AF410824

Role of miRNAs in Cancer—Analysis of Their Targetome

Role of miRNAs in Cancer—Analysis of Their Targetome

Editor

Alfons Navarro

MDPI • Basel • Beijing • Wuhan • Barcelona • Belgrade • Manchester • Tokyo • Cluj • Tianjin

Editor
Alfons Navarro
Human Anatomy
University of Barcelona
Barcelona
Spain

Editorial Office
MDPI
St. Alban-Anlage 66
4052 Basel, Switzerland

This is a reprint of articles from the Special Issue published online in the open access journal *Cancers* (ISSN 2072-6694) (available at: www.mdpi.com/journal/cancers/special_issues/miRNAs_Cancer).

For citation purposes, cite each article independently as indicated on the article page online and as indicated below:

LastName, A.A.; LastName, B.B.; LastName, C.C. Article Title. *Journal Name* **Year**, *Volume Number*, Page Range.

ISBN 978-3-0365-1686-8 (Hbk)
ISBN 978-3-0365-1685-1 (PDF)

Contents

About the Editor

Alfons Navarro

Alfons Navarro, PhD (University of Barcelona, 2008) is a molecular biologist with a longstanding interest in non-coding RNAs and exosomes in cancer. In 2010, he did a post-doctoral research at the Comprehensive Cancer Center at Ohio State University (Carlo M. Croce Lab). In 2016, he obtained a MSc in Bioinformatics and Biostatistics from the Open University of Catalonia. He has authored more than 70 scientific articles (cited more than 4490 times, H-index: 33) and several book chapters. He has participated in more than 14 financed projects and he has been editor in several journals. Since 2012, he has worked as a Professor in Human Anatomy and Embryology at University of Barcelona.

Preface to "Role of miRNAs in Cancer—Analysis of Their Targetome"

MicroRNAs (miRNAs) can be considered the best reference in the world of non-coding RNAs. They are the best studied and most cancer-related non-coding RNA species, despite their short nucleotide length of about 25 nucleotides. Several functions have been described for miRNAs, but the most relevant is the negative regulation of the translation process of their target mRNAs—based on sequence complementarity mechanisms. Although the first evidence linking miRNAs to cancer comes from 2005, nowadays not all the targets related to their functions in the process of tumorogenesis have been elucidated. The specific miRNA-targetome changes from tumor to tumor and is highly correlated with the transcriptomic profile of each cell, which makes its identification very difficult. The present book aims to add some light to this topic by reviewing the main known miRNA-gene interactions in the context of tumor development and metastasis, as well as describing new relationships, not known to date. The present book is addressed to all authors interested in knowing the interactions between miRNAs and target genes during the tumorogenesis process. It was written by many authors around a common theme—the role of miRNAs in cancer trough the analysis of their targetome. It is beyond the scope of this book to review the miRNA-targetome for every cancer. However, it is the intent of this editor to include a broad selection of cancer models. The contributing authors have written six original research papers and eight valuable reviews that expand our knowledge on the role of miRNAs in tumorogenesis. I am most grateful to all the expert contributors for bringing their expertise and experience together in this book and I hope you will find it a helpful contribution towards the deciphering of the miRNA-targetomes in cancer development and metastasis.

Alfons Navarro
Editor

Editorial

Deciphering miRNA–Target Relationships to Understand miRNA-Mediated Carcinogenesis

Alfons Navarro [1,2]

1   Molecular Oncology and Embryology Laboratory, Human Anatomy Unit, Faculty of Medicine and Health Sciences, University of Barcelona, IDIBAPS, 08036 Barcelona, Spain; anavarroponz@ub.edu; Tel.: +34-934-021-903
2   Thoracic Oncology Unit, Hospital Clinic, 08036 Barcelona, Spain

Citation: Navarro, A. Deciphering miRNA–Target Relationships to Understand miRNA-Mediated Carcinogenesis. *Cancers* **2021**, *13*, 2415. https://doi.org/10.3390/cancers13102415

Received: 6 May 2021
Accepted: 7 May 2021
Published: 17 May 2021

Publisher's Note: MDPI stays neutral with regard to jurisdictional claims in published maps and institutional affiliations.

We now accept that the non-coding part of the genome is essential for fine-tuning most cellular functions, and that its deregulation drives carcinogenesis [1]. However, the first link between non-coding RNAs and cancer was identified only 16 years ago, where the microRNAs (miRNAs) miR-15/16 were identified as tumor suppressors in CLL by Croce's group [2]. From that stepping stone multiple types of miRNAs first, and other non-coding RNAs later have been identified and related to all stages of tumor development, from tumor initiation to dissemination during the metastasis process [3]. Now the non-coding RNAs group has been so enlarged that people classify them in two main groups according to their size, where those of less than 200 nucleotides are call small non-coding RNAs, in contrast to long non-coding RNAs. The best studied non-coding RNA sequences of our genome, in part because they were the first formally related to cancer, are miRNAs, which belong to the small non-coding RNA group with a sequence length of 15–25 nt. The first miRNA, lin-4, was discovered in 1993 by V. Ambros' group in *C. elegans* [4], but it took eight years to discover the second miRNA, let-7a, which really boosted miRNA discovery because in contrast to lin-4, homologous sequences of let-7a were discovered in most species including humans [5]. Since the finding of let-7a, the identification of new miRNAs has increased considerably. In fact, as of 2021, we know 2693 mature miRNAs in humans (according to miRBase v22.1). Several functions have been described for miRNAs (reviewed in [6]), but the most relevant one is the posttranscriptional regulation of protein-coding genes by binding to target mRNAs in a sequence-dependent way and inhibiting its translation. Therefore, to identify the functions of the dysregulated miRNAs, it is necessary to decipher their target genes, their specific targetome. But this task becomes complex, because each miRNA can target thousands of mRNAs and at the same time one mRNA is targeted by multiple different miRNAs. To increase the complexity, we know that the miRNA–target interactions depend on the cellular context and the specific associated transcriptome, which generate a cell type-dependent targetome for each miRNA that is also variable according to tumor types. Thousands of targets have been predicted for each miRNA using different bioinformatic algorithms such as TargetScan, miRanda or Pictar, which are mainly based in the identification of miRNA–target sequence complementarities, and a high number of them have been experimentally validated. However, an important gap exists yet between the number of pathological identified miRNAs and their real targets, since their complete targetome remains to be elucidated [7]. Recently, the miRNA–target network became more entangled with the addition of other non-coding RNAs, such as long non-coding RNAs or circRNAs, and with the fact that miRNAs not only regulate the transcriptome of the parental cell but also other cell types since they can be released outside through extracellular vesicles acting on other cellular transcriptomes different from the parental cell.

In this Special Issue of *Cancers* called "Role of miRNAs in Cancer—Analysis of Their Targetome" we collected 14 articles (5 original Articles, 1 Communication, 7 reviews and

1 Perspective). In the following lines I am going to summarize the main results of the articles included in the Special Issue. The original articles included move from the study of the role of individual miRNAs and their targets in different tumors, to the identification of new diagnostic/prognostic biomarkers in tissue and liquid biopsy.

Hunter et al. [8] studied the role of two *COX-2*-activated miRNAs, miR-526b and miR-655, in the control of angiogenesis and lymphomagenesis in an estrogen receptor positive breast cancer model. In the MCF7 cell line, these miRNAs by targeting *PTEN* induced *HIF1A*, which increased *VEGF* secretion. Moreover, the overexpression of these miRNAs enhanced the expression of *VEGFA/C/D*, *COX-2* and *LYVE1*. When HUVEC cells were cultured with the supernatant of overexpressing cells, VEGF and EP4 receptors augmented, and higher migration and higher tube formation was observed in HUVEC cells. In contrast, the use of *COX-2* inhibitors and *EP* antagonists reversed the phenotype. They also observed a correlation in patient samples between these miRNAs and angiogenesis and lymphomagenesis markers.

Borkowska et al. [9] studied the potential role as diagnostic biomarker of eight selected miRNAs (miR-10a, miR-20a, miR-21, miR-103, miR-130b, miR-145, miR-182 and miR-205) in 55 patients with bladder cancer and 30 controls. They concluded that the best diagnostic signature included miR-20a, miR-205 and miR-145, which discriminated bladder cancer from healthy controls.

Quin et al. [10] generated a sophisticated prognostic score in clear cell renal carcinoma patients, combining miRNAs and their target genes (TCGA data). The miRNA analysis initially identified eight miRNAs impacting patient outcome. The correlation with the expression data allowed the identification of mRNA signatures also impacting prognosis that explained the role of the identified miRNAs. To increase the efficiency of the miRNA score, the authors identified transcription factors associated with the miRNA signature and generated a combined score including miR-365b-3p, miR-223-3p, miR-1269a, miR-144-5p, miR-183-5p, miR-335-3p, *TFAP2A*, *KLF5*, *IRF1*, *MYC* and *IKZF1* which had the highest impact in patient survival. They concluded that transcription factors and miRNAs can cooperatively regulate oncogenesis and impact prognosis in clear cell renal carcinoma patients.

Niu et al. [11] by small RNAseq identified miR-378a-3p as an upregulated miRNA impacting proliferation in Burkitt lymphoma cell lines. After validating its overexpression in patient tissues, they performed targetome analysis by Ago2-RIP-Chip. The bioinformatic analysis of the results allowed them to identify 63 potential targets when they inhibited the miRNA in ST486 cells, and 20 targets when they overexpressed it. The authors focused on *MYCBP*, *CISH*, *BCR*, *TUB1C*, *FOXP1*, *MNT*, *IRAK4* and the lncRNA *JPX* for validation with luciferase assays finally confirming *FOXP1*, *MNT*, *IRAK4* and the lncRNA *JPX* as real miR-378-3p targets.

Cuscino et al. [12] identified eight novel miRNAs in osteosarcoma by analyzing the cellular and exosomal RNA from the cell lines SAOS-2, MG-63 and U-2 OS. The validation in tissue and plasma samples from osteosarcoma patients showed that seven miRNAs were detected in all samples, and five where significantly upregulated in plasma samples. The in silico analysis of the miRNA–targets revealed several KEGG pathways linked to cancer, suggesting that the novel miRNAs identified could have a role in osteosarcoma pathogenesis.

Nguyen et al. [13] made a combined article that simultaneously reviewed the literature and at the same time provided their own data on blood miRNAs as biomarkers for radiotherapy response in pancreatic cancer. Using mice harboring pancreatic tumors, they obtained by small RNAseq a miRNA signature associated with the presence of this tumor. Then, they treated the mice with radiotherapy (a single 5Gy dose) and at 24 h plasma was collected for miRNA analysis. A miRNA signature that included 20 downregulated miRNAs and upregulated one (miR-184) was identified. The role of these miRNAs and their targets were revised showing that several of them have been previously related to radioresistance in other tumor models, but only a few of them have been previously associated with pancreatic cancer radioresistance.

The review articles included in the Special Issue cover various topics from the known miRNA targetome in specific tumors, including glioblastoma, thyroid or adrenocortical cancer, to the analysis of miRNAs involved in the metastasis process in general or focusing in the targetome of epithelial mesenchymal transition (EMT) genes such as SNAIL or the involvement of miRNAs in the crosstalk between tumor and immune system cells. Moreover, specific targetomes such as miR-361 or viral-associated miRNA targetome have been explored.

Xu et al. [14] concentrated on a unique miRNA, miR-361, which can be considered a tumor suppressor miRNA because its targets are mostly oncogenes. They summarized the main reasons for miR-361 downregulation, which include DNA hypermethylation, transcriptional inhibition, sponging by lncRNAs and gene deletion. The cellular and extracellular targetome of this miRNA explains the aggressive tumor phenotype observed when the miRNA is lost and its potential utility as diagnostic, prognostic or therapeutic biomarker.

Sole et al. [15] nicely summarized the miRNAs involved in the metastasis process. They organized the miRNAs into three groups: metastasis-promoting miRNAs, metastasis-suppressing miRNAs, and metastasis associated circulating blood miRNAs. They exhaustively listed the miRNAs associated with metastasis in different tumors together with their known targets.

Skrzypek et al. [16] also examined metastasis but centering on the *SNAIL* transcription factor, which is involved in EMT regulation and metastasis. The authors reviewed not only miRNAs, but also lncRNAs and circRNAs that either regulate *SNAIL* levels or that are regulated by *SNAIL*.

Gallo et al. [17] examined the role of viral (mainly herpesvirus) miRNAs and their cellular targets in the tumorigenesis process. The article listed the known miRNAs from EBV, HPV, KSHV/HHV8, HBV and MCPyV and their cellular targetomes involved in both the own viral cycle control and the induction of tumorigeneses.

In glioblastoma multiforme, the methylation status of the *MGMT* gene classifies patients into unmethylated and methylated, and the last group is treated with totemozolomide-based chemotherapy. Kirstein et al. [18] discussed the potential role of miRNAs targeting *MGMT* as therapeutic tools for unmethylated patients, which are resistance totemozolomide-based chemotherapy. They examined whether miRNAs inhibiting *MGMT* could enhance response to this line of treatment.

Chehade et al. [19] reviewed miRNAs associated with adrenocortical cancer, which is a rare but aggressive malignancy. They revised from differentially expressed miRNAs in tumor tissue to differentially expressed circulating miRNAs, and their utility as diagnostic/prognostic biomarkers. Moreover, they revised the known targetome of adrenocortical cancer miRNAs focusing on relevant disease-associated pathways such a p53 pathway, mTOR or Wnt/B-catenin.

Tabatabaeian et al. [20] presented a systematic review on miRNAs, lncRNAs and circRNAs in thyroid cancer. They summarized the most relevant non-coding RNAs involved in the tumorigenesis process and their role as diagnostic, prognostic or therapeutic biomarkers. Of note, they highlighted the list of ongoing clinical trials which includes some of these non-coding RNA biomarkers.

Lastly, Cho et al. [21] briefly reviewed the miRNA targetome involved in the interaction between tumor and immune cells, stressing the miRNAs responsible for inhibiting the anti-tumor immune response. They focused on two main groups: miRNAs released in exosomes and participating in the crosstalk between tumor cells, mesenchymal cells and immune cells (macrophages, CD4+ T cells or CD8+ T cells); and cellular miRNAs regulating the levels of immunomodulatory proteins such as CD47, IDO1 or PD-L1 in the tumor cell. Moreover, the authors discussed the main techniques used for study of the miRNA targetome such as CLIP-seq or CLASH, that are necessary to fully understand the role of miRNAs in tumorigenesis.

In summary, this Special Issue includes miRNA-related articles that add new information to the miRNA–target interaction in different tumor models and review some of the most recent information on the tumorigenic targetome. However, more efforts are still needed to decipher the critical targets—both coding and non-coding RNAs—of the miRNAs involved in cancer to identify their contribution to the malignant transformation and metastasis process.

Funding: This work was supported by grants from the Ministry of Economy, Industry, and Competition, Agencia Estatal de Investigación co-financed with the European Union FEDER funds SAF2017-88606-P (AEI/FEDER, UE).

Conflicts of Interest: The author declares no conflict of interest.

References

1. Anastasiadou, E.; Jacob, L.S.; Slack, F.J. Non-coding RNA networks in cancer. *Nat. Rev. Cancer* **2018**, *18*, 5–18. [CrossRef] [PubMed]
2. Calin, G.A.; Ferracin, M.; Cimmino, A.; Di Leva, G.; Shimizu, M.; Wojcik, S.E.; Iorio, M.V.; Visone, R.; Sever, N.I.; Fabbri, M.; et al. A MicroRNA Signature Associated with Prognosis and Progression in Chronic Lymphocytic Leukemia. *N. Engl. J. Med.* **2005**, *353*, 1793–1801. [CrossRef] [PubMed]
3. Slack, F.J.; Chinnaiyan, A.M. The Role of Non-coding RNAs in Oncology. *Cell* **2019**, *179*, 1033–1055. [CrossRef] [PubMed]
4. Lee, R.C.; Feinbaum, R.L.; Ambros, V. The *C. elegans* heterochronic gene lin-4 encodes small RNAs with antisense complementarity to lin-14. *Cell* **1993**, *75*, 843–854. [CrossRef]
5. Pasquinelli, A.E.; Reinhart, B.J.; Slack, F.; Martindale, M.Q.; Kuroda, M.I.; Maller, B.; Hayward, D.C.; Ball, E.E.; Degnan, B.; Müller, P.; et al. Conservation of the sequence and temporal expression of let-7 heterochronic regulatory RNA. *Nature* **2000**, *408*, 86–89. [CrossRef]
6. Bartel, D.P. MicroRNAs: Target Recognition and Regulatory Functions. *Cell* **2009**, *136*, 215–233. [CrossRef]
7. Navarro, A. Argonaute-crosslinking and immunoprecipitation deciphers the liver miR-122 targetome. *Non-Coding RNA Investig.* **2017**, *1*, 23. [CrossRef]
8. Hunter, S.; Nault, B.; Ugwuagbo, K.C.; Maiti, S.; Majumder, M. Mir526b and Mir655 promote tumour associated angio-genesis and lymphangiogenesis in breast cancer. *Cancers* **2019**, *11*, 938. [CrossRef]
9. Borkowska, E.M.; Konecki, T.; Pietrusiński, M.; Borowiec, M.; Jabłonowski, Z. MicroRNAs which can prognosticate aggressiveness of bladder cancer. *Cancers* **2019**, *11*, 1551. [CrossRef]
10. Qin, S.; Shi, X.; Wang, C.; Jin, P.; Ma, F. Transcription Factor and miRNA Interplays Can Manifest the Survival of ccRCC Patients. *Cancers* **2019**, *11*, 1668. [CrossRef]
11. Niu, F.; Dzikiewicz-Krawczyk, A.; Koerts, J.; De Jong, D.; Wijenberg, L.; Hernandez, M.F.; Slezak-Prochazka, I.; Winkle, M.; Kooistra, W.; Van Der Sluis, T.; et al. MiR-378a-3p Is Critical for Burkitt Lymphoma Cell Growth. *Cancers* **2020**, *12*, 3546. [CrossRef]
12. Cuscino, N.; Raimondi, L.; De Luca, A.; Carcione, C.; Russelli, G.; Conti, L.; Baldi, J.; Conaldi, P.G.; Giavaresi, G.; Gallo, A. Gathering Novel Circulating Exosomal microRNA in Osteosarcoma Cell Lines and Possible Implications for the Disease. *Cancers* **2019**, *11*, 1924. [CrossRef]
13. Nguyen, L.; Schilling, D.; Dobiasch, S.; Raulefs, S.; Franco, M.S.; Buschmann, D.; Pfaffl, M.W.; Schmid, T.E.; Combs, S.E. The Emerging Role of miRNAs for the Radiation Treatment of Pancreatic Cancer. *Cancers* **2020**, *12*, 3703. [CrossRef]
14. Xu, D.; Dong, P.; Xiong, Y.; Yue, J.; Ihira, K.; Konno, Y.; Kobayashi, N.; Todo, Y.; Watari, H. MicroRNA-361: A Multifaceted Player Regulating Tumor Aggressiveness and Tumor Microenvironment Formation. *Cancers* **2019**, *11*, 1130. [CrossRef]
15. Solé, C.; Lawrie, C.H. MicroRNAs and Metastasis. *Cancers* **2020**, *12*, 96. [CrossRef]
16. Skrzypek, K.; Majka, M. Interplay among SNAIL Transcription Factor, MicroRNAs, Long Non-Coding RNAs, and Circular RNAs in the Regulation of Tumor Growth and Metastasis. *Cancers* **2020**, *12*, 209. [CrossRef]
17. Gallo, A.; Miceli, V.; Bulati, M.; Iannolo, G.; Contino, F.; Conaldi, P.G. Viral miRNAs as Active Players and Participants in Tumorigenesis. *Cancers* **2020**, *12*, 358. [CrossRef]
18. Kirstein, A.; Schmid, T.E.; Combs, S.E. The role of miRNA for the treatment of MGMT unmethylated glioblastoma multiforme. *Cancers* **2020**, *12*, 1099. [CrossRef]
19. Chehade, M.; Bullock, M.; Glover, A.; Hutvagner, G.; Sidhu, S. Key MicroRNA's and Their Targetome in Adrenocortical Cancer. *Cancers* **2020**, *12*, 2198. [CrossRef]
20. Tabatabaeian, H.; Yang, S.P.; Tay, Y. Non-Coding RNAs: Uncharted Mediators of Thyroid Cancer Pathogenesis. *Cancers* **2020**, *12*, 3264. [CrossRef]
21. Cho, S.; Tai, J.W.; Lu, L.-F. MicroRNAs and Their Targetomes in Tumor-Immune Communication. *Cancers* **2020**, *12*, 2025. [CrossRef] [PubMed]

cancers

Article

Mir526b and Mir655 Promote Tumour Associated Angiogenesis and Lymphangiogenesis in Breast Cancer

Stephanie Hunter, Braydon Nault [†], Kingsley Chukwunonso Ugwuagbo [†], Sujit Maiti and Mousumi Majumder *

Department of Biology, Brandon University, 3rd Floor, John R. Brodie Science Centre, 270–18th Street, Brandon, MB R7A 6A9, Canada
* Correspondence: majumderm@brandonu.ca
† These authors contributed equally to this work.

Received: 11 June 2019; Accepted: 29 June 2019; Published: 4 July 2019

Abstract: MicroRNAs (miRNAs) are small endogenously produced RNAs, which regulate growth and development, and oncogenic miRNA regulate tumor growth and metastasis. Tumour-associated angiogenesis and lymphangiogenesis are processes involving the release of growth factors from tumour cells into the microenvioronemnt to communicate with endothelial cells to induce vascular propagation. Here, we examined the roles of cyclo-oxygenase (COX)-2 induced miR526b and miR655 in tumour-associated angiogenesis and lymphangiogenesis. Ectopic overexpression of miR526b and miR655 in poorly metastatic estrogen receptor (ER) positive MCF7 breast cancer cells resulted in upregulation of angiogenesis and lymphangiogenesis markers vascular endothelial growth factor A (VEGFA); VEGFC; VEGFD; COX-2; lymphatic vessel endothelial hyaluronan receptor-1 (LYVE1); and receptors *VEGFR1, VEGFR2,* and *EP4*. Further, miRNA-high cell free conditioned media promoted migration and tube formation by human umbilical vein endothelial cells (HUVECs), and upregulated *VEGFR1, VEGFR2,* and *EP4* expression, showing paracrine stimulation of miRNA in the tumor microenvironment. The miRNA-induced migration and tube formation phenotypes were abrogated with *EP4* antagonist or PI3K/Akt inhibitor treatments, confirming the involvement of the *EP4* and PI3K/Akt pathway. Tumour supressor gene *PTEN* was found to be downregulated in miRNA high cells, confirming that it is a target of both miRNAs. *PTEN* inhibits hypoxia-inducible factor-1 (HIF1α) and the PI3K/Akt pathway, and loss of regulation of these pathways through *PTEN* results in upregulation of VEGF expression. Moreover, in breast tumors, angiogenesis marker *VEGFA* and lymphangiogenesis marker *VEGFD* expression was found to be significantly higher compared with non-adjacent control, and expression of miR526b and miR655 was positively correlated with *VEGFA, VEGFC, VEGFD, CD31,* and *LYVE1* expression in breast tumour samples. These findings further strengthen the role of miRNAs as breast cancer biomarkers and *EP4* as a potential therapeutic target to abrogate miRNA-induced angiogenesis and lymphangiogenesis in breast cancer.

Keywords: miR526b; miR655; breast cancer; angiogenesis; lymphangiogenesis; *EP4*; PI3K/Akt

---

1. Introduction

Breast cancer is the deadliest and most prevalent cancer among women, being responsible for the greatest number of cancer-related deaths among women worldwide [1]. In many cancers, including human breast cancer, cyclo-oxygenase (COX)-2 enzyme is found to have higher than normal expression [2]. Specifically, upregulation of COX-2 is correlated with breast cancer disease progression, metastasis, and poor patient survival [3,4]. COX-2 is responsible for the production of the inflammatory molecule, prostaglandin E2 (PGE2). Production of PGE2 by COX-2 results in binding of PGE2 to four G-protein coupled PGE receptors, EP1-4, each of which have distinct signalling pathways [5]. EP1 couples with Gq, EP3 couples with Gi, and EP2 and *EP4* couple with Gs. Additionally, *EP4* stimulates non-canonical pathways PI3K/Akt and ERK, which are associated with cell survival and migration [6]. PGE2 induces early vascular maturation and angiogenesis in vertebrates by upregulation of VEGFs and PGE2 receptors [7]. Overproduction of PGE2 and activation of *EP4* receptor results in many tumourigenesis-promoting phenotypes such as inactivation of host anti-tumour immune cells, enhanced tumour cell migration and invasion, stem-like cell (SLC) induction, and tumour-associated angiogenesis and lymphangiogenesis [6,8]. Overexpression of COX-2 in two poorly metastatic MCF7 (COX-2 low, human epidermal growth factor receptor 2 (HER-2) negative, progesterone receptor (PR) positive, estrogen receptor (ER) positive) and SK-BR-3 (COX-2 low, HER-2 high, ER negative) breast cancer cell lines has been shown to induce aggressive breast cancer phenotypes and promote metastasis, which could be abrogated with *EP4* antagonist treatment. Moreover, MCF7 cells show lower ER expression with COX-2 overexpression, and COX-2 overexpression caused a change in gene and microRNA (miRNA) expression in MCF7 cells [8].

MicroRNAs are a class of endogenously produced, short non-coding RNAs that can down regulate gene expression of target messenger RNA (mRNA) at the post-transcriptional level by partial or complete complimentary base pairing [9]. Abnormal expression of miRNAs has been well highlighted in various types of cancer, including breast cancer. Previously, two COX-2 upregulated miRNAs, miR526b and miR655, have been identified and established as oncogenic miRNAs in human breast cancer [10,11]. The roles of both miR526b and miR655 have been implicated in many hallmarks of cancer, including driving primary tumour growth, induction of stem-like cells (SLCs) phenotype, epithelial-to-mesenchymal transition (EMT), invasion and migration, as well as distant metastasis when tested in vivo in mouse models [10,11]. Moreover, both miRNAs target a transcription factor, *CPEB2A* gene, which was recently validated as a tumor suppressor [12]. However, the potential roles of miR526b and miR655 in tumour-associated angiogenesis and lymphangiogenesis in breast cancer have not yet been investigated.

Angiogenesis and lymphangiogenesis are processes involving the formation of new blood or lymph vessels from pre-existing vasculature, both of which progress through the proliferation, migration, and maturation of nearby blood or lymph vessel endothelial cells [13,14]. Angiogenesis is an essential biological process that is fundamental for development, reproduction, and wound repair; however, this process is also considered a major hallmark of cancer [15]. As tumour growth can only reach 1–2 mm without sufficient blood supply, angiogenesis is essential during the uncontrolled growth of tumours for supply of sufficient oxygen and nutrients [15,16]. Similarly, lymphangiogenesis is an essential biological process that has also been implicated in many cancers, including breast cancer. The initial sites of metastasis in breast cancer are often the regional lymph nodes, and the migration of tumour cells to these sites is facilitated by lymphangiogenesis [17]. Vascular endothelial growth factors (VEGF) play a key role in both angiogenesis and lymphangiogenesis. Specifically, VEGFA is a major mediator of angiogenesis, through binding with VEGF receptors (VEGFR)1 and VEGFR2, leading to proangiogenic activity and migration of endothelial cells [18]. VEGFC and VEGFD are the primary ligands regulating lymphangiogenesis through binding of VEGFR3 on lymphatic endothelial cells; however, both ligands also have a weak affinity for VEGFR2, thus in part activating angiogenesis [18]. CD31 is also a well known stimulator of angiogenesis specifically involved in cell to cell interactions necessary for the organization of blood endothelial cells [19]. Moreover, lymphatic vessel endothelial hyaluronan receptor-1 (LYVE-1) [5,14] is a marker of lymphangiogenesis.

The majority of breast cancer patients are estrogen receptor (ER) positive and treated with tamoxifen [1]. Thus, in the present study, we investigated miR526b and miR655 and their potential roles in the process of breast cancer tumour-associated angiogenesis and lymphangiogenesis with an emphasis on ER-positive breast cancer, and we selected MCF7 as the breast cancer model to overexpress miRNAs. In vitro studies involving cell migration and capillary-like tube formation assays were conducted with human umbilical vein endothelial cells (HUVECs) and cell-free conditioned media collected from miR526b and miR655 overexpressing breast cancer cell lines [10,11]. Furthermore, the involvement of the COX2, *EP4*, and PI3K/Akt signalling pathways was investigated during angiogenesis assays by either COX-2 inhibition, *EP4* receptor antagonism, or PI3K/Akt pathway inhibition. We establish that overexpression of miR526b and miR655 is linked with tumour-associated angiogenesis and lymphangiogenesis in vitro. Our study also suggests that this stimulation occurs via the activation of the *EP4* receptor and subsequent activation of the PI3K/Akt pathway. In support of these findings, we used an in situ model to examine the relationship of miR526b and miR655 expression in human breast cancer tissues with expression of angiogenesis and lymphangiogenesis markers. Previously, we have shown that high expression of miR526b or miR655 in human breast cancer tissue is associated with reduced breast cancer patient survival [10,11]. Our present results demonstrate that both miR526b and miR655 expression is positively correlated with established angiogenesis and lymphangiogenesis markers in human breast cancer. Overall, our study establishes the roles of miR526b and miR655 in human breast cancer tumour-associated angiogenesis and lymphangiogenesis.

2. Results

2.1. Over-Expression of miR526b and miR655 in Poorly Metastatic (ER Positive) MCF7 Breast Cancer Cell Line Results in Upregulation of Angiogenesis and Lymphangiogenesis Markers

RNA was extracted from various passages of 90% confluent MCF7, MCF7-miR526b, and MCF7-miR655 cell lines and reverse transcribed into cDNA. The cDNA was then used to quantify the expression of known markers of lymphangiogenesis and angiogenesis. Relative gene expression fold change analysis was performed to compare the miRNA high cell lines, MCF7-miR526b and MCF7-miR655, to miRNA-low MCF7 cells. It was found that lymphangiogenesis marker, *VEGFD*, was significantly upregulated in MCF7-miR526b and MCF7-miR655 cell lines, while *VEGFC* and *LYVE-1* were marginally upregulated in MCF7-miR526b and significantly upregulated in MCF7-miR655 (Figure 1A). Furthermore, angiogenesis marker, *VEGFA*, was found to be significantly upregulated in both miRNA high cell lines (Figure 1A). Further, we extracted total protein from MCF7, MCF7-miR526b, and MCF7-miR655 cells for the quantification of VEGFA, VEGFC, VEGFD, COX-2, and LYVE-1 markers. MCF7-miR655 overexpressing cell line showed high expression of all VEGF markers and COX-2 expression at the protein level, while MCF7-miR526b showed significantly higher expression of VEGFC and VEGFD expression (Figure 1B,C). Whole Western blot data with corresponding molecular weights are presented in Figures S1 and S2.

Expression of receptors *VEGFR1* and *EP4* was found to be significantly upregulated in both the MCF7-miR526b and MCF7-miR655 cell lines (Figure 2A). *VEGFR2* was found to be marginally upregulated in MCF7-miR526b and MCF7-miR655 cell lines; however, this was not statistically significant (Figure 2A).

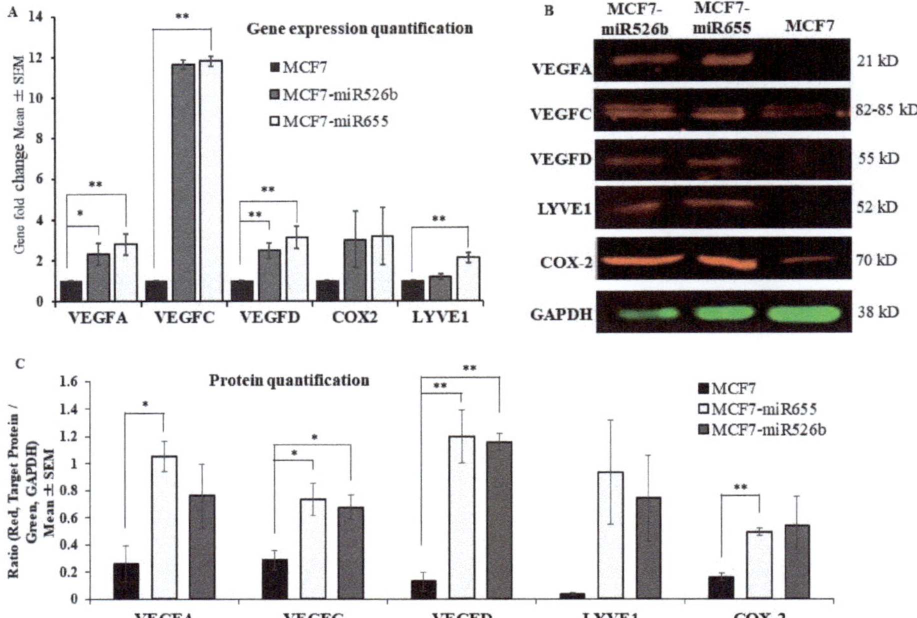

Figure 1. Overexpression of miR526b and miR655 in MCF7 cell line results in upregulation of angiogenesis and lymphangiogenesis markers. (**A**) Quantitative real time polymerase chain reaction (qRT-PCR) shows angiogenic and lymphangiogenic markers at the mRNA level with significant positive fold changes in the miRNA overexpressed cell lines, MCF7-miR526b and MCF7-miR655, compared with the MCF7 cell line. (**B**) The Western blot analysis shows a larger presence of angiogenesis and lymphangiogenesis markers at the protein level in the MCF7-miR655 and MCF7-miR526b cell lines compared with the MCF7 cell line. Green: control marker Glyceraldehyde 3-phosphate dehydrogenase (GAPDH), red: target proteins, either VEGFA or VEGFC or VEGFD or LYVE1 or COX-2. (**C**) Quantitative analysis of Western blot shows increased levels of angiogenesis and lymphangiogenesis markers. RT-PCR and Western blot quantitative data are presented as the mean ± SEM of triplicate replicates; * $p < 0.05$, ** $p < 0.01$. LYVE-1—lymphatic vessel endothelial hyaluronan receptor-1; VEGF—vascular endothelial growth factors; COX-2—cyclo-oxygenase 2.

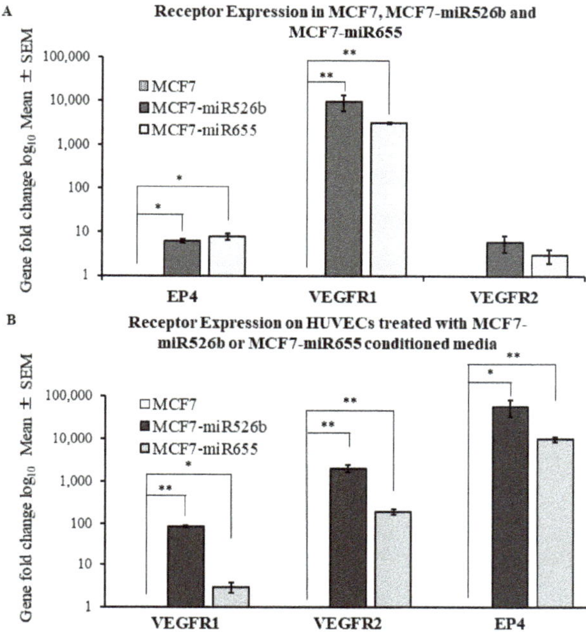

Figure 2. Overexpression of miR526b and miR655 in MCF7 or treatment of human umbilical vein endothelial cells (HUVECs) with miRNA-conditioned media results in upregulation of angiogenesis and lymphangiogenesis receptor markers. (**A**) Quantitative RT-PCR analysis shows significant upregulation of *prostaglandin E2 receptor 4 (EP4)*, *VEGF receptor 1 (VEGFR1)*, and *VEGFR2* expression in the MCF7-miR526b and MCF7-miR655 cell lines, when compared with MCF7. (**B**) Quantitative RT-PCR analysis shows the HUVECs treated with MCF7-miR526b or MCF7-miR655 conditioned media have greater receptor gene expression compared with HUVECs treated with MCF7 conditioned media. RT-PCR quantitative data are presented as the mean ± SEM of triplicate replicates; * $p < 0.05$, ** $p < 0.01$.

2.2. Human Umbilical Vein Endothelial Cells (HUVECs) Treated with MCF7-miR526b and MCF7-miR655 Conditioned Media Show Higher Expression of VEGF and EP4 Receptors

HUVECs grown to 90% confluency were treated overnight with conditioned media collected from MCF7-miR526b and MCF7-miR655 cell cultures. Total RNA was extracted from treated HUVECs and reverse transcribed to cDNA. TaqMan gene expression assay comparing the expression of *VEGFR1*, *VEGFR2*, and *EP4* was conducted. By virtue, HUVECs show very high expression of VEGFR1 and VEGFR2; here, we observed that miRNA-conditioned media further induced expression, showing a significant upregulation of *VEGFR2* in comparison with HUVECs treated with MCF7 conditioned media (Figure 2B). Moreover, expression of *VEGFR1* was found to be significantly upregulated in HUVECs treated with MCF7-miR526b conditioned media and marginally upregulated in those treated with MCF7-miR655 conditioned media (Figure 2B). HUVECs express a low level of *EP4* [20]; here, we observed a very significant upregulation of *EP4* expression in HUVECs treated with MCF7-miR655 conditioned media, and marginal upregulation following treatment with MCF7-miR526b conditioned media (Figure 2B).

2.3. Cancer Cell Conditioned Media Induces Migration and Tube Formation of HUVEC Cells

To examine the in vitro role of miR526b and miR655 in angiogenesis, we tested the cell migration capacity of HUVECs in cell-free conditioned media collected from MCF7, MCF7-miR526b, and MCF7-miR655 cell lines. Here, HUVEC cells were seeded and grown in a 24-well plate and a scratch wound was made with a 2 µL pipette tip. Cell-free conditioned from MCF7-miR526b and MCF7-miR655 cell lines were found to result in a significant increase in HUVEC migration during wound healing compared with the MCF7 conditioned media at the 24 h time point (Figure 3A,B). Cell migration images of the positive and negative controls, along with the experimental conditions at other time points, are provided in Figure S3.

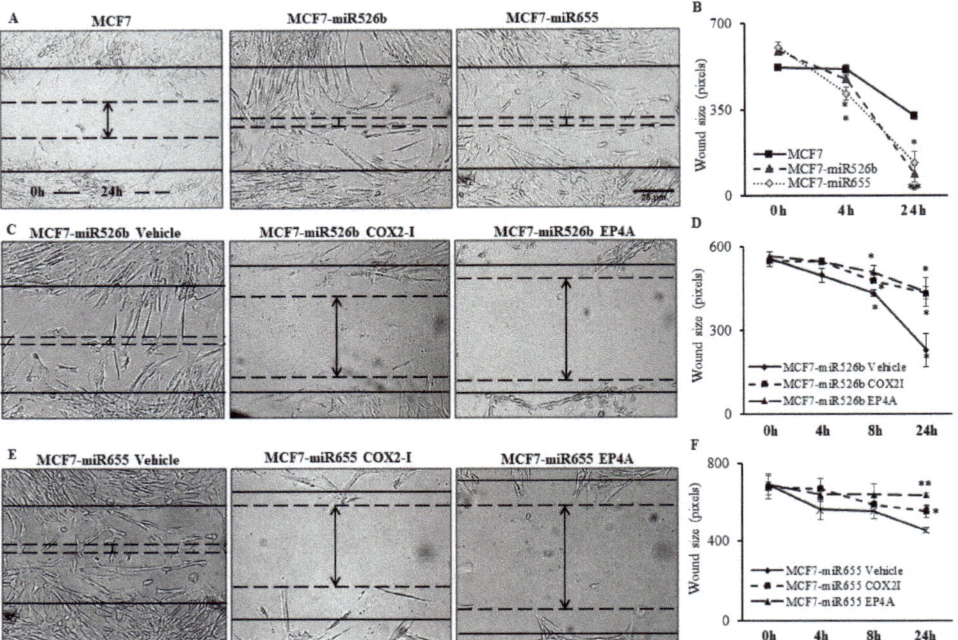

Figure 3. MCF7-miR526b and MCF7-miR655 conditioned media promotes cellular migration of HUVECs and inhibition of the COX2/*EP4* signaling pathway abrogates these phenotypes. Baseline scratches represented by black line; wound size at 24 h represented by dashed line. (**A**) Images of migration assay with conditioned media from MCF7, MCF7-miR526b, and MCF7-miR655 at the 24 h time point. (**B**) Quantitative data representing wound size per time point in conditioned media from MCF7, MCF7-miR526b, or MCF7-miR655. (**C**) Images of migration assay with conditioned media from MCF7-miR526b with the addition of vehicle, COX2-I, or EP4A, at the 24 h time point. (**D**) Quantitative data representing wound size per time point in conditioned media from MCF7-miR526b with the addition of either vehicle, COX2-I, or *EP4*A. (**E**) Images of migration assay with conditioned media from MCF7-miR655 with the addition of vehicle, COX2-I, or *EP4*A, at the 24 h time point. (**F**) Quantitative data representing wound size per time point in conditioned media from MCF7-miR655 with the addition of either vehicle, COX2-I, or *EP4*A. Data shown as mean ± SEM of three biological replicates, including three experimental replicates per biological replicate; * $p < 0.05$, ** $p < 0.01$.

To further investigate the effects of miR526b and miR655 on the angiogenesis potential of HUVECs, a tube formation assay was performed. The tube formation assay performed with growth factor reduced Matrigel is an established surrogate of angiogenesis in vitro. Cultured HUVECs were harvested and resuspended in MCF7, MCF7-miR526b, or MCF7-miR655 conditioned media, and seeded in a Matrigel-coated 24-well plate. Tube formation was observed and recorded at 0–24 h with image capturing. MCF7-miR655 conditioned media significantly stimulated an increase in the formation of both tubes and branching points at the 24 h time point when compared with the MCF7 conditioned media (Figure 4A–C), while MCF7-miR526b conditioned media resulted in a marginal increase of tubes, and a very significant increase in branching points at the 24 h time-point (Figure 4A–C). Tube formation images of the positive and negative controls, along with the experimental conditions at other time points, are provided in Figure S4.

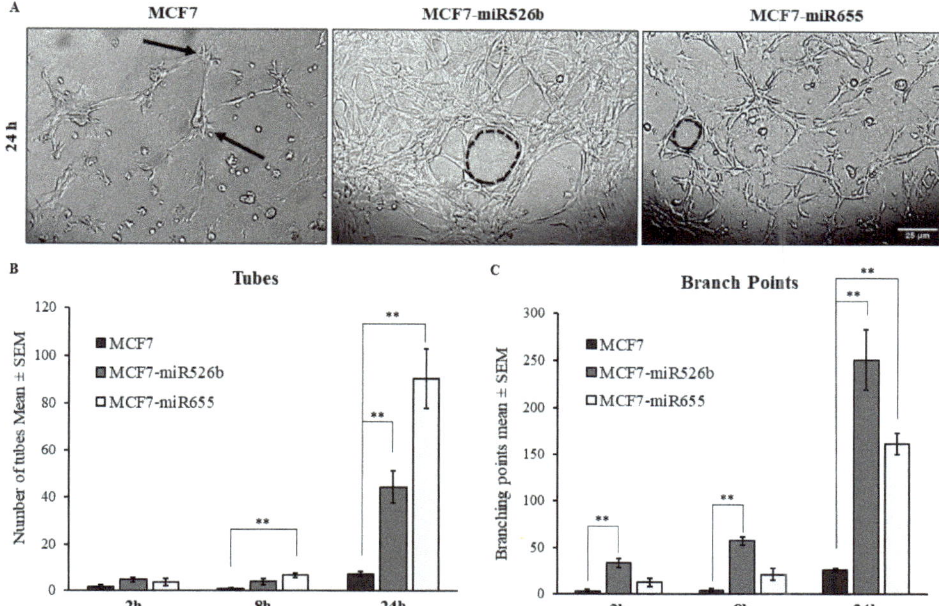

Figure 4. Overexpression of miR526b and miR655 results in an increase of tube formation of HUVECs. (**A**) Visual representation of tube formation of HUVECs at the 24 h time point in conditioned media from MCF7, MCF7-miR526b, or MCF7-miR655 cell line. Both tubes (dotted circle) and branching points (arrows) were greater in MCF7-miR526b or MCF7-miR655 conditioned media compared with MCF7 conditioned media. (**B**) Quantitative data represent number of tubes per time point, per condition. (**C**) Quantitative data represent number of branching points per time point, per condition. Quantitative data presented as the mean ± SEM of three biological replicates, including three experimental replicates per biological replicate; ** $p < 0.01$.

2.4. Treatments with COX2 Inhibitor (COX-I) and EP4 Antagonist (EP4A) Significantly Inhibits miRNA Induced Functions

PGE2 is the major product of COX-2 enzyme activity. The activation of *EP4* receptor by binding of PGE2 results in activation of the PI3K/Akt signalling pathway and is associated with promotion of tumour cell migration and angiogenesis [6]. To examine whether the stimulatory actions of MCF7-miR526b and MCF7-miR655 is the result of involvement of COX-2 activity or *EP4*-signalling, we tested both the migration (Figure 3) and tube formation (Figure 5) phenotypes of HUVECs stimulated with miRNA-conditioned media, with the addition of either COX-2 inhibitor (COX2-I, NS398), *EP4* antagonist (*EP4A*, ONO-AE3-208), or PI3K/Akt pathway inhibitor (Wortmannin, WM).

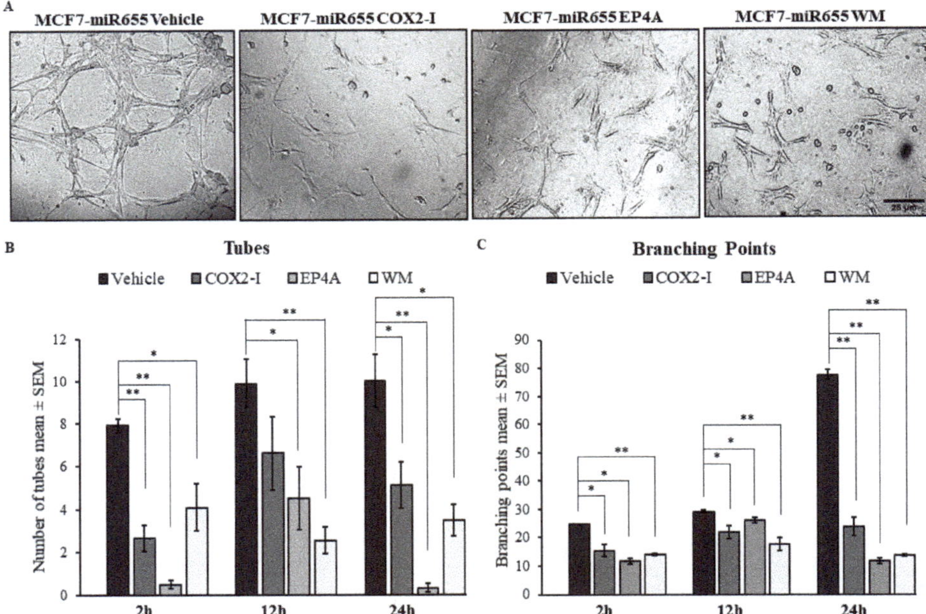

Figure 5. Treatment with COX2-I, *EP4A,* or PI3K pathway inhibitor (Wortmannin, WM) abrogates the tube formation stimulation abilities of MCF7-miR655 conditioned media. (**A**) Images of all treatments; miRNA-conditioned media with vehicle, COX2-I, *EP4A,* or PI3K inhibitor (WM) conditions at the 24 h time point. (**B**) Quantitative analysis of number of tubes formed per time point, per condition. (**C**) Quantitative analysis of branching points per time point, per condition. Quantitative data presented as the mean ± SEM of three biological replicates, including three experimental replicates per biological replicate; * $p < 0.05$, ** $p < 0.01$.

2.4.1. Inhibition of Cell Migration

Specifically, for MCF7-miR526b conditioned media, the addition of COX2-I or *EP4A* significantly inhibited migration of HUVECs at the 8 h and 24 h time points, when compared with the vehicle. (images at 24 h are shown in Figure 3C, and quantification in Figure 3D). Moreover, for MCF7-miR655 conditioned media, addition of COX2-I and *EP4A* significantly inhibited HUVEC migration at 24 h (images at 24 h are shown in Figure 3E and quantification provided in Figure 3F). Additional time points of migration of HUVECs with COX2-I or *EP4A* are presented in Figure S5.

2.4.2. Inhibition of Tube Formation

The addition of COX2-I, *EP4A,* or WM to MCF7-miR655 conditioned media significantly reduced the number of both tubes and branching points formed by HUVECs at the 24 h timepoint (images at 24 h are shown in Figure 5A and quantification provided in Figure 5B,C). While COX2-I could inhibit miR655-conditioned media induced tube formation and branching formation, the addition of either *EP4A* or WM resulted in complete inhibition tube formation of HUVECs as early as 2 h, and we observed the same trend in all other time points compared with the vehicle. Quantitative data are presented for 2 h, 12 h, and 24 h (Figure 5B,C), and images are presented only at 24 h (Figure 5A). Images captured at other time points are provided in Figure S6.

2.5. Expression of miR526b and miR655 in Human Breast Tumour Tissue Correlated with Angiogenesis and Lymphangiogenesis Markers

We investigated miRNA expression *in situ* with human breast tumour tissue retrieved from Ontario Institute of Cancer Research (OICR) Tumour Bank to examine the relationship of miR526b and miR655 expression with expression of angiogenesis and lymphangiogenesis markers. We used 105 tumour tissue samples and 20 non-cancerous tissues. Demographic data of the sample are provided in Table 1 and a description of the sample has been published previously [8,10,11,21]. The majority of the tumor samples used in this study were ER and PR positive and HER-2 negative; with only 10 triple negative breast cancer samples. Moreover, this set has no stage IV tumor and only a few stage I tumor samples. Taqman gene expression (quantitative real time polymerase chain reaction; qRT-PCR) analysis was performed on cDNA produced from each sample and the relative fold change of mRNA was measured to compare control and tumour tissues. We found that the tumour samples illustrated higher expression of *VEGFA* (29.2 fold) than the control group (Figure 6A). The expression of both miR526b and miR655 in this sample set was quantified and published earlier [10,11]. Here, we evaluated the correlation between miRNA expression and *VEGFA* expression in tumour samples. Both miRNAs show a positive correlation with *VEGFA*, with miR526b having a correlation coefficient of R = 0.338 ($p < 0.001$) (Figure 6B), and miR655 having a correlation coefficient of R = 0.425 ($p < 0.0001$) (Figure 6C). Therefore, in both cases, *VEGFA* expression increases as miRNA expression increased. We previously reported the expression of angiogenesis marker *CD31* in this same sample set [21]. Here, we show that the expression of both miR526b and miR655 was significantly correlated with *CD31* expression, showing a positive correlation of R = 0.463 ($p < 0.0001$) for miR655 (Figure 6D) and R = 0.526 ($p < 0.0001$) for miR526b (Figure 6E).

Table 1. Demography, tobacco exposures, tumour grade, and hormone status of the control and tumor samples used in this study. Patient demography of human tissue biopsy samples illustrated. Samples were age matched; majority of the sample members are female. Also, alcohol consumption, tobacco exposure, and hormone receptor status were quantified, none was significantly different. Sample description has been published previously [8–10,21].

Subjects		Control n = 20 (%)	Cancer n = 105 (%)
Sex	Male	0	3 (2.8)
	Female	20 (100)	102 (97.2)
Age Distribution (years)	Range	52–87	27–92
Age (years)	Mean ± SD	66 ± 11	64 ± 12
Smoking Habit	Smokers	1 (5)	3 (2.8)
	Pack Year (PY)	40	56 ± 11
Alcohol Consumption	Social/Occasional Drinker	5 (25)	29 (27.62)
	Regular Drinker	0	3 (2.8)
Estrogen Receptor (ER) Status	Positive		80 (76)
	Negative		19(18)
Progesteron Receptor Status (PR) Status	Positive		66 (62.9)
	Negative		33 (31)
Human epidermal growth factor receptor 2 (HER2) Status	Positive		21 (20)
	Negative		68 (64.8)
ER, PR, HER2 (Triple) Negative	Negative		10 (9.5)

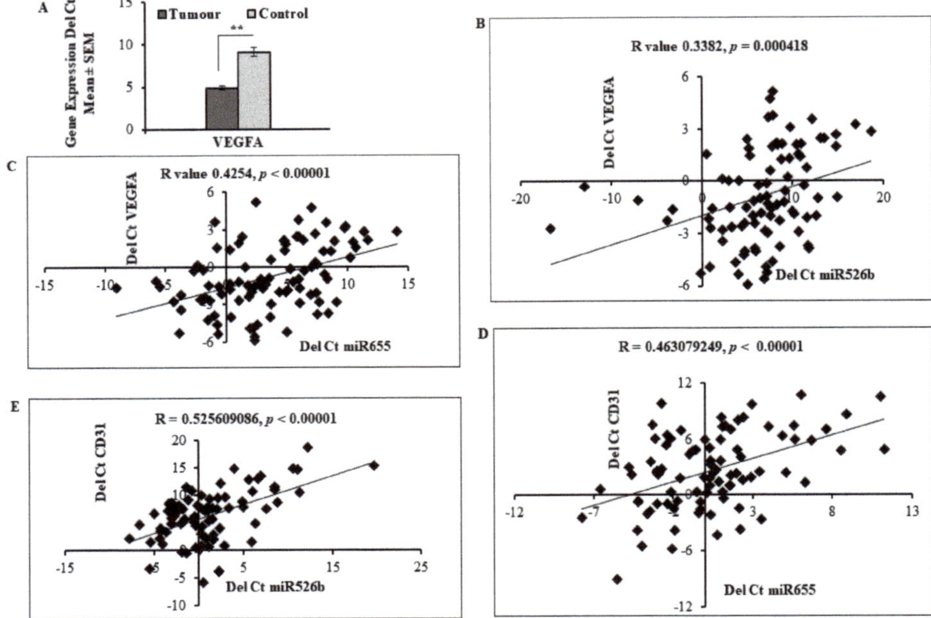

Figure 6. miR526b and miR655 expression is positively correlated with angiogenesis and vascular markers in human breast tumours. (**A**) qRT-PCR analysis of angiogenesis marker *VEGFA* mRNA expression in control (adjacent non-tumour) and tumoural tissues. Data are represented as mean ± SD. (******) indicates significant differences ($p < 0.01$). (**B,C**) miR526b and miR655 expression is positively correlated with *VEGFA* in primary breast cancer samples; Pearson's coefficient indicates positive, but moderate correlations. (**D,E**) miR526b and miR655 expression level is positively correlated with *CD31* in primary breast cancer samples. Pearson's coefficient suggests moderate correlation between the two variables.

We have previously measured lymphangiogenesis markers *VEGFC* and *LYVE-1* expression in these tumour samples [21]. For the first time here, we report that *VEGFD* expression is significantly high in tumour samples compared with control tissue (Figure 7A). Expression of miR526b and miR655 in human tumour tissue was significantly correlated with lymphangiogenesis markers *VEGFC*, *VEGFD*, and *LYVE-1* (Figure 7). Specifically, miR526b and miR655 were very significantly correlated with *VEGFD*. Not only did *VEGFD* show significantly higher expression in tumour samples (Figure 7A), a strong positive correlation of R = 0.7652 ($p < 0.00001$) with miR526b (Figure 7B) and R = 0.933 ($p < 0.00001$) with miR655 (Figure 7C) is observed. Furthermore, miR526b had a positive correlation coefficient of R = 0.5286 ($p < 0.00001$) with *VEGFC* (Figure 7D), and miR655 had a coefficient of R = 0.6053 ($p < 0.00001$) with *VEGFC* (Figure 7E). Expression of *LYVE-1* was also positively correlated with both miRNAs, showing a positive correlation coefficient (R = 0.5256, $p < 0.00001$) for miR526b in Figure 7F as well as for miR655 (Figure 7G) (R = 0.3437, $p = 0.001279$).

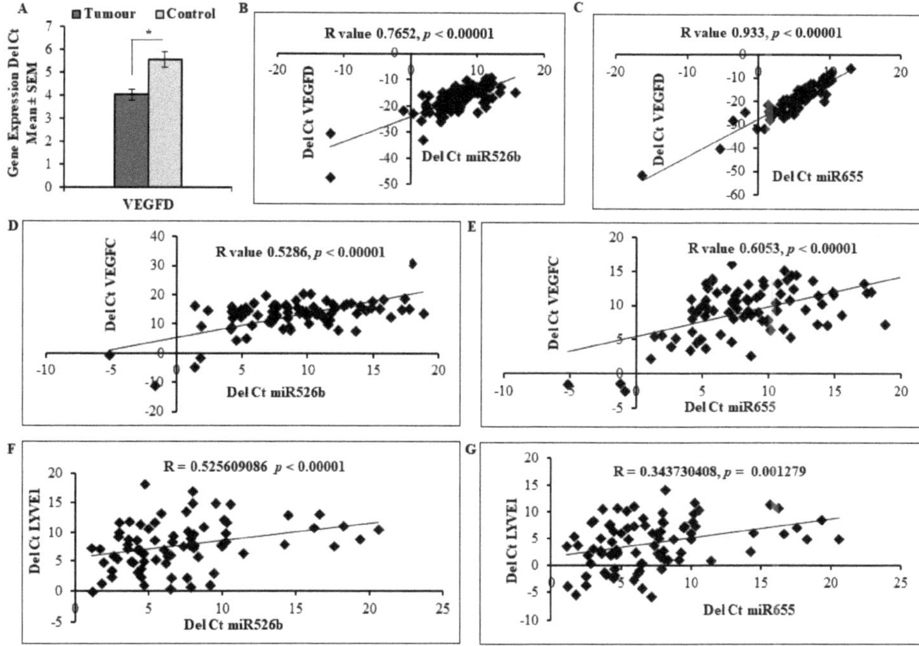

Figure 7. miR526b and miR655 expression is positively correlated with lymphangiogenesis markers in human breast tumours. (**A**) qRT-PCR analysis of *VEGFD* mRNA expression in control (adjacent non-tumour) and tumoural tissues. Data are represented as a mean ± SD. (*) indicates significant differences ($p < 0.05$). (**B**,**C**) miR526b and miR655 expression is positively correlated with the *VEGFD* in primary breast cancer samples; Pearson's coefficient indicates a very strong positive correlation. (**D**,**E**) miR526b and miR655 expression level is positively correlated with another lymphangiogenesis marker *VEGFC* in primary breast cancer samples. Pearson's coefficient suggests a positive correlation between the two variables. (**F**,**G**) Finally, *LYVE1* mRNA expression is positively correlated with miR526b and miR655 expression in tumour samples. Pearson's coefficient suggests a moderate correlation between the two variables.

Further, we investigated if there is any difference in the distribution of high miRNA expression across hormone status (Table 2). We subdivided miRNA expression according to low ($+\Delta$Ct value) and very high miRNA expression ($-\Delta$Ct value) in various tumor stages and different hormone receptor status. Both miRNA high samples were higher in proportion in ER and PR positive and HER2 negative samples; however, none of these distributions were statistically significant.

Table 2. Percentage of samples that show higher expression of the two miRNAs (negative delta Ct values) in various stages of tumour and hormonal receptors status. We conducted a Z score analysis by comparing proportions. The analysis shows the proportion of miR526b high expression samples were more in tumor grade II and III; however, this was not significant. For miR655, there was no difference in the distribution of high expression across tumor grades recorded. A higher proportion of samples showing high miR526b and miR655 expression can be seen in ER positive samples; however, the difference is not statistically significant, although larger sample size may increase significance. Other receptor status also shows no significant difference between presence or absence of individual receptor.

Tumour Grade and High miRNA Expression in Cancer Samples			
Tumour Grade	*n* (%)	**miR-526b High** *n* (%)	**miR-655 High** *n* (%)
I (low-well differentiated)	7 (6.7)	0	2 (28.5)
II (intermediate-moderately differentiated)	26 (24.76)	2 (7.69)	7 (26.9)
III (high-poorly differentiated)	63 (60)	5 (7.94)	13 (20.6)
X (Unknown)	9 (8.57)	0	2 (28.6)
Tumour Receptor Status and High miRNA Expression in Cancer Samples			
Receptor Status of Cancer Samples	*n* (%)	**miR-526b High** *n* (%)	**miR-655 High** *n* (%)
ER Status — Positive	40 (38.1)	6 (15)	10 (25)
ER Status — Negative	20 (19)	0	4 (20)
PR Status — Positive	33 (31.4)	3 (9.1)	8 (24.2)
PR Status — Negative	33 (31.4)	2 (6.1)	7 (21.2)
HER2 Status — Positive	22 (21)	2 (9.1)	8 (36.4)
HER2 Status — Negative	68 (64.8)	4 (5.9)	13 (19.1)
ER,PR, HER2 (Triple) Negative — Negative	10 (9.5)	0	0

3. Discussion

In this study, the role of miR526b and miR655 in breast tumour-associated angiogenesis and lymphangiogenesis was investigated. The roles of specific miRNA in tumor-associated angiogenesis and lymphangiogenesis have been well highlighted. In a study by Cascico et al., oncogenic miR-20b was shown to be involved in the regulation of VEGF in breast cancer cells by targeting HIF-1α [22], while the role of miR-10b as an angiogenic regulator was validated in a study by Liu et al. [23]. Expression of VEGFs has been found to be associated with other miRNAs involved in the regulation of angiogenesis and lymphangiogenesis, such as miR-182 [24] and miR-20a [25], which were both found to cause an upregulation of VEGFA; in contrast, another study has highlighted the role of miR-128 as a tumour suppressor miRNA, which results in the reduction of both VEGFA and VEGFC expression [26]. Further, other studies have shown that down regulation of tumour suppressive miR-126 [27] and miR-128 [26] results in the promotion of tumour-associated lymphangiogenesis, which was assessed by the subsequent reduction of a lymphangiogenic marker LYVE1. Therefore, miRNA can directly or indirectly regulate both angiogenesis and lymphangiogenesis by regulating expression of angiogenesis and lymphangiogenesis markers.

We have previously shown that COX-2 overexpression in ER-positive MCF7 breast cancer cells significantly increased expression of miR526b and miR655 [10,11]. Interestingly, overexpression of these two miRNAs in poorly metastatic and ER positive MCF7 cells promotes aggressive breast cancer phenotypes [10,11]. Recently, a common target of both miRNAs *CPEB2* has been validated as a tumor suppressor gene in breast cancer [12]. However, there was no direct report on these two miRNAs and their roles in breast cancer-associated angiogenesis and lymphangiogenesis. miR526b and miR655 can upregulate COX-2 expression via the NFKB pathway [10,11]. It is very well known that COX-2 induces angiogenesis and lymphangiogenesis through production of VEGFC and VEGFD, and upregulation of PI3K/Akt signalling and COX-2 overexpression can also induce LYVE-1 over expression in mouse breast tumours [28,29]. We have previously shown low expression of miR526b and miR655 in ER receptor positive cell lines MCF7 and T47D, and that these miRNA have the highest expression in triple negative breast cancer cells MDAMB231, MCF7-COX2, and Hs578T, with a relative correlation of miRNA expression with COX-2 expression [11]. Thus, we chose to overexpress miR526b and miR655 in an ER positive, poorly metastatic breast cancer cell MCF7 to investigate miRNA gain of functions.

In this study, we show that miR526b and miR655 overexpression results in significant upregulation of angiogenesis and lymphangiogenesis markers, specifically, VEGFA, VEGFC, VEGFD, COX-2, and LYVE-1. Furthermore, the expression of angiogenesis and lymphangiogenesis receptors in miRNA high cells was measured to test the autocrine regulation of miRNA in tumor associated angiogenesis. Although VEGFR1 and VEGFR2 receptors are primarily expressed on endothelial cells, previous studies have reported VEGFR expression in breast cancer cells, establishing the involvement of a VEGF–VEGFR autocrine loop [30]. VEGFR1 expression in breast cancer cells might promote tumour growth and metastasis, and has been established as an unfavourable indicator of progression in breast cancer patients [31]. Moreover, a VEGFR2 autocrine signalling loop has been established in breast cancer cell lines, and has been shown to activate MAP kinase pathways [32]. Here, we show that miR526b and miR655 overexpression in MCF7 cell line results in an extremely significant upregulation of *VEGFR1* expression at the mRNA level. *VEGFR2* expression was also found to be marginally upregulated at the mRNA level in MCF7-miR526b and MCF7-miR655. These results suggest that the upregulation of VEGFA, VEGFC, and VEGFD in MCF7-miR526b and MCF7-miR655 results in the production and release of these ligands into the tumour microenvironment, which may feed into an autocrine loop on breast cancer cells to activate VEGFR1 and VEGFR2 expression. Furthermore, it has been previously shown that high expression of COX-2, and thus overproduction of PGE2, leads to overexpression of miR526b and miR655 in breast cancer [10,11]. Production of PGE2 by COX-2 results in upregulation of PGE2 receptors EPs. Specifically, binding of PGE2 to receptor *EP4* stimulates non-canonical pathways PI3K/Akt and ERK [6], and is associated with tumour-associated angiogenesis and lymphangiogenesis [6,7,14,33,34]. Here, expression of *EP4* receptor was found to be significantly upregulated in MCF7-miR526b and MCF7-miR655 cells, suggesting that PGE2 released into the tumour microenvironment is involved in autocrine signalling through *EP4* receptor on breast cancer cells.

In this study, we also tested the paracrine potential of miR526b and miR655. Other reports have demonstrated the effects of miRNA-overexpressing cell lines and their involvement with angiogenesis in vitro through secretions of stimulatory proteins or by co-culture with endothelial cells such as miR-155 [35], miR-494 [36], and miR-182 [24], which were all established to promote endothelial tube formation and migration. Similarly, we investigated the paracrine potential of MCF7-miR526b and MCF7-miR655 cell line secretions and tested the effects on HUVEC cell tube formation and migration potential. Here, we observed that the conditioned media collected from miR526b and miR655 overexpressing breast cancer cell lines stimulates both tube formation and migration of endothelial cells. Moreover, we found *VEGFR1* and *VEGFR2* to be upregulated in HUVEC cells following treatment with miRNA overexpressing cell conditioned media, and that expression of *EP4* receptor was found to be upregulated in HUVECs treated with miRNA high cell line conditioned media. Because *EP4* is a main receptor that regulates COX-2/PGE2 induced functions [8,10,11,28,29,34], this led us to further investigate the involvement of this *EP4* signalling pathway in miRNA induced angiogenesis and lymphangiogenesis. To test *EP4* signalling mechanisms, we used a specific COX-2I, *EP4A* or PI3K/Akt inhibitor along with miRNA overexpressing cell conditioned media. *EP4A* could significantly abrogate HUVECs' migration and tube formation, however, the PI3K/Akt inhibitor completely blocked these phenotypes, suggesting that miR526b and miR655 induce angiogenesis and lymphangiogenesis via *EP4*/PI3K/Akt pathways. Thus, miR526b and miR655 regulate angiogenesis through the production of VEGFs and their subsequent release into the tumour microenvironment for both paracrine and autocrine signalling.

To further investigate the translational involvement of miR526b and miR655 in tumour associated angiogenesis and lymphangiogenesis, we examined the expression of angiogenesis and lymphangiogenesis markers in situ. The majority of the tumor samples used in this study are ER and PR positive and HER2 negative, with only a few triple negative breast cancer samples. We have previously shown that in this set of human tumour samples both miR526b and miR655 expression is high [10,11]. We found proportionately more samples with high miRNA expression in the ER positive samples, but the distribution was not statistically significantly, which could be because of the fact that we have only a few sample with very high expression. We have also previously shown expression of *CD31*, *VEGFC*, and *LYVE1* to be high in this tumour set [21]. Here, we show that in malignant breast samples, expression of *VEGFA* and *VEGFD* is significantly high compared with the disease-free control samples. Moreover, we show that expression of both miR526b and miR655 is positively correlated with expression of angiogenesis (*CD31* and *VEGFA*) and lymphangiogenesis markers (*VEGFC*, *VEGFD*, and *LYVE1*) in human breast cancer tissue. We found the strongest correlation of miRNA expression with *VEGFD* expression in tumor samples, which confirms our cell line findings of high expression of VEGFD in miRNA high cell lines. This observation is also supported by other studies showing a correlation of miRNA with angiogenic markers, including the positive correlation of miR-20a expression in breast tumour tissue with *VEGFA* expression [25]. Further, a study by He et al. found that tumour suppressive miR-186 was negatively correlated with expression of *VEGFC* in tumour tissue collected from bladder cancer patients [37].

The exact mechanisms by which miR526b and miR655 stimulate angiogenesis and lymphangiogenesis remain unknown. However, we have shown that COX-2 overexpression results in overexpression of both miR526b and miR655, and further results in overproduction of inflammatory PGE2. We have also shown that miRNA overexpression induces COX-2 expression via NFKB pathway [10,11]. Therefore, miRNA overexpressing cell lines are high in both production of COX-2 and secretions of PGE2. Furthermore, with bioinformatics analysis, we have previously reported that *PTEN* (phosphatase and tensin homolog) is a target of both miR526b and miR655 [10]. Here, we validated that *PTEN* is indeed a direct target of miR526b and miR655, as it is significantly down regulated in both MCF7-miR526b and MCF7-miR655 cell lines compared with MCF7 (Figure 8A). *PTEN* is also a known target of other established miRNAs, such as miR-494, which has been reported by Mao et al. to target *PTEN* in response to hypoxic conditions to promote angiogenesis and tumour growth in non-small lung cancer [36]. *PTEN* acts as a tumour suppressor gene both by inhibiting the PI3K/Akt pathway, and by acting as a negative regulator of HIF-1α (hypoxia-inducible factor-1) [38,39]. In turn, HIF-1α is a transcription factor known to promote the transcription of many angiogenesis-associated genes, including VEGFs [40]. Although we did not measure HIF-1α expression in our samples, we speculate that both miR526b and miR655 target this pathway, resulting in VEGF secretions into the tumour microenvironment, thus facilitating tumour-associated angiogenesis and lymphangiogenesis. The proposed pathway is presented in Figure 8B. A similar report in triple negative breast cancer was shown, in which *PTEN* downregulation resulting in cell proliferation via the PI3K pathway was shown [41].

Here, we established the roles of miR526b and miR655 as promoter and regulator of breast cancer angiogenesis and lymphangiogenesis using an ER positive breast cancer cell model MCF7 and showed that miRNA expression is high in ER positive breast cancer samples and highly correlated with angiogenesis and lymphangiogenesis markers in breast cancer. Further investigation into the roles of these miRNAs in the triple negative breast cancer model and incorporation of more tumor samples of various hormonal receptor status and tumor stages would give greater insight into the mechanisms of these miRNA across differential subtypes of breast cancer.

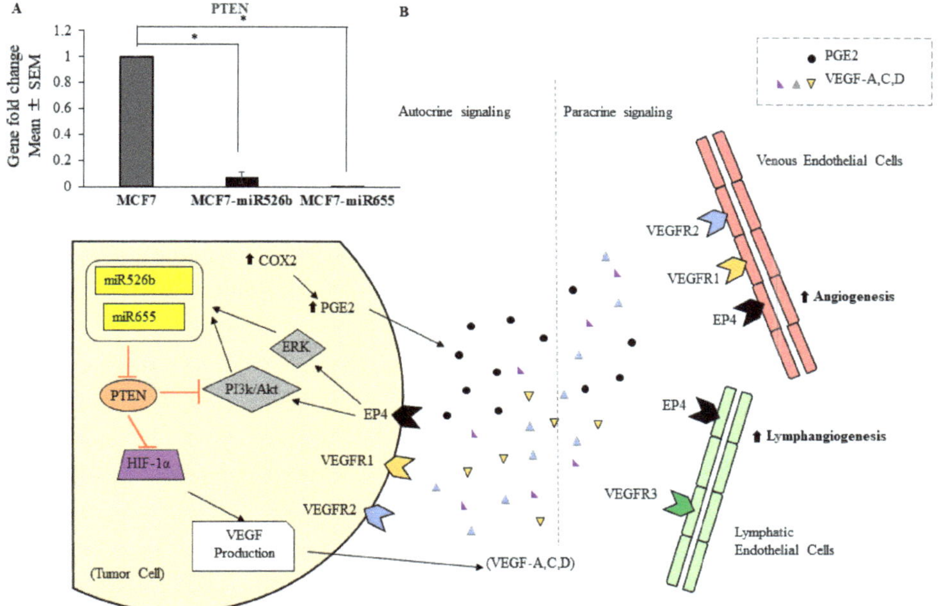

Figure 8. (**A**) miR526b and miR655 overexpressing cells showing significant down regulation of tumour suppressor gene *phosphatase and tensin homolog (PTEN)* mRNA expression. Data are represented as mean ± SD.; * ($p < 0.05$). (**B**) The proposed pathway of tumour associated angiogenesis and lymphangiogenesis promotion by miR655 and miR526b. miRNA over expressing cells are high in COX-2 expression, resulting in production of PGE2 and its subsequent release into the tumour microenvironment. PGE2 can signal through *EP4* in both an autocrine and paracrine fashion. Signalling through *EP4* on tumour cells stimulates ERK and PI3k/Akt signalling pathways, resulting in the upregulation of miR526b and miR655. Subsequently, miR526b and miR655 target tumour suppressor gene *PTEN*. Thus, the inhibitory effect of *PTEN* on hypoxia-inducible factor-1 (HIF-1a), as well as the PI3K/Akt pathway, is prevented by upregulation of these miRNAs. Without regulation, transcription factor HIF-1a promotes VEGFA, C, D transcription, and thus an overproduction of VEGFs and their subsequent release into the tumour microenvironment for both autocrine and paracrine signalling. VEGF molecules will bind to various VEGF receptor (VEGFR1, VEGFR2) molecules present on vascular endothelial cells (VECs) and lymphatic endothelial cells (LECs) to promote angiogenesis and lymphangiogenesis, respectively. In the absence *PTEN*'s regulation on PI3k/Akt pathway within the tumour cell, growth and proliferation is promoted.

4. Materials and Methods

4.1. Ethics Statements

Brandon University Research Ethics Committee approves this study (#21986, 21 April 2017). The human tissues used in this project were obtained by Dr. Peeyush K Lala at the University of Western Ontario (UWO) from the Ontario Institute for Cancer Research (OICR) repository (created on the basis of donor consent) following approval of human ethics by the Ethics Review Board of the OICR and UWO. Total RNA and miRNA were extracted using Qiagen (Qiagen, Toronto, ON, Canada) RNA and miRNA extraction kits followed by cDNA synthesis at UWO, and a portion of cDNA of all samples were transferred to Dr. Mousumi Majumder at Brandon University following a material transfer agreement.

4.2. Cell Culture

Human breast cancer cell line MCF7 was purchased from American Type Culture Collection (ATCC, Rockville, MD, USA). Stable miRNA overexpressing MCF7-miR526b and MCF7-miR655 cell lines were established as previously described [9,10]. MCF7, MCF7-miR526b, and MCF7-miR655 cells were all grown in minimal essential medium (MEM) (Life Technologies, Thermofisher, Ottawa, ON, Canada) supplemented with 10% fetal bovine serum (FBS) and 1% Penstrep. Furthermore, MCF7-miR526b and MCF7-miR655 cell lines were sustained with Geneticin (Life Technologies Thermofisher, Ottawa, ON, Canada) at 200 ng/mL.

HUVECs were purchased from Life Technologies and grown in Medium 200 (GIBCO, ON) supplemented with low serum growth supplement (LSGS) kit (GIBCO, Toronto, ON, Canada) containing 2% FBS, hydrocortisone (1 µg/mL), human epidermal growth factor (10 ng/mL), basic fibroblast growth factor (3 ng/mL), and heparin (10 µg/mL). All cell lines were maintained in a humidified incubator at 37 °C with 5% CO_2.

4.3. Collection of Conditioned Media

MCF7, MCF7-miR526b, and MCF7-miR655 cell lines were grown in complete serum supplemented media until 90% confluent. Cells were then washed with 1X phosphate buffered saline (PBS) to remove any trace of serum. The cells were then starved with basal MEM (serum-free) for 12 h prior to collection of media. Cell free supernatant was then collected for HUVEC functional assays.

4.4. Drugs and Chemicals

NS398 (COX-2 inhibitor) was purchased from Cayman Chemical (Ann Arbor, MI, USA). ONO-AE3-208 (selective *EP4* antagonist, *EP4A*) was a gift from ONO Pharmaceuticals, Osaka, Japan. Wortmannin (WM), an irreversible PI3K inhibitor purchased from Sigma-Aldrich (Saint Louis, MO, USA), Dr. Lala generously shared 1 mM of each drug with us. For all treatments in vitro, DMSO (vehicle) served as the control.

4.5. Tube Formation Assay

The assay was carried out as previously described [21,34], using a 24-well plate. HUVECs were resuspended in either non-supplemented, serum free Medium 200 to serve as a negative control; MCF7 conditioned media, MCF7-miR526b conditioned media, or MCF7-miR655 conditioned media as experimental conditions; or complete serum supplemented Medium 200 to serve as the positive control. Each condition was tested in triplicate (experimental replicates) and repeated three times (biological replicates). HUVECs with each condition were then seeded on a growth factor reduced Matrigel (BD Biosciences, Mississauga, ON, Canada) coated 24-well plate. Matrigel was prepared using a 1:2 ratio with one part Matrigel and two parts un-supplemented Medium 200. Tube formation was examined at different time intervals, and images were obtained with a Nikon inverted microscope. Quantification of tubes and branching points was carried out using NIH Image J software (Version 64, NIH, Bethesda, MD, USA). To test the involvement of COX-2, *EP4* receptor, and the PI3K/Akt pathway, tube formation assay was repeated with cancer cell conditioned media along with 20 µM NS398 or 50 µM ONO-AE3-208. To confirm the involvement of the PI3K/Akt signalling pathway, we also used 10 µM WM, an irreversible P13K/Akt inhibitor.

4.6. HUVEC Migration Assay

HUVECs were grown in LSGS supplemented Medium 200 in a T75 flask, then harvested and re-suspended in supplemented Medium 200, after which 300 µL of harvested cells was added in a 24-well attachment plate and maintained until 90% confluency. The surface of each well was scratched with a 2 µL sterile micropipette tip and cells were washed with PBS. Each condition was then applied to the wells. Basal Medium 200 served as the negative control, and serum supplemented Medium

200 acted as the positive control. MCF7, MCF7-miR526b, and MCF7-miR655 conditioned media were the experimental conditions. A total of 300 µL of the respective condition was added per well. To test the involvement of the *EP4* and the PI3K/Akt signalling pathway, 300 µL of cell-free conditioned media from MCF7-miR526b or MCF7-miR655 was added to the wells. In another set of experiments along with cancer cell conditioned media, additionally either 20 µM NS398, 50 µM ONO-AE3-208, or 10 µM WM was added, and DMSO served as a control to test the involvement of COX-2, *EP4*, and the PI3K/Akt signalling pathway in cell migration. The progress of HUVEC migration and photos of the scratch wound size were captured using an inverted microscope at differing time intervals, and NIH ImageJ software was used to measure the width of the wound in pixels.

4.7. Quantitative Real-Time PCR (qRT-PCR)

Total RNA was extracted from MCF7, MCF7-miR526b, and MCF7-miR655 cell lines using miRNeasy Mini Kit (Qiagen, Toronto, ON, Canada) and reverse transcribed using the TaqMan microRNA and mRNA cDNA Reverse Transcription Kit (Applied Biosystems, Waltham, MA, USA). The TaqMan MiRNA or gene expression assays was used for quantitative PCR. Two control markers, *Beta-actin* (Hs01060665_g1) and *RPL5* (Hs03044958_g1) expression was quantified using RT-PCR and used to normalize the expression of the following angiogenesis and lymphangiogenesis markers using relative analysis: *VEGFA* (Hs00900055_m1), *VEGFC* (Hs01099203_m1), *FIGF* (Hs01128659_m1), *PTGS2* (Hs00153133_m1), *LYVE1* (Hs00272659_m1), *CD31* (Hs01065279_m1), and *PTGER4* (Hs00168761_m1). Moreover, expression of tumour suppressor gene *PTEN* (Hs0082981_s1) was examined. Gene expression was measured using delta CT values to obtain the fold change, as described earlier [42].

For quantification of receptor expression on HUVECs, HUVECs were grown in a 6-well plate until confluent. HUVECs were then treated with MCF7, MCF7-miR526b, or MCF7-miR655 conditioned media for 12 h. HUVECs were then trypsinized and collected for RNA extraction and cDNA synthesis. Quantitative qPCR was carried out with *FLT1* (Hs01052961_m1) (*VEGFR1*), *KDR* (Hs00964383_g1) (*VEGFR2)*, and *PTGER4* (Hs00911700) (*EP4*) as described above. We also measured *VEGFR1, VEGFR2,* and *EP4* in MCF7, MCF7-miR526b, and MCF7-miR655 cell lines.

4.8. Western Blot

Cancer cells were treated with M-PER® Mammalian Protein Extraction Reagent (Thermo Scientific, Rockford, IL, USA), HALT Protease Inhibitor Cocktail (Thermo Scientific), and Phosphatase Inhibitor Cocktail (Thermo Scientific) to extract total protein. A total of 15–20 µg of total protein were electrophoresed per well on a SDS-polyacrylamide gel; transferred onto Immobilon-FL PVDF membranes (Millipore, Billerica, MA, USA); and further incubated with the following primary antibodies: VEGFA (sc-507), VEGFC (sc-1881), VEGFD (sc-13085), LYVE1 (sc-28190), and COX-2 (sc-1747), using antibodies (1:500 dilutions) from Santa Cruz Biotechnology (Santa Cruz Biotechnology, Santa Cruz, CA, USA). Monoclonal GAPDH antibody (MAB374) was from Millipore, Billerica, MA, USA. After blocking with primary antibodies, overnight blots were probed with a mixture of IRDye polyclonal secondary antibodies (LI-COR Biosciences, Lincoln, NE, USA). Images were read with an Odyssey infrared imaging system (LI-COR Biosciences).

4.9. Human Breast Cancer Tissue Samples

Frozen human breast tumour ($n = 105$) and control ($n = 20$) tissue samples were obtained previously from the Ontario Tumour Bank with the demographic description shown in Table 1. Qiagen miRNeasy mini kit was used to extract mRNA or miRNA from tissue samples, followed by cDNA synthesis using cDNA Reverse Transcription Kit (Life Technologies, Applied Biosystems, cat # 4368814, ON, Canada) To examine the potential correlations between miRNA and lymphangiogenesis or angiogenesis markers, tissue sample cDNA was screened using qRT-PCR. Expression of *miR526b* (Hs03304873_pri) and *miR655* (Hs03296227_pri); angiogenesis markers *VEGFA* (Hs00900055_m1) and *CD31* (Hs01065279_m1);

and lymphangiogenesis markers *LYVE1* (Hs00272659_m1), *VEGFC* (Hs01099203_m1), and *VEGFD* (Hs01128659_m1) were all examined.

4.10. Statistical Analysis

Statistical calculations were performed using GraphPad Prism software version 5 (GraphPad Software, San Diego, CA, USA). All parametric data were analyzed with one-way analysis of variance (ANOVA) followed by Tukey–Kramer or Dunnett post-hoc comparisons. Student's *t*-test was used when comparing two datasets and Pearson's coefficient was employed to assess statistical correlations. We used Z-score to compare the proportion of miRNA high expression in various tumor grades and ER, PR, and HER2 status positive and negative samples in the tumors. Statistically relevant differences between means were accepted at $p < 0.05$.

5. Conclusions

The roles of angiogenesis and lymphangiogenesis in breast cancer tumour growth and metastasis have been well established. Breast cancer metastasis requires that primary tumour cells possess the ability to enter the blood or lymph vessels and travel to secondary sites [43]. This is greatly facilitated by angiogenesis and lymphangiogenesis [44,45]. Overall, our study establishes the involvement of microRNA (miR526b and miR655) in tumour-associated angiogenesis and lymphangiogenesis, specifically in ER positive breast cancer. These findings further establish the involvement of these miRNAs in breast cancer metastasis and their potential as future breast cancer biomarkers. Our study also validates the involvement of PGE2 signalling through *EP4* receptor, and the subsequent PI3K/Akt pathways in these processes, further validating the use of *EP4* receptor antagonists as a potential therapeutic target in COX2-high breast cancer patients.

Supplementary Materials: The following are available online at http://www.mdpi.com/2072-6694/11/7/938/s1, Figure S1: Western blot analysis of VEGFs in cancer cell lines, Figure S2: Western blot analysis of COX-2 and LYVE1 in cancer cell lines, Figure S3: MCF7-miR526b and MCF7-miR655 conditioned media promotes cellular migration of HUVECs, Figure S4: Secretions collected from miR526b and miR655 overexpressing cells results in an increase of tube formation of HUVECs, Figure S5: HUVEC migration assays. Figure S6: HUVEC tube formation assays.

Author Contributions: Concept, project design, and supervision: M.M.; Experiments: M.M., K.C.U., B.N., and S.H.; Data Analysis: M.M., K.C.U., B.N., S.H., and S.M.; Figures and Image Data Processing: B.N., K.C.U., S.M., and S.H.; Manuscript writing: S.H., B.N., K.C.U., and M.M.

Funding: This research is funded by NSERC-Discovery grant and Brandon University Research Committee grants to Mousumi Majumder. Stephanie Hunter and Braydon Nault are recipients of NSERC-USRA scholarships.

Acknowledgments: The authors thank Bernadette Ardelli at Brandon University (BU) for giving us access to dissection microscope and the Rotor-gene PCR machine and Vincent Chen at BU for kindly allowing access to the PCR machine. We acknowledge the assistance of Patrice Colville at BU with cell culturing, RNA extraction, and qRT-PCR. We thank Ling Liu at UWO for helping us with Western blot development. We sincerely thank Peeyush K Lala at UWO for providing us with human tumour tissue from the Ontario Tumour Bank and for sharing NS398, *EP4A*, and WM with us.

Conflicts of Interest: The authors declare no conflict of interest.

References

1. *Global Health Risks: Mortality and Burden of Disease Attributable to Selected Major Risks*; World Health Organization: Geneva, Switzerland, 2009; ISBN 9789241563871.

2. Singh-Ranger, G.; Salhab, M.; Mokbel, K. The Role of Cyclooxygenase-2 in Breast Cancer: Review. *Breast Cancer Res. Treat.* **2008**, *109*, 189–198. [CrossRef] [PubMed]

3. Howe, L.R. Inflammation and Breast Cancer. Cyclooxygenase/Prostaglandin Signaling and Breast Cancer. *Breast Cancer Res. BCR* **2007**, *9*, e210. [CrossRef] [PubMed]

4. Ristimaki, A.; Sivula, A.; Lundin, J.; Lundin, M.; Salminen, T.; Haglund, C.; Joensuu, H.; Isola, J. Prognostic Significance of Elevated Cyclooxygenase-2 Expression in Breast Cancer. *Cancer Res.* **2002**, *62*, e632.

5. Kawahara, K.; Hohjoh, H.; Inazumi, T.; Tsuchiya, S.; Sugimoto, Y. Prostaglandin E2-Induced Inflammation: Relevance of Prostaglandin E Receptors. *BBA Mol. Cell Biol. Lipids* **2015**, *1851*, 414–421. [CrossRef] [PubMed]

6. Majumder, M.; Nandi, P.; Omar, A.; Ugwuagbo, K.C.; Lala, P.K. EP4 as a Therapeutic Target for Aggressive Human Breast Cancer. *Int. J. Mol. Sci.* **2018**, *19*, 1019. [CrossRef]

7. Ugwuagbo, K.C.; Maiti, S.; Omar, A.; Hunter, S.; Nault, B.; Northam, C.; Majumder, M. Prostaglandin E2 Promotes Embryonic Vascular Development and Maturation in Zebrafish. *Biol. Open* **2019**, *8*, bio039768. [CrossRef]

8. Majumder, M.; Xin, X.; Liu, L.; Tutunea-Fatan, E.; Rodriguez-Torres, M.; Vincent, K.; Postovit, L.; Hess, D.; Lala, P.K. COX-2 Induces Breast Cancer Stem Cells Via EP4/PI3K/AKT/NOTCH/WNT Axis. *Stem Cells* **2016**, *34*, 2290–2305. [CrossRef]

9. Singh, R.; Mo, Y. Role of microRNAs in Breast Cancer. *Cancer Biol. Ther.* **2013**, *14*, 201–212. [CrossRef]

10. Majumder, M.; Landman, E.; Liu, L.; Hess, D.; Lala, P.K. COX-2 Elevates Oncogenic miR-526b in Breast Cancer by EP4 Activation. *Mol. Cancer Res.* **2015**, *13*, 1022–1033. [CrossRef]

11. Majumder, M.; Dunn, L.; Liu, L.; Hasan, A.; Vincent, K.; Brackstone, M.; Hess, D.; Lala, P.K. COX-2 Induces Oncogenic Micro RNA miR655 in Human Breast Cancer. *Sci. Rep.* **2018**, *8*, e3. [CrossRef]

12. Tordjman, J.; Majumder, M.; Amirr, M.; Hasan, A.; Hess, D.; Lala, P.K. Tumor suppressor role of cytoplasmic polyadenylation element binding protein 2 (CPEB2) in human mammary epithelial cells. *BMC Cancer* **2019**, *19*, e561. [CrossRef] [PubMed]

13. Urbich, C.; Kuehbacher, A.; Dimmeler, S. Role of microRNAs in Vascular Diseases, Inflammation, and Angiogenesis. *Cardiovasc. Res.* **2008**, *79*, 581–588. [CrossRef] [PubMed]

14. Lala, P.K.; Nandi, P.; Majumder, M. Roles of Prostaglandins in Tumor-Associated Lymphangiogenesis with Special Reference to Breast Cancer. *Cancer Metastasis Rev.* **2018**, *37*, 1–16. [CrossRef] [PubMed]

15. Otrock, Z.K.; Mahfouz, R.A.R.; Makarem, J.A.; Shamseddine, A.I. Understanding the Biology of Angiogenesis: Review of the most Important Molecular Mechanisms. *Blood Cells Mol. Dis.* **2007**, *39*, 212–220. [CrossRef] [PubMed]

16. Lou, W.; Liu, J.; Gao, Y.; Zhong, G.; Chen, D.; Shen, J.; Bao, C.; Xu, L.; Pan, J.; Cheng, J.; et al. MicroRNAs in Cancer Metastasis and Angiogenesis. *Oncotarget* **2017**, *8*, e115787. [CrossRef]

17. Ran, S.; Volk, L.; Hall, K.; Flister, M.J. Lymphangiogenesis and Lymphatic Metastasis in Breast Cancer. *Pathophysiology* **2010**, *17*, 229–251. [CrossRef] [PubMed]

18. Shibuya, M. Vascular Endothelial Growth Factor (VEGF) and its Receptor (VEGFR) Signaling in Angiogenesis: A Crucial Target for Anti- and Pro-Angiogenic Therapies. *Genes Cancer* **2011**, *2*, 1097–1105. [CrossRef]

19. Park, S.; Sorenson, C.M.; Sheibani, N. PECAM-1 Isoforms, eNOS and Endoglin Axis in Regulation of Angiogenesis. *Clin. Sci.* **2015**, *129*, 217–234. [CrossRef]

20. Rao, R.; Redha, R.; Macias-Perez, I.; Su, Y.; Hao, C.; Zent, R.; Breyer, M.D.; Pozzi, A. Prostaglandin E2-EP4 Receptor Promotes Endothelial Cell Migration Via ERK Activation and Angiogenesis in Vivo. *J. Biol. Chem.* **2007**, *282*, 16959–16968. [CrossRef]

21. Tutunea-Fatan, E.; Majumder, M.; Xin, X.; Lala, P.K. The Role of CCL21/CCR7 Chemokine Axis in Breast Cancer-Induced Lymphangiogenesis. *Mol. Cancer* **2015**, *14*, 35. [CrossRef]

22. Cascio, S.; D'Andrea, A.; Ferla, R.; Surmacz, E.; Gulotta, E.; Amodeo, V.; Bazan, V.; Gebbia, N.; Russo, A. miR-20b Modulates VEGF Expression by Targeting HIF-1 Alpha and STAT3 in MCF-7 Breast Cancer Cells. *J. Cell. Physiol.* **2010**, *224*, 242–249. [PubMed]

23. Liu, X.; Guan, Y.; Wang, L.; Niu, Y. MicroRNA-10b Expression in Node-Negative Breast Cancer-Correlation with Metastasis and Angiogenesis. *Oncol. Lett.* **2017**, *14*, 5845–5852. [CrossRef] [PubMed]

24. Chiang, C.; Chu, P.; Hou, M.; Hung, W. MiR-182 Promotes Proliferation and Invasion and Elevates the HIF-1α-VEGF-A Axis in Breast Cancer Cells by Targeting FBXW7. *Am. J. Cancer Res.* **2016**, *6*, 1785. [PubMed]

25. Luengo-Gil, G.; Gonzalez-Billalabeitia, E.; Perez-Henarejos, S.A.; Manzano, E.N.; Chaves-Benito, A.; Garcia-Martinez, E.; Garcia-Garre, E.; Vicente, V.; Ayala La Peña, F. Angiogenic Role of miR-20a in Breast Cancer. *PLoS ONE* **2018**, *13*, e0194638. [CrossRef] [PubMed]

26. Hu, J.; Cheng, Y.; Li, Y.; Jin, Z.; Pan, Y.; Liu, G.; Fu, S.; Zhang, Y.; Feng, K.; Feng, Y. microRNA-128 Plays a Critical Role in Human Non-Small Cell Lung Cancer Tumourigenesis, Angiogenesis and Lymphangiogenesis by Directly Targeting Vascular Endothelial Growth Factor-C. *Eur. J. Cancer* **2014**, *50*, 2336–2350. [CrossRef] [PubMed]

27. Sasahira, T.; Kurihara, M.; Bhawal, U.K.; Ueda, N.; Shimomoto, T.; Yamamoto, K.; Kirita, T.; Kuniyasu, H. Downregulation of miR-126 Induces Angiogenesis and Lymphangiogenesis by Activation of VEGF-A in Oral Cancer. *Br. J. Cancer* **2012**, *107*, 700–706. [CrossRef] [PubMed]

28. Xin, X.; Majumder, M.; Girish, G.V.; Mohindra, V.; Maruyama, T.; Lala, P.K. Targeting COX-2 and *EP4* to Control Tumor Growth, Angiogenesis, Lymphangiogenesis and Metastasis to the Lungs and Lymph Nodes in a Breast Cancer Model. *Lab. Investig.* **2012**, *92*, 1115–1128. [CrossRef]

29. Majumder, M.; Xin, X.; Liu, L.; Girish, G.V.; Lala, P.K. Prostaglandin E2 Receptor *EP4* as the Common Target on Cancer Cells and Macrophages to Abolish Angiogenesis, Lymphangiogenesis, Metastasis, and Stem-like Cell Functions. *Cancer Sci.* **2014**, *105*, 1142–1151. [CrossRef]

30. Perrot-Applanat, M.; Di Benedetto, M. Autocrine Functions of VEGF in Breast Tumor Cells. *Cell Adh. Migr.* **2012**, *6*, 547–553. [CrossRef]

31. Ning, Q.; Liu, C.; Hou, L.; Meng, M.; Zhang, X.; Luo, M.; Shao, S.; Zuo, X.; Zhao, X. Vascular Endothelial Growth Factor Receptor-1 Activation Promotes Migration and Invasion of Breast Cancer Cells through Epithelial-Mesenchymal Transition. *PLoS ONE* **2013**, *8*, e65217. [CrossRef]

32. Weigand, M.; Hantel, P.; Kreienberg, R.; Waltenberger, J. Autocrine Vascular Endothelial Growth Factor Signalling in Breast Cancer. Evidence from Cell Lines and Primary Breast Cancer Cultures in Vitro. *Angiogenesis* **2005**, *8*, 197–204. [CrossRef] [PubMed]

33. Timoshenko, A.V.; Chakraborty, C.; Wagner, G.F.; Lala, P.K. COX-2-Mediated Stimulation of the Lymphangiogenic Factor VEGF-C in Human Breast Cancer. *Br. J. Cancer* **2006**, *94*, 1154–1163. [CrossRef] [PubMed]

34. Nandi, P.; Girish, G.V.; Majumder, M.; Xin, X.; Tutunea-Fatan, E.; Lala, P.K. PGE2 Promotes Breast Cancer-Associated Lymphangiogenesis by Activation of *EP4* Receptor on Lymphatic Endothelial Cells. *BMC Cancer* **2017**, *17*, e11. [CrossRef] [PubMed]

35. Kong, W.; He, L.; Richards, E.J.; Challa, S.; Xu, C.X.; Permuth-Wey, J.; Lancaster, J.M.; Coppola, D.; Sellers, T.A.; Djeu, J.Y.; et al. Upregulation of miRNA-155 Promotes Tumour Angiogenesis by Targeting VHL and is Associated with Poor Prognosis and Triple-Negative Breast Cancer. *Oncogene* **2014**, *33*, 679–689. [CrossRef] [PubMed]

36. Mao, G.; Liu, Y.; Fang, X.; Liu, Y.; Fang, L.; Lin, L.; Liu, X.; Wang, N. Tumor-Derived microRNA-494 Promotes Angiogenesis in Non-Small Cell Lung Cancer. *Angiogenesis* **2015**, *18*, 373–382. [CrossRef]

37. He, X.; Ping, J.; Wen, D. MicroRNA-186 Regulates the Invasion and Metastasis of Bladder Cancer Via Vascular Endothelial Growth Factor, C. *Exp. Ther. Med.* **2017**, *14*, 3253–3258. [CrossRef] [PubMed]

38. Li, N.; Meng, X.; Bao, Y.; Wang, S.; Li, T. Evidence for the Involvement of COX-2/VEGF and *PTEN*/Pl3K/AKT Pathway the Mechanism of Oroxin B Treated Liver Cancer. *Pharmacogn. Mag.* **2018**, *14*, 207–213.

39. Forsythe, J.A.; Jiang, B.H.; Iyer, N.V.; Agani, F.; Leung, S.W.; Koos, R.D.; Semenza, G.L. Activation of Vascular Endothelial Growth Factor Gene Transcription by Hypoxia-Inducible Factor 1. *Mol. Cell. Biol.* **1996**, *16*, 4604–4613. [CrossRef]

40. Hirota, K.; Semenza, G.L. Regulation of Angiogenesis by Hypoxia-Inducible Factor 1. *Crit. Rev. Oncol. Hematol.* **2006**, *59*, 15–26. [CrossRef]

41. Khan, F.; Esnakula, A.; Ricks-Santi, L.J.; Zafar, R.; Kanaan, Y.; Naab, T. Loss of *PTEN* in high grade adnaced triple negative breast ductal cancers in African American women. *Pathol. Res. Pract.* **2018**, *5*, 673–678. [CrossRef]

42. Livak, K.J.; Schmittgen, T.D. Analysis of Relative Gene Expression Data using Real-Time Quantitative PCR and the 2 (-Delta Delta C (T)) Method. *Methods* **2001**, *25*, 402–408. [CrossRef] [PubMed]

43. Xie, H.Y.; Shao, Z.M.; Li, D.Q. Tumor Microenvironment: Driving Forces and Potential Therapeutic Targets for Breast Cancer Metastasis. *Chin. J. Cancer* **2017**, *36*, e36. [CrossRef] [PubMed]

44. Li, S.; Li, Q. Cancer Stem Cells, Lymphangiogenesis, and Lymphatic Metastasis. *Cancer Lett.* **2014**, *357*, 438–447. [CrossRef] [PubMed]

45. Weis, S.M.; Cheresh, D.A. Tumor Angiogenesis: Molecular Pathways and Therapeutic Targets. *Nat. Med.* **2011**, *17*, 1359–1370. [CrossRef] [PubMed]

Article

MicroRNAs Which Can Prognosticate Aggressiveness of Bladder Cancer

Edyta Marta Borkowska [1,*], Tomasz Konecki [2], Michał Pietrusiński [1], Maciej Borowiec [1] and Zbigniew Jabłonowski [2]

[1] Medical University of Lodz, Chair of Laboratory and Clinical Genetics Department of Clinical Genetics, 251 Pomorska Street, 92-213 Lodz, Poland; Michal.pietrusinski@umed.lodz.pl (M.P.); maciej.borowiec@umed.lodz.pl (M.B.)

[2] Medical University of Lodz, 1st Clinic of Urology, 113 Żeromski Street, 90-549 Lodz, Poland; Tomasz.konecki@umed.lodz.pl (T.K.); zbigniew.jablonowski@umed.lodz.pl (Z.J.)

* Correspondence: edyta.borkowska@umed.lodz.pl; Tel.: +48 42 272 53 58

Received: 21 September 2019; Accepted: 9 October 2019; Published: 14 October 2019

Abstract: Bladder cancer (BC) is still characterized by a very high death rate in patients with this disease. One of the reasons for this is the lack of adequate markers which could help determine the biological potential of the tumor to develop into its invasive stage. It has been found that some microRNAs (miRNAs) correlate with disease progression. The purpose of this study was to identify which miRNAs can accurately predict the presence of BC and can differentiate low grade (LG) tumors from high grade (HG) tumors. The study included 55 patients with diagnosed bladder cancer and 30 persons belonging to the control group. The expression of seven selected miRNAs was estimated with the real-time PCR technique according to miR-103-5p (for the normalization of the results). Receiver operating characteristics (ROC) curves and the area under the curve (AUC) were used to evaluate the feasibility of using selected markers as biomarkers for detecting BC and discriminating non-muscle invasive BC (NMIBC) from muscle invasive BC (MIBC). For HG tumors, the relevant classifiers are miR-205-5p and miR-20a-5p, whereas miR-205-5p and miR-182-5p are for LG (AUC = 0.964 and AUC = 0.992, respectively). NMIBC patients with LG disease are characterized by significantly higher miR-130b-3p expression values compared to patients in HG tumors.

Keywords: Bladder cancer; microRNA; genetic marker; progression

1. Introduction

Bladder cancer (BC) is characterized by the high rate of non-muscle invasive BC (NMIBC) at the moment of diagnosis (75–80%) [1,2]. Transitional cell carcinoma (TCC) constitutes the majority of the urothelial carcinoma of the bladder. There are two described alternative molecular pathways of developing BC, characterized by different genetic changes and different biological potentials. The first alternative includes changes of papillary and an always non-invasive character, while the other alternative can be either papillary or non-papillary and is often invasive (into the lamina propria—T1 stage; or muscularis propria—T2 stage) [3,4]. Patients suffering from muscle invasive BC (MIBC) at the moment of the initial diagnosis are treated with radical cystectomy (RC). This is not the optimal solution, as patients' quality of life after RC is low and a high rate of relapse and death has been observed within a short period of time after operation [5,6]. As far as patients with NMIBC are concerned, it is impossible to predict which of them will have disease progression. In consequence, they undergo systematic cystoscopy examinations aimed at assessing the disease development stage. This also decreases patients' quality of life (time in hospital, stress, uncertainty connected with another examination) and generates enormous costs for the health care [7].

MicroRNAs (miRNAs) are known to be dysregulated in bladder cancer (BC) and implicated in the pathogenesis of the development of bladder tumors mostly via their influence on genes involved in two molecular pathways, specifically the gene which codes fibroblast growth factor receptor 3 (FGFR3) and the gene which codes tumor protein 53 (TP53). Numerous miRNA studies have identified histological grade and stage (pT) classification-dependent miRNA expression and have proven the existence of miRNAs alterations related to the two divergent pathways found in the development of NMIBC and MIBC [8,9]. Only a few studies have analyzed miRNA as a prognostic and predictive biomarker [10–12]. Each miRNA can have multiple targets, and changes in their expression profile could have a magnified effect on cellular phenotype. The previously published studies emphasize the possible prognostic potential of some miRNAs to predict progression and disease specific or overall survival in BC patients. Unfortunately, none of these miRNAs are used in routine practice. This in the result of quite a few factors: Using different platforms for assessing marker expressions, using various biological samples (tissue or cell lines) secured in different ways (paraffin, RNAlater, freezing), and using various normalization methods and reference genes [13]. Some analyses are based on relative expression and others are based on absolute expression. Finally, these factors also include the lack of a control group. That is why we decided to evaluate the expression of selected miRNAs in an adequately selected group of both NMIBC and MIBC patients characterized by the high rate of observed progression.

In tumors, downregulated miRNAs are considered to be tumor suppressor candidates, whereas miRNAs with increased expression may play a promotional role in cancer progression. Potential BC suppressors include miR-100, miR-99a, miR-202, and miR-30a. Some miRNAs, including miR-145-5p (locus on chromosome 5), miR-195, and miR-199a-5p have been shown to inhibit the proliferation of or induce the apoptosis of BC cells [14]. MiR-145-5p appears to play a key role as a tumor suppressor by targeting N-cadherin and its downstream effector matrix metalloproteinase-9 (MMP9), and it is the most frequently reported downregulated miRNA in BC. MiR-205-5p (locus on chromosome 1), miR-182-5p (locus on chromosome 7), mir-130b-3p (locus on chromosome 22), miR-10a-5p and miR-21-5p (loci on chromosome 17), and miR-20a-5p (locus on chromosome 13) are mainly overexpressed in BC tissue. They promote proliferation, migration, and invasion, and they inhibit BC cells apoptosis. The potential target/regulator for miR-130b-3p and miR-205-5p is the *PTEN* gene (phosphatase and tensin homolog); for miR-182-5p, it is the *SMAD4* gene (drosophila protein, mothers against decapentaplegic homolog 4); and for miR-10a-5p, it is the *FGFR3* gene [15]. miR-21-5p overexpression is related to *TP53* inactivation, invasion, and tumor progression. It has been seen to be simultaneously upregulated in the tissue, plasma and urinary exosomes of BC patients, but its role needs further elucidation. However, there are still conflicting results regarding the function of miRNAs in publications, so, for our analysis, we chose a panel of the best described miRNAs for BC and the miRNAs connected with genes or chromosomes whose genetic alterations are well documented in pathogenesis BC [14,15].

2. Materials and Methods

The tested group consisted of tumor tissue samples stored in the tissue bank in the Clinical Genetics Department, the Chair of Clinical and Laboratory Genetics, Medical University of Lodz. The tumor tissues were obtained during the TURBT (transurethral resection of bladder tumor) examination at the Urology Ward of the University Clinical Hospital Military Memorial Medical Academy in Lodz. Official permission to conduct the tests was granted by the Bioethics Advisory Commission at Lodz Medical University, No. RNN/62/15/ KE/M, and the patients signed consent forms. The tumors selected for RNA isolation were submerged in an RNA later solution (Sigma) and stored at −20 °C before isolation time. The tests were carried out on a group of 55 patients with diagnosed bladder cancer. The clinical and pathological characteristics of the cohorts are summarized in Table 1 and Supplemental Table S1. All the tumors were of urothelial origin. Only samples with more than 60% tumor content were included in the study. The age range was 44–88 with an average age of 72.8. The majority of the patients were male (45/55–81.7%). Nineteen patients (34.55%) in the group suffered from non-invasive bladder cancer in stage Ta, and 18 patients (32.75%) were in stage T1. The remaining 18 patients (32.75%) were diagnosed with invasive bladder cancer in stage T2. Tumor stage was determined according to the 2002 UICC TNM classification, and histological grading was assessed in accordance with the World Health Organization/International Society of Urological Pathology criteria of 2004 [16,17]. A progressive disease was defined as a disease that had progressed to stage T2 or higher, the development of nodal or distant metastases, or death. The control group consisted of 30 patients admitted to the urology ward. They underwent control cystoscopy aimed at confirming or excluding tumor changes in the bladder. The examination did not reveal any tumor changes.

A MirVana™ miRNA Isolation Kit (Life Technologies, Cat No. 1560, Foster City, CA, USA) was used to isolate microRNA from the frozen tumor tissues. The whole procedure was carried out in accordance with the instructions of the producers. Briefly: 1 mL of Lysis/Binding buffer was added to each sample (1 mL per 0.1 g of tissue) and homogenized. After that, 100 µL of miRNA Homogenate Additive was added to sample and incubated for 10 min on ice. Next, 1100 µL of acid-phenol:chloroform was mixed with the sample and centrifuged (5 min at 10,000× *g*). The aqueous phase was transferred to a fresh tube and vortexed with 200 µL of 100% ethanol. A lysate/ethanol mixture was pipetted onto the filter cartridge and centrifuged (15 seconds at 10,000× *g*). The filtrate was collected, and the step was repeated. After that, 400 µL of 100% ethanol was added to the filtrate, pipetted onto new filter cartridge, and centrifuged in the same condition. Two washing steps were conducted: (1) 700 µL of miRNA Wash Solution 1 was applied to the filter cartridge, and (2) 500 µL of Wash Solution 2 and 3 were applied to the filter cartridge; this was repeated twice (at each step, samples were centrifuged for 15 seconds at 10,000× *g*). In the last step, 70 µL of the preheated (95 °C) elution solution was applied to the filter cartridge, which was then spun for 30 seconds at 16,000 g. The collected eluate was stored at −20 °C. An additional DNase and digestion step was performed. The obtained microRNA concentrations were monitored using the spectrophotometric method on the NanoDrop® ND-1000 instrument (NanoDrop Technologies, Wilmington, DE, USA). The purity measurement of the obtained extracts used the relationships A260/230 and A260/280. It is accepted that for good quality nucleic acids, these relationships are, respectively, 1.8–2.2 and 1.8–2.0. The measurement results of the samples selected for further analysis met the required criteria. The purity of the samples was also verified using a Qubit microRNA Assay Kit (Invitrogen, Cat No. Q32880). For reverse transcription, 10 ng of RNA was taken. MiRNAs (hsa-mir-10a, hsa-mir-20a, hsa-mir-21, hsa-mir-130b, hsa-mir-145, hsa-mir-182, hsa-mir-205, and hsa-mir-103) for 55 samples were reverse transcribed using a TaqMan MicroRNA Reverse Transcription Kit (Applied Biosystems Cat No. 4366596) and a 50 nM pool of miRNA specific stem loop primers (Applied Biosystems Cat No. 4427975; details and ID of assays specified in Supplemental Table S2) following the manufacturer's protocol (100 mM dNTPs 0,15 µL, MultiScribe™ Reverse Transcriptase, 50 U/µL 1 µL, 10× Reverse Transcription Buffer 1 µL, RNase Inhibitor, 20 U/µL 0,19 µL, and nuclease-free Water 4.16 µL). The reaction mixtures were incubated at 16 °C for 30 min, at 42 °C for 30 min, and at 85 °C for 5 min (Applied Biosystems microAmp Optical 96-well reaction plate Cat No. N8010560, Micro Amp optical adhesive film Cat No. 4311971), and then the products of the reaction were stored at −20 °C until use. Purity and quantity were verified using a Qubit dsDNA HS assay kit (Invitrogen, Cat No. Q32851).

Table 1. Differences in expression level of selected miRNAs in the patients' group according to clinicopathological parameters.

		FCmiR-145		p-Value	FCmiR-21		p-Value	FCmiR-182		p-Value	Abnormal Expression 1		p-Value	Abnormal Expression 2		p-Value
Clinicopathological Parameters		HE n (%)	LE n (%)		HE n (%)	LE n (%)		HE n (%)	LE n (%)		Yes n (%)	No n (%)		Yes n (%)	No n (%)	
Total	55															
Sex																
Female		4 (7.27%)	6 (10.91%)		3 (5.45%)	7 (12.73%)		5 (9.09%)	5 (9.09%)		7 (12.73%)	3 (5.45%)		3 (5.45%)	7 (12.73%)	
Male		26 (47.27%)	19 (34.55%)	0.503 (Y)	9 (16.36%)	36 (65.45%)	0.787 (Y)	26 (47.27%)	19 (34.55%)	0.923 (Y)	34 (61.82%)	11 (20%)	0.971 (Y)	20 (36.36%)	25 (45.45%)	0.629 (Y)
Age at Diagnosis																
<60		2 (3.64%)	4 (7.27%)		1 (1.82%)	5 (9.09%)		6 (10.91%)	0 (0%)		6 (10.91%)	0 (0%)		2 (3.64%)	4 (7.27%)	
>60		28 (50.91%)	21 (38.18%)	0.502 (Y)	11 (20%)	38 (69.09%)	0.841 (Y)	25 (45.45%)	24 (43.64%)	0.064 (Y)	35 (63.64%)	14 (25.45%)	0.308 (Y)	21 (38.18%)	28 (50.91%)	0.994 (Y)
Smoking Status																
Yes		23 (41.82%)	23 (41.82%)		9 (16.36%)	37 (67.27%)		26 (47.27%)	20 (36.36%)		34 (61.82%)	12 (21.82%)		17 (30.91%)	29 (52.73%)	
No		7 (12.73%)	2 (3.64)	0.244 (Y)	3 (5.45%)	6 (10.91%)	0.636 (Y)	5 (9.09%)	4 (7.27%)	0.753 (Y)	7 (12.73%)	2 (3.64%)	0.861 (Y)	6 (10.91%)	3 (5.45%)	0.199 (Y)
Occupatinal Exposure																
Yes		21 (38.18%)	19 (34.55%)		6 (10,91%)	34 (61.82%)		21 (38.18%)	19 (34.55%)		28 (50.91%)	12 (21.82%)		15 (27.27%)	25 (45.45%)	
No		9 (16.36%)	6 (10.91%)	0.622 (V)	6 (10.91%)	9 (16.36%)	0.102 (Y)	10 (18.18%)	5 (9.09%)	0.349	13 (23.64%)	2 (3.64%)	0.359 (Y)	8 (14.55%)	7 (12.73%)	0.293 (V)
Tumour Stage																
Ta		9 (16.36%)	10 (18.18%)		1 (1.82%)	18 (32.73%)		11 (20%)	8 (14.55%)		14 (25.45%)	5 (9.09%)		6 (10.91%)	13 (23.64%)	
T1		10 (18.18%)	8 (14.55%)		6 (10.91%)	12 (21.82%)		9 (16.36%)	9 (16.36%)		13 (23.64%)	5 (9.09%)		8 (14.55%)	10 (18.18%)	
T2		11 (20%)	7 (12.73%)	0.699	5 (9.09%)	13 (23.64%)	0.089	11 (20%)	7 (12.73%)	0.786	14 (25.45%)	4 (7.27%)	0.924	9 (16.36%)	9 (16.36%)	0.505

Table 1. *Cont.*

Clinicopathological Parameters	*FCmiR-145* HE n (%)	LE n (%)	*p*-Value	*FCmiR-21* HE n (%)	LE n (%)	*p*-Value	*FCmiR-182* HE n (%)	LE n (%)	*p*-Value	Abnormal Expression 1 Yes n (%)	No n (%)	*p*-Value	Abnormal Expression 2 Yes n (%)	No n (%)	*p*-Value
Grade															
high grade	13 (23.64%)	9 (16.36%)		5 (9.09%)	17 (30.91%)		12 (21.82%)	10 (18.18%)		16 (29.09%)	6 (10.91%)		11 (20%)	11 (20%)	
low grade	17 (30.91%)	16 (29.09)	0.580	7 (12.73%)	26 (47.27%)	0.841 (Y)	19 (34.55%)	14 (25.45%)	0.826 (V)	25 (45.45%)	8 (14.55%)	0.802 (V)	12 (21.82%)	21 (38,18%)	0.319 (V)
Recurrence															
Yes	13 (23.64%)	13 (23.64%)		3 (5.45%)	23 (41.82%)		16 (29.09%)	10 (18.18%)		21 (38.18%)	5 (9.09%)		9 (16.36%)	17 (30.91%)	
No	17 (30.91%)	12 (21.82%)	0.521	9 (16.36%)	20 (36.36%)	0.083 (V)	15 (27.27%)	14 (25.45%)	0.463	20 (36.36%)	9 (16.36%)	0.320 (V)	14 (25.45%)	15 (27.27%)	0.305
Progression															
Yes	17 (30.91%)	13 (23.64%)		7 (12.73%)	23 (41.82%)		16 (29.09%)	14 (25.45%)		21 (38.18%)	9 (16.36%)		14 (25.45%)	16 (29.09%)	
No	13 (23.64%)	12 (21.82%)	0.729	5 (9.09%)	20 (36.36%)	0.767 (V)	15 (27.27%)	10 (18.18%)	0.619	20 (36.36%)	5 (9.09%)	0.401 (V)	9 (16.36%)	16 (29.09%)	0.424
Death															
Yes	10 (18.18%)	7 (12.73%)		3 (5.45%)	14 (25.45%)		9 (16.36%)	8 (14.55%)		11 (20%)	6 (10.91%)		9 (16.36%)	8 (14.55%)	
No	20 (36.36%)	18 (32.73%)	0.673 (V)	9 (16.36%)	29 (52.73%)	0.882 (Y)	22 (40%)	16 (29.09%)	0.734	30 (54.55%)	8 (14.55%)	0.432 (Y)	14 (25.45%)	24 (43.64%)	0.267 (V)

Y-test chi-squared with Yeats corrections; V-test V-squared.

Real-time polymerase chain reactions (rt-PCR) were performed on CFX96 (BioRad, Hercules, CA, USA) including related documentation with regard to the specific items of MIQE guidelines (Supplemental Table S3) [18]. Each sample was run in duplicate at a final volume of 18 µL containing 10 µL of TaqMan 2× Universal PCR Master mix II with no UNG (Applied Biosystems Cat No. 4440040), 7 µL of nuclease free water, and 1 µL of TaqMan® Small RNA Assay (20×). Each PCR included no template control, and all of them were negative. The reaction was heated to 90 °C for 10 min, 55 °C for 2 min, and 72 °C for 2 min, followed by 50 cycles. The mean threshold cycle value (Ct) was used for downstream analyses. miR-103-5p was chosen as an endogenous control. The ΔΔCt method, also defined as the comparative method, was applied in order to mark the expression level of the examined microRNAs [19]. This method is based on mathematical calculations that enable us to determine the relative difference in the expression level of the tested marker between unknown samples and the reference. The first stage consists of the analysis of the marked Ct (the cycle at which the fluorescence level reaches a certain amount/threshold) in the amplification reaction of the examined microRNAs and control microRNA for both the tested and the control groups. The calculated expression level of each patient was normalized against the endogenous control, which was miR-103a-5p [20]. After that, the difference of the tested and control microRNAs (ΔCt) was calculated for individual samples. The calculations were made for both the unknown and control samples.

(ΔCt) (tested group) = Ct miRNA target – Ct miRNA reference

(ΔCt) (control group) = Ct miRNA target – Ct miRNA reference

Next, ΔΔCt was calculated for each sample:

ΔΔCt = ΔCt (tested samples) – ΔCt (median of the control group)

The calculation of the normalized value of the relative expression level (FC) of the tested marker in the tested sample against the control sample was made as follows:

$$FC = 2^{-\Delta\Delta Ct} \tag{1}$$

The $2^{-\Delta\Delta Ct}$ method assumes a uniform PCR amplification efficiency of 100% across all samples. In our study, the efficiency was between 98.9% and 100%.

3. Data Analysis

The statistical calculations were made using the program STATISTICA 13, Stat-Soft Inc. The differential miRNA expression between bladder cancer cases and controls was determined using Student's t-statistics. In fact, the distribution of variables differed from the standard normal distribution; therefore, non-parametric tests were applied. The analysis of the unrelated variables was made with the Mann–Whitney U test. The value $p < 0.05$ was accepted as the threshold of statistical difference or correlation significance. Kaplan–Meier analyses with a long-rank test and Cox regression were performed for overall survival time (OS), time to recurrence, and time to progression. The discriminating capacity of miRNAs was assessed by a receiver operating characteristics (ROC) analysis.

4. Results

In the first stage of the analysis, the relationship between the abnormal expression of selected microRNAs and other clinical parameters was examined. The raw data from the Ct for individual miRNAs were recalculated for fold change (FCmiR) (Supplemental Fata File SF1, Supplemental Table S4). In the case of miR-205-5p, all the patients were classified into the reduced expression (low expression—LE) group, while in the case of miR-130b-3p, miR-20a-5p and miR-10a-5p, all the patients were classified into the increased expression (high expression—HE) group. These miRNAs did not differentiate the patients according to clinicopathological parameters; therefore, only FCmiR145, FCmiR21 and FCmiR182 were selected for further analysis. Table 1 presents the probability values (p) of the relevant statistics used to make conclusions regarding the existence of relationships between individual variables. The analysis was performed depending on the fulfilled assumptions by the a classic Chi^2 test, *V*-square test (*V*), or with Yates's correction (*Y*). There were not any significant correlations observed. Additionally, we did not observe any significant differences for the division of the tested group into Expression 1 (when at least one of the miRNAs indicates abnormal expression) and Expression 2 (when at least two of the analyzed miRNAs show changes). The results are presented in Table 1.

The next step was comparing differences between the level of expression in different stages (TaT1and T2) or grades (LG and HG tumors). The question that was sought next was whether the selected miRNAs could be prognostic classifiers for patients at different stages or grades of cancer. For this purpose, the patients were divided into two groups: 0—patients with stage T2 or higher; 1—patients with stage Ta or T1. The results are presented in Table 2 and Figure 1. The parametric t-test tested the one-sided hypothesis that miRNA for TaT1 < miRNA for T2 and above. The remaining *p*-values were read from the Mann–Whitney U test, which compares distributions (medians). This test is less powerful than the t-test, but it is the only one for non-normal distributions. NMIBC patients (TaT1 in our study) with an LG disease were characterized by significantly higher miR-130b-3p expression values compared to patients with HG tumors. If we consider patients with the LG disease, miR-205-5p, miR-182-5p and miR-20a-5p differentiated this group with BC in stage TaT1 from patients in a higher stage (*p* < 0.05). If we focus on a group of patients in HG, it is miR-130b-3p which best differentiated patients in terms of stage.

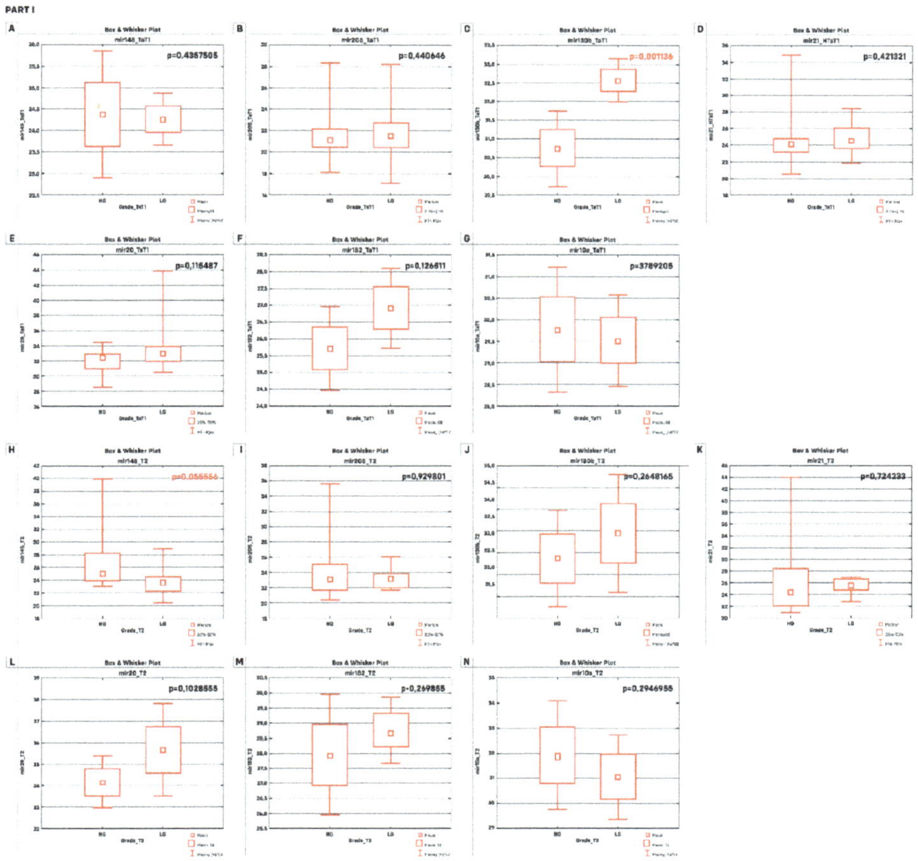

Figure 1. *Cont.*

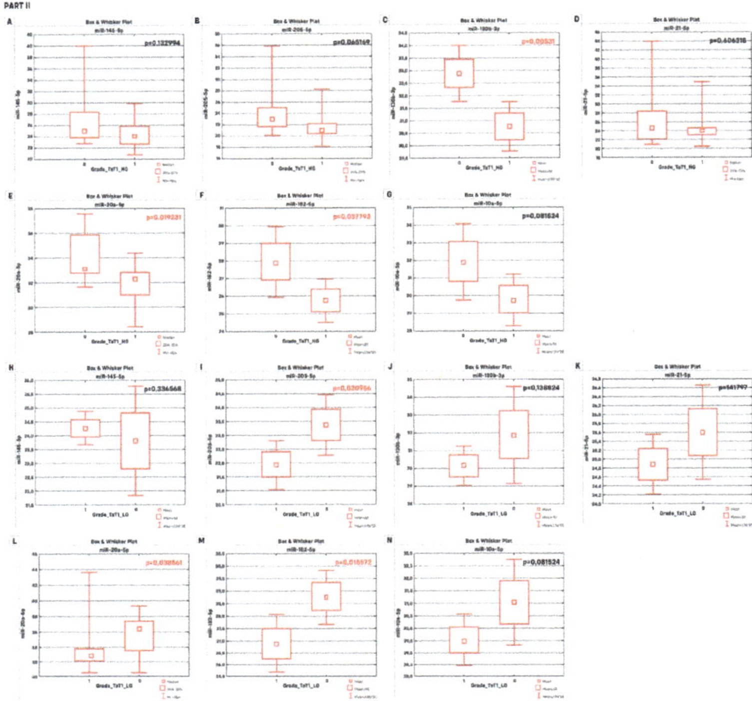

Figure 1. Differences in expression level for patients: Group 0—BC patients in T2; Group 1—BC patients in stage TaT1 or in low or high grade group (low grade (LG) or high grade (HG)). Part I presents results for the differentiation of patients in terms of grade in TaT1 (**A–G**) and T2 (**H–N**) groups. Only miR-130b-3p differentiated patients in stage TaT1 according to grade (miR-145-5p was close to significance). Part II presents results for the differentiation of patients in terms of stage in high (**A–G**) or low grade (**H–N**) groups. If we take patients with LG, the differentiating miRNAs in terms of stage were miR-205-5p, miR-20a-5p and miR-182-5p. In the case of patients in the HG group, miR-20a-5p, miR-205-5p and miR-182-5p always had significantly lower values of expression in patients in the TaT1 stage of the disease compared to patients in the T2 stage. (**A**) and (**H**) miR-145-5p; (**B**) and (**I**) miR-205-5p; (**C**) and (**J**) miR-130b-3p; (**D**) and (**K**) miR-21-5p; (**E**) and (**L**) miR-20a-5p; (**F**) and (**M**) miR-182-5p; and (**G**) and (**N**) miR-10a-5p. *p*-values in read are significant (*p* < 0.05).

To assess the clinical relevance of all miRNAs, a Kaplan–Meier analysis with the log-rank test and Cox regression analyses were performed for overall survival, recurrence-free survival, and progression-free survival (results presented in Table 3 and Figure 2). We did not find any significant differences. Univariate Cox regression was performed to assess the factors that affect the risk of progression, recurrence or death (results presented in Table 4). It has been shown that the older patients are, the higher the risk of death (increasing each year by 30%). In addition, the risk of death for people with stage T2 of the disease is more than six times higher than for patients in stage Ta or T1. People with disease progression are nine times more likely to die. The increase in the expression of miR-205-5p, miR-145-5p and miR-21-5p makes the risk of death higher by 13%, 0.03%, and 0.009%, respectively (Table 4). These percentages result from the interpretation of the hazard ratio (HR) of important risk parameters, for which the Cox regression analysis for OS had *p* < 0.05. The risk of recurrence in patients in stage Ta is over two times higher than people in stage T1 or T2. In the group of patients with recurrence, the death rate is four times lower. The increased expression of miR-20a-5p and miR-182-5p heightens the risk of recurrence by 5% and 6%, respectively. More advanced age increases the risk of progression (by 7% each year). This risk is more than three times higher for people in the T2 stage compared to people in the Ta or T1 stages.

Table 2. Differences in expression in A) stage TaT1 and T2 according to different grade and B) low and higf grade tumors according to stage of BC.

A)	TaT1 p-value	T2 p-value
miR-145-5p	0.4357505 *	0.055556
miR-205-5p	0.440646	0.929801
miR-130b-3p	**0.001136 ***	0.2648165 *
miR-21-5p	0.421321	0.724233
miR-20a-5p	0.115487	0.1028555 *
miR-182-5p	0.126511 *	0.269855 *
miR-10a-5p	0.3987205 *	0.2946955 *
B)	HG p-value	LG p-value
miR-145-5p	0.132994	0.336568 *
miR-205-5p	0.065169	**0.030956 ***
miR-130b-3p	**0.00531 ***	0.138824 *
miR-21-5p	0.606318	0.141797 *
miR-20a-5p	**0.019231**	**0.038561**
miR-182-5p	**0.037793 ***	**0.015572 ***
miR-10a-5p	0.06102 *	0.081524 *

Results obtained with Mann-Whitney U test (p-values without *) and parametric t test (p-values with *). Bold face represents p-value <0.05.

Table 3. Kaplan-Meier analysis for overall survival, time to recurrence and time to progression in the patients' group.

		Kaplan-Meier Analysis					
		Overall Survival		Recurrence		Progression	
	Overall n (%)	Rate	Log-Rank Value	Rate	Log-Rank Value	Rate	Log-Rank Value
Total	55						
FCmiR-145							
HE	30	10		13		17	
LE	25	7	0.6992	13	0.5745	13	0.9267
FCmiR-21							
HE	12	3		3		7	
LE	43	14	0.7390	23	0.1789	7	0.7993
FCmiR-182							
HE	31	9		16		16	
LE	24	8	0.6576	10	0.4189	14	0.5976
Total	55						
Abnormal expression 1							
Yes	41	11		21		21	
No	14	6	0.2875	5	0.3499	9	0.2847
Abnormal expression 2							
Yes	23	9		9		14	
No	32	8	0.2551	17	0.6881	16	0.5205

LE—low expression; HE—high expression.

Table 4. Univaraite Cox regression analysis of potential predictor variables and overall survival, time to recurrence and time to progression in the group of patients (n = 55).

	Overall Survival				Time to Recurrence				Time to Progression			
	Beta	HR (95% CI)	*p*-value	*p*-value for Chi^2	Beta	HR (95% CI)	*p*-value	*p*-value for Chi^2	Beta	HR (95% CI)	*p*-value	*p*-value for Chi^2
Gender	−0.57	0.56 (0.13–2.47)	0.448	0.415	−0.816	0.44 (0.13–1.47)	0.184	0.142	0.029	1.03 (0.42–2.42)	0.948	0.948
Age at diagnosis	0.266	1.30 (1.17-1.45)	**0.000**	0.000	-0.002	0.997 (0.99-1.005)	0.526	0.524	0.068	1.07 (1.02-1.12)	**0.0034**	0.003
Stage												
Ta–T1&T2	−0.013	0.98 (0.36–2.67)	0.97	0.98	0.738	2.09 (0.97–4.53)	**0.0607**	0.061	−1.388	0.25 (0.09–0.65)	**0.005**	0.0013
Ta&T1–T2	1.821	6.17 (2.25–16.89)	**0.0004**	0.00028	−0.766	0.46 (0.16–1.35)	0.159	0.126	1.109	3.03 (1.46–6.26)	**0.0027**	0.0034
Occupatinal Exposure	1.12	3.08 (0.7–13.47)	0.135	0.087	−0.669	0.51 (0.23–1.13)	0.097	0.108	0.874	2.39 (0.91–6.28)	0.075	0.052
Grade	2.85	17.36 (3.89–77.41)	**0.00018**	0.000	−1.615	0.19 (0.06–0.66)	**0.008**	0.001	1.775	5.89 (2.59–13.38)	**0.00002**	0.00001
Smoking Status	0.37	1.45 (0.33–6.37)	0.619	0.603	0.101	1.11 (0.38–3.21)	0.852	0.85	−0.07	0.93 (0.36–2.43)	0.885	0.886
Recurrence	−1.28	0.28 (0.09–0.86)	**0.026**	0.015					−2.229	0.107 (0.04–0.28)	**0.000008**	0.00000
Progression	2.16	8.67 (1.97–8.13)	**0.004**	0.00031	−1.717	0.18 (0.07–0.48)	**0.0006**	0.00008				
FCmiR-145	0.0003	1.0003 (1.00009–1.0006)	**0.0069**	0.038	−0.019	0.98 (0.93–1.03)	0.393	0.099	0.0001	1.0001 (0.99–1.0003)	0.243	0.321
FCmiR-205	0.12	1.13 (1.03–1.24)	**0.0089**	0.045	−0.167	0.85 (0.36–1.96)	0.697	0.521	0.046	1.05 (0.97–1.13)	0.233	0.311
FCmiR-130b	0.0003	0.99 (0.99–1.00)	0.466	0.398	0.0003	1.0003 (0.99–1.0007)	0.131	0.176	−0.0002	0.99 (0.99–1.00)	0.484	0.437
FCmiR-21	0.00009	1.00009 (1.000025–1.00015)	**0.0069**	0.038	0.0004	1.0000006 (0.98–1.006)	0.145	0.156	0.00003	1.00003 (0.99–1.00008)	0.259	0.336
FCmiR-20a	−0.00013	0.999 (0.999–1.0)	0.412	0.177	0.000002	1.000002 (1.0–1.000003)	**0.031**	0.097	−0.00013	0.999 (0.999–1.0)	0.412	0.177
FCmiR-182	−0.034	0.966 (0.87–1.07)	0.529	0.172	0.0006	1.0006 (0.00004–1.001)	**0.035**	0.104	−0.0009	0.999 (0.995–1.002)	0.599	0.243
FCmiR-10a	−0.0004	0.999 (0.997–1.001)	0.672	0.47	−0.0004	0.999 (0.998–1.0007)	0.505	0.301	0.0003	1.0003 (0.999–1.0006)	0.129	0.218
Abnormal Expression 1	−0.5328	0.587 (0.217–1.588)	0.294	0.309	0.4376	1.549 (0.583–4.11)	0.379	0.358	−0.4315	0.649 (0.297–1.42)	0.279	0.295
Abnormal Expression 2	0.5419	1.719 (0.663–4.459)	0.265	0.265	−0.1626	0.85 (0.378–1.91)	0.694	0.691	0.2274	1.255 (0.612–2.575)	0.535	0.536

Bold face representing *p*-values <0.05; CI—Coincidence Interval; HR—Hazard Ratio.

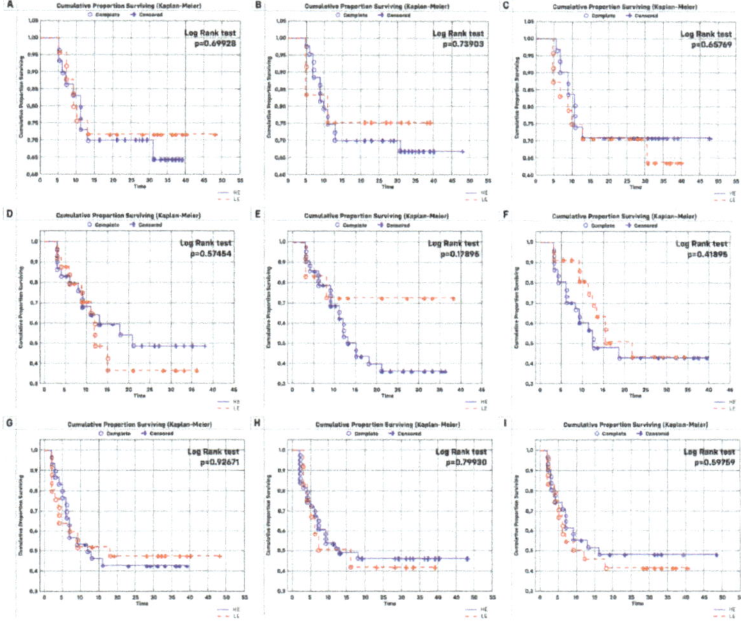

Figure 2. Kaplan–Meier plots for patients' group divided into low expression (LE) and high expression (HE): Cancer-specific survival (**A**) miR-145-5p, (**B**) miR-21-5p, and (**C**) miR-182-5p; recurrence free-survival (**D**) miR-145-5p, (**E**) miR-21-5p, and (**F**) miR-182-5p; and progression-free survival (**G**) miR-145-5p, (**H**) miR-21-5p, and (**I**) mir-182-5p.

The next step of the analysis was based on results of the area under receiver-operating characteristic curves (ROC). Based on the data from 55 patients with BC and from 30 patients of the control group, an attempt was made to find out which miRNAs, among the selected ones, are the best potential cancer classifier. The conclusion about the significant influence of individual miRNAs on the classification of patients was formed using the multivariable logistic regression model (a logistic regression model with many explanatory variables). The results are presented in Figure 3, Table 5 and Supplemental Table S5. The Mann–Whitney U test showed that the distribution of miR-130b-3p was not significantly different for high grade (HG), Ta, and TaT1 patients. Only for the low grade (LG) group did all miRNAs have significantly different distributions compared to the control group. Figures 3 and 4 present the results. Mir-205-5p seems to be a good classifier for LG and HG patients and also for Ta and T1 stages. Logistic regression assessed with a backward elimination approach resulted in a pattern of three miRNAs (miR-205-5p, miR-20a-5p and miR-182-5p). For HG, the relevant classifiers are miR-205-5p and miR-20a-5p, which gave an AUC = 0.964, whereas low LG miR-205-5p and miR-182-5p gave an AUC = 0.992. The model classifies HG as well as BC. The results are presented in Figure 4 and Table 6.

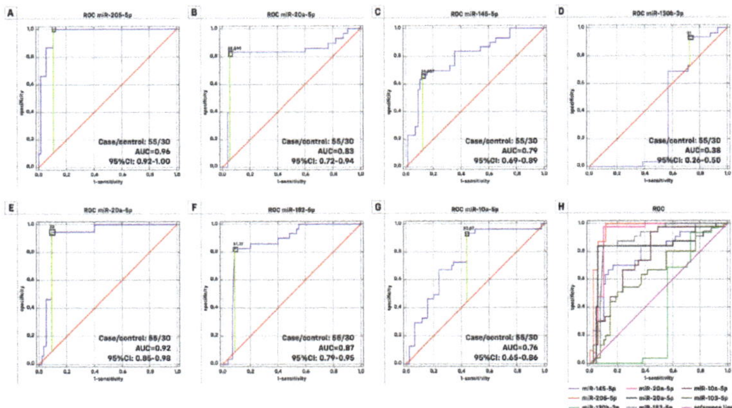

Figure 3. The receiver operating characteristics (ROC) curves for BC prediction using the expression level of (**A**) miR-205-5p, (**B**) miR-20a-5p, (**C**) miR-145-5p, (**D**) miR-130b-3p, (**E**) miR-21-5p, (**F**) miR-182-5p, (**G**) miR-10a-5p and (**H**) summary of curves for all miRNAs. The best classifiers are miR-205-5p, miR-20a-5p and miR-145-5p, as these could significantly discriminate BC patients from the control group by an AUC higher than 0.9; $p < 0.05$.

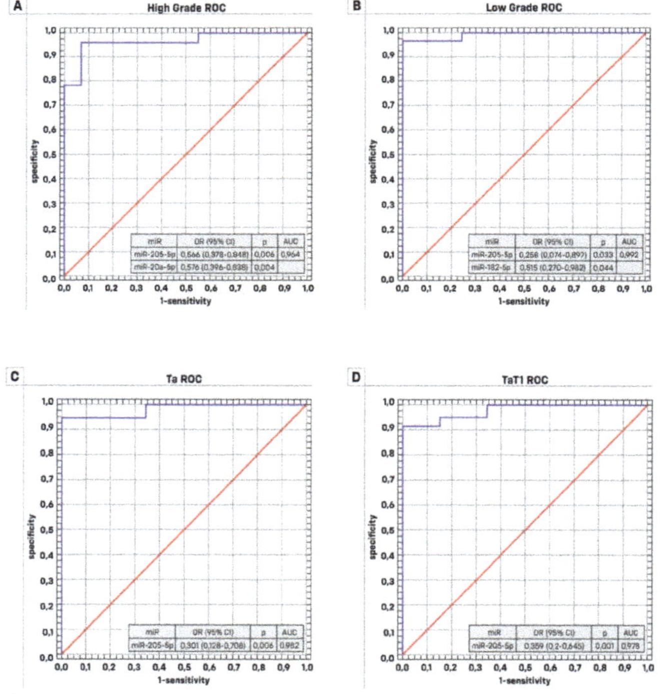

Figure 4. Multivariate logistic model of one or two signature microRNAs (miRNAs): (**A**) For high grade tumors, miR-205-5p and miR-20a-5p are the best classifiers; AUC = 0.964. (**B**) For low grade tumors, miR-205-5p and miR-182-5p are the best classifiers; AUC = 0.992. (**C**) for Ta and (**D**) for TaT1, mir-205-5p is the best classifier; AUC = 0.982 and AUC = 0.978 respectively. (OR—odds ratio; CI—confidence interval; AUC—area under curve).

Table 5. Results of Mann-Withney U test in bladder cancer (BC) group and in subgrups divided according to grade or stage.

Mann Whitney U Test	BC Group	Subgroups			
	p-value	HG *p*-value	LG *p*-value	Ta *p*-value	TaT1 *p*-value
miR-145-5p	**0.000005**	**0.003612**	**0.000002**	**0.000026**	**0.000001**
miR-205-5p	**0.000000**	**0.00000**	**0.00000**	**0.000000**	**0.000000**
miR-130b-3p	0.073733	0.770102	**0.011493**	0.257699	0.479923
miR-21-5p	**0.000000**	**0.000004**	**0.000024**	**0.000000**	**0.000000**
miR-20-5p	**0.000000**	**0.000001**	**0.000001**	**0.000003**	**0.000001**
miR-182-5p	**0.000000**	**0.000009**	**0.00000**	**0.000001**	**0.000000**
miR-10a-5p	**0.000048**	**0.014889**	**0.000016**	**0.000009**	**0.000004**

Bold face represents *p* value <0.05; HG—high grade; LG—low grade.

Table 6. ROC characteristics for subgroups of patients with BC (HG-high grade, LG-low grade, Ta stage, TaT1 stages) and control group.

	HG (Case/Control = 22/30)			LG (Case/Control = 33/30)		
ROC Characteristics	AUC	95% Cl	Significance *p*	AUC	95% Cl	Significance *p*
miR-145-5p	0.732	0.591–0.873	**0.0013**	0.833	0.731–0.936	**0.0001**
miR-205-5p	0.941	0.860–1.000	**0.0001**	0.981	0.955–1.000	**0.0001**
miR-130b-3p	0.475	0.287–0.663	0.7964	0.313	0.167–0.458	**0.0115**
miR-21-5p	0.851	0.717–0.984	**0.0001**	0.936	0.866–1.000	**0.0001**
miR-20a-5p	0.87	0.761–0.978	**0.0001**	0.801	0.675–0.927	**0.0001**
miR-182-5p	0.841	0.703–0.976	**0.0001**	0.895	0.809–0.980	**0.0001**
miR-10a-5p	0.696	0.545–0.846	**0.0109**	0.807	0.698–0.916	**0.0001**
	Ta (case/control = 19/30)			TaT1 (case/control = 37/30)		
ROC Characteristics	AUC	95% Cl	Significance *p*	AUC	95% Cl	Significance *p*
miR-145-5p	0.842	0.734–0.950	**0.0001**	0.83	0.728–0.932	**0.0001**
miR-205-5p	0.982	0.947–1.000	**0.0001**	0.978	0.950–1.000	**0.0001**
miR-130b-3p	0.401	0.205–0.597	0.3236	0.448	0.300–0.596	0.493
miR-21-5p	0.939	0.837–1.000	**0.0001**	0.925	0.849–1.000	**0.0001**
miR-20a-5p	0.872	0.764–0.980	**0.0001**	0.83	0.713–0.946	**0.0001**
miR-182-5p	0.888	0.777–0.998	**0.0001**	0.902	0.821–0.983	**0.0001**
miR-10a-5p	0.858	0.738–0.978	**0.0001**	0.817	0.714–0.920	**0.0001**

AUC—Area Under Curve; CI—Coincidence Interval; Bold face representing *p*-values <0.05; ROC—Receiver-operating Characteristcs.

5. Discussion

The progression in bladder cancer is a complex and multifactorial process [21,22]. In oncology, histopathological examination is still the most important method to determine the diagnosis and classification of tumors; however, current prognosticators such as the tumor grade, stage, size, and multifocality do not accurately reflect clinical outcomes and have limited usefulness for a reliable risk-adjusted therapy decision. At present, there are not enough good markers that could be used as tools to support screening, detecting or monitoring the disease [23,24]. miRNA is an "attractive candidate" as a potential diagnostic and prognostic biomarker, not only due to its high level of stability in body tissues and fluids but also due to its ability to be quantified in relatively easy and cheap techniques like real-time PCR [25,26]. Various miRNAs have been identified as important targets in bladder cancer development, but the large number of different expression profiling platforms such as microarrays, miRCURY ready to use PCR, TaqMan Human MicroRNA Probes, and different reference genes used for normalization are the reason that the results are not comparable and it is difficult to put miRNAs into clinical practice. Therefore, obtaining reliable, not biased miRNA expression data is crucial for selecting clinically useful markers.

It is estimated that over 30% of the protein-coding genes in human cells are controlled by miRNAs. One type of miRNA can even control the expression of hundreds of target genes, and one gene can be controlled by numerous miRNAs. These molecules are regarded as the "key" ones in the gene regulatory network. MiRNAs are involved in many significant biological processes, such as apoptosis, proliferation, cell diversification, and oncogenesis. In this study, we compared the expression of selected miRNAs in non-malignant and malignant bladder tissue, and we identified three down-regulated ones (miR-205-5p, miR-182-5p, and mir-145-5p) and two up-regulated (miR-20a-5p and miR-130b-3p). In previous studies, all of them have been found to be differently expressed in malignant bladder tissue (mainly the underexpression of miR-145-5p and the overexpression of the others), but in this study, the normalization of expression data was performed using miR-103-5p as a reference [25,26]. An endogenous control, in relation to which we normalized the results of other miRNAs, should, as a rule, show stability in a given tissue. In reality, this is very difficult to achieve, and different groups of researchers choose different controls and obtain different results due to these controls. Ratert at al. confirmed that using RNU6B and RNU48 could lead to seriously biased results regarding miRNA expression analysis [27]. Peltier and Latham found that some miRNAs (including miR-106a and miR-191) were the most consistently expressed across different human tissues [28]. They also observed that RNU6 and RNA5S were the least stable. Hofbauer et al. used two endogenous controls in their research, RNU48 and miR-103-5p, and they achieved satisfactory results [29]. In our research, we followed the results of others, including the possibility of the use of miR-103-5p as an endogenous control in commercial sets (Exiqon, Vedbaek Denmark). The studies of Boisen et al. and Parvaee showed that mir-103-5p expression assessed in isolation from formalin-fixed parafin-embedded (FFPE) cancer tissue was the most stable reference miRNA in colorectal (CRC), pancreatic (PC), and intestinal type gastric cancer [30,31].

We subdivided the tumor samples in terms of the low and high grade diseases. The comparison of the miRNAs revealed four significant differentially expressed miRNAs (miR-205-5p, miR-130b-3p, miR-20a-5p, and miR-182-5p). Several studies have implicated miRNAs as prognostic markers for BC. As already shown in previous studies, miR-205-5p expression in normal and tumor samples seems to be coordinated with the *mir-8* family. Lenherr found abnormal expression between progressors and non-progressors for several miRNAs including miR-205-5p and miR-20a-5p. Some of the known targets of miR-205-5p include *ZEB1/2*, *PTEN*, and *VEGFA* [32]. The downregulation of miR-205-5p has been linked to the epithelial-mesenchymal transition (EMT) and has been significantly associated with progression in non-muscle invasive BC. However, the results obtained by different research groups are not consistent due to factors that were already-mentioned in the introduction (differences in the chosen methods). Contrary to that, Dip et al. observed that miR-205-5p was overexpressed in pT2–3 stages of BC [33]. In their study, miR-10a-5p overexpression was associated with shorter disease-free and disease specific survival. Ecke et al. did not confirm the statistical significance for differences in the expression of miRNA-205-5p between non-malignant and BC samples, but they detected a statistically significant reduction in the expression of miR-130b-3p (the best discriminator, also shown in our research) [34]. miR-145-5p overexpression inhibited cell proliferation and migration in BC [35]. Moreover, the downregulation of mir-145-5p was found to be directly targeting the *TAGLN2* gene (its increased expression promoted cell proliferation and migration). Li et al. also confirmed the correlation between the overexpression of miR-145-5p and poor survival [36]. Unfortunately, we failed to achieve such correlation. Inamoto et al. also confirmed the deregulation of miR-145-5p expression and its association with the aggressive phenotype, but they showed its protective effect. miR-145-5p expression was significantly lower in BC samples and cell lines compared to those in normal bladder tissue [37]. Pignot et al. observed that most of the examined miRNAs were deregulated in the same way in the two types of bladder cancer, irrespective of the pathological stage [38]. In their study, miR-182-5p was downregulated and was found to be related to tumor aggressiveness (associated with both recurrence-free and overall survival in univariate analysis). In our study, the high expression of mir-182-5p and mir-20a-5p correlated with the risk of disease recurrence (Table 4; risk higher by 0.06%

and 0.0002%, respectively). Urquidi et al. identified a few miRNA set classifiers for predicting the presence of bladder cancer (25 miRNAs, 20 miRNAs, 15 miRNAs, and 10 miRNAs), but none of them included those ones which we found in our study [39]. The authors note that these biomarkers were correlated with the presence of BC, but their association with clinical variables was much less evident. In our opinion, different sets of miRNAs can be suggested as prognostic biomarkers (three: miR-9, mir-183, and mir-200b; two: miR-143 and miR-145); however, until now, only one study had verified the examined miRNAs as independent markers [27]. Ecke et al. identified miR-199a-3p and miR-214-3p as independent prognostic biomarkers for the prediction of overall survival (OS) in MIBC patients after radical cystectomy (RC). They used a combination of four miRNAs (miR-101, miR-125a, miR-148b, and miR-151-5p) or three miRNAs (miR-148b, miR-181b, and miR-874) as endogenous controls. The study was carried out in a formalin-fixed, paraffin-embedded (FFPE) tissue specimen. These markers were not evaluated by us. Ecke at al. also analyzed the expression of miR-205-5a, but they did not confirm its usefulness. It needs to be stressed, however, that the marking was done in FFPE, whereas our tests were carried out in fresh, frozen tumor tissue. Armstrong's results for matched tumor and bio-fluids in BC showed that there is an overlap between the expression of miRNAs in different bio-specimen sources, but overexpression in all three kinds of the biological samples has only been observed for two tested miRNAs (miR-4454 and miR-21) [40]. No correlation has been observed between expression in tumors and plasma exosomes (using the NanoString nCounter microRNA assay technique). In their review, Lee et al. showed a correlation in the changes of the expression of miRNAs isolated from bladder cancer tissues and urine (in multiple results) for only 14 miRNAs, including miR-145, miR-182, and miR-205 [15]. On the other hand, Baumgart et al. observed that nine miRNAs were consistently differently expressed in both invasive cells and their secreted exosomes, but the remaining six miRNAs were only dysregulated in exosomes [41]. The NanoString technique has its advantages, as it does not require the application of any nucleoid acids. However, it is expensive and hardly available. A real-time PCR technique is available, but any obtained result is affected by many factors, such as the kind of tissue, the way of normalization, and the way of analysis.

Receiver operating characteristics analyses showed a good ability to discriminate between non-malignant and malignant tissues for the investigated miRNAs. Based on binary logistic regression using the backward elimination approach, the optimal combination for discriminating healthy people from BC patients is miR-205-5p, miR-20a-5p and miR-182-5p (AUC > 0.9; $p < 0.05$). Lv et al. Egawa et al. and Liu at al. also confirmed that miR-130b-3p could play a critical role in the development and progression of bladder cancer [42–44]. Fang et al. found an miR-205-5p area under the receiver-operating characteristics curve value of 0.950 for discriminating BC patients from healthy people and a value of 0.668 for discriminating MIBC from NMIBC [10]. The log-rank test and univariate and multivariate Cox regression analyses did not indicate that high miR-205-5p expression in NMIBC patients was associated with cancer specific survival.

We faced some limitations in our study, one of which was a relatively small group of patients. We only tried to use the samples that were characterized by adequate amounts of tumor cells. It is not easy to obtain a large group of patients with bladder cancer progression who can provide biological material for tests, as such cases constitute the minority in this disease. The applied study technique is relatively cheap and easy. Thus, it could be used for examining chosen markers on a daily basis.

6. Conclusions

This study follows the strategy "from top to bottom," which means choosing the phenotype of patients (histopathological characteristics and survival) and evaluating the molecular markers of such a phenotype. The goal for the future is the opposite course of analysis, which is "from genotype to phenotype." Based on the detection of diagnostically and prognostically significant differences between normal and cancer samples, we could assess the biological potential of the tumor and its aggressiveness. As a result, we could enable the choice of appropriate therapeutic measures, tailored to individual patients; this is personalized medicine. Finally, we could lengthen a patient's life and improve its quality without offering a radical treatment if it is not necessary. The implementation of miR-205-5p, miR-20a-5p and miR-130b-3p into routine practice can be an alternative to screening or the follow up of treatment effects. Such analyses can help in the search of non-invasive markers, especially since they can also be evaluated in urine or plasma. Our findings could be of clinical importance, but the results should be validated in a bigger group.

Supplementary Materials: The following are available online at http://www.mdpi.com/2072-6694/11/10/1551/s1, Table S1: Characteristics of the patients group. Table S2: Details of TaqMan MicroRNA assays. Table S3: Experimental details of the rt-PCR analyses according to the checklist of the MIQE Giudelines (minimum information for the publication of real-time quantitative PCR experiments) Table S4: Summary statistics of the normalized expression data of miRNAs in the three clinical sample groups (BC, NMIBC, MIBC) Table S5: ROC characteristics for A) selected miRNAs B) between AUC of different miRNAs Data File SF1: Raw data-expression calculation.

Author Contributions: Conceptualization, E.M.B. and T.K.; Data curation, E.M.B., T.K. and Z.J.; Funding acquisition, Z.J.; Investigation, M.P.; Methodology, E.M.B. and M.P.; Project administration, M.B.; Resources, T.K. and Z.J.; Software, E.M.B.; Supervision, E.M.B. and M.B.; Validation, E.M.B.; Writing – original draft, E.M.B.; Writing – review & editing, E.M.B. and Z.J.

Funding: This work was supported by Medical University Grant No502-03/5-138-02/502-54-146 and Grant No. 502-03/5-138-02/502-54-133.

Acknowledgments: The authors are grateful to all participants who consented to be included in the study. The authors would like to thank Aneta Zglińska-Pietrzak for her assistance in the field of statistical analysis and Magdalena Kaźmierczak for the English correction of the manuscript.

Conflicts of Interest: The authors declare no conflict of interest.

References

1. Cumberbatch, M.G.K.; Jubber, I.; Black, P.C.; Esperto, F.; Figueroa, J.D.; Kamat, A.M.; Kiemeney, L.; Lotan, Y.; Pang, K.; Silverman, D.T.; et al. Epidemiology of Bladder Cancer: A Systematic Review and Contemporary Update of Risk Factors in 2018. *Eur. Urol.* **2018**, *74*, 784–795. [CrossRef]

2. Hurst, C.D.; Knowles, M.A. Bladder cancer: Multi-omic profiling refines the molecular view. *Nat. Rev. Clin. Oncol.* **2018**, *15*, 203–204. [CrossRef]

3. Klaassen, Z.; Kamat, A.M.; Kassouf, W.; Gontero, P.; Villavicencio, H.; Bellmunt, J.; Van Rhijn, B.W.; Hartmann, A.; Catto, J.W.; Kulkarni, G.S. Treatment Strategy for Newly Diagnosed T1 High-grade Bladder Urothelial Carcinoma: New Insights and Updated Recommendations. *Eur. Urol.* **2018**, *74*, 597–608. [CrossRef]

4. Lin-Brande, M.; Pearce, S.M.; Ashrafi, A.N.; Nazemi, A.; Burg, M.L.; Ghodoussipour, S.; Miranda, G.; Djaladat, H.; Schuckman, A.; Daneshmand, S. Assessing the Impact of Time to Cystectomy for Variant Histology of Urothelial Bladder Cancer. *Urology* **2019**. [CrossRef]

5. Psutka, S.P.; Barocas, D.A.; Catto, J.W.F.; Gore, J.L.; Lee, C.T.; Morgan, T.M.; Master, V.A.; Necchi, A.; Rouprêt, M.; Boorjian, S.A. Staging the Host: Personalizing Risk Assessment for Radical Cystectomy Patients. *Eur. Urol. Oncol.* **2018**, *1*, 292–304. [CrossRef]

6. Palou, J.; Brausi, M.; Catto, J.W.F. Management of Patients with Normal Cystoscopy but Positive Cytology or Urine Markers. *Eur. Urol. Oncol.* **2019**. [CrossRef]

7. Sloan, F.A.; Yashkin, A.P.; Akushevich, I.; Inman, B.A. The Cost to Medicare of Bladder Cancer Care. *Eur. Urol. Oncol.* **2019**. [CrossRef]

8. Guancial, E.A.; Bellmut, J.; Yeh, S.; Rosenberg, J.E.; Berman, D.M. The evolving understanding of microRNA in bladder cancer. *Urol. Oncol.* **2014**, *32*, e31–e41. [CrossRef]

9. Li, Q.; Wang, H.; Peng, H.; Huang, Q.; Huyan, T.; Huang, Q.; Yang, H.; Shi, J. MicroRNAs: Key Players in Bladder Cancer. *Mol. Diagn. Ther.* **2019**. [CrossRef]

10. Fang, Z.; Dai, W.; Wang, X.; Chen, W.; Shen, C.; Ye, G.; Li, L. Circulating miR-205: A promising biomarker for detection and prognosis evaluation of bladder cancer. *Tumor. Biol.* **2016**, *37*, 8075–8082. [CrossRef]

11. Miah, S.; Dudziec, E.; Dryton, R.M.; Zlotta, A.R.; Morgan, S.L.; Rosario, D.J.; Hamdy, F.C.; Catto, J.W.F. An evaluation of urinary micro RNA reveals a high sensitivity for bladder cancer. *Br. J. Cancer* **2012**, *107*, 123–128. [CrossRef]

12. Zhou, B.; Guo, R. Integrative analysis of genomic and clinical data reveals intrinsic characteristics of bladder urothelial carcinoma progression. *Genes* **2019**, *10*, 464. [CrossRef]

13. Jensen, S.G.; Lamy, P.; Rasmussen, M.H.; Ostenfeld, M.S.; Dyrskjøt, L.; Ørntoft, T.F.; Andersen, C.L. Evaluation of two commercial global miRNA expression profiling platforms for detection of less abundant miRNAs. *BMC Genom.* **2011**, *12*, 435. [CrossRef]

14. Enokida, H.; Yoshino, H.; Matsushita, R.; Nakagawa, M. The role of microRNAs in bladder cancer. *Investig. Clin. Urol.* **2016**, *57* (Suppl. 1), S60–S76. [CrossRef]

15. Lee, J.Y.; Ryu, D.S.; Kim, W.J.; Kim, S.J. Aberrantly expressed microRNAs in the context of bladder tumorigenesis. *Investig. Clin. Urol.* **2016**, *57* (Suppl. 1), S52–S59. [CrossRef]

16. Sobin, L.H.; Gospodarowicz, M.K.; Wittekind, C.; IUAC (Eds.) *TNM Classification of Malignant Tumors*, 6th ed.; Springer: Berlin/Heidelberg, Germany, 2002.

17. Compérat, E.M.; Burger, M.; Gontero, P.; Mostafid, A.H.; Palou, J.; Rouprêt, M.; van Rhijn, B.W.G.; Shariat, S.F.; Sylvester, R.J.; Zigeuner, R.; et al. Grading of Urothelial Carcinoma and The New "World Health Organisation Classification of Tumours of the Urinary System and Male Genital Organs 2016". *Eur. Urol. Focus* 2018. [CrossRef]

18. Bustin, S.A.; Benes, V.; Garson, J.A.; Huggett, J.; Kubista, M.; Mueller, R.; Nolan, T.; Pfaffl, M.W.; Shipley, G.L.; Vandesompele, J.; et al. The MIQE guidelines: Minimum information for publication of quantitative real-time PCR experiments. *Clin. Chem.* **2009**, *55*, 611–622. [CrossRef]

19. Xiayu, R.; Huang, X.; Zhou, Z.; Lin, X. An improvement of the 2v(-delta delta CT) method for quantitative real-time polymerase chain reaction data analysis. *Biostat. Bioinform. Biomath.* **2013**, *3*, 71–85.

20. Schwarzenbach, H.; da Silva, A.M.; Calin, G.; Pantel, K. Data normalization strategies for microRNA quantification. *Clin. Chem.* **2015**, *61*, 1333–1342. [CrossRef]

21. van Kessel, K.E.M.; van der Keur, K.A.; Dyrskjøt, L.; Algaba, F.; Welvaart, N.Y.; Beukers, W.; Segersten, U.; Keck, B.; Maurer, T.; Simic, T.; et al. Molecular Markers Increase Precision of the European Association of Urology Non-Muscle-Invasive Bladder Cancer Progression Risk Groups. *Clin. Cancer Res.* **2018**, *24*, 1586–1593. [CrossRef]

22. Bruchbacher, A.; Soria, F.; Hassler, M.; Shariat, S.F.; D'Andrea, D. Tissue biomarkers in nonmuscle-invasive bladder cancer: Any role in clinical practice? *Curr. Opin. Urol.* **2018**, *28*, 584–590. [CrossRef]

23. Soria, F.; Krabbe, L.M.; Todenhöfer, T.; Dobruch, J.; Mitra, A.P.; Inman, B.A.; Gust, K.M.; Lotan, Y.; Shariat, S.F. Molecular markers in bladder cancer. *World J. Urol.* **2019**, *37*, 31–40. [CrossRef]

24. Miyake, M.; Owari, T.; Hori, S.; Fujimoto, K. Significant lack of urine-based biomarkers to replace cystoscopy for the surveillance of non-muscle invasive bladder cancer. *Transl. Androl. Urol.* **2019**, *8* (Suppl. 3), S332–S334. [CrossRef]

25. Lee, J.Y.; Yun, S.J.; Jeong, P.; Piao, X.-M.; Kim, Y.-H.; Kim, J.; Subramaniyam, S.; Byun, Y.J.; Kang, H.W.; Seo, S.P.; et al. Identification of differentially expressed miRNAs and miRNA-targeted genes in bladder cancer. *Oncotarget* **2018**, *9*, 27656–27766. [CrossRef]

26. Ratert, N.; Meyer, H.A.; Jung, M.; Lioudmer, P.; Mollenkopf, H.; Wagner, I.; Miller, K.; Kilic, E.; Erbersdobler, A.; Weikert, S.; et al. miRNA profiling identifies candidate miRNAs for bladder cancerdiagnosis and clinical outcome. *J. Mol. Diagn.* **2013**, *15*. [CrossRef]

27. Ratert, N.; Meyer, H.A.; Jung, M.; Mollenkopf, H.-J.; Wagner, I.; Miller, K.; Kilic, E.; Erbersdobler, A.; Weikert, S.; Jung, K. Reference miRNAs for miRNAome analysis of urothelial carcinomas. *PLoS ONE* **2012**, *7*, e39309. [CrossRef]

28. Peltier, H.J.; Latham, G.J. Normalization of microRNA expression levels in quantitative RT-PCR assays: Identification of suitable reference RNA targets in normal and cancerous human solid tissues. *RNA* **2008**, *14*, 844–852. [CrossRef]

29. Hofbauer, S.L.; de Martino, M.; Lucca, I.; Haitel, A.; Susani, M.; Shariat, S.F.; Klatte, T. A urinary microRNA (miR) signature for diagnosis of bladder cancer. *Urol. Oncol.* **2018**, *36*, 531.e1–531.e8. [CrossRef]

30. Boisen, M.K.; Dehlendorff, C.; Linnemann, D.; Schultz, N.A.; Jensen, B.V.; Hogdal, E.V.S.; Johansen, J.S. MicroRNA expression in formalin-fixed paraffin-embedded cancer tissue: Identifying reference microRNAs and variability. *BMC Cancer* **2015**, *15*, 1024. [CrossRef]

31. Parvaee, P.; Sarmadian, H.; Khansarinejad, B.; Amini, M.; Mondanizadeh, M. Plasma level of microRNAs, miR-107, miR-194 and miR-210 as potential biomarkers for diagnosis intestinal-type gastric cancer in human. *Asian Pac. J. Cancer Prev.* **2019**, *20*, 1421–1426. [CrossRef]

32. Lenherr, S.; Tsai, S.; Neto, B.S.; Sullivan, T.B.; Cimmino, C.B.; Logvinenko, T.; Gee, J.; Huang, W.; Libertino, J.A.; Summerhayes, I.C.; et al. MicroRNA expression profile identifies high grade, non-muscle-invasive bladder tumors at elevated risk to progress to an invasive phenotype. *Genes* **2017**, *8*, 77. [CrossRef]

33. Dip, N.; Reis, S.T.; Timoszczuk, L.S.; Viana, N.I.; Piantino, C.B.; Morais, D.R.; Moura, C.M.; Abe, D.K.; Silva, I.A.; Srougi, M.; et al. Stage, grade and behavior of bladder urothelial carcinoma defined by the microRNA expression profile. *J. Urol.* **2012**, *188*, 1951–1956. [CrossRef]

34. Ecke, T.H.; Stier, K.; Weickmann, S.; Zhao, Z.; Buckendahl, L.; Stephan, C.; Kilic, E.; Jung, K. miR-199a-3p and miR-214-3p improve the overall survival prediction of muscle-invasive bladder cancer patients after radical cystectomy. *Cancer Med.* **2017**, *6*, 2252–2262. [CrossRef]

35. Zhang, H.; Jiang, M.; Liu, Q.; Han, Z.; Zhao, Y.; Ji, S. mir-145-5p inhibits the proliferation and migration of bladder cancer cells by targeting TAGLN2. *Oncol. Lett.* **2018**, *16*, 6355–6360. [CrossRef]

36. Li, D.; Hao, X.; Song, Y. An integrated analysis of key microRNAs, regulatory pathways and clinical relevance in bladder cancer. *Onco Targets Ther.* **2018**, *11*, 3075–3085. [CrossRef]

37. Inamoto, T.; Uehara, H.; Akao, Y.; Ibuki, N.; Komura, K.; Takahara, K.; Takai, T.; Uchimoto, T.; Saito, K.; Tanda, N.; et al. A Panel of MicroRNA Signature as a Tool for Predicting Survival of Patients with Urothelial Carcinoma of the Bladder. *Dis. Markers* **2018**, *2018*, 5468672. [CrossRef]

38. Pignot, G.; Cizeron-Clairac, G.; Vacher, S.; Susini, A.; Tozlu, S.; Vieillefond, A.; Zerbib, M.; Lidereau, R.; Debre, B.; Amsellem-Ouazana, D. microRNA expression profile in a large series of bladder cancer tumors: Identyfication of 3-miRNA signature associated with aggressiveness of muscle-invasive bladder cancer. *Int. J. Cancer* **2013**, *132*, 2479–2491. [CrossRef]

39. Urquidi, V.; Netherton, M.; Gomes-Giacoia, E.; Serie, D.J.; Eckel-Passow, J.; Rosser, C.J.; Goodison, S. A microRNA biomarker panel for the non-invasive detection of bladder cancer. *Oncotarget* **2016**, *7*, 66290–86299. [CrossRef]

40. Armstromg, D.A.; Green, B.B.; Seigne, J.D.; Schned, J.D.; Marsit, C.J. MicroRNA molecular profiling from matched tumor and bio-fluids in bladder cancer. *Mol. Cancer* **2015**, *14*, 194. [CrossRef]

41. Baumgart, S.; Holters, S.; Ohlmann, C.H.; Bohle, R.; Stockle, M.; Ostenfeld, M.S.; Dyrskjøt, L.; Junker, K.; Heinzelmann, J. Exosome of invasive urothelial carcinoma cells are characterized by a specific miRNA expression signature. *Oncotarget* **2017**, *8*, 58278–58291. [CrossRef]

42. Egawa, H.; Jingushi, K.; Hirono, T.; Ueda, Y.; Kitae, K.; Nakata, W.; Fujita, K.; Uemura, M.; Nonomura, N.; Tsujikawa, K. The. miR-130 family promotes cell migration and invasion inbladder cancerthrough FAK and Akt phosphorylation by regulating PTEN. *Sci. Rep.* **2016**, *6*, 20574. [CrossRef]

43. Lv, M.; Zhong, Z.; Chi, H.; Huang, M.; Jiang, R.; Chen, J. Genome-Wide Screen of miRNAs and Targeting mRNAs Reveals the Negatively Regulatory Effect of miR-130b-3p on PTEN by PI3K and Integrin β1 Signaling Pathways in Bladder Carcinoma. *Int. J. Mol. Sci.* **2016**, *18*, 78. [CrossRef]

44. Liu, X.; Kong, C.; Zhang, Z. miR-130bpromotesbladder cancer cell proliferation, migration and invasion by targeting VGLL4. *Oncol. Rep.* **2018**, *39*, 2324–2332. [CrossRef]

Article

Transcription Factor and miRNA Interplays Can Manifest the Survival of ccRCC Patients

Shijie Qin, Xuejia Shi, Canbiao Wang, Ping Jin * and Fei Ma *

Laboratory for Comparative Genomics and Bioinformatics, College of Life Science, Nanjing Normal University, Nanjing 210046, China; 191201005@stu.njnu.edu.cn (S.Q.); 181202063@stu.njnu.edu.cn (X.S.); 191202061@stu.njnu.edu.cn (C.W.)
* Correspondence: jinping@njnu.edu.cn (P.J.); mafei01@tsinghua.org.cn (F.M.)

Received: 28 September 2019; Accepted: 24 October 2019; Published: 28 October 2019

Abstract: Clear cell renal cell carcinoma (ccRCC) still remains a higher mortality rate in worldwide. Obtaining promising biomakers is very crucial for improving the diagnosis and prognosis of ccRCC patients. Herein, we firstly identified eight potentially prognostic miRNAs (hsa-miR-144-5p, hsa-miR-223-3p, hsa-miR-365b-3p, hsa-miR-3613-5p, hsa-miR-9-5p, hsa-miR-183-5p, hsa-miR-335-3p, hsa-miR-1269a). Secondly, we found that a signature containing these eight miRNAs showed obviously superior to a single miRNA in the prognostic effect and credibility for predicting the survival of ccRCC patients. Thirdly, we discovered that twenty-two transcription factors (TFs) interact with these eight miRNAs, and a signature combining nine TFs (*TFAP2A, KLF5, IRF1, RUNX1, RARA, GATA3, IKZF1, POU2F2,* and *FOXM1*) could promote the prognosis of ccRCC patients. Finally, we further identified eleven genes (hsa-miR-365b-3p, hsa-miR-223-3p, hsa-miR-1269a, hsa-miR-144-5p, hsa-miR-183-5p, hsa-miR-335-3p, *TFAP2A, KLF5, IRF1, MYC, IKZF1*) that could combine as a signature to improve the prognosis effect of ccRCC patients, which distinctly outperformed the eight-miRNA signature and the nine-TF signature. Overall, we identified several new prognosis factors for ccRCC, and revealed a potential mechanism that TFs and miRNAs interplay cooperatively or oppositely regulate a certain number of tumor suppressors, driver genes, and oncogenes to facilitate the survival of ccRCC patients.

Keywords: ccRCC; prognostic biomarker; miRNA; transcription factor; interplay

---

1. Introduction

Clear cell renal cell carcinoma (ccRCC) is the most common malignant tumor subtype of kidney cancer [1], which still remains a higher mortality rate in worldwide [2]. At present, the main treatment method on ccRCC patients is early resection, but its curative effect and prognosis are not very good for these terminally ccRCC patients [3]. Currently, above 30–50% of ccRCC patients have missed the best surgical opportunity due to the lack of early clinical symptoms [3]. Therefore, acquiring new molecular biomarkers not only are urgently needed for establishing clinically stratifying system to improve the diagnostic efficiency of ccRCC patients, but are of great clinical value for effectively improving the management and treatment strategy of ccRCC patients.

MicroRNAs (miRNAs), a class of small noncoding RNAs with about 22 nucleotides, can negatively regulate gene expression at the post-transcriptional level by influencing the mRNA degradation and/or translational efficiency involved in manifold biological processes [4–7]. In recent years, many studies have revealed that miRNAs not only can act as oncogenes or tumor suppressor genes involve in tumorigenesis and progressions of various cancers [8], but can serve as valuable boimarkers for the detection and prognosis of cancer patients [9–15]. Thus, the current study has been focused on finding novel miRNAs as effective prognostic predictors for the overall survival of cancer patients [16–20].

At present, some studies have suggested that a lot of miRNAs could act as oncogenic miRNAs or tumor suppressor miRNAs to participate in the tumorigenesis and progressions of ccRCC, and they might serve as diagnosic and prognosic biomarkers for ccRCC patients [21–24]. However, several studies have shown that some miRNAs may play complete contradictor roles in diagnosic and prognosic effects for ccRCC, such as miR-99a [25,26], miR-106a [27,28], miR-125b [21,29], miR-144 [30,31], miR-203 [32,33], and miR-378 [34,35], which might greatly limit their applications into the clinical diagnosis and prognosis for ccRCC patients. Therefore, we should further study the functional roles of miRNAs as potentially diagnostic and prognostic biomarkers for ccRCC, whilst it is also very necessary for expanding the screening of new reliable miRNAs as diagnostic and prognostic biomarkers for ccRCC.

Nowadays, a number of studies are mainly focused on the fuctional mechanism of miRNAs that regulate their targets to cause the occurrence and development of ccRCC. However, how are miRNAs themselves regulated in the occurrence of ccRCC, which is still largely unknown. Intriguingly, currently, many studies have shown that transcription factors (TFs) can regulate miRNA expressions, and miRNAs may also regulate TF expressions in gene regulatory networks [36–39], and TFs and miRNAs interplay could precisely modulate gene expressions to maintain cell homeostasis [37,40,41]. Therefore, considering the complexity of TFs and miRNAs interplay mediating gene expression, substantial works should be further made in elucidating the mechanism that TFs and miRNAs interplay drives the occurrence and development of ccRCC, and finding TFs and miRNAs as novel diagnostic and prognostic biomarkers in the survival of ccRCC patients.

Considering that miRNAs often carry out their functions through fine-tuning the expression of their target genes, and multiple miRNAs can also synergistically or antagonistically regulate one or more target genes to control the strength and duration of cell response [42,43]. Here, we firstly screened eight potentially prognostic miRNAs based on RNA-seq and clinical information from the TCGA database. We next combined the eight miRNAs as an integrative prognostic predictor to evaluate the prognostic efficiency. This result showed that the prognostic efficiency and credibility of the eight-miRNA signature significantly outperformed a single miRNA, which implies that the synergistical regulation of miRNAs plays key roles in the tumorigenesis and progression of ccRCC. Subsequently, we utilized target prediction software and KEGG enrichment analysis to find that these eight miRNAs mainly control a certain number of oncogenic and onco-suppressive genes from some crucially cancer-related pathways improving the survival of ccRCC patients. Additionally, we found that twenty-two TFs could interact with eight miRNAs based on deepCAGE, TransmiR v2.0, and MirWalk3.0 database [44–46]. To further reveal the possible molecular mechanism that TFs and miRNAs interplay facilitates the survival of ccRCC patients, we further constructed a interplay network of TFs and miRNAs and the network analysis revealed that the interplay between twenty-two TFs and eight miRNAs could synergistically control the expression of oncogenes, driver genes, and tumor suppressor genes to involve in the regulation of tumorigenesis and progression of ccRCC. Finally, we performed cox regression analysis to identify eleven genes as a eleven-gene signature, including six miRNAs and five TFs and the prognostic effect and the credibility of the eleven-gene signature also obviously outperformed the eight-miRNA signature and the nine-TF signature. Taken together, our findings not only revealed a novel possible mechanism that TFs and miRNAs interplay could regulate cooperatively oncogenes, driver genes, and tumor suppressor genes to facilitate the survival of ccRCC patients, but also identified some new potential prognostic factors and therapeutic targets for ccRCC patients.

2. Results

2.1. Identification of miRNAs as Potential Prognostic Biomarkers

In this work, we found 110 differentially expressed miRNAs between 480 ccRCC tissues and 68 paracancerous tissues, including 50 up-regulated and 60 down-regulated miRNAs (Figure S1A). Here, to identify these prognostic miRNAs for predicting the overall survival in ccRCC patients, we grouped 480 patients with at least 90 days into high- and low-expression groups according to the median expression level of each of 110 differentially expressed miRNAs. We next performed survival analysis to find that 37 miRNAs might be associated with the survival of ccRCC patients (*p*-value < 0.05) (Table S1). Next, we performed univariate analysis and screened these top 21 miRNAs with a *p*-value < 0.001 (Figure 1, Table S2) for multivariate stepwise cox regression analysis to further determine independently prognostic miRNA biomarkers for ccRCC patients. We finally obtained eight potential prognostic miRNAs, including four down-regulated miRNAs (hsa-miR-9-5p, hsa-miR-1269a, hsa-miR-183-5p, hsa-miR-335-3p) and four up-regulated miRNAs (hsa-miR-365b-3p, hsa-miR-223-3p, hsa-miR-144-5p, hsa-miR-3613-5p), which might be involved in the survival of ccRCC patients. We divided 480 ccRCC patients into high-risk and low-risk groups according to the median univariate cox risk score of each of these eight miRNAs to further detect the association between these identified eight miRNAs and the overall survival of ccRCC patients. Kaplan-Meier survival analysis and log-rank test indicated that, in all eight independent miRNA cohorts, ccRCC patients with high-risk groups exhibited the overall survival more badly than low-risk groups (all cohorts *p*-value < 0.01) (Figure S2). Interestingly, the range of the AUC value for eight miRNAs was about 0.6~0.7 (Figure S3), which indicated that the established prognosis model has a very good prognosis effect. Overall, these findings suggested that any of eight miRNAs might act as a possible prognostic biomarker for the survival of ccRCC patients.

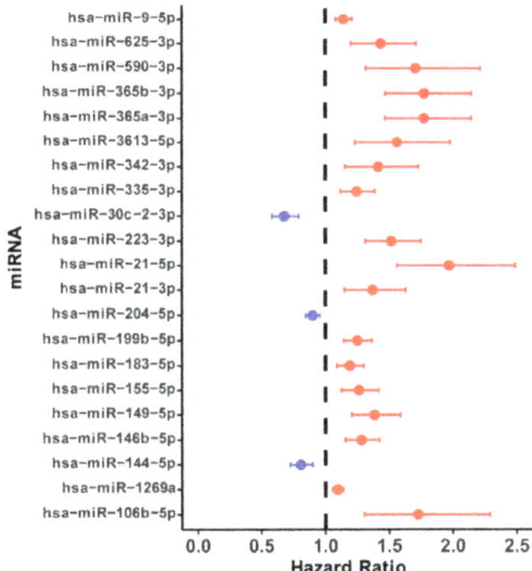

Figure 1. **The forest plots of hazard ratios (HR) of top 21 most significant survival associated miRNAs in ccRCC (*p*-value < 0.001).** Red represents the risk factors (HR > 1) and blue represents protective factors (HR < 1).

2.2. *The Exprssion Level of Eight Prognostic miRNAs is Associated with the Survival of ccRCC Patients*

To explore whether the eight miRNAs could be used as potential diagnostic biomarkers for distinguishing patients with ccRCC from controls. Here, we used survival curves to assess the association between the expression level of these eight miRNAs and the overall survival of ccRCC patients. We divided ccRCC patients into high-expression and low-expression miRNA groups according to the median expression level of each of these eight miRNAs. This result indicated that, in all eight independent miRNA cohorts, ccRCC patients with high-expression miRNA groups exhibited a worse overall survival rate than the low-expression miRNA groups, except for hsa-miR-144-5p (Figure 2). These results seemed to indicate that these highly expressed hsa-miR-9-5p, hsa-miR-1269a, hsa-miR-183-5p, hsa-miR-335-3p, hsa-miR-365b-3p, hsa-miR-223-3p, and hsa-miR-3613-5p might be associated with a poor prognosis for ccRCC patients. It is noteworthy that the highly expressed hsa-miR-144-5p was associated with better overall survival (Figure 2), which suggested that hsa-miR-144-5p should be a good prognostic factor for ccRCC patients. In addition, we further calculated the association between the expression level of eight miRNAs and patient's clinical diagnostic factors, respectively. Our results showed that eight prognostic miRNAs were significantly associated with T stage, M stage, G stage, and pathologic stage (Figure S4, Table S3), implying that these eight miRNAs might be involved in tumorigenesis and progression of ccRCC and could be served as prognostic biomarkers for the survival of ccRCC patients.

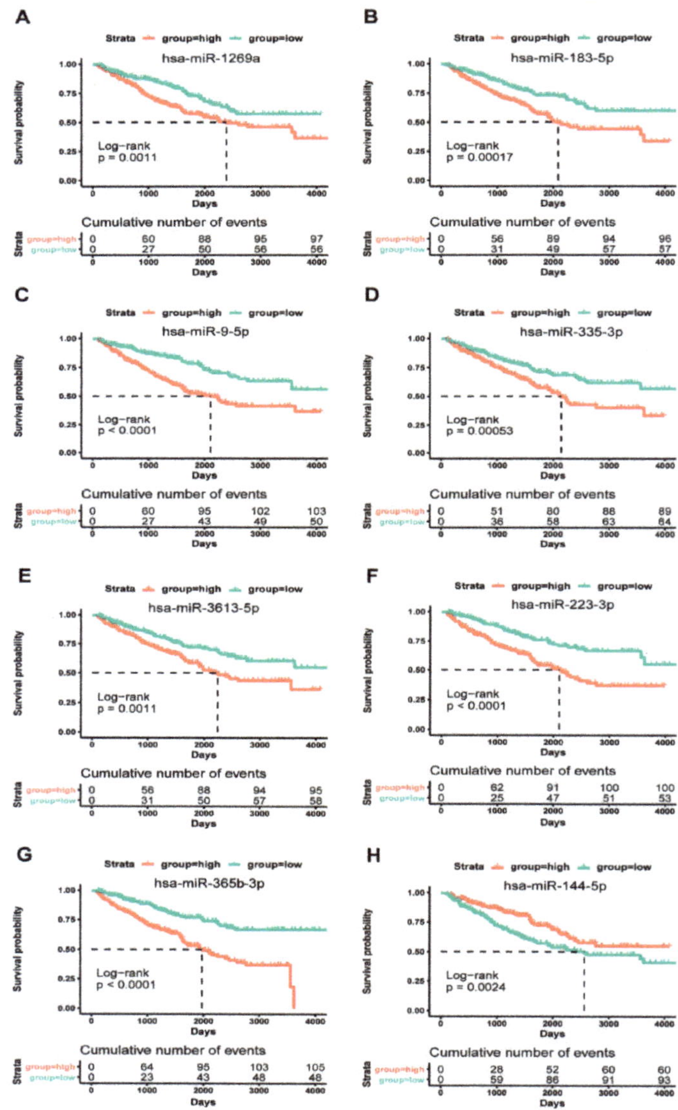

Figure 2. Kaplan Meier survival based on the expression level of eight miRNAs. Overall survival curves for high-expression and low-expression ccRCC patient cohorts. (**A**): hsa-miR-1269a; (**B**): hsa-miR-183-5p; (**C**): hsa-miR-9-5p; (**D**): hsa-miR-335-3p; (**E**): hsa-miR-3613-5p; (**F**): hsa-miR-223-3p; (**G**): hsa-miR-365b-3p; and, (**H**): hsa-miR-144-5p.

2.3. Prognostic Value of Combined Eight miRNAs as a Signature in ccRCC Patients

Considering multiple miRNAs could synergistically or antagonistically regulate one or more target genes to control cell fate. Herein, we further combined these eight miRNAs as an integrative prognostic predictor. The 480 patients were divided into low-risk group and high-risk group and then subjected to survival analysis. Our results showed that there was a significant difference in the overall survival between the two risk groups, and the high-risk group had more worse overall survival than the low-risk group ($p < 0.0001$, Figure 3A). In addition, the ROC curve based on the eight-miRNA signature also, respectively, showed an average 3, 5, and 10 year AUC for 0.762, 0.747, and 0.746 (Figure 3B). Interestingly, the concordance index (0.7305) of the combined prognostic model of the eight-miRNA signature was higher than that of all single miRNA (Table S4), whereas the Akaike information criterion (1622.2491) of the combined prognostic model of the eight-miRNA signature was lower than that of all single miRNA (Table S4), which indicated that the prognostic effect and credibility of the eight-miRNA signature were clearly superior to all single miRNA. These findings suggested that the eight-miRNA signature could act as a prognostic biomarker for promoting the survival of ccRCC patients.

Cox proportional hazard regression analysis was further used to characterize the impact of various clinical factors on the overall survival of ccRCC patients (Table 1). Age, gender, tumor size, metastasis, pathologic stage, neoplasm histologic grade, and the combined eight miRNAs signature were coded as continuous variables. As shown in Table 1, the univariate analysis showed that all factors, except for gender, might act as prognostic indicators for ccRCC patients. However, the multivariate analysis indicated that only age and the eight-miRNA signature can be used as independent prognostic indicators for ccRCC patients. This result revealed that the eight-miRNA signature could not only could serve as an independent prognostic factor for overall survival of ccRCC patients, but also act as an effective risk stratification indicator for ccRCC patient diagnosis.

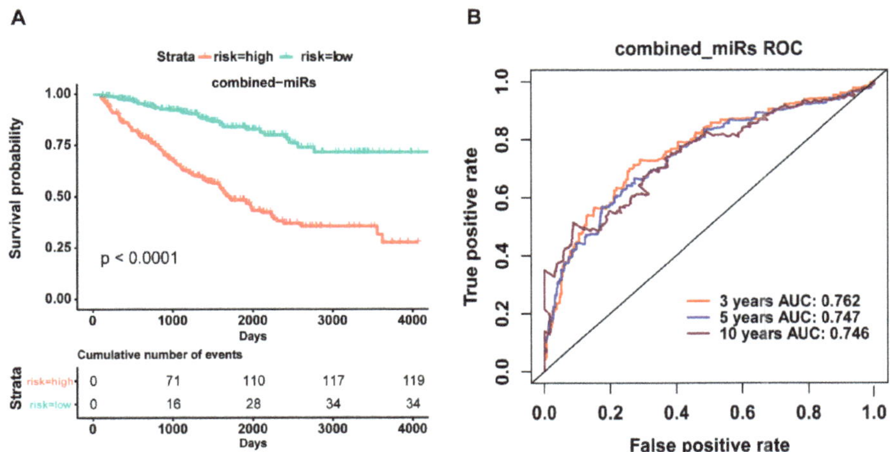

Figure 3. Kaplan Meier survival and receiver operating characteristic (ROC) curves based on the riskscore of the eight-miRNA signature. (**A**): Overall survival curves of high-risk and low-risk based on the eight-miRNA signature model. (**B**): Receiver operating characteristic (ROC) curves for high and low risk from the eight-miRNA signature model.

Table 1. Univariate and multivariate Cox regression analysis of the overall survival for clinical factors and risk of the combined eight prognostic miRNAs as a signature.

Variables	Univariate Analysis		Multivariate Analysis	
	Hazard Ratio (95% CI)	*p*-Value	Hazard Ratio (95% CI)	*p*-Value
Age	1.029 (1.014–1.043)	<0.001	1.028 (1.012–1.044)	<0.001
Gender	0.920 (0.660–1.284)	0.624	0.832 (0.590–1.172)	0.292
Tumor_pathologic_T	1.855 (1.556–2.211)	<0.001	0.816 (0.525–1.269)	0.367
Metastasis_pathologic_M	4.671 (3.369–6.475)	<0.001	1.514 (0.744–3.078)	0.253
Pathologic_stage_Stage	1.884 (1.634–2.172)	<0.001	1.605 (0.994–2.590)	0.053
Histologic_grade_G	2.238 (1.800–2.783)	<0.001	1.275 (0.996–1.633)	0.054
The eight-miRNA signature	4.009 (2.725–5.898)	<0.001	2.666 (1.768–4.020)	<0.001

Age, gender, tumor stage, metastasis pathologic, pathologic stage, histologic grade, and the eight-miRNA signature were coded as continuous variable. Specifically, pathologic stage was coded as I = 1, II = 2, III = 3, IV = 4. Tumor stage was coded as T1 = 1, T2 = 2, T3 = 3, T4 = 4. Histologic grade was coded as G1 = 1, G2 = 2, G3 = 3, G4 = 4.

2.4. Function Roles of Eight Prognostic miRNAs in ccRCC

To reveal the functional role of eight prognostic miRNAs in ccRCC, we first identified 3672 differentially expressed genes between 480 ccRCC tissues and 68 paracancerous tissues, including 2057 up-regulated genes (of which 116 transcription factors) and 1012 down-regulated genes (of which 61 transcription factors) (Figure S1B). Next, we obtained 357 down-regulated targets of 4 up-regulated miRNAs (hsa-miR-365b-3p, hsa-miR-223-3p, hsa-miR-144-5p, hsa-miR-3613-5p) and 1012 up-regulated targets of 4 down-regulated miRNAs (hsa-miR-9-5p, hsa-miR-1269a, hsa-miR-183-5p, hsa-miR-335-3p).

To elucidate the function of these target genes of these eight miRNAs in ccRCC, we further performed KEGG pathway analysis using clusterProfiler R package. We found that these significantly up-regulated target genes of four down-regulated miRNAs were widely involved in cancer-related signaling pathways, such as MAPK, Ras, NF-kappa B, Chemokine and Cytokine-cytokine receptor (Figure S5A). These results suggested that the interplay among multiple signaling pathways might synergistically mediate the occurrence and development of ccRCC.

We further picked out these up-regulated target genes of four down-regulated miRNAs from these main cancer signaling pathways to construct the miRNA-gene regulation network (Figure 4A). As shown in Figure 4A, some up-regulated target genes were regulated by more than one down-regulated miRNA, implying that the cooperative regulation of multiple miRNAs might play key roles in the initiation and progression of ccRCC. In addition, the protein-protein interaction (PPI) network was also constructed for these up-regulated target genes of four down-regulated miRNAs using the STRING database, which demonstrated a close interaction within these target genes (Figure 4B). Here, a node with ≥ 20 degrees is defined as a hub gene, thus we found 30 hub genes (Table S5). We next used multivariate cox regression analysis for 30 hub genes to further select prognosis-related genes for ccRCC patients. We found ten potential prognostic genes, including eight tumor suppressor and/or driver genes (*VEGFA, CCND1, BAX, IL7R, SHC1, FLT1, IL7,* and *JAK3*) [47–51], as well as two chemokines (*CXCL9* and *CXCL10*) [50]. Additionally, we combined the above ten hub genes as a signature to perform survival analysis. The result indicated that the low-risk group has a better overall survival rate than the high-risk group (*p* < 0.0001, Figure 4C), whilst the ROC curve also demonstrated that the ten-hub gene signature had a better prognostic effect and credibility for ccRCC patients with an average 3, 5 and 10 year AUC for 0.728, 0.751 and 0.796, respectively (Figure 4D). Taken together, our present findings suggested a possible molecular mechanism that the down-regulated expressed hsa-miR-9-5p, hsa-miR-1269a, hsa-miR-183-5p, and hsa-miR-335-3p might cooperatively up-regulate the expression level of numerous tumor suppressor and/or driver genes from some cancer-related pathways to improve the overall survival of ccRCC patients.

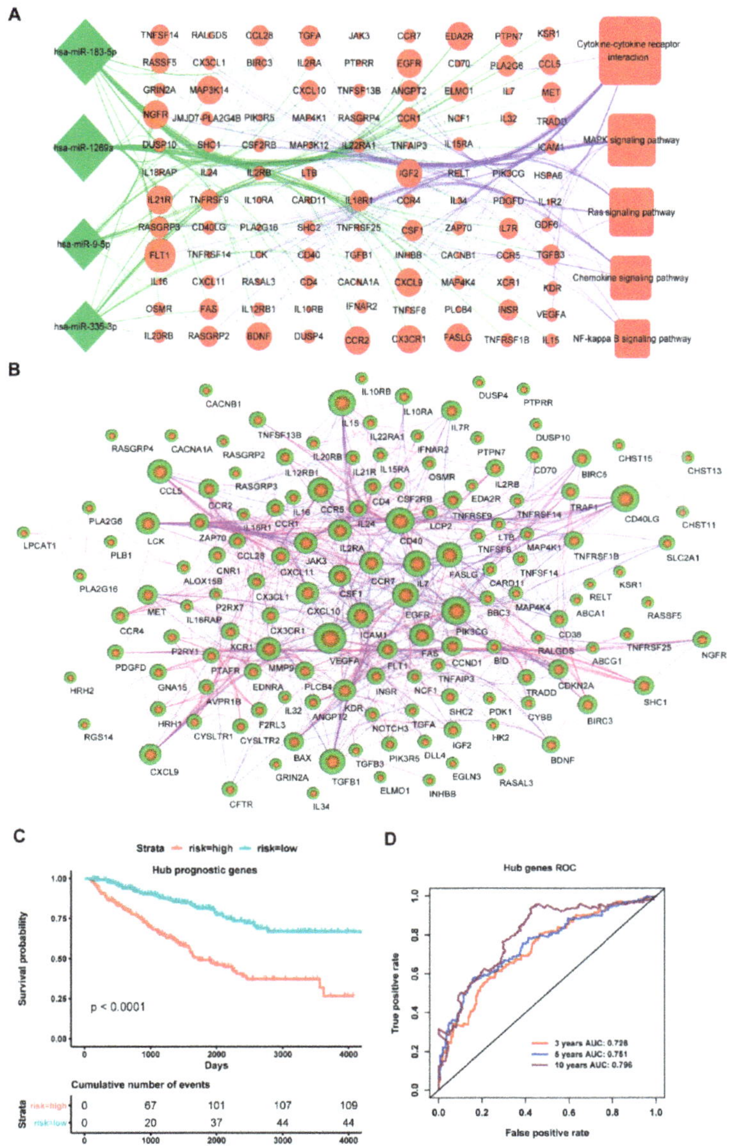

Figure 4. The cancer signaling pathway and protein network analysis of these up-regulated targets and the prognostic model of the ten-hub gene signature. (**A**): Target genes of four down-regulated miRNAs involve in the network map of cancer-related signaling pathways. The green represents the down-regulated expression and the red represents up-regulated expression; (**B**): Target protein interaction network of four down-regulated miRNAs. The blue line means low credibility and the purple line means high credibility; (**C**): Overall survival curves for high-risk and low-risk groups based on the ten-hub gene model; and, (**D**): ROC curves for high-risk and low-risk based on the ten-hub gene model.

Compared to four down-regulated miRNAs, these down-regulated target genes of four up-regulated miRNAs were less enriched in cancer-associated signaling pathways (Figure S5B). However, we further analyzed these target genes of three up-regulated miRNAs (hsa-miR-365b-3p, hsa-miR-223-3p, hsa-miR-3613-5p), finding that many genes of their target genes are involved in some cancer-associated signaling pathways (Figure 5A). Therefore, here, we reused multivariate cox regression analysis to screen potential prognostic genes for ccRCC patients. The result showed that five genes (*PRKCA, ADORA1, PPARGC1A, KL, GNG7*) could be identified as potentially prognostic factors, and they have also been suggested as tumor suppressor genes [49]. Interestingly, the survival analysis indicated that the five-gene signature could significantly stratify ccRCC patients into a high- and low-risk group ($p < 0.0001$, Figure 5B), and the AUC value of an average 3, 5 and 10 year is 0.696, 0.698, and 0.708, respectively (Figure 5C). Overall, we proposed a possibly functional mechanism that three highly expressed miRNAs (hsa-miR-365b-3p, hsa-miR-223-3p, and hsa-miR-3613-5p) might synergistically down-regulate the expression of many tumor suppressor genes to decrease the survival of ccRCC patients.

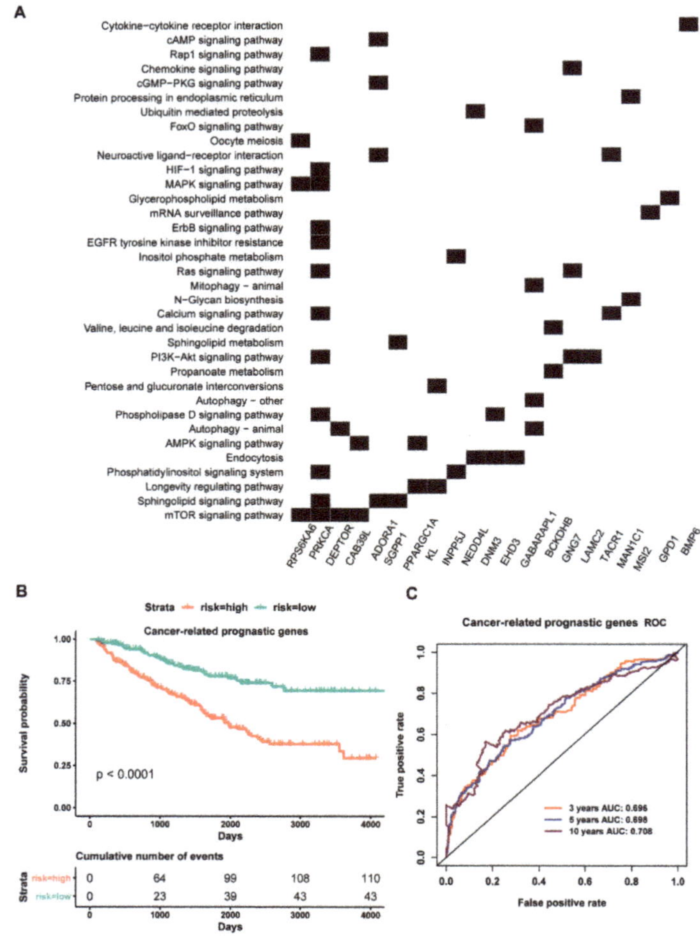

Figure 5. **The heat map of down-regulated targets of three up-regulated miRNAs involve in cancer-related signaling pathway and the prognostic model of the five-gene signature.** (**A**): The heat map of these down-regulated targets regulated by has-miR-3613-5p, has-miR-223-3p, and has-miR-365b-3p involved in cancer-related signaling pathways; (**B**): Overall survival curves for high-risk and low-risk groups based on the prognostic model of a five-gene signature; (**C**): ROC curves for high and low risk from the prognostic model of a five-gene signature.

Interestingly, our above results have indicated that the highly expressed hsa-miR-144-5p could act as a good prognostic factor for ccRCC patients. How does the highly expressed hsa-miR-144-5p facilitate the survival of ccRCC patients? Thus, we carried out a in-depth analysis for these target genes of hsa-miR-144-5p. Intriguingly, we found that hsa-miR-144-5p could regulate 55 genes, and, in particular, twelve genes of them have been reported as oncogenes or driver genes (Table S6) [49]. Thus, we further used multivariate cox regression analysis for twelve oncogenes and driver genes to screen potential prognostic factors for ccRCC patients. Consequently, these five genes (*MAGI3, CDKL1, CDH1, PPM1K, MSI2*) could be served as potential prognostic factors. We further combined the five genes as an integrative prognostic predictor for survival analysis. These results showed that the high-risk group had a worse overall survival than the low-risk group ($p < 0.0001$, Figure 6A), and the AUC value of an average 3, 5 and 10 year is 0.659, 0.676 and 0.759, respectively, based on the ROC curve (Figure 6B). Collectively, our present findings implied that the highly expressed hsa-miR-144-5p might facilitate the overall survival of ccRCC patients through down-regulating the expression level of some certain oncogenes and/or driver genes.

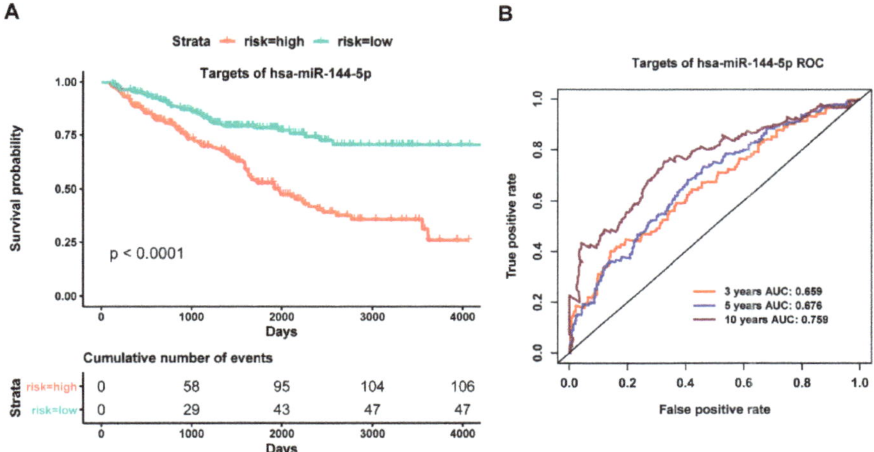

Figure 6. Kaplan Meier survival and ROC curves based on the risk score of five prognostic targets of has-miR-144-5p. (**A**): Overall survival curves for high-risk and low-risk groups based on the prognostic model of a five-gene signature; (**B**): ROC curves for high and low risk from the prognostic model of a five-gene signature.

2.5. The Interplay Network Between Twenty-Two TFs and Eight miRNAs

To explore the interplay between TFs and eight prognostic miRNAs, we further used TransmiR2.0 to predict these TFs that may regulate the eight prognostic miRNAs. Here, we found six down-regulated TFs (KLF5, SREBF2, TFAP2A, HIF1A, GATA3, and GATA2), which could regulate three down-regulated miRNAs (hsa-miR-9-5p, hsa-miR-183-5p, and hsa-miR-335-3p), but no TF was found for regulating hsa-miR-1269a (Figure S6A). Whilst, 16 up-regulated TFs (CEBPA, E2F1, FOXM1, FLI1, HEY1, IKZF1, IRF1, MEF2C, MYC, POU2F2, POU5F1, PRDM1, RARA, RUNX1, RUNX3, and TCF4) were also found to regulate four up-regulated miRNAs (hsa-miR-365b-3p, hsa-miR-223-3p, hsa-miR-144-5p, hsa-miR-3613-5p) (Figure S6B). Furthermore, we also predicted these twenty-two TFs regulated by the eight prognostic miRNAs. These detailed TF-miRNA regulatory pairs were outlined in the Table S7. Based on the TF-miRNA pairs, we further constructed a TF-miRNA interplay network (Figure 7), which showed that these up-regulated TFs, such as MYC, IKZF1, and IRF1 could up-regulate the expression level of hsa-miR-223-3p and hsa-miR-365b-3p to inhibit the expression of some TFs, such as GATA2 and SREBF2, then reducing the expression of hsa-miR-183-5p and hsa-miR-335-3p to up-regulate tumor suppressor gene expression (Figure 7). Whilst these down-regulated TFs, such as KLF5, SREBF2, TFAP2A, HIF1A, GATA3, and GATA2, could also down-regulate the expression level of hsa-miR-9-5p, hsa-miR-183-5p, and hsa-miR-335-3p to up-regulate their target TF expression, such as E2F1, RUNX1, and RUNX3, and then up-regulated the expression of hsa-miR-365b-3p, hsa-miR-223-3p, hsa-miR-144-5p, and hsa-miR-3613-5p to down-regulate the expression of some tumor suppressor genes or oncogenes (Figure 7). Specially, TF and miRNA carry out opposing functions. Therefore, our study seemed to imply that the interplay between twenty-two TFs and eight prognostic miRNAs might precisely control the expression of oncogenes, driver genes, and tumor suppressor genes to facilitate the survival of ccRCC patients.

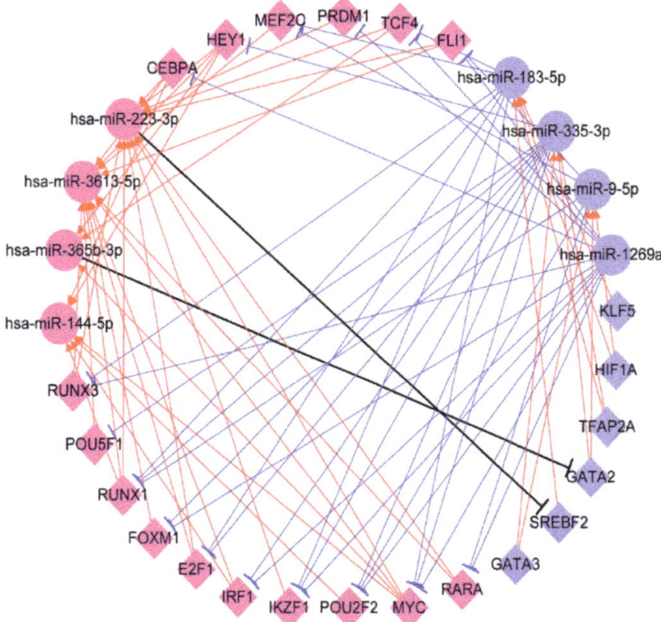

Figure 7. **The interplay network between twenty-two transcription factors and eight miRNAs.** The circle represents the miRNA and the diamond represents the transcription factor. The blue represents a down-regulation expression and the purple represents an up-regulation expression. The sharp arrow represents activation and the flat arrow represents inhibition.

2.6. Prognostic Value of the Combined Nine TFs as a Signature in ccRCC Patients

Based on these above results, we further used multivariate cox regression analysis for these above twenty-two TFs to determine independent prognostic TFs for ccRCC patients. Herein, we identified nine potential prognostic TFs (*TFAP2A, KLF5, IRF1, RUNX1, RARA, GATA3, IKZF1, POU2F2,* and *FOXM1*) that could act as prognostic factors for ccRCC patients. We next combined these nine TFs as a signature to detect the prognostic effect for ccRCC patients. The 480 patients were divided into low-risk group and high-risk group and subjected to survival analysis. Our results showed that the high-risk group had worse overall survival than the low-risk group ($p < 0.0001$, Figure 8A). In addition, the ROC curve based on the nine-TF pool also, respectively, showed an average 3, 5, and 10 year AUC for 0.721, 0.748, and 0.780 (Figure 8B), indicating that the nine-TF signature had very good prognosis effect and credibility for the survival of ccRCC patients. These findings suggested that the nine-TF signature could serve as a prognostic biomarker for improving the survival of ccRCC patients.

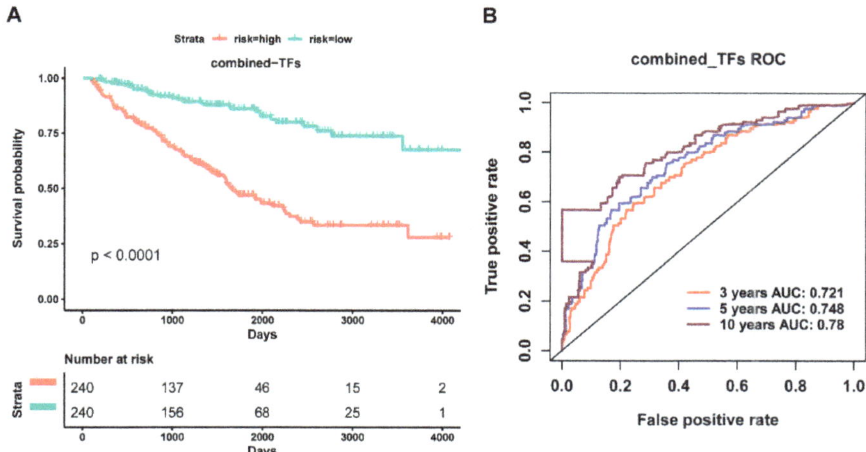

Figure 8. Kaplan Meier survival and ROC curves based on the risk score of the combined nine transcription factors as a signature. (**A**): Overall survival curves for high-risk and low-risk groups based on the prognostic model of a nine-TF signature; (**B**): ROC curves for high and low risk from the prognostic model of a nine-TF signature.

2.7. Clinical Value of TFs and miRNAs Interplay as a Prognostic Signature in ccRCC Patients

To further reveal the role of the interaction between TFs and miRNAs in the overall survival of ccRCC patients, we hypothesized that the interaction between TFs and miRNAs is likely to improve prognosis. Thus, we next performed multivariate analysis for above twenty-two TFs and eight miRNAs to identify prognostic TFs and miRNAs for ccRCC patients. We ultimately screened eleven potential prognostic factors, including hsa-miR-365b-3p, hsa-miR-223-3p, hsa-miR-1269a, hsa-miR-144-5p, hsa-miR-183-5p, hsa-miR-335-3p, *TFAP2A*, *KLF5*, *IRF1*, *MYC*, *IKZF1*. We further combined the above eleven genes as a signature for survival analysis. We found that the high-risk group had worse overall survival than the low-risk group ($p < 0.0001$, Figure 9A). Whilst the ROC curve based on the eleven-gene signature also, respectively, showed an average 3, 5, and 10 year AUC values for 0.777, 0.771, and 0.785 (Figure 10B). Strikingly, the concordance index (0.7552) of the combined prognostic model of the eleven-gene signature was more higher than that of the eight-miRNA signature (0.7305) and the nine-TF signature (0.7281), and the Akaike information criterion (1606.0516) of the combined prognostic model of the eleven-gene signature was lower than that of the eight-miRNA signature (1622.2941) and the nine-TF signature (1632.5955) (Table S8), which indicated that the prognostic effect and the credibility of the eleven-gene signature were better than both the eight-miRNA signature and the nine-TF signature did. These findings suggested that the eleven-gene signature could act as a prognostic factor for the overall survival of ccRCC patients.

Here, we also further used Cox proportional hazard regression analysis to characterize the impact of various clinical factors on overall survival of ccRCC patients (Table 2). Age, gender, tumor-pathologic, metastasis pathologic, pathological stage, neoplasm histologic grade, and the eleven-gene signature were coded as continuous variables. The univariate analysis showed that all factors, except for gender, might serve as prognostic indicators for ccRCC patients (Table 2). Notably, the multivariate analysis demonstrated that only age and the eleven-gene signature could be used as independent prognostic indicators for ccRCC patients (Table 2). Taken together, our present results revealed that the interaction between TFs and miRNAs might have very important effects on the overall survival of ccRCC patients, and the eleven-gene signature might serve as an independent factor to improve the prognosis for ccRCC patients.

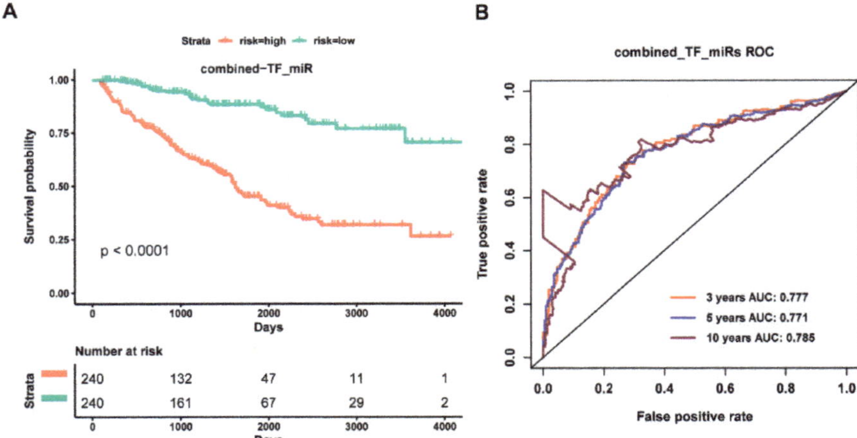

Figure 9. **Kaplan Meier survival and ROC curves based on the risk score of the combined five transcription factors and six miRNAs as a signature.** (**A**): Overall survival curves for high-risk and low-risk groups based on the prognostic model of an eleven-gene signature; and, (**B**): ROC curves for high and low risk from the prognostic model of an eleven-gene signature.

Table 2. Univariate and multivariate Cox regression analysis of overall survival for clinical factors and risk of the combined five prognostic transcription factors and six miRNAs as a signature.

Variables	Univariate Analysis		Multivariate Analysis	
	Hazard Ratio (95% CI)	*p*-Value	Hazard Ratio (95% CI)	*p*-Value
Age	1.029 (1.014–1.043)	<0.001	1.028 (1.011–1.044)	<0.001
Gender	0.920 (0.6595–1.284)	0.624	0.949 (0.673–1.338)	0.765
Tumor_pathologic_T	1.855 (1.556–2.211)	<0.001	0.840 (0.539–1.311)	0.443
Metastasis_pathologic_M	4.671 (3.369–6.475)	<0.001	1.707 (0.830–3.513)	0.146
Pathologic_stage_Stage	1.884 (1.634–2.172)	<0.001	1.410 (0.865–2.300)	0.178
Histologic_grade_G	2.238 (1.800–2.783)	<0.001	1.420 (1.118–1.803)	0.765
The eleven-gene signature	4.349 (2.943–6.428)	<0.001	2.590 (1.706–3.930)	<0.001

Age, gender, tumor stage, metastasis pathologic, pathologic stage, histologic grade and the eleven-gene signature were coded as continuous variable. Specifically, pathologic stage was coded as I = 1, II = 2, III = 3, IV = 4. Tumor stage was coded as T1 = 1, T2 = 2, T3 = 3, T4 = 4. Histologic grade was coded as G1 = 1, G2 = 2, G3 = 3, G4 = 4.

Overall, herein we propose a potential molecular mechanism that the interplay between TFs and miRNAs facilitates the overall survival of ccRCC patients (Figure 10). On the one hand, down-regulated expressed TFs could down-regulate some miRNA expression to further up-regulate tumor suppressor gene expression, whilst the down-regulated miRNAs could also up-regulate TF expression to further up-regulate the expression of tumor suppressor miRNA to down-regulate the expression of some oncogenes to improve prognosis for ccRCC. On the other hand, up-regulated TFs could up-regulate miRNA expression to inhibit the expression of some tumor suppressor genes, whilst the up-regulated miRNAs could also down-regulate the expression of some TFs to further down-regulate the expression of some miRNAs to up-regulate the expression of some tumor suppressor genes, which might promote ccRCC development. Taken together, our results seemed to reveal that the interplay between TFs and miRNAs might synergistically regulate the expression of a certain number of oncogenes, driver genes, and tumor suppressor genes to improve prognosis for ccRCC patients in transcriptional and post-transcriptional levels.

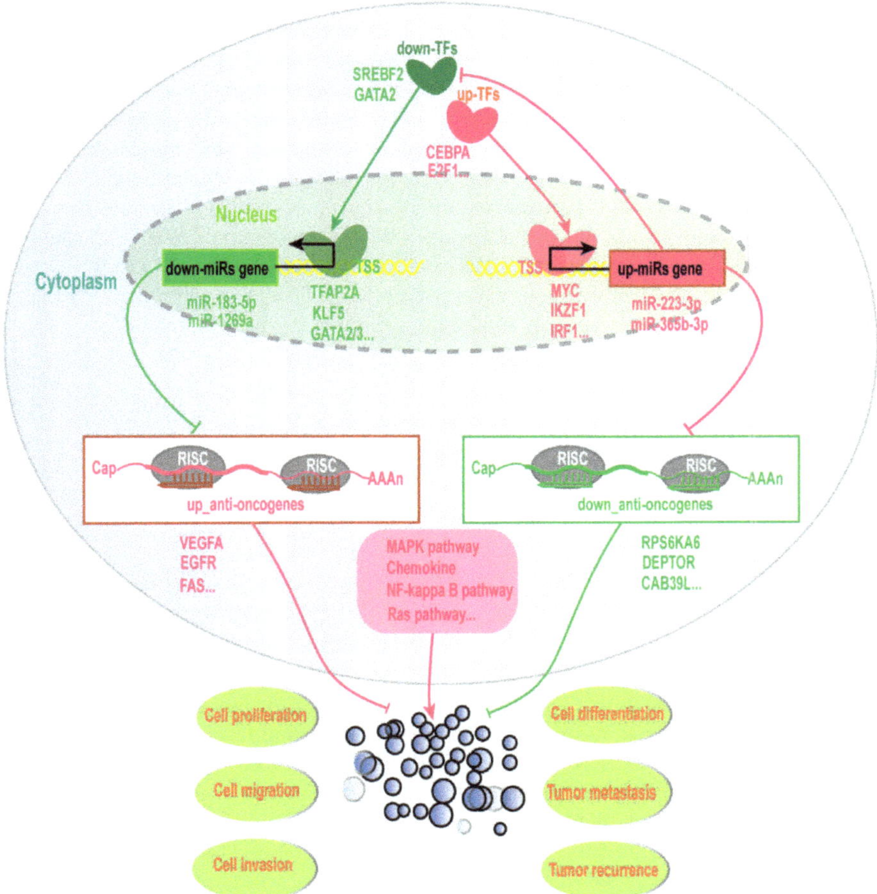

Figure 10. The molecular mechanism of the interplay between transcription factors and miRNAs improving the prognosis of ccRCC patients. The red represents an up-regulation of gene expression and the green represents down-regulation of gene expression. The sharp arrow represents activation and the flat arrow represents inhibition.

3. Discussion

At present, a number of miRNAs have been suggested as diagnosic and prognosic biomarkers for ccRCC patients, but, since a single miRNA mainly fine-tunes gene expression to execute its function, and the functional effect of a single miRNA is relatively weak, which might result in a lack of miRNA application into the clinical diagnosis and prognosis for ccRCC patients. Many studies have shown that miRNAs are more likely to regulate a certain number of gene expressions to control cell fate [52,53]. Remarkably, several studies have revealed that TFs can regulate miRNA expressions, whilst miRNAs may also regulate TF expressions, and TFs and miRNAs interplay can precisely modulate gene expression in transcriptional and post-transcriptional levels [36,45,54–57]. However, the regulatory landscape by miRNAs and TFs interplay is still largely unknown in the tumorigenesis and progression of ccRCC up to now. Especially, the interplay network of miRNA and TF has not yet been systemically studied in ccRCC. Therefore, in this current work, we have integratedly analyzed miRNA, TF, and mRNA profilings, and identified some potential diagnostic and prognostic factors participated in the survival of ccRCC patients, as well as revealed a possible molecular mechanism that miRNA and TF interplay can serve as an effective prognostic factor to facilitate the survival of ccRCC patients.

In this study, we have identified eight potentially diagnostic and prognostic miRNAs that could significantly distinguish the survival and pathological stratification for ccRCC patients. Among them, four down-regulated miRNAs (hsa-miR-9-5p, hsa-miR-1269a, hsa-miR-183-5p, hsa-miR-335-3p) could serve as good prognostic factors for clinical application of ccRCC patients, conversely the three up-regulated hsa-miR-365b-3p, hsa-miR-223-3p and hsa-miR-3613-5p of them might been acted as poor prognostic factors for ccRCC patients (Figure 2). At present, a few studies have reported that hsa-miR-183-5p can promote canceration by targeting SRSF2 in renal cancer [58], miR-335 could inhibit the proliferation and invasion of clear renal cells through suppressing Bcl-w [59], hsa-miR-365b-3p is poorly associated with ccRCC patient survival [60] and hsa-miR-9-5p is associated with the development and risk of renal cancer recurrence [61]. Although studies on the function roles of the other three miRNAs in ccRCC are still less reported to date, some other studies have revealed that hsa-miR-1269a could function as an onco-miRNA in NSCLC via down-regulating its target SOX6 [62], as well as that miR-1269a could promote colorectal cancer (CRC) metastasis by targeting Smad7 and HOXD10 [63]. The highly expressed hsa-miR-3613 as a oncogene could inhibit apoptosis via the down-regulation of target *APAF1* in human neuroblastoma BE(2)-C cells, as well as serve as a potential prognostic biomarker for pancreatic adenocarcinoma [64]. Some recent studies have indicated that miR-223-3p could down-regulate aNHEJ expression to result in synthetic lethality in human BRCA1-deficient cancers [65], and also act as an oncogenic miRNA in colon cancer through regulating EMT and PRDM1 [66]. These previous studies have revealed that the high expression of the seven miRNAs could down-regulate the respective tumor suppressor genes or driver genes involved in different malignant tumor occurrences. Especially, the same miRNA might regulate different target genes involved in specific tumor formation, which suggests that the different roles of same miRNA might depend on different cell microenvironment and cancer types. Interestingly, our present study has indicated that the seven aforementioned miRNAs mainly regulate tumor suppressor and/or driver genes to involve in ccRCC. Thus, we suggest that the seven miRNAs could served as potentially diagnostic and prognostic factors that may be due to the four down-regulated miRNAs that could restore or elevate the expression level of a certain number of tumor suppressor or driver genes to improve the survival of ccRCC patients, whereas the three up-regulated miRNAs might decrease the expression level of many tumor suppressor genes to result in the worse survival of ccRCC patients. Particularly, our present findings have shown that the highly expressed hsa-miR-144-5p could reduce the expression level of multiple oncogenic genes to promote the survival for ccRCC prognosis. A previous study has also shown that the hsa-miR-144-5p could serve as a tumor-suppressor gene for inhibiting cell growth and arresting cells in the G1 phase in renal cancer [30], which is also agreement with our present result. Taken together, we urge that hsa-miR-223-3p, hsa-miR-365b-3p, hsa-miR-3613-5p, hsa-miR-9-5p, hsa-miR-183-5p, hsa-miR-335-3p, and hsa-miR-1269a could serve as oncogenic miRNAs, as well as the hsa-miR-144-5p could act as a tumor suppressor miRNA for ccRCC patient diagnosis, but it is still further experimental verification.

A single miRNA molecule is known to carry out its function through fine-tuning the expression of target genes [42,43]. Therefore, it is hard to say that the expression of a single gene regulated by a single miRNA could significantly impact the proliferation and migration of cancer cell. Here, our findings seem to imply that single miRNA might need to regulate the expression of a lot of genes involved in the progression of ccRCC, suggesting that a single miRNA might be more suitable for ccRCC diagnosis and prognosis than a single gene. Specially, multiple miRNAs prefer synergistically or antagonistically regulating one or more target genes to control the strength and duration of cell response [42,43]. Thus, the combined multi-miRNAs as diagnosic and prognosic factors might be more suitable for the clinical application of ccRCC patients than a single miRNA. In the present study, just as we wish the prognostic effect and the credibility of the eight-miRNA signature are clearly superior to a single miRNA (Figure 3, Figure S3 and Table S4), implying that the cooperative regulation of multi-miRNAs might play very important roles in tumorigenesis and progression of ccRCC.

In the present work, we have found that eight prognostic miRNAs could interplay with twenty-two TFs. Many studies have revealed that the interplay between miRNA and TFs play key roles in establishing and maintaining cell phenotype [67–69]. Notably, in the gene regulatory network, TF and miRNA interplay could constitute positive or negative feedback loops to execute similar and opposing functions, which can precisely control the regulation of gene expression to reduce noise and maintain cell homeostasis [40,70,71]. Therefore, while considering the complexity of TFs and miRNAs interplay regulating gene expression, we further constructed an interplay network of TF-miRNA and propose a potential molecular mechanism that the interaction between TFs and miRNAs facilitates the survival of ccRCC patients (Figure 10). As shown in Figure 10, down-regulated transcription factors, such as KLF5 and GATA2, can down-regulate miR-183-3p and miR-1269a to up-regulated FAS and other tumor suppressor genes, whilst down-regulated miR-183-3p and miR-1269a can also up-regulate IKZF1 and IRF1 to down-regulated downstream targets. Additionally, up-regulated MYC and IKZF1 can activate miR-223-3p and miR-365b-3p to down-regulate RPS6KA6, DEPTOR, and other tumor suppressor genes, whilst miR-223-3p and miR-365b-3p can also down-regulate TFAP2A and GATA2 to down-regulate miR-183-5p and miR-1269a to up-regulated FAS and VEGFA. Remarkably, some previous reports showed that the transcription factor GATA2 could activate the expression of miR-194 to promote this distant metastasis of prostate cancer by inhibiting SOCS2 [72], and the transcription factor KLF5 could also promote the expression of miR-145, miR-124, and miR-183 by binding to their promoter involved in the progression of invasive pituitary adenoma [73], as well as the transcription factor TFAP2C could promote lung tumorigenesis and aggressiveness through it activating miR-183 and miR-33a-mediated cell cycle regulation [74]. Especially, the interplay of MYC and hsa-miR-144 has been reported in chronic myelogenous leukemia cell K562 [75], and the transcription factor E2F1 could also up-regulate miR-224/452 expressions to inhibit the expression of TXNIP to drive EMT in malignant melanoma [36], as well as miR-3188, as a tumour suppressor, could control the nasopharyngeal carcinoma proliferation and chemosensitivity through a mechanism where FOXO1 modulated a positive feedback loop of mTOR-p-PI3K/AKT-c-JUN [38]. These above results supported our conclusion that the interplays between transcription factors and miRNAs might play very important roles in the prognosis of ccRCC patients. Thus, based on the importance of TF and miRNA interplay in gene expression regulation, we ultimately screened six miRNAs (hsa-miR-365b-3p, hsa-miR-223-3p, hsa-miR-1269a, hsa-miR-144-5p, hsa-miR-183-5p, hsa-miR-335-3p) and five TFs (TFAP2A, KLF5, IRF1, MYC, IKZF1) as an integrative prognostic predictor. Interestingly, the prognostic effect and the credibility of the combined six miRNAs and five TFs signature are better than both the eight-miRNA signature and the nine-TF signature. The reason might be that not only TFs can regulate the expression of multiple target genes, including miRNAs, but also miRNAs can fine-tune multiple gene expressions, including TFs, as well as their closely coordinated regulations control cell homeostasis. This also suggests why TFs and miRNAs interplay is effective as a clinical prognostic factor for ccRCC patients. Of course, all of these regulatory pairs predicted by bioinformatics and data integration are still to be further verified experimentally.

4. Materials and Method

4.1. Data Sources and Pre-Processing

All KIRC sample RNA-seq data of mRNA and miRNA isoform and corresponding clinical information were downloaded from the The Cancer Genome Atlas (TCGA) database (https://portal.gdc.cancer.gov/, Springer Netherlands, Bethesda, MD, USA). The samples were filtered based on survival days greater than three months and simultaneous possessing mRNA and miRNA expression data. Based on the AnimalTFDB3.0 (http://bioinfo.life.hust.edu.cn/AnimalTFDB/, Oxford University Press, Hubei, China.) and Ensemble (http://asia.ensembl.org/index.html, Oxford University Press, Cambridgeshire, UK.) annotations, we identified 19,780 coding genes, of which 1400 were transcription factors. The expression level of 2,104 miRNA matures was obtained after miRNA isoform alignment. These lower expressed genes (sum(cpm) < 1) were removed and these genes expressed in at least 50% of the sample were retained.

4.2. Differentially Expressed Gene Analysis

Difference gene expression analysis between all tumor and paracancerous tissues was performed while using the edgeR package with filter parameters |log2FC| > 1 and p.adjust < 0.05. Similarly, differentially expressed miRNA was screened by limma package with criteria: |log2FC| > 1, p.adjust < 0.05. In addition, the mRNA and miRNA expression profiles were further respectively converted to log2 (normalized value + 1) and log2 (RPKM + 1) to be used for the next operation.

4.3. Survival Analysis and Prognosis Model Establishment

The samples of ccRCC with OS > 90 days were selected for survival analysis. Firstly, according to the median expression of miRNAs, batch survival analysis was performed to screen out the significantly differentially expressed miRNAs that are associated with survival. Subsequently, the univariate Cox regression was used to further assess the miRNAs related to survival. Only those miRNAs with a p-value < 0.001 were selected as candidate biomarker miRNAs. Finally, these candidate miRNAs were subjected to multivariate cox regression to determine independent prognostic marker miRNAs and calculate the risk value constructing prediction model for each miRNA. Based on the above results, the time-dependent receiver operating characteristic (ROC) curve was drawn using the Survival ROC R package, and the classification model was evaluated according to the area under the curve (AUC). In addition, we also analyzed the concordance index (C-index) and the Akaike information criterion (AIC). The C-index represents the consistency of the probability of the actual occurrence of the outcome and the probability of the prediction, and the AIC represents a standard for measuring the goodness of statistical model fitting. The prognostic signature of individual calculated according to the risk values of each marker miRNA and combined miRNAs. Next, the patient were divided into high and low risk groups according to the median risk score, and the survival analysis curve was then performed to check significant difference of patients in two groups over time.

4.4. miRNA Target Prediction and Target Function Analysis

These target genes of prognostic miRNAs were predicted by integrating Mirwalk3.0 (http://mirwalk. umm.uni-heidelberg.de/, Public Library of Science, Mannheim, Germany.) and a negative correlation between miRNA and mRNA expression. The Kyoto Encyclopedia of Genes and Genomes (KEGG) analysis was performed using the clusterProfiler package [76] with the filtration standard: p.adjust < 0.05. The PPI network from target genes was derived from the STING database (https://string-db.org/, Oxford University Press, Zurich, Switzerland).

4.5. Prediction of Transcription Factors Regulating miRNAs

The TransmiR2.0 (http://www.cuilab.cn/transmir, Oxford University Press, Beijing, China.) has collected the human regulatory pair of TFs regulating miRNAs (TFs-miRNAs) based on accurate transcriptional start site (TSS) of miRNA and ChIP-seq sequencing as well as experimental validation. Combining the TFs-miRNAs regulatory pair with the expression positive correlation of miRNAs and TFs, and TFs that may regulate these biomarker miRNAs were screened.

4.6. Data Statistics and Visualization

All data analysis was performed using R software (version 3.5.1, R Core Team, Vienna, Austria). Cytoscape software [77] (version 3.6.1, Cold Spring Harbor Laboratory Press, Washington, WA, USA.) were used to visualize the network. The survival curve was plotted using Kaplan Meier function, and the difference significance was evaluated by log-rank test.

5. Conclusions

In this work, we have identified eight prognostic miRNAs. Among them, seven miRNAs (hsa-miR-223-3p, hsa-miR-365b-3p, hsa-miR-3613-5p hsa-miR-9-5p, hsa-miR-183-5p, hsa-miR-335-3p, hsa-miR-1269a) can serve as potential oncogenes, whereas hsa-miR-144-5p might act as a tumor suppressor gene for ccRCC diagnosis. In addition, the eleven-gene signature (hsa-miR-365b-3p, hsa-miR-223-3p, hsa-miR-1269a, hsa-miR-144-5p, hsa-miR-183-5p, hsa-miR-335-3p, *TFAP2A*, *KLF5*, *IRF1*, *MYC*, *IKZF1*) can sever as an effective prognostic predictor to significantly improve the overall survival of ccRCC patients. Especially, our study has revealed a possible molecular mechanism that TFs and miRNAs interplay can cooperatively regulate the expression of oncogenes, driver genes, and tumor suppressor genes to facilitate the survival of ccRCC patients. Thus, our findings not only provide a new insight into the mechanism that TFs and miRNAs interplay control the tumorigenesis and progression of ccRCC, but also identify several novel diagnostic and prognostic biomarkers as well as potential therapeutic targets that are very crucial for making individualized therapeutic strategies of ccRCC patients.

Supplementary Materials: The following are available online at http://www.mdpi.com/2072-6694/11/11/1668/s1, Figure S1: The volcano map of differentially expressed genes and miRNAs. The volcano map of differentially expressed miRNA (A) and mRNA (B). The red point represents up-regulated expression and the green point represents down-regulated expression with statistical significance (p-value < 0.05), Figure S2: Kaplan-Meier survival based on the risk score of eight miRNAs. Overall survival curves of high-risk and low-risk groups for hsa-miR-1269a (A), hsa-miR-183-5p (B), hsa-miR-9-5p (C), hsa-miR-335-3p (D), hsa-miR-3613-5p (E), hsa-miR-223-3p (F), hsa-miR-365b-3p (G), hsa-miR-144-5p (H), Figure S3: ROC curves based on the risk score of eight miRNAs. ROC curves for high and low risk of hsa-miR-1269a (A), hsa-miR-183-5p (B), hsa-miR-9-5p (C), hsa-miR-335-3p (D), hsa-miR-3613-5p (E), hsa-miR-223-3p (F), hsa-miR-365b-3p (G), hsa-miR-144-5p (H), Figure S4: The correlation between the expression level of eight prognostic miRNAs and clinical factors. The correlation heat map between clinical parameters and prognostic miRNA expression levels in ccRCC. The red box represents a p-value < 0.05; the green box represents a p-value > 0.05, Figure S5: The KEGG signaling pathway analysis of up-regulated and down-regulated targets. Signaling pathways of KEGG enrichment of up-regulated targets(A) and down-regulated targets(B), Figure S6: Transcription factors regulating eight miRNAs. Six down-regulated transcription factors down-regulate three down-regulated miRNAs (A); sixteen up-regulate transcription factors up-regulate four up-regulated miRNAs (B). The red represents up-regulation of gene expression and the blue represents down-regulation of gene expression. Table S1: These 37 miRNAs are related to ccRCC survival, Table S2: These top 21 miRNAs are significant associated with ccRCC survival, Table S3: The correlation between prognostic miRNA expression levels and clinical factors, Table S4: The concordance index (C-index) and Akaike information criterion (AIC) of combined miRNAs and single miRNA model, Table S5: These top 30 hub genes from PPI network. Table S6: Targets of has-miR-144-5p. Table S7: TF-miRNA regulation pairs, Table S8. The C-index and AIC of TFs and miRNAs.

Author Contributions: F.M. and P.J. designed this study and edited this manuscript; S.Q. and X.S. acquired and analyzed the data; S.Q. and P.J. wrote the original draft; P.J. supervised this project; C.W. collected literatures. All authors read and approved the manuscript.

Funding: This research was funded by grants from the National Natural Science Foundation of China (No. 31970477) and the Natural Science Foundation of Jiangsu Province (No. BK20191368).

Acknowledgments: We thank our colleagues for their suggestions and criticisms on the manuscript.

Conflicts of Interest: The authors declare no conflict of interest.

References

1. Choudhary, S.; Sudarshan, S.; Choyke, P.L.; Prasad, S.R. Renal cell carcinoma: Recent advances in genetics and imaging. *Semin. Ultrasound CT MR* **2009**, *30*, 315–325. [CrossRef] [PubMed]

2. Siegel, R.L.; Miller, K.D.; Jemal, A. Cancer statistics, 2018. *CA A Cancer J. Clin.* **2018**, *68*, 7–30. [CrossRef] [PubMed]

3. Bellido, J.A.; Palou, J.; Hubner, M.; Pascual, M.; Sagrista, R.; Martinez, E.; Martinez, J.; Rosales, A.; Villavicencio, H. [Early ultrasound detection of renal tumors in patients with end stage renal disease in dialysis]. *Arch. Esp. Urol.* **2007**, *60*, 079–083. [PubMed]

4. Herrera-Carrillo, E.; Berkhout, B. Dicer-independent processing of small RNA duplexes: Mechanistic insights and applications. *Nucleic Acids Res.* **2017**, *45*, 10369–10379. [CrossRef] [PubMed]

5. Lee, R.C.; Ambros, V. An extensive class of small RNAs in Caenorhabditis elegans. *Science* **2001**, *294*, 862–864. [CrossRef] [PubMed]

6. Hansen, T.B. Detecting Agotrons in Ago CLIPseq Data. *Methods Mol. Biol.* **2018**, *1823*, 221–232. [CrossRef] [PubMed]

7. Bartel, D.P. MicroRNAs: Target recognition and regulatory functions. *Cell* **2009**, *136*, 215–233. [CrossRef]

8. Cho, W.C. OncomiRs: The discovery and progress of microRNAs in cancers. *Mol. Cancer* **2007**, *6*, 60. [CrossRef]

9. Yang, N.; Ekanem, N.R.; Sakyi, C.A.; Ray, S.D. Hepatocellular carcinoma and microRNA: New perspectives on therapeutics and diagnostics. *Adv. Drug Deliv. Rev.* **2015**, *81*, 62–74. [CrossRef]

10. Klingenberg, M.; Matsuda, A.; Diederichs, S.; Patel, T. Non-coding RNA in hepatocellular carcinoma: Mechanisms, biomarkers and therapeutic targets. *J. Hepatol.* **2017**, *67*, 603–618. [CrossRef]

11. Wu, W.L.; Wang, W.Y.; Yao, W.Q.; Li, G.D. Suppressive effects of microRNA-16 on the proliferation, invasion and metastasis of hepatocellular carcinoma cells. *Int. J. Mol. Med.* **2015**, *36*, 1713–1719. [CrossRef] [PubMed]

12. Meng, F.; Henson, R.; Wehbe-Janek, H.; Ghoshal, K.; Jacob, S.T.; Patel, T. MicroRNA-21 regulates expression of the PTEN tumor suppressor gene in human hepatocellular cancer. *Gastroenterology* **2007**, *133*, 647–658. [CrossRef] [PubMed]

13. Wong, C.C.; Wong, C.M.; Tung, E.K.; Au, S.L.; Lee, J.M.; Poon, R.T.; Man, K.; Ng, I.O. The microRNA miR-139 suppresses metastasis and progression of hepatocellular carcinoma by down-regulating Rho-kinase 2. *Gastroenterology* **2011**, *140*, 322–331. [CrossRef] [PubMed]

14. Xu, Y.; Xia, F.; Ma, L.; Shan, J.; Shen, J.; Yang, Z.; Liu, J.; Cui, Y.; Bian, X.; Bie, P.; et al. MicroRNA-122 sensitizes HCC cancer cells to adriamycin and vincristine through modulating expression of MDR and inducing cell cycle arrest. *Cancer Lett.* **2011**, *310*, 160–169. [CrossRef]

15. Zhai, W.; Li, S.; Zhang, J.; Chen, Y.; Ma, J.; Kong, W.; Gong, D.; Zheng, J.; Xue, W.; Xu, Y. Sunitinib-suppressed miR-452-5p facilitates renal cancer cell invasion and metastasis through modulating SMAD4/SMAD7 signals. *Mol. Cancer* **2018**, *17*, 157. [CrossRef]

16. Mishra, P.J. MicroRNAs as promising biomarkers in cancer diagnostics. *Biomark. Res.* **2014**, *2*, 19. [CrossRef]

17. Yang, X.; Zhang, X.F.; Lu, X.; Jia, H.L.; Liang, L.; Dong, Q.Z.; Ye, Q.H.; Qin, L.X. MicroRNA-26a suppresses angiogenesis in human hepatocellular carcinoma by targeting hepatocyte growth factor-cMet pathway. *Hepatology* **2014**, *59*, 1874–1885. [CrossRef]

18. Yang, F.; Li, Q.J.; Gong, Z.B.; Zhou, L.; You, N.; Wang, S.; Li, X.L.; Li, J.J.; An, J.Z.; Wang, D.S.; et al. MicroRNA-34a targets Bcl-2 and sensitizes human hepatocellular carcinoma cells to sorafenib treatment. *Technol. Cancer Res. Treat.* **2014**, *13*, 77–86. [CrossRef]

19. Hua, S.; Lei, L.; Deng, L.; Weng, X.; Liu, C.; Qi, X.; Wang, S.; Zhang, D.; Zou, X.; Cao, C.; et al. miR-139-5p inhibits aerobic glycolysis, cell proliferation, migration, and invasion in hepatocellular carcinoma via a reciprocal regulatory interaction with ETS1. *Oncogene* **2018**, *37*, 1624–1636. [CrossRef]

20. Youssef, Y.M.; White, N.M.; Grigull, J.; Krizova, A.; Samy, C.; Mejia-Guerrero, S.; Evans, A.; Yousef, G.M. Accurate molecular classification of kidney cancer subtypes using microRNA signature. *Eur. Urol.* **2011**, *59*, 721–730. [CrossRef]

21. Braga, E.A.; Fridman, M.V.; Loginov, V.I.; Dmitriev, A.A.; Morozov, S.G. Molecular Mechanisms in Clear Cell Renal Cell Carcinoma: Role of miRNAs and Hypermethylated miRNA Genes in Crucial Oncogenic Pathways and Processes. *Front. Genet.* **2019**, *10*, 320. [CrossRef] [PubMed]

22. Zhang, Z.; Sun, H.; Dai, H.; Walsh, R.M.; Imakura, M.; Schelter, J.; Burchard, J.; Dai, X.; Chang, A.N.; Diaz, R.L.; et al. MicroRNA miR-210 modulates cellular response to hypoxia through the MYC antagonist MNT. *Cell Cycle* **2009**, *8*, 2756–2768. [CrossRef] [PubMed]

23. Pan, Y.; Hu, J.; Ma, J.; Qi, X.; Zhou, H.; Miao, X.; Zheng, W.; Jia, L. MiR-193a-3p and miR-224 mediate renal cell carcinoma progression by targeting alpha-2,3-sialyltransferase IV and the phosphatidylinositol 3 kinase/Akt pathway. *Mol. Carcinog.* **2018**, *57*, 1067–1077. [CrossRef] [PubMed]

24. Fedorko, M.; Pacik, D.; Wasserbauer, R.; Juracek, J.; Varga, G.; Ghazal, M.; Nussir, M.I. MicroRNAs in the pathogenesis of renal cell carcinoma and their diagnostic and prognostic utility as cancer biomarkers. *Int. J. Biol. Mark.* **2016**, *31*, e26–e37. [CrossRef]

25. Cui, L.; Zhou, H.; Zhao, H.; Zhou, Y.; Xu, R.; Xu, X.; Zheng, L.; Xue, Z.; Xia, W.; Zhang, B.; et al. MicroRNA-99a induces G1-phase cell cycle arrest and suppresses tumorigenicity in renal cell carcinoma. *BMC Cancer* **2012**, *12*, 546. [CrossRef]

26. Oliveira, R.C.; Ivanovic, R.F.; Leite, K.R.M.; Viana, N.I.; Pimenta, R.C.A.; Junior, J.P.; Guimaraes, V.R.; Morais, D.R.; Abe, D.K.; Nesrallah, A.J.; et al. Expression of micro-RNAs and genes related to angiogenesis in ccRCC and associations with tumor characteristics. *BMC Urol.* **2017**, *17*, 113. [CrossRef]

27. Xiang, W.; He, J.; Huang, C.; Chen, L.; Tao, D.; Wu, X.; Wang, M.; Luo, G.; Xiao, X.; Zeng, F.; et al. miR-106b-5p targets tumor suppressor gene SETD2 to inactive its function in clear cell renal cell carcinoma. *Oncotarget* **2015**, *6*, 4066–4079. [CrossRef]

28. Pan, Y.J.; Wei, L.L.; Wu, X.J.; Huo, F.C.; Mou, J.; Pei, D.S. MiR-106a-5p inhibits the cell migration and invasion of renal cell carcinoma through targeting PAK5. *Cell Death Dis.* **2017**, *8*, e3155. [CrossRef]

29. Jin, L.; Zhang, Z.; Li, Y.; He, T.; Hu, J.; Liu, J.; Chen, M.; Gui, Y.; Chen, Y.; Lai, Y. miR-125b is associated with renal cell carcinoma cell migration, invasion and apoptosis. *Oncol. Lett.* **2017**, *13*, 4512–4520. [CrossRef]

30. Xiang, C.; Cui, S.P.; Ke, Y. MiR-144 inhibits cell proliferation of renal cell carcinoma by targeting MTOR. *J. Huazhong Univ. Sci. Technol. Med. Sci. = Hua zhong Ke Ji Da Xue Xue Bao. Yi Xue Ying De Wen Ban = Huazhong Keji Daxue Xuebao. Yixue Yingdewen Ban* **2016**, *36*, 186–192. [CrossRef]

31. Xiao, W.; Lou, N.; Ruan, H.; Bao, L.; Xiong, Z.; Yuan, C.; Tong, J.; Xu, G.; Zhou, Y.; Qu, Y.; et al. Mir-144-3p Promotes Cell Proliferation, Metastasis, Sunitinib Resistance in Clear Cell Renal Cell Carcinoma by Downregulating ARID1A. *Cell. Physiol. Biochem. Int. J. Exp. Cell. Physiol. Biochem. Pharmacol.* **2017**, *43*, 2420–2433. [CrossRef]

32. Xu, M.; Gu, M.; Zhang, K.; Zhou, J.; Wang, Z.; Da, J. miR-203 inhibition of renal cancer cell proliferation, migration and invasion by targeting of FGF2. *Diagn. Pathol.* **2015**, *10*, 24. [CrossRef] [PubMed]

33. Hu, G.; Lai, P.; Liu, M.; Xu, L.; Guo, Z.; Liu, H.; Li, W.; Wang, G.; Yao, X.; Zheng, J.; et al. miR-203a regulates proliferation, migration, and apoptosis by targeting glycogen synthase kinase-3beta in human renal cell carcinoma. *Tumour Biol. J. Int. Soc. Oncodevelopmental Biol. Med.* **2014**, *35*, 11443–11453. [CrossRef] [PubMed]

34. Fedorko, M.; Stanik, M.; Iliev, R.; Redova-Lojova, M.; Machackova, T.; Svoboda, M.; Pacik, D.; Dolezel, J.; Slaby, O. Combination of MiR-378 and MiR-210 Serum Levels Enables Sensitive Detection of Renal Cell Carcinoma. *Int. J. Mol. Sci.* **2015**, *16*, 23382–23389. [CrossRef] [PubMed]

35. Redova, M.; Poprach, A.; Nekvindova, J.; Iliev, R.; Radova, L.; Lakomy, R.; Svoboda, M.; Vyzula, R.; Slaby, O. Circulating miR-378 and miR-451 in serum are potential biomarkers for renal cell carcinoma. *J. Transl. Med.* **2012**, *10*, 55. [CrossRef]

36. Knoll, S.; Furst, K.; Kowtharapu, B.; Schmitz, U.; Marquardt, S.; Wolkenhauer, O.; Martin, H.; Putzer, B.M. E2F1 induces miR-224/452 expression to drive EMT through TXNIP downregulation. *EMBO Rep.* **2014**, *15*, 1315–1329. [CrossRef]

37. Song, W.; Wang, L.; Wang, L.; Li, Q. Interplay of miR-21 and FoxO1 modulates growth of pancreatic ductal adenocarcinoma. *Tumour Biol. J. Int. Soc. Oncodevelopmental Biol. Med.* **2015**, *36*, 4741–4745. [CrossRef]

38. Zhao, M.; Luo, R.; Liu, Y.; Gao, L.; Fu, Z.; Fu, Q.; Luo, X.; Chen, Y.; Deng, X.; Liang, Z.; et al. miR-3188 regulates nasopharyngeal carcinoma proliferation and chemosensitivity through a FOXO1-modulated positive feedback loop with mTOR-p-PI3K/AKT-c-JUN. *Nat. Commun.* **2016**, *7*, 11309. [CrossRef]

39. Zhou, J.; Wang, K.C.; Wu, W.; Subramaniam, S.; Shyy, J.Y.; Chiu, J.J.; Li, J.Y.; Chien, S. MicroRNA-21 targets peroxisome proliferators-activated receptor-alpha in an autoregulatory loop to modulate flow-induced endothelial inflammation. *Proc. Natl. Acad. Sci. USA* **2011**, *108*, 10355–10360. [CrossRef]

40. Su, Y.; Chen, C.; Guo, L.; Du, J.; Li, X.; Liu, Y. Ecological Balance of Oral Microbiota Is Required to Maintain Oral Mesenchymal Stem Cell Homeostasis. *Stem Cells* **2018**, *36*, 551–561. [CrossRef]

41. Shan, J.; Feng, L.; Sun, G.; Chen, P.; Zhou, Y.; Xia, M.; Li, H.; Li, Y. Interplay between mTOR and STAT5 signaling modulates the balance between regulatory and effective T cells. *Immunobiology* **2015**, *220*, 510–517. [CrossRef]

42. Qin, S.; Ma, F.; Chen, L. Gene regulatory networks by transcription factors and microRNAs in breast cancer. *Bioinformatics* **2015**, *31*, 76–83. [CrossRef]

43. Ng, W.L.; Chen, G.; Wang, M.; Wang, H.; Story, M.; Shay, J.W.; Zhang, X.; Wang, J.; Amin, A.R.; Hu, B.; et al. OCT4 as a target of miR-34a stimulates p63 but inhibits p53 to promote human cell transformation. *Cell Death Dis.* **2014**, *5*, e1024. [CrossRef]

44. Sticht, C.; De La Torre, C.; Parveen, A.; Gretz, N. miRWalk: An online resource for prediction of microRNA binding sites. *PLoS ONE* **2018**, *13*, e0206239. [CrossRef]

45. Tong, Z.; Cui, Q.; Wang, J.; Zhou, Y. TransmiR v2.0: An updated transcription factor-microRNA regulation database. *Nucleic Acids Res.* **2019**, *47*, D253–D258. [CrossRef]

46. Marsico, A.; Huska, M.R.; Lasserre, J.; Hu, H.; Vucicevic, D.; Musahl, A.; Orom, U.; Vingron, M. PROmiRNA: A new miRNA promoter recognition method uncovers the complex regulation of intronic miRNAs. *Genome Biol.* **2013**, *14*, R84. [CrossRef]

47. Ko, J.; Lee, Y.H.; Hwang, S.Y.; Lee, Y.S.; Shin, S.M.; Hwang, J.H.; Kim, J.; Kim, Y.W.; Jang, S.W.; Ryoo, Z.Y.; et al. Identification and differential expression of novel human cervical cancer oncogene HCCR-2 in human cancers and its involvement in p53 stabilization. *Oncogene* **2003**, *22*, 4679–4689. [CrossRef]

48. Ito, K.; Maruyama, Z.; Sakai, A.; Izumi, S.; Moriishi, T.; Yoshida, C.A.; Miyazaki, T.; Komori, H.; Takada, K.; Kawaguchi, H.; et al. Overexpression of Cdk6 and Ccnd1 in chondrocytes inhibited chondrocyte maturation and caused p53-dependent apoptosis without enhancing proliferation. *Oncogene* **2014**, *33*, 1862–1871. [CrossRef]

49. Lever, J.; Zhao, E.Y.; Grewal, J.; Jones, M.R.; Jones, S.J.M. CancerMine: A literature-mined resource for drivers, oncogenes and tumor suppressors in cancer. *Nat. Methods* **2019**, *16*, 505–507. [CrossRef]

50. Rabant, M.; Amrouche, L.; Morin, L.; Bonifay, R.; Lebreton, X.; Aouni, L.; Benon, A.; Sauvaget, V.; Le Vaillant, L.; Aulagnon, F.; et al. Early Low Urinary CXCL9 and CXCL10 Might Predict Immunological Quiescence in Clinically and Histologically Stable Kidney Recipients. *Am. J. Transplant. Off. J. Am. Soc. Transplant. Am. Soc. Transplant. Surg.* **2016**, *16*, 1868–1881. [CrossRef]

51. Rawal, R.M.; Joshi, M.N.; Bhargava, P.; Shaikh, I.; Pandit, A.S.; Patel, R.P.; Patel, S.; Kothari, K.; Shah, M.; Saxena, A.; et al. Tobacco habituated and non-habituated subjects exhibit different mutational spectrums in head and neck squamous cell carcinoma. *3 Biotech.* **2015**, *5*, 685–696. [CrossRef]

52. Rupp, L.J.; Brady, B.L.; Carpenter, A.C.; De Obaldia, M.E.; Bhandoola, A.; Bosselut, R.; Muljo, S.A.; Bassing, C.H. The microRNA biogenesis machinery modulates lineage commitment during alphabeta T cell development. *J. Immunol.* **2014**, *193*, 4032–4042. [CrossRef]

53. Chavali, S.; Bruhn, S.; Tiemann, K.; Saetrom, P.; Barrenas, F.; Saito, T.; Kanduri, K.; Wang, H.; Benson, M. MicroRNAs act complementarily to regulate disease-related mRNA modules in human diseases. *RNA* **2013**, *19*, 1552–1562. [CrossRef]

54. Zhang, B.; Zhang, Z.; Xia, S.; Xing, C.; Ci, X.; Li, X.; Zhao, R.; Tian, S.; Ma, G.; Zhu, Z.; et al. KLF5 activates microRNA 200 transcription to maintain epithelial characteristics and prevent induced epithelial-mesenchymal transition in epithelial cells. *Mol. Cell. Biol.* **2013**, *33*, 4919–4935. [CrossRef]

55. Ellwanger, D.C.; Leonhardt, J.F.; Mewes, H.W. Large-scale modeling of condition-specific gene regulatory networks by information integration and inference. *Nucleic Acids Res.* **2014**, *42*, e166. [CrossRef]

56. Yang, Y.; Zhou, L.; Lu, L.; Wang, L.; Li, X.; Jiang, P.; Chan, L.K.; Zhang, T.; Yu, J.; Kwong, J.; et al. A novel miR-193a-5p-YY1-APC regulatory axis in human endometrioid endometrial adenocarcinoma. *Oncogene* **2013**, *32*, 3432–3442. [CrossRef]

57. Zhang, X.; Zhang, X.; Wang, T.; Wang, L.; Tan, Z.; Wei, W.; Yan, B.; Zhao, J.; Wu, K.; Yang, A.; et al. MicroRNA-26a is a key regulon that inhibits progression and metastasis of c-Myc/EZH2 double high advanced hepatocellular carcinoma. *Cancer Lett.* **2018**, *426*, 98–108. [CrossRef]

58. Sokol, E.; Kedzierska, H.; Czubaty, A.; Rybicka, B.; Rodzik, K.; Tanski, Z.; Boguslawska, J.; Piekielko-Witkowska, A. microRNA-mediated regulation of splicing factors SRSF1, SRSF2 and hnRNP A1 in context of their alternatively spliced 3′UTRs. *Exp. Cell Res.* **2018**, *363*, 208–217. [CrossRef]

59. Wang, K.; Chen, X.; Zhan, Y.; Jiang, W.; Liu, X.; Wang, X.; Wu, B. miR-335 inhibits the proliferation and invasion of clear cell renal cell carcinoma cells through direct suppression of BCL-W. *Tumour Biol. J. Int. Soc. Oncodev. Biol. Med.* **2015**, *36*, 6875–6882. [CrossRef]

60. Tian, Q.; Sun, H.F.; Wang, W.J.; Li, Q.; Ding, J.; Di, W. miRNA-365b promotes hepatocellular carcinoma cell migration and invasion by downregulating SGTB. *Future Oncol.* **2019**, *15*, 2019–2028. [CrossRef]

61. Hildebrandt, M.A.; Gu, J.; Lin, J.; Ye, Y.; Tan, W.; Tamboli, P.; Wood, C.G.; Wu, X. Hsa-miR-9 methylation status is associated with cancer development and metastatic recurrence in patients with clear cell renal cell carcinoma. *Oncogene* **2010**, *29*, 5724–5728. [CrossRef]

62. Jin, R.H.; Yu, D.J.; Zhong, M. MiR-1269a acts as an onco-miRNA in non-small cell lung cancer via down-regulating SOX6. *Eur. Rev. Med. Pharmacol. Sci.* **2018**, *22*, 4888–4897. [CrossRef]

63. Bu, P.; Wang, L.; Chen, K.Y.; Rakhilin, N.; Sun, J.; Closa, A.; Tung, K.L.; King, S.; Kristine Varanko, A.; Xu, Y.; et al. miR-1269 promotes metastasis and forms a positive feedback loop with TGF-beta. *Nat. Commun.* **2015**, *6*, 6879. [CrossRef]

64. Nowak, I.; Boratyn, E.; Durbas, M.; Horwacik, I.; Rokita, H. Exogenous expression of miRNA-3613-3p causes APAF1 downregulation and affects several proteins involved in apoptosis in BE(2)-C human neuroblastoma cells. *Int. J. Oncol.* **2018**, *53*, 1787–1799. [CrossRef]
65. Srinivasan, G.; Williamson, E.A.; Kong, K.; Jaiswal, A.S.; Huang, G.; Kim, H.S.; Scharer, O.; Zhao, W.; Burma, S.; Sung, P.; et al. MiR223-3p promotes synthetic lethality in BRCA1-deficient cancers. *Proc. Natl. Acad. Sci. USA* **2019**, *116*, 17438–17443. [CrossRef]
66. Chai, B.; Guo, Y.; Cui, X.; Liu, J.; Suo, Y.; Dou, Z.; Li, N. MiR-223-3p promotes the proliferation, invasion and migration of colon cancer cells by negative regulating PRDM1. *Am. J. Transl. Res.* **2019**, *11*, 4516–4523.
67. Liu, N.; Liu, X.; Zhou, N.; Wu, Q.; Zhou, L.; Li, Q. Gene expression profiling and bioinformatics analysis of gastric carcinoma. *Exp. Mol. Pathol.* **2014**, *96*, 361–366. [CrossRef]
68. Hu, H.; Du, L.; Nagabayashi, G.; Seeger, R.C.; Gatti, R.A. ATM is down-regulated by N-Myc-regulated microRNA-421. *Proc. Natl. Acad. Sci. USA* **2010**, *107*, 1506–1511. [CrossRef]
69. Agrawal, R.; Garg, A.; Benny Malgulwar, P.; Sharma, V.; Sarkar, C.; Kulshreshtha, R. p53 and miR-210 regulated NeuroD2, a neuronal basic helix-loop-helix transcription factor, is downregulated in glioblastoma patients and functions as a tumor suppressor under hypoxic microenvironment. *Int. J. Cancer* **2018**, *142*, 1817–1828. [CrossRef]
70. Wei, C.; Li, L.; Kim, I.K.; Sun, P.; Gupta, S. NF-kappaB mediated miR-21 regulation in cardiomyocytes apoptosis under oxidative stress. *Free Radic. Res.* **2014**, *48*, 282–291. [CrossRef]
71. Groenendyk, J.; Peng, Z.; Dudek, E.; Fan, X.; Mizianty, M.J.; Dufey, E.; Urra, H.; Sepulveda, D.; Rojas-Rivera, D.; Lim, Y.; et al. Interplay between the oxidoreductase PDIA6 and microRNA-322 controls the response to disrupted endoplasmic reticulum calcium homeostasis. *Sci. Signal.* **2014**, *7*, ra54. [CrossRef] [PubMed]
72. Das, R.; Gregory, P.A.; Fernandes, R.C.; Denis, I.; Wang, Q.; Townley, S.L.; Zhao, S.G.; Hanson, A.R.; Pickering, M.A.; Armstrong, H.K.; et al. MicroRNA-194 Promotes Prostate Cancer Metastasis by Inhibiting SOCS2. *Cancer Res.* **2017**, *77*, 1021–1034. [CrossRef] [PubMed]
73. Yang, W.; Xu, T.; Qiu, P.; Xu, G. Caveolin-1 promotes pituitary adenoma cells migration and invasion by regulating the interaction between EGR1 and KLF5. *Exp. Cell Res.* **2018**, *367*, 7–14. [CrossRef] [PubMed]
74. Kang, J.; Kim, W.; Lee, S.; Kwon, D.; Chun, J.; Son, B.; Kim, E.; Lee, J.M.; Youn, H.; Youn, B. TFAP2C promotes lung tumorigenesis and aggressiveness through miR-183- and miR-33a-mediated cell cycle regulation. *Oncogene* **2017**, *36*, 1585–1596. [CrossRef] [PubMed]
75. Liu, L.; Wang, S.; Chen, R.; Wu, Y.; Zhang, B.; Huang, S.; Zhang, J.; Xiao, F.; Wang, M.; Liang, Y. Myc induced miR-144/451 contributes to the acquired imatinib resistance in chronic myelogenous leukemia cell K562. *Biochem. Biophys. Res. Commun.* **2012**, *425*, 368–373. [CrossRef]
76. Yu, G.; Wang, L.G.; Han, Y.; He, Q.Y. clusterProfiler: An R package for comparing biological themes among gene clusters. *Om. A J. Integr. Biol.* **2012**, *16*, 284–287. [CrossRef]
77. Shannon, P.; Markiel, A.; Ozier, O.; Baliga, N.S.; Wang, J.T.; Ramage, D.; Amin, N.; Schwikowski, B.; Ideker, T. Cytoscape: A software environment for integrated models of biomolecular interaction networks. *Genome Res.* **2003**, *13*, 2498–2504. [CrossRef]

Article

MiR-378a-3p Is Critical for Burkitt Lymphoma Cell Growth

Fubiao Niu [1], Agnieszka Dzikiewicz-Krawczyk [2], Jasper Koerts [1], Debora de Jong [1], Laura Wijenberg [1], Margot Fernandez Hernandez [1], Izabella Slezak-Prochazka [3], Melanie Winkle [1], Wierd Kooistra [1], Tineke van der Sluis [1], Bea Rutgers [1], Miente Martijn Terpstra [4], Klaas Kok [4], Joost Kluiver [1] and Anke van den Berg [1,*]

[1] Departments of Pathology and Medical Biology, University Medical Center Groningen, University of Groningen, 9700 RB Groningen, The Netherlands; f.niu@umcg.nl (F.N.); j.a.koerts@umcg.nl (J.K.); d.de.jong03@umcg.nl (D.d.J.); l.wijenberg@student.rug.nl (L.W.); margot.fernandezhz@udlap.mx (M.F.H.); m.winkle@protonmail.com (M.W.); w.kooistra@umcg.nl (W.K.); t.van.der.sluis@umcg.nl (T.v.d.S.); b.rutgers@umcg.nl (B.R.); j.l.kluiver@umcg.nl (J.K.)

[2] Institute of Human Genetics, Polish Academy of Sciences, 60-479 Poznan, Poland; krawczyk@man.poznan.pl

[3] Biotechnology Centre, Silesian University of Technology, 44-100 Gliwice, Poland; izabella.slezak-prochazka@polsl.pl

[4] Department of Genetics, University Medical Center Groningen, University of Groningen, 9700 RB Groningen, The Netherlands; m.m.terpstra@umcg.nl (M.M.T.); k.kok@umcg.nl (K.K.)

* Correspondence: a.van.den.berg01@umcg.nl; Tel.: +31-050-361-1476

Received: 16 October 2020; Accepted: 25 November 2020; Published: 27 November 2020

Simple Summary: MicroRNAs (miRNAs) are small RNAs that regulate expression of specific target genes. We observed elevated levels of miR-378a-3p in Burkitt lymphoma (BL) and studied its role in the pathogenesis of BL. Inhibition of miR-378a-3p reduced growth of BL cells, confirming its significance in BL. Identification of BL specific target genes of miR-378a-3p revealed four candidates. For two of them, MNT and IRAK4, miR-378a-dependent regulation was confirmed at the protein level. Overexpression of MNT and IRAK4 in BL cell lines resulted in a similar effect as observed upon miR-378a-3p inhibition, suggesting their involvement in the growth regulatory role of miR-378a-3p.

Abstract: MicroRNAs (miRNAs) are small RNA molecules with important gene regulatory roles in normal and pathophysiological cellular processes. Burkitt lymphoma (BL) is an MYC-driven lymphoma of germinal center B (GC-B) cell origin. To gain further knowledge on the role of miRNAs in the pathogenesis of BL, we performed small RNA sequencing in BL cell lines and normal GC-B cells. This revealed 26 miRNAs with significantly different expression levels. For five miRNAs, the differential expression pattern was confirmed in primary BL tissues compared to GC-B cells. MiR-378a-3p was upregulated in BL, and its inhibition reduced the growth of multiple BL cell lines. RNA immunoprecipitation of Argonaute 2 followed by microarray analysis (Ago2-RIP-Chip) upon inhibition and ectopic overexpression of miR-378a-3p revealed 63 and 20 putative miR-378a-3p targets, respectively. Effective targeting by miR-378a-3p was confirmed by luciferase reporter assays for MAX Network Transcriptional Repressor (MNT), Forkhead Box P1 (FOXP1), Interleukin 1 Receptor Associated Kinase 4 (IRAK4), and lncRNA Just Proximal To XIST (JPX), and by Western blot for IRAK4 and MNT. Overexpression of IRAK4 and MNT phenocopied the effect of miR-378a-3p inhibition. In summary, we identified miR-378a-3p as a miRNA with an oncogenic role in BL and identified IRAK4 and MNT as miR-378a-3p target genes that are involved in its growth regulatory role.

Keywords: Burkitt lymphoma; miR-378a-3p; cell growth; microRNA

1. Introduction

Burkitt lymphoma (BL) is one of the fastest growing human tumors with a cell doubling time of about 24 h. BL mainly affects children and young adults but can also occur at a later age [1]. The tumor cells are derived from germinal center B (GC-B) cells and usually carry the hallmark translocation involving *MYC* and the immunoglobulin heavy or light chain loci which results in high expression of MYC [2,3].

MicroRNAs (miRNAs) are a class of short noncoding RNAs of about 22 nucleotides. They modulate gene expression at the post-transcriptional level by translational inhibition or by inducing mRNA degradation [4,5]. MiRNAs regulate a wide range of cellular processes, including cell cycle, proliferation, and apoptosis, and they are important determinants of B-cell development and maturation [6]. A widespread deregulation of miRNA expression has been observed in all B-cell lymphoma subtypes [7].

We and others identified distinct miRNA expression patterns in BL and demonstrated the central role of MYC in regulating miRNA levels [8–12]. Functional studies showed crucial roles for the miR-17~92 cluster, miR-28, miR-150, and miR-155 as either oncogenic or tumor suppressor miRNAs in the pathogenesis of BL [9,13–19]. Nevertheless, the role of most of the deregulated miRNAs in BL remains to be explored.

In this study, we carried out small RNA-sequencing in BL cell lines and normal GC-B cells and subsequently focused on downstream functional experiments for miR-378a-3p. We show for the first time that this miRNA is upregulated in BL and confirmed its regulation by MYC. Further analysis indicated that miR-378a-3p is essential for the growth of BL cells. Using a combination of genome-wide target gene identification, luciferase reporter assays, and Western blot upon modulating miR-378a-3p levels, we confirmed targeting of Interleukin 1 Receptor Associated Kinase 4 (IRAK4) and MAX Network Transcriptional Repressor (MNT) by miR-378a-3p. Overexpression of these two genes phenocopied the effect of miR-378a-3p inhibition on growth of BL cells.

2. Results

2.1. MiRNA Expression Profiling in BL and GC-B Cells

An overview of the total number of reads and percentages of mapped reads per sample is given in Table S1. The top 10 most abundantly expressed miRNAs accounted for 73% of all reads in BL and for 71% in GC-B cells (total GC-B cells sorted on the basis of a CD20$^+$IgD$^-$CD38$^+$ or IgD$^-$CD138$^-$CD3$^-$CD10$^+$ phenotype). Seven of the top 10 most abundantly expressed miRNAs were shared between BL and GC-B cells (Figure 1A). Twenty-six miRNAs were significantly differentially expressed between BL and GC-B cells, including eight miRNAs upregulated in BL and 18 downregulated (Figure 1B). qRT-PCR validation on the same set of samples confirmed differential expression for six out of eight selected miRNAs (Figure 1C and Figure S1). Of the six validated miRNAs, miR-378a-3p levels were increased in BL relative to GC-B cells, while expression levels of miR-28-5p, miR-155-5p, miR-363-3p, miR-221-3p, and miR-222-3p were decreased. Further expression analysis in primary BL tissue samples and GC-B cells confirmed the differential expression for five of the six miRNAs, excluding miR-221-3p (Figure 1D).

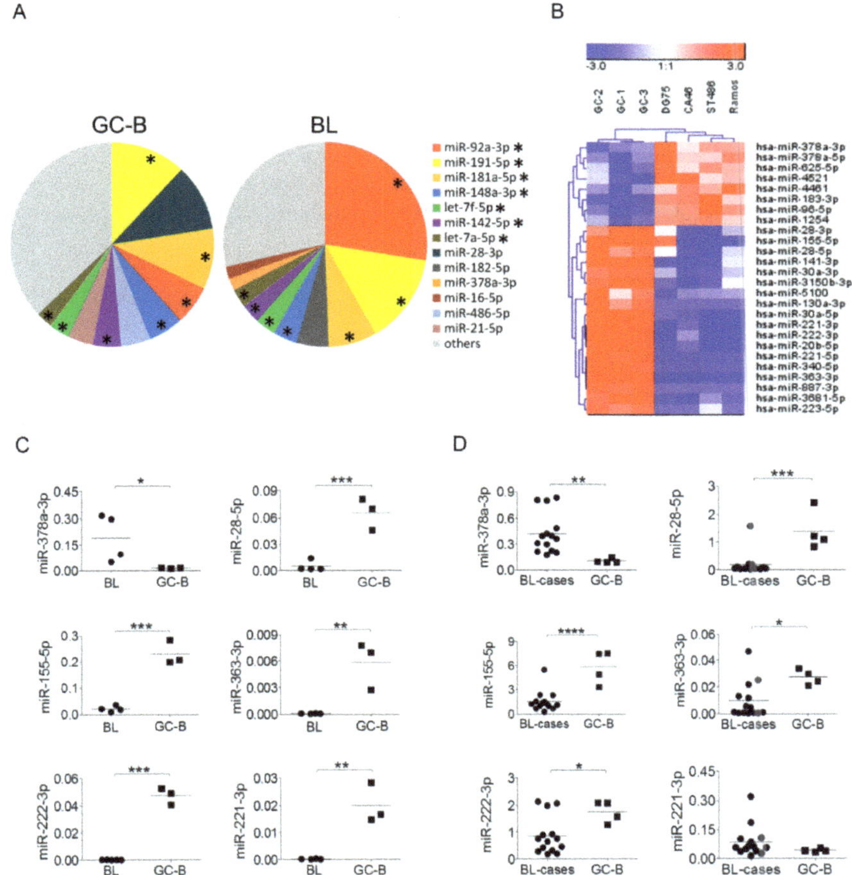

Figure 1. Deregulated expression patterns of microRNAs (miRNAs) in Burkitt lymphoma (BL) compared to germinal center B (GC-B) cells. (**A**) Overview of the top 10 most abundantly expressed miRNAs in Burkitt lymphoma (BL) and normal germinal center B (GC-B) cells as determined by small RNA-sequencing. Asterisks indicate miRNAs present in the top 10 of both BL and GC-B cells. (**B**) Heatmap of miRNAs significantly differentially expressed between BL and GC-B cells. (**C**) qRT-PCR validation results for six of the eight tested miRNAs with significantly differential expression between BL cell lines and GC-B cells. MiRNA expression levels were normalized to Small Nucleolar RNA C/D Box 44 (SNORD44). (**D**) The differential expression pattern was confirmed for five of the six tested miRNAs when BL tissues and GC-B cells were compared. MiRNA expression levels were normalized to Small Nucleolar RNA C/D Box 49 (SNORD49). Significant differences were calculated using an unpaired *t*-test. * $p < 0.05$, ** $p < 0.01$, *** $p < 0.001$, and **** $p < 0.0001$

2.2. MYC-Induced miR-378a-3p Controls BL Cell Growth

We selected miR-378a-3p for further functional analysis, because it was the only significantly upregulated miRNA with a high expression level in BL. Previous studies demonstrated that miR-378a-3p is induced by MYC in human mammary epithelial cells [20]. We assessed the regulatory role of MYC in B cells using the P493-6 B-cell model that has a tetracycline-repressible MYC allele [21]. Our results showed that this miRNA is also induced by MYC in B cells (Figure 2A).

To explore the role of miR-378a-3p in growth of BL cells, we inhibited miR-378a-3p using a lentiviral miRNA inhibition construct (mZip-378a-3p) in four BL cell lines and followed cell growth in a GFP competition assay. Compared to the negative control (mZip-SCR), a significant decrease in the number of GFP-positive cells was observed in three (ST486, CA46, and DG75) out of four BL cell lines (Figure 2B). No relationship was observed between the reduction in the percentage of GFP$^+$ cells and the level of miR-378a-3p expression (Figure S2). Together, our data indicate that inhibition of miR-378a-3p is disadvantageous for BL cells, suggesting miR-378a-3p is indispensable for growth of BL cells.

Overexpression of miR-378a in ST486 using a lentiviral miRNA overexpression construct (pCDH-378a) resulted in a ~47-fold increase in miR-378a-3p level. In a GFP competition assay, miR-378a overexpression had no effects on cell growth, probably due to the already high endogenous levels.

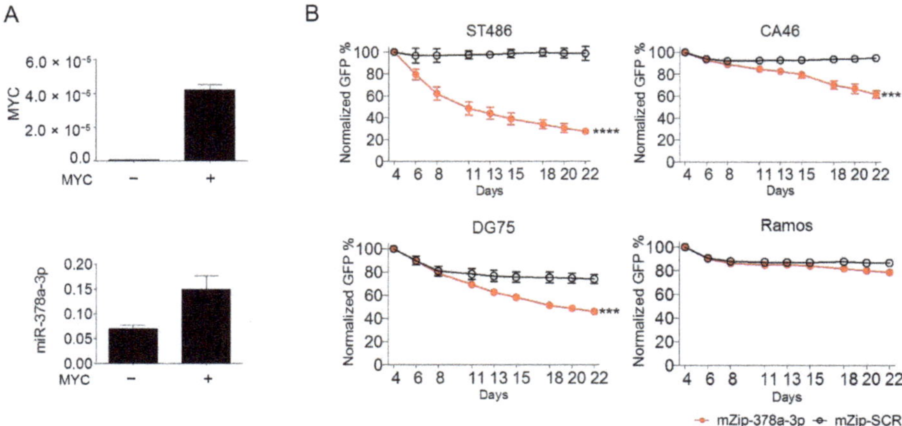

Figure 2. MYC-induced miR-378a-3p is essential for BL cell growth. (**A**) Levels of MYC and miR-378a-3p in tetracycline treated ("−", MYC-off) and non-treated ("+", MYC-on) P493-6 B-cells. MYC levels were normalized to RNA Polymerase II Subunit A (POLR2A). MiRNA levels were normalized to SNORD44. (**B**) Green fluorescent protein (GFP) growth competition assay upon miR-378a-3p inhibition in four Burkitt lymphoma (BL) cell lines. The miR-378a inhibitor (mZip-378a-3p) and the scrambled control (mZip-SCR) were stably transduced in BL cells using a lentiviral vector, which co-expresses GFP. The GFP percentage was measured for 18 days, and the GFP percentage at the first day of measurement (day 4) was set to 100%. All assays were performed in triplicate. Significant differences were calculated using a mixed model analysis. *** $p < 0.001$ and **** $p < 0.0001$.

2.3. Identification of miR-378a-3p Targets

To identify miR-378a-3p target genes, we performed Ago2-RIP-Chip upon miR-378a-3p inhibition and overexpression in ST486 cells (Figure S3A,B). Efficient pulldown of Ago2-containing RISC and miRNAs was confirmed by qRT-PCR for miR-378a-3p and the unrelated highly expressed miR-181a-5p (Figure S3C,D) and by Western blot for Ago2 protein (Figures S3E and S4). The number of Ago2 immunoprecipitation (IP)-enriched probes was similar in all four conditions, ranging between 6.3% and 9.8% of the probes with consistent expression levels (Table 1).

Table 1. Number of genes in the miRNA targetome upon miR-378a-3p overexpression (pCDH) and inhibition (mZip).

IP/T Ratio	pCDH ($n = 9233$)			mZip ($n = 8944$)		
	EV	378a	378a/EV	SCR	mZip-378a-3p	SCR/mZip-378a-3p
≥2	611	586	20	741	878	63
≥4	117	117	2	171	196	4
≥8	25	18	0	53	34	0

EV = pCDH-EV, 378a = pCDH-378a, SCR = mZip-SCR, IP = Ago2 immunoprecipitated fraction, T = total fraction.

A total of 22 probes corresponding to 20 genes showed a ≥2-fold increased IP enrichment upon miR-378a overexpression compared to empty vector control infected cells (Figure 3A and Table 2). Nine of the 20 genes (45%) had at least one putative miR-378a-3p binding site (7mer-A1, 7mer-m8, or/and 8mer). Six of them, i.e., MAX Network Transcriptional Repressor (MNT), Heat Shock Protein Family B (Small) Member 1 (HSPB1), Interleukin 1 Receptor Associated Kinase 4 (IRAK4), Cyclin K (CCNK), Cyclin Dependent Kinase Inhibitor 2A (CDKN2A), and Ring Finger Protein 34 (RNF34), had a Gene Ontology (GO) term related to cell growth, apoptosis, and/or cell cycle. Upon miR-378a-3p inhibition, 74 probes, corresponding to 63 genes, showed a ≥2-fold decreased IP enrichment compared to the negative control infected cells (Figure 3B and Table 3). Nineteen of these 63 genes (30%) contained at least one putative miR-378a-3p binding site, including Cytokine Inducible SH2 Containing Protein (CISH), BCR Activator Of RhoGEF And GTPase (BCR), Tubulin Alpha 1c (TUBA1C), SWI/SNF Related, Matrix Associated, Actin Dependent Regulator Of Chromatin, Subfamily A, Member 4 (SMARCA4), and Forkhead Box P1 (FOXP1) with a GO term related to cell growth, apoptosis, or cell cycle. One of the target genes, i.e., MYC Binding Protein (MYCBP), was identified with both experimental set-ups.

Table 2. Identified targets of miR-378a-3p upon overexpression.

Gene	Transcript ID	IP/T Ratio			miR-378a-3p Binding Site			Growth-Related GO
		EV *	378a	FC	5′UTR	CDS	3′UTR	
IRAK4	ENST00000613694	1.0	4.6	4.6		8m		yes
CDKN2A	ENST00000304494	1.0	4.0	4.0				yes
JPX	ENST00000415215	1.5	5.8	3.9			8m **	
PLGRKT	ENST00000223864	1.0	3.0	3.0	8m			
TMEM245	ENST00000374586	1.0	2.9	2.9			7m8/8m	
TOMM6	ENST00000398884	1.5	4.1	2.7				
CDK1	ENST00000395284	1.0	2.6	2.6				yes
FAM117A	ENST00000240364	1.4	3.6	2.6				
WDR83OS	ENST00000596731	1.3	3.2	2.5	7m8			
CCNK	ENST00000389879	1.0	2.4	2.4		8m		yes
MYCBP	ENST00000397572	2.3	5.4	2.3			7mA1	
RNF34	ENST00000392465	1.3	3.0	2.3				yes
UBC	ENST00000339647	1.0	2.3	2.3				yes
POP4	ENST00000585603	1.0	2.3	2.3				
MNT	ENST00000174618	1.2	2.7	2.3		7m8	7mA1	yes
HSPB1	ENST00000248553	1.8	3.9	2.2	7mA1			yes
INAFM1	ENST00000552360	1.3	2.8	2.2				
PCNA	ENST00000379160	1.0	2.1	2.1				yes
SP100	ENST00000264052	1.2	2.5	2.1				
RPP25L	ENST00000297613	1.2	2.4	2.0				

* IP/T ratios in pCDH-EV (EV) were set to 1.0 in cases where the ratios were <1.0. ** The binding site in the noncoding RNA is listed in the 3′-UTR column. 7mA1 = 7mer-A1, 7m8 = 7mer-m8, 8m = 8mer, FC = fold change, 378a = pCDH-378a, IP = Ago2 immunoprecipitated fraction, T = Total fraction, 5′UTR = 5′ untranslated region, CDS = coding sequence, 3′UTR = 3′ untranslated region, GO = gene ontology.

Table 3. Identified targets of miR-378a-3p upon inhibition.

Gene	Transcript ID	IP/T Ratio			miR-378a-3p Binding Site			Growth-Related GO
		mZip-378a-3p *	SCR	FC	5′UTR	CDS	3′UTR	
DYNLRB1	ENST00000357156	1.0	6.4	6.4				
VPS18	ENST00000220509	8.5	43.4	5.1				
NAPA-AS1	ENST00000594367	1.2	5.2	4.3			7mA1 **	
C11orf95	ENST00000433688	1.0	4.2	4.2				
HOMEZ	ENST00000357460	1.2	4.3	3.5				
TMEM79	ENST00000405535	1.7	5.7	3.4				
FOXP1	ENST00000493089	4.5	14.9	3.3		7m8/8m		yes
PCIF1	ENST00000372409	1.0	3.2	3.2		7m8		
ATP6V0C	ENST00000330398	1.0	3.1	3.1				
CISH	ENST00000348721	2.1	6.6	3.1			8m	yes
lnc-FOXB1-8	lnc-FOXB1-8:1	1.0	3.3	3.1				
MT1B	ENST00000334346	1.1	3.5	3.1				yes

Table 3. *Cont.*

Gene	Transcript ID	IP/T Ratio			miR-378a-3p Binding Site			Growth-Related GO
		mZip-378a-3p *	SCR	FC	5′UTR	CDS	3′UTR	
lnc-EGLN1-1	lnc-EGLN1-1:6	1.0	2.9	2.9				
NELFA	ENST00000382882	1.0	3.0	2.9				
BCR	ENST00000305877	1.0	2.8	2.8		7m8	7m8	yes
MT1L	ENST00000565768	1.2	3.2	2.8				
FAT3	ENST00000409404	1.0	2.7	2.7		8m	7mA1	
TRAF3IP2-AS1	ENST00000525151	1.0	2.7	2.7				
C22orf39	ENST00000611555	1.0	2.6	2.6				
KRTCAP2	ENST00000295682	1.6	4.1	2.6				
LINC01122	ENST00000427421	11.2	29	2.6			7mA1 **	
ACTG1P20	ENSG00000241547	1.3	3.2	2.5				
lnc-KRTAP5-10-1	lnc-KRTAP5-10-1:1	1.5	3.5	2.4			7m8 **	
MT1E	ENST00000306061	1.1	2.8	2.4				yes
NUDT19	ENST00000397061	1.0	2.5	2.4			7mA1	
PRDX4	ENST00000379341	1.0	2.4	2.4				yes
TMEM258	ENST00000537328	1.1	2.6	2.4				
XLOC_l2_005952	TCONS_l2_00011050	6.6	15.6	2.4		7m8		
BTG3	ENST00000629582	1.6	3.7	2.3				yes
EVI5L	ENST00000270530	1.3	3.0	2.3				
LINC01534	ENST00000433232	1.0	2.3	2.3				
lnc-ADA-1	lnc-ADA-1:2	1.0	2.3	2.3				
lnc-ZNF431-4	lnc-ZNF431-4:1	1.1	2.6	2.3				
MT1A	ENST00000290705	2.5	5.6	2.3				yes
CSRP2	ENST00000311083	2.7	5.9	2.2				
FYCO1	ENST00000296137	1.0	2.2	2.2			7m8	
HYAL3	ENST00000336307	2.0	4.3	2.2				
KCNQ1	ENST00000632153	1.0	2.2	2.2				
lnc-RP11-158I9.5.1-2	TCONS_00019776	1.5	3.3	2.2				
PCNX	ENST00000304743	3.8	8.3	2.2		7mA1/8m		
PRSS36	ENST00000268281	1.4	2.8	2.2				
SMARCA4	ENST00000344626	1.0	2.2	2.2		7m8		yes
TNRC6C	ENST00000335749	8.5	18.3	2.2		7m8		
TOLLIP	ENST00000317204	1.3	2.8	2.2			7mA1	
ARF4	ENST00000303436	1.0	2.1	2.1				yes
ATG4D	ENST00000309469	1.4	2.8	2.1				yes
CSE1L	ENST00000262982	1.2	2.5	2.1				yes
lnc-PCF11-1	lnc-PCF11-1:12	1.5	3.2	2.1				
MARS	ENST00000262027	1.7	3.6	2.1				
NANS	ENST00000210444	1.0	2.1	2.1				
ORMDL2	ENST00000243045	1.1	2.3	2.1				
PFKFB2	ENST00000367080	1.4	3.0	2.1		7m8	7mA1	
PGM2L1	ENST00000298198	1.8	3.7	2.1				
PTPN23	ENST00000265562	1.5	3.1	2.1				
TSACC	ENST00000368255	1.5	3.1	2.1				
TUBA1C	ENST00000301072	1.1	2.4	2.1		7mA1		yes
XLOC_l2_013031	TCONS_l2_00024809	6.5	13.7	2.1				
C15orf61	ENST00000342683	1.3	2.6	2.0			8m	
LAT	ENST00000360872	1.0	2.0	2.0				
LOC101929494	N/A	1.6	3.2	2.0				
MYCBP	ENST00000397572	1.5	3.1	2.0			7mA1	
NDRG4	ENST00000394279	1.1	2.2	2.0				yes
TUBE1	ENST00000368662	1.1	2.2	2.0				yes

* IP/T ratios in mZip-378a-3p were set to 1.0 if <1.0. ** Binding sites on noncoding RNAs are listed in 3′-UTR column. 7mA1 = 7mer-A1, 7m8 = 7mer-m8, 8m = 8mer, FC = fold change, SCR = mZip-SCR, N/A = not available, IP = Ago2 immunoprecipitated fraction, T = total fraction, 5′UTR = 5′ untranslated region, CDS = coding sequence, 3′UTR = 3′ untranslated region, GO = gene ontology.

2.4. IRAK4 and MNT Are Involved in the Function of miR-378a-3p

For further analysis, we selected MYCBP that was identified in both approaches and six candidates that had at least one 7mer-A1, 7mer-m8, or an 8mer and a GO term related to cell growth, apoptosis, or/and cell cycle (CISH, BCR, TUBA1C, FOXP1, MNT, and IRAK4). We also included the lncRNA JPX, as it showed a strong enrichment upon miR-378a-3p overexpression and contained an 8-mer seed binding site (Figure 3C).

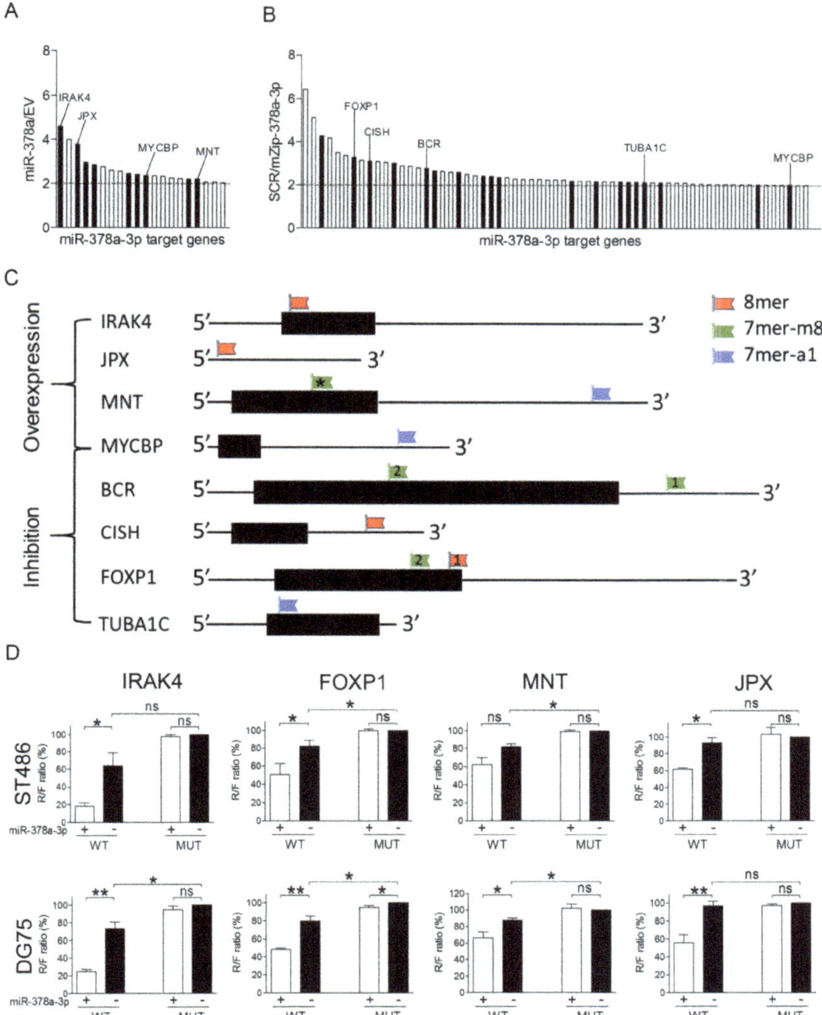

Figure 3. Identification and validation of miR-378a-3p targets. MiR-378a-3p targets identified by Ago2-RIP-Chip upon (**A**) miR-378a overexpression relative to empty vector (EV) and (**B**) scrambled vector relative to miR-378a-3p inhibition (SCR/mZip-378a). The black bars indicate genes with miR-378a-3p seed binding sites. (**C**) Schematic representations of the eight genes selected for luciferase reporter assay validation. Black boxes indicate positions of the open reading frames (ORFs). Positions and types of miR-378a-3p binding sites are indicated relative to the ORF. The binding site in MNT indicated by an asterisk was not tested. (**D**) Luciferase reporter assay results upon co-transfection of ST486 and DG75 cells with the Psi-check-2 construct containing the wildtype (WT) or mutated (MUT) miR-378a-3p binding sites from the selected genes and either an miR-378a-3p mimic or a negative control mimic. Significant differences were calculated using a paired *t*-test. * $p < 0.05$, ** $p < 0.01$, ns = not significant.

To confirm targeting of the eight selected genes by miR-378a-3p, we carried out luciferase reporter assays for 10 putative miR-378a-3p binding sites in ST486 and DG75. This revealed a strong reduction in the *Renilla*/firefly ratio for four of the miR-378a-3p binding sites in four genes (IRAK4, FOXP1 (site

1), MNT, and JPX) (Figure S5). To further confirm specific binding by miR-378a-3p, we generated constructs with mutations in these four miR-378a-3p binding sites. For IRAK4 (trend), FOXP1, and MNT (both significant), wildtype binding sites showed lower *Renilla*/firefly ratios compared to the mutated binding sites (black bars in Figure 3D), indicating binding of endogenous miR-378a-3p to these sequences. Upon miR-378a-3p overexpression a significantly reduced *Renilla*/firefly ratio was observed for the wildtype, but not for the mutated target sites, confirming efficient and specific targeting (Figure 3D).

To validate the regulatory role of miR-378a-3p on endogenous IRAK4, FOXP1, and MNT protein levels, we analyzed protein levels upon modulation of miR-378a-3p levels in ST486 and DG75 cells (Figure 4 and Figure S6). For IRAK4, this revealed a significant decrease upon miR-378a-3p overexpression and a significant increase upon miR-378a-3p inhibition in DG75. In ST486, the same pattern was observed albeit not significant. MNT levels were significantly decreased upon miR-378a-3p overexpression in DG75 and ST486. Increased MNT levels upon miR-378a-3p inhibition were observed only in DG75 cells although the increase was not significant. No significant differences in FOXP1 expression were observed in either cell line.

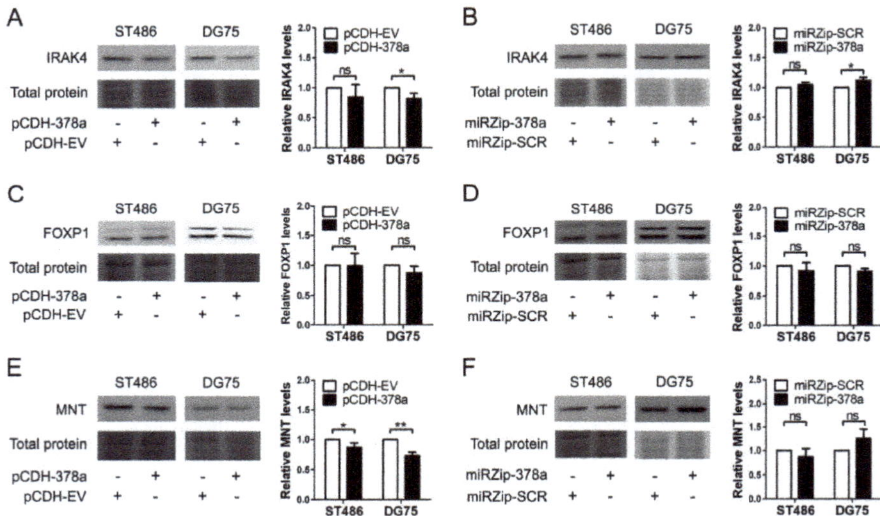

Figure 4. Analysis of the effect of modulating miR-378a-3p levels upon the protein levels of the target genes. Representative examples of the effect of miR-378a overexpression on IRAK4 (**A**), FOXP1 (**C**), and MNT (**E**). Representative examples of the effect of miR-378a-3p inhibition on IRAK4 (**B**), FOXP1 (**D**), and MNT (**F**). cells. Graphs show the quantification of the protein levels of three to four independent infections. For each protein, the part of the total protein lane corresponding to the position of the protein band is shown as an indication for protein loading. Protein levels were quantified relative to the total protein amount as measured in the complete lane. Uncropped blots can be found in Figure S6. Significant differences were calculated using a paired *t*-test. * $p < 0.05$, ** $p < 0.01$, ns = not significant.

Next, we performed gain-of-function experiments to determine whether overexpression of the validated targets IRAK4 and MNT could phenocopy the effect of miR-378a-3p inhibition. Overexpression of IRAK4 in DG75 cells resulted in a 74% decrease in GFP$^+$ cells in 11 days (Figure 5). Overexpression of MNT resulted in a >90% decrease in GFP$^+$ cells in 11 days in DG75. Thus, our results indicate that overexpression of IRAK4 and MNT could phenocopy the effect of miR-378a-3p inhibition on growth of BL cells. Altogether, our results suggest that growth of BL cells depends on high miR-378a-3p levels through regulating IRAK4 and MNT.

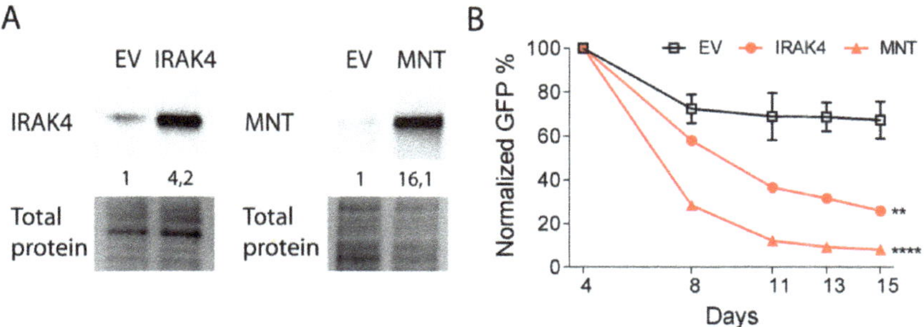

Figure 5. Overexpression of IRAK4 and MNT phenocopies the effect of miR-378a-3p inhibition in DG75 cells. (**A**) Validation of the IRAK4 and MNT overexpression in DG75. An empty vector (EV) was used as a control. For each protein, the part corresponding to the position of the protein band is shown as control of protein loading. Protein levels were quantified relative to the total protein amount in the complete lanes. Uncropped blots can be found in Figure S7. (**B**) GFP competition assays upon overexpression of IRAK4 and MNT resulted in a strong decrease in GFP$^+$ cells over time, while no or only a mild effect was observed for the EV control. The GFP percentages were measured for 11 days, and the GFP percentage as measured on day 4 was set to 100%. Assays were performed in duplicate. Significant differences between IRAK4 and MNT overexpression and EV control were calculated using a mixed model analysis. ** $p < 0.01$ and **** $p < 0.0001$.

3. Discussion

In this study, we identified 26 miRNAs differentially expressed in BL cell lines compared to GC-B cells. For five of the miRNAs, deregulated expression levels were confirmed in both BL cell lines and primary tissues. Among them, miR-378a-3p is MYC-induced, highly abundant (top 10 within BL), and overexpressed in BL compared to GC-B cells. Inhibition of miR-378a-3p showed a negative effect on BL cell growth. In a genome-wide Ago2-RIP-Chip analysis, 20 and 63 genes were identified as the potential targets of miR-378a-3p upon miR-378a-3p overexpression and inhibition, respectively. MNT and IRAK4 were confirmed as novel targets of miR-378a-3p in BL, and their overexpression phenocopied the effect of miR-378a-3p on BL cell growth.

Six of the 26 identified differentially expressed miRNAs were reported to be deregulated in BL by Oduor et al. in a previous study [8]. These six included miR-155-5p, miR-221-3p, miR-222-3p, and miR-28-5p which we validated by qRT-PCR in BL cell lines and tissue samples. Nine additional miRNAs were proven to be differentially expressed in BL compared to other B-cell lymphomas [9–12,20]. Thus, our small RNA sequencing data confirmed some of the previously identified deregulated miRNAs in BL. Moreover, we showed for the first time that miR-378a-3p is an MYC-induced and significantly upregulated miRNA in BL. Since the role of miR-378a-3p in BL was not studied before, we focused on this miRNA for further functional analysis.

Inhibition of miR-378a-3p resulted in a strong reduction of BL cell growth, suggesting a possible oncogenic role of miR-378a-3p in BL. The effect on growth upon miR-378a-3p was most pronounced in ST486. DG75 and CA46 showed intermediate phenotypes while no significant effect was observed in Ramos. There was no obvious relationship between endogenous miR-378a-3p levels and the decrease in percentage of GFP$^+$ cells upon miR-378a-3p inhibition. Given the complex interactions between miRNAs and target genes, i.e., multiple targets per miRNA and multiple miRNAs per target, the differences in the observed phenotypes might be related to endogenous levels of other miRNAs or target genes. Previous studies have shown opposite roles of miR-378a-3p in different cancer types. MiR-378a-3p was reported to inhibit growth or promote apoptosis and, thus, act as a tumor suppressor in colorectal cancer, lung cancer, ovarian cancer, prostate cancer, and rhabdomyosarcoma [22–26]. In contrast, in line with our findings, miR-378a-3p was shown to promote proliferation and reduce apoptosis in gastric cancer, nasopharyngeal carcinoma, colorectal cancer, and acute myeloid leukemia [27–30].

Using an unbiased genome-wide experimental approach, we identified 83 putative miR-378a-3p target genes. Luciferase reporter assays confirmed targeting of four out of eight selected genes, i.e., the protein-coding genes MNT, IRAK4, and FOXP1, and the lncRNA JPX. Targeting of the endogenous *MNT* and *IRAK4* transcripts by miR-378a-3p was confirmed at the protein level, albeit with limited effects. For FOXP1, we could not confirm targeting by miR-378a-3p at the protein level, while we did not follow up potential targeting of endogenous JPX transcripts in BL. Moreover, we showed that overexpression of MNT and IRAK4 strongly inhibited the growth of DG75 cells. Together, these data suggest that the effect of miR-378a-3p might at least in part be dependent on targeting MNT and IRAK4. The apparent discrepancy between the limited effect of miR-378a-3p on the levels of these proteins and the relatively strong phenotype on growth is in line with the general thought that miRNAs most often do not work as on/off switches but rather fine-tune the expression of their targets [5]. A combined moderate effect on MNT, IRAK4, and possible others might explain the strong effect on cell growth. Further experiments in additional cell lines using, e.g., phenotype rescue approaches with more precise control of the overexpression levels are required to confirm the relevance of these targets for the miR-378a-3p-induced phenotype.

Previous studies showed a dual role of MNT in tumorigenesis. On the one hand, MNT was reported as a facilitator of MYC-driven T-cell proliferation and survival [31]. In line with this, a study in Eμ-MYC mice showed that reduced MNT levels reduced tumorigenesis [32], suggesting that MNT is indispensable for MYC-driven oncogenesis. However, in most studies, MNT acted as a tumor suppressor and was a functional antagonist of MYC by repressing its activities related to cell cycle, proliferation, and apoptosis [33,34]. Loss of MNT in mouse embryonic fibroblasts (MEFs) phenocopied the effect of MYC overexpression [35–37]. These findings are in line with our results and suggest that high levels of miR-378a-3p could promote BL tumorigenesis by reducing MNT levels, thereby enabling MYC to execute its oncogenic effects in BL.

IRAK4 plays an essential role in the Toll-like receptor (TLR) pathway [38], which mediates inflammatory signals in B cells and causes activation of nuclear factor kappa-light-chain-enhancer of activated B cells (NF-κB). The TLR pathway is hyperactive in mantle cell lymphoma and diffuse large B-cell lymphoma (DLBCL), and activation of NF-κB promotes B-cell survival and proliferation [39–41]. Depletion of IRAK4 showed a negative effect on NF-κB activity and autocrine IL-6/IL-10 engagement of the Janus Kinase (JAK)-Signal Transducer and Activator of Transcription 3 (STAT3) pathway, reducing survival of DLBCL cells [42,43]. Despite the pro-survival role of NF-κB in DLBCL, activation of NF-κB has been reported to be disadvantageous in MYC positive BL consistent with our data, supporting a role of miR-378a-3p-dependent repression of IRAK4 in limiting the activation of NF-κB [44,45].

Although we could not show that FOXP1 protein levels are affected by miR-378a-3p, we cannot exclude that expression changes in FOXP1 are more subtle and not captured by our experimental set-up. FOXP1 is a member of the forkhead box (Fox) transcription factor family and a regulator of early B-cell development [46]. Interestingly FOXP1 expression is downregulated during the normal GC reaction. In BL, FOXP1 levels are comparable to GC-B cells, but lower than the level in other B-cell lymphomas [47,48]. The potential relevance of maintaining low FOXP1 in BL is further supported by the finding that aberrant expression of FOXP1 cooperates with (constitutive) NF-κB activity [49], which might be disadvantageous for the survival of BL. The relevance of the long noncoding RNA JPX in the phenotype induced by miR-378a-3p inhibition remains to be elucidated. JPX is an activator of X Inactive Specific Transcript (XIST) and acts as a molecular switch for X chromosome inactivation [50]. JPX was reported to act as an oncogene in ovarian cancer and non-small-cell lung cancer by promoting cell proliferation, invasion, and migration [51,52]. In hepatocellular carcinoma, JPX-dependent induction of XIST suppresses hepatocellular carcinoma progression by binding to the miR-155-5p oncomiR [53]. The potential role of miR-378a-3p in regulating FOXP1 and JPX levels need to be further established.

4. Materials and Methods

4.1. BL Cell Lines, Germinal Center (GC) B Cells, and BL Patient Material

BL cell lines were purchased from American Type Culture Collection (ATCC/LGC Standards, Molsheim Cedex, France) (ST486 and Ramos) and German Collection of Microorganisms and Cell Cultures (DSMZ, Braunschweig, Germany) (CA46 and DG75). BL cells were cultured at 37 °C under an atmosphere containing 5% CO_2 in Roswell Park Memorial Institute (RPMI)-1640 medium (Cambrex Biosciences, Walkersville, MD, USA) supplemented with 2 nM ultra-glutamine, 100 U/mL penicillin, 0.1 mg/mL streptomycin, and 10% (CA46, DG75, and Ramos) or 20% (ST486) fetal bovine serum (Sigma-Aldrich, Zwijndrecht, The Netherlands). P493-6 B-cells were cultured as described previously [54]. We routinely confirmed cell line identity using the PowerPlex® 16HS System (Promega, Leiden, The Netherlands) and absence of mycoplasma contamination.

GC-B cells and frozen BL tissue sections were obtained previously as described in [9,54,55]. GC-B cells (defined as $CD20^+IgD^-CD38^+$, $n = 6$ or $IgD^-CD138^-CD3^-CD10^+$, $n = 1$) were sorted from routinely removed tonsil specimens of children. Specifically, for small RNA seq experiments, fluorescence-activated cell sorter (FACS)-sorted ($CD20^+IgD^-CD38^+$, $n = 2$) and magnetic-activated cell sorter MACS-sorted ($IgD^-CD138^-CD3^-CD10^+$, $n = 1$) GC-B cells were used as controls, while FACS-sorted ($CD20^+IgD^-CD38^+$, $n = 4$) GC-B cells were used for qRT-PCR validation experiments. Written permission for the use of the tonsil tissues to isolate GC-B cells was obtained from the parents of the children. The study protocol was consistent with international ethical guidelines (the Declaration of Helsinki and the International Conference on Harmonization Guidelines for Good Clinical Practice). According to the Medical ethics review board of the University Medical Center Groningen our studies fulfilled requirements for patient anonymity and were in accordance with their regulations. The Medical ethics review board waives the need for approval if rest material is used under law in the Netherlands, and waives the need for informed consent when patient anonymity is assured (BL tissue samples).

4.2. RNA Isolation

RNA was isolated using miRNeasy Mini or Micro kit (Qiagen, Hiden, Germany) according to the manufacturer's instructions. RNA concentration was measured by a NanoDropTM 1000 Spectrophotometer (Thermo Fisher Scientific Inc., Waltham, MA, USA) and integrity was evaluated on a 1% agarose gel.

4.3. Small RNA Library Preparation and Sequencing

Firstly, 1–2 µg of total RNA from four BL cell lines and three samples of GC-B cells was used to generate small RNA libraries using Truseq Small RNA Sample Preparation Kit and TruSeq small RNA indices (Illumina, San Diego, CA, USA). Sequencing was performed on an Illumina 2000 HiSeq platform. After removal of 3′- and 5′-adaptor sequences from the raw reads using the CLC Genomics Workbench (CLC Bio, Cambridge, MA, USA), sequencing data were analyzed using miRDeep version 2.0 (Max Delbrück Center for Molecular Medicine in the Helmholtz Association, https://www.mdc-berlin.de/8551903/en) [56] and annotated against miRbase version 21 (http://www.mirbase.org) [57] allowing one nucleotide mismatch. Read counts of miRNAs with the same mature miRNA sequence were merged. Total read counts per sample were normalized to 1,000,000. For statistical analysis, we included all unique miRNAs with at least 50 read counts in all seven samples. Genesis software v1.7.6 (Institute for Genomics and Bioinformatic Graz, Graz, Austria) was used to generate the heat map. The small RNA sequencing data were deposited in the Gene Expression Omnibus database (http://www.ncbi.nlm.nih.gov/geo; accession number GSE92616).

4.4. qRT-PCR

For validation of small RNA sequencing results, we selected differentially expressed miRNAs with expression $\log_2$ reads per million (RPM) >8 in BL or GC-B cells. We selected the most optimal

Taqman assay on the basis of isoform abundance as observed in the small RNA sequencing data (Table S2). The miRNA expression levels were analyzed using Taqman miRNA quantitative PCR assays (Thermo Fisher Scientific Inc.) in a multiplexed fashion as described previously [58]. Cycle crossing point (Cp) values were determined with Light Cycler 480 software version 1.5.0 (Roche, Basel, Switzerland). Relative expression levels of miRNAs to the housekeeping gene (SNORD44 or SNORD49) were determined by calculating $2^{-\Delta Cp}$ ($\Delta Cp = Cp_{miRNA} - Cp_{house-keeping\ gene}$). MYC transcript levels were analyzed as indicated previously [54].

4.5. Lentiviral Constructs, Transduction, and GFP Competition Assay

Lentiviral constructs to inhibit (mZip-378a-3p) or overexpress (pCDH-miR-378a) miR-378a-3p were purchased from System Biosciences (Palo Alto, CA, USA). A nontargeting mZip-scrambled (SCR) and an empty vector pCDH (EV) construct were used as negative controls. Lentiviral pLV-EGFP:T2A:Puro-EF1A IRAK4, MNT, and empty vector control constructs were purchased from VectorBuilder (Chicago, IL, USA). Lentiviral particles were produced in HEK-293T cells by calcium phosphate precipitation transfection using a third-generation packaging system as described previously [55]. Briefly, HEK-293T cells were seeded in six-well plates and grown until ~80% confluence. A plasmid mix consisting of 15 µL of CaCl$_2$ (2.5 M), 1 µg of pMSCV-VSV-G, 1 µg of pRSV.REV, 1 µg of pMDL-gPRRE, 2 µg of lentiviral vector, and 150 µL of 2× HEPES buffered saline (HBS) was prepared to transfect the HEK-293T cells. The virus was harvested and filtered using a 0.45 µm filter 48 h after transfection. The virus was either used directly or stored at −80 °C.

For GFP competition assays, BL cell lines were infected with the mZip-378a-3p and the negative control mZip-SCR in three biological replicates per construct, aiming at an infection efficiency of 20% to 50% GFP$^+$ cells on day 4. The percentage of GFP$^+$ cells was monitored by flow cytometry (BD Biosciences, San Jose, CA, USA) three times per week for a total period of 18 days. For the IRAK4 and MNT GFP competition assays, two biological replicates per construct were performed, and the percentage of GFP$^+$ cells was monitored for 11 days.

4.6. AGO2-IP Procedure

Immunoprecipitation (IP) of the Ago2-containing RNA-induced silencing complex (RISC) (Ago2-IP) procedure was done as described previously [59]. To identify miR-378a-3p target genes, we applied the Ago2-RIP-Chip on BL cells infected with lentiviral miR-378a-3p inhibition or overexpression constructs. For both constructs, a parallel infection with appropriate control constructs (nontargeting or empty vector) was performed. We aimed at a high infection percentage and harvested the cells at day 5, either directly or after sorting to reach a GFP$^+$ percentage >95% for inhibition and >85% for overexpression. For each AGO2-IP experiment, we started with ~30 million cells. RNA was isolated from the Ago2-IP and total (T) fractions. Efficiency of the Ago2-IP procedure was confirmed by qRT-PCR for miR-378a-3p and miR-181a and by Western blot for the Ago2 protein.

4.7. Western Blotting

Infected cells were harvested and lysed in lysis buffer (#9803, Cell Signaling Technology, Danvers, MA, USA) supplemented with protease inhibitor. After centrifugation at 14,000 rpm for 10 min (4 °C), supernatants were collected, and protein concentrations were measured using the bicinchoninic acid (BCA) Protein Assay Kit (Thermo Fisher Scientific Inc., Waltham, MA, USA) according to the manufacturer's protocol. Then, 20 µg of protein was separated on a polyacrylamide gel and transferred to a nitrocellulose membrane followed by incubation overnight at 4 °C with primary antibodies, anti-Ago2 (2E12-1C9, Abnova, Taipei, Taiwan), anti-MNT (#A303-626A, Bethyl Lab, Montgomery, TX, USA), anti-IRAK4 (ab32511, Cambridge, UK), and anti-FOXP1 (#2005, Cell Signaling Technology, Danvers, MA, USA)). All antibodies were diluted 1000× in 5% milk in Tris-buffered saline with Tween-20 (TBST). After incubation with the secondary and tertiary (for MNT, IRAK4, and FOXP1) antibodies and with the enhanced chemiluminescence (ECL) substrate (Thermo Fisher Scientific

Inc.), chemiluminescence was detected. Ago2 was quantified with Image Lab 4.0.1 software (BioRad, Hercules, CA, USA), and MNT, IRAK4, and FOXP1 were quantified using ImageJ (NIH, Bethesda, MD, USA). MNT, IRAK4, and FOXP1 protein levels were normalized relative to the total protein amount in the complete lane.

4.8. Microarray Analysis

About 50 ng RNA of both the total (T) and the IP fractions were labeled and hybridized on an Agilent gene expression microarray (AMADID no.: 072363, SurePrint G3 Human Gene Exp v3 array kit, Agilent Technologies, Santa Clara, CA, USA). The microarray contained 58,341 probes against coding and noncoding transcripts. The procedure and data analysis were performed as previously described [54]. Briefly, after complementary RNA (cRNA) synthesis and amplification, labeling was done with cyanine 3-CTP (Cy3) or cyanine 5-CTP (Cy5) using the LowInput QuickAmp Labeling kit (catalog no.: 0006322867). Equal amounts of Cy3- or Cy5-labeled cRNA samples were mixed and hybridized on the microarray slide overnight. Raw data were quantile normalized without baseline transformation using GeneSpring GX 12.5 software (Agilent Technologies). Probes were selected for further analysis if they were flagged present in all samples, expressed in the 25th to 100th percentile in at least half of the total (T) fractions, and showed consistent expression in the duplicate measurements (<2-fold change). The average signals of replicates were used to calculate the IP/T ratio, and probes with a ≥ 2-fold enrichment in the IP fraction as compared to total (T) fraction were considered as potential miRNA targets.

For pCDH-378a transduced cells, we next assessed miR-378a-3p targets by identifying probes that were enriched at least ≥2-fold higher in miR-378a-overexpressing cells as compared to empty vector (pCDH-EV). For mZip-378a-3p transduced cells, we assessed miR-378a-3p targets by identifying probes with at least ≥2-fold higher enrichment in mZip-SCR transduced cells as compared to mZip-378a-3p cells. Gene expression microarray data are deposited in the Gene Expression Omnibus database (http://www.ncbi.nlm.nih.gov/geo; accession number: GSE141691).

4.9. Identification of miR-378a-3p Seed Sites and Gene Ontology (GO) Terms

For all Ago2-RIP-Chip identified targets of miR-378a-3p, we used a Perl script to search for 7mer-A1, 7mer-m8, and 8mer [60] miR-378a-3p seed sites in 5′-untranslated region (UTR), coding sequence (CDS), and 3′-UTR). Ensemble transcript isoforms were selected on the basis of a refseq identifier (ID) conversion using Biomart (https://www.ensembl.org/biomart). For lncRNA transcripts without an Ensembl ID, we used the LNCipedia or a XLOC/TCONS (BROAD Institute, Cambridge, MA, USA) transcript ID. The growth-related Gene Ontology (GO) terms (proliferation, cell cycle, and apoptosis) for selected genes were retrieved from the Ensembl database (https://www.ensembl.org/biomart).

4.10. Cloning of miRNA Binding Sites and Luciferase Reporter Assay

We adapted the psi-Check-2 vector (Promega, Madison, WI, USA) to remove the predicted miR-378a-3p target site (7mer-A1) in the open reading frame of the *Renilla* luciferase gene. The binding site was mutated by changing two nucleotides in the seed region without affecting the amino-acid sequence. This was accomplished by substituting the 460 nt long fragment between the EcoRV and XhoI sites of the psi-Check-2 vector (Figure S8; Integrated DNA Technologies, Leuven, Belgium). Effective *Renilla* luciferase production independent of miR-378a-3p levels was confirmed before cloning putative binding sites of target genes.

Ten potential miRNA binding sites of eight miR-378a-3p target genes were ordered as 58-mer oligo duplexes (Integrated DNA technologies) and cloned into the *Xho*I and *Not*I restriction sites of the modified luciferase reporter vector (Table S3). For the binding sites with a positive result in the first luciferase reporter assay, mutant controls were generated by cloning oligo duplexes with mutations in three nucleotides in the seed region. The reporter vectors with miR-378a-3p wildtype or mutated binding sites were co-transfected with 10 μM of either miR-378a pre-miRNA (Cat. NO.:

AM17100, Ambion, Austin, TX, USA) or control oligos (Cat. NO.: AM17111, Ambion) to ST486 and DG75 cells using an Amaxa nucleofector device (program A23) and the Amaxa Cell Line Nucleofector Kit V (Cat NO.: VACA-1003) (Amaxa, Gaithersburg, MD, USA). Cells were harvested 24 h after transfection. *Renilla* and firefly luciferase activity in cell lysates was measured using a Dual-Luciferase Reporter Assay System (Promega). Each experiment was measured in duplicate and results were averaged per experiment. For each construct, the luciferase assay was performed in three independent biological replicates.

4.11. Statistical Analysis

MiRNAs significantly differentially expressed in the small RNA-seq profiling were identified with a moderated t-test and Benjamini–Hochberg correction for multiple testing using the GeneSpring GX software (version 12.5, Agilent Technologies, Santa Clara, CA, USA). For confirmation of differentially expressed miRNAs by qRT-PCR, we used the nonparametric Mann–Whitney U-test (GraphPad Software Inc., San Diego, CA, USA). Statistical analysis of GFP competition assays was performed as described previously [55]. Briefly, the percentage of mZip-378a-3p infected cells at day 4 was set to 100% and compared to percentages in the control over time using a mixed model, with time and the interaction of time and miRNA construct type as a fixed effect and the measurement repeat within miRNA construct type as a random effect in SPSS (22.0.0.0 version, IBM, Armonk, NY, USA). For the luciferase reporter assay, significance was calculated on the basis of the *Renilla*-to-firefly (RL/FL) luciferase ratios between experimental samples and negative controls using a paired t-test (GraphPad Software Inc.). MNT, IRAK4, and FOXP1 protein levels were normalized to total protein loading as visualized with Ponceau S staining. The average change in protein levels of three to four experiments was calculated relative to pCDH-EV and miRZip-SCR controls, which were set to 1. Significance was calculated using a one-tailed paired t-test (GraphPad Software Inc.).

5. Conclusions

In conclusion, we identified 26 miRNAs differentially expressed between BL cells and GC-B cells and confirmed deregulated expression of five out of eight miRNAs in both BL cell lines and tissue samples. For one of the differentially and highly abundant (top 10 in BL) miRNAs, miR-378a-3p, we showed a negative effect on BL cell growth upon inhibition. Overexpression of two experimentally proven miR-378a-3p targets, i.e., IRAK4 and MNT, phenocopied the effect observed upon miR-378a-3p inhibition. Together, our data show a critical role for miR-378a-3p in promoting BL cell growth and suggest that this involves controlling IRAK4 and MNT levels.

Supplementary Materials: The following are available online at http://www.mdpi.com/2072-6694/12/12/3546/s1, Figure S1: Validation of the differential expression pattern observed in the BL cell lines relative to GC-B cells by qRT-PCR, Figure S2: miR-378a-3p levels based on small RNA-seq data derived from individual BL cell lines and germinal center B-cells, Figure S3: Validation of Ago2-IP procedure on ST486 cells upon inhibition and overexpression of miR-378a-3p, Figure S4: Uncropped western blot results for Ago2 protein, Figure S5: Results of the luciferase reporter assays for 10 putative miR-378a-3p binding sites identified in eight Ago-2 IP-enriched genes in (A) ST486 and (B), Figure S6: Uncropped Western blot results for IRAK4, FOXP1, and MNT upon miR-378a-3p inhibition or overexpression, Figure S7: Uncropped Western blot results for IRAK4 and MNT upon lentiviral overexpression, Figure S8: Sequence of the minigene used to generate a luciferase reporter vector without the putative miR-378a-3p binding site in the *Renilla* gene, Table S1: Small RNA sequencing read summary, Table S2: Taqman miRNA assays used for qRT-PCR validation, Table S3: Oligos for cloning miR-378a-3p binding sites from selected genes, wildtype, and with mutations in miR-378a-3p seed. Green letters indicate miR-378a-3p binding sites and red letters indicate the mismatches.

Author Contributions: Conceptualization, A.D.-K., J.K. (Joost Kluiver), K.K., and A.v.d.B.; methodology, J.K. (Jasper Koerts), D.d.J., L.W., M.F.H., W.K., T.v.d.S., I.S.-P., M.W., M.M.T., and B.R.; software, F.N., A.D.-K., M.M.T., and J.K. (Joost Kluiver); validation, F.N., J.K. (Jasper Koerts), and D.d.J.; formal analysis, F.N., A.D.-K., and J.K. (Joost Kluiver); investigation, F.N., I.S.-P., A.D.-K., J.K. (Joost Kluiver), and A.v.d.B.; resources, J.K. (Joost Kluiver) and A.v.d.B.; data curation, A.D.-K. and J.K. (Joost Kluiver); writing—original draft preparation, F.N., A.D.-K., J.K. (Joost Kluiver), I.S.-P., K.K., and A.v.d.B.; writing—review and editing, F.N., A.D.-K., J.K. (Joost Kluiver), and A.v.d.B.; visualization, F.N.; supervision, A.D.-K., J.K. (Joost Kluiver), and A.v.d.B.; project administration, F.N.; funding acquisition, F.N. All authors have read and agreed to the published version of the manuscript.

Funding: This research was funded by the China Scholarship Council (CSC), grant number 601317 (awarded to F.N.) and by a grant from the National Science Center, Poland (2015/19/D/NZ1/03443 to I.S.P.).

Acknowledgments: We thank the UMCG Genomics Coordination center, the UG Center for Information Technology, and their sponsors BBMRI-NL & TarGet for storage and computing infrastructure.

Conflicts of Interest: The authors declare no conflict of interest.

References

1. Burkitt, D.P. Etiology of Burkitt's lymphoma—An alternative hypothesis to a vectored virus. *J. Natl. Cancer Inst.* **1969**, *42*, 19–28. [PubMed]
2. Zech, L.; Haglund, U.; Nilsson, K.; Klein, G. Characteristic chromosomal abnormalities in biopsies and lymphoid-cell lines from patients with Burkitt and non-Burkitt lymphomas. *Int. J. Cancer J. Int. Cancer* **1976**, *17*, 47–56. [CrossRef] [PubMed]
3. Victora, G.D.; Dominguez-Sola, D.; Holmes, A.B.; Deroubaix, S.; Dalla-Favera, R.; Nussenzweig, M.C. Identification of human germinal center light and dark zone cells and their relationship to human B-cell lymphomas. *Blood* **2012**, *120*, 2240–2248. [CrossRef] [PubMed]
4. Selbach, M.; Schwanhausser, B.; Thierfelder, N.; Fang, Z.; Khanin, R.; Rajewsky, N. Widespread changes in protein synthesis induced by microRNAs. *Nature* **2008**, *455*, 58–63. [CrossRef]
5. Baek, D.; Villen, J.; Shin, C.; Camargo, F.D.; Gygi, S.P.; Bartel, D.P. The impact of microRNAs on protein output. *Nature* **2008**, *455*, 64–71. [CrossRef]
6. Ameres, S.L.; Zamore, P.D. Diversifying microRNA sequence and function. *Nat. Rev. Mol. Cell Biol.* **2013**, *14*, 475–488. [CrossRef]
7. Lawrie, C.H. MicroRNAs in hematological malignancies. *Blood Rev.* **2013**, *27*, 143–154. [CrossRef]
8. Oduor, C.I.; Kaymaz, Y.; Chelimo, K.; Otieno, J.A.; Ong'echa, J.M.; Moormann, A.M.; Bailey, J.A. Integrative microRNA and mRNA deep-sequencing expression profiling in endemic Burkitt lymphoma. *BMC Cancer* **2017**, *17*, 761. [CrossRef]
9. Robertus, J.L.; Kluiver, J.; Weggemans, C.; Harms, G.; Reijmers, R.M.; Swart, Y.; Kok, K.; Rosati, S.; Schuuring, E.; van Imhoff, G.; et al. MiRNA profiling in B non-Hodgkin lymphoma: A MYC-related miRNA profile characterizes Burkitt lymphoma. *Br. J. Haematol.* **2010**, *149*, 896–899. [CrossRef]
10. Lenze, D.; Leoncini, L.; Hummel, M.; Volinia, S.; Liu, C.G.; Amato, T.; De Falco, G.; Githanga, J.; Horn, H.; Nyagol, J.; et al. The different epidemiologic subtypes of Burkitt lymphoma share a homogenous micro RNA profile distinct from diffuse large B-cell lymphoma. *Leukemia* **2011**, *25*, 1869–1876. [CrossRef]
11. Hezaveh, K.; Kloetgen, A.; Bernhart, S.H.; Mahapatra, K.D.; Lenze, D.; Richter, J.; Haake, A.; Bergmann, A.K.; Brors, B.; Burkhardt, B.; et al. Alterations of microRNA and microRNA-regulated messenger RNA expression in germinal center B-cell lymphomas determined by integrative sequencing analysis. *Haematologica* **2016**, *101*, 1380–1389. [CrossRef] [PubMed]
12. Di Lisio, L.; Sanchez-Beato, M.; Gomez-Lopez, G.; Rodriguez, M.E.; Montes-Moreno, S.; Mollejo, M.; Menarguez, J.; Martinez, M.A.; Alves, F.J.; Pisano, D.G.; et al. MicroRNA signatures in B-cell lymphomas. *Blood Cancer J.* **2012**, *2*, e57. [CrossRef] [PubMed]
13. Bartolome-Izquierdo, N.; de Yebenes, V.G.; Alvarez-Prado, A.F.; Mur, S.M.; Lopez Del Olmo, J.A.; Roa, S.; Vazquez, J.; Ramiro, A.R. miR-28 regulates the germinal center reaction and blocks tumor growth in preclinical models of non-Hodgkin lymphoma. *Blood* **2017**, *129*, 2408–2419. [CrossRef] [PubMed]
14. Chen, S.; Wang, Z.; Dai, X.; Pan, J.; Ge, J.; Han, X.; Wu, Z.; Zhou, X.; Zhao, T. Re-expression of microRNA-150 induces EBV-positive Burkitt lymphoma differentiation by modulating c-Myb in vitro. *Cancer Sci.* **2013**, *104*, 826–834. [CrossRef] [PubMed]
15. Dzikiewicz-Krawczyk, A.; Kok, K.; Slezak-Prochazka, I.; Robertus, J.L.; Bruining, J.; Tayari, M.M.; Rutgers, B.; de Jong, D.; Koerts, J.; Seitz, A.; et al. ZDHHC11 and ZDHHC11B are critical novel components of the oncogenic MYC-miR-150-MYB network in Burkitt lymphoma. *Leukemia* **2017**, *31*, 1470–1473. [CrossRef]
16. Kluiver, J.; Haralambieva, E.; de Jong, D.; Blokzijl, T.; Jacobs, S.; Kroesen, B.J.; Poppema, S.; van den Berg, A. Lack of BIC and microRNA miR-155 expression in primary cases of Burkitt lymphoma. *Genes Chromosom. Cancer* **2006**, *45*, 147–153. [CrossRef]
17. Sandhu, S.K.; Fassan, M.; Volinia, S.; Lovat, F.; Balatti, V.; Pekarsky, Y.; Croce, C.M. B-cell malignancies in microRNA Emu-miR-17~92 transgenic mice. *Proc. Natl. Acad. Sci. USA* **2013**, *110*, 18208–18213. [CrossRef]

18. Teng, G.; Hakimpour, P.; Landgraf, P.; Rice, A.; Tuschl, T.; Casellas, R.; Papavasiliou, F.N. MicroRNA-155 is a negative regulator of activation-induced cytidine deaminase. *Immunity* **2008**, *28*, 621–629. [CrossRef]

19. Dorsett, Y.; McBride, K.M.; Jankovic, M.; Gazumyan, A.; Thai, T.H.; Robbiani, D.F.; Di Virgilio, M.; San-Martin, B.R.; Heidkamp, G.; Schwickert, T.A.; et al. MicroRNA-155 suppresses activation-induced cytidine deaminase-mediated Myc-Igh translocation. *Immunity* **2008**, *28*, 630–638. [CrossRef]

20. Feng, M.; Li, Z.; Aau, M.; Wong, C.H.; Yang, X.; Yu, Q. Myc/miR-378/TOB2/cyclin D1 functional module regulates oncogenic transformation. *Oncogene* **2011**, *30*, 2242–2251. [CrossRef]

21. Pajic, A.; Spitkovsky, D.; Christoph, B.; Kempkes, B.; Schuhmacher, M.; Staege, M.S.; Brielmeier, M.; Ellwart, J.; Kohlhuber, F.; Bornkamm, G.W.; et al. Cell cycle activation by c-myc in a burkitt lymphoma model cell line. *Int. J. Cancer* **2000**, *87*, 787–793. [CrossRef]

22. Li, H.; Dai, S.; Zhen, T.; Shi, H.; Zhang, F.; Yang, Y.; Kang, L.; Liang, Y.; Han, A. Clinical and biological significance of miR-378a-3p and miR-378a-5p in colorectal cancer. *Eur. J. Cancer* **2014**, *50*, 1207–1221. [CrossRef] [PubMed]

23. Megiorni, F.; Cialfi, S.; McDowell, H.P.; Felsani, A.; Camero, S.; Guffanti, A.; Pizer, B.; Clerico, A.; De Grazia, A.; Pizzuti, A.; et al. Deep Sequencing the microRNA profile in rhabdomyosarcoma reveals down-regulation of miR-378 family members. *BMC Cancer* **2014**, *14*, 880. [CrossRef] [PubMed]

24. Wang, M.; Sun, X.; Yang, Y.; Jiao, W. Long non-coding RNA OIP5-AS1 promotes proliferation of lung cancer cells and leads to poor prognosis by targeting miR-378a-3p. *Thorac. Cancer* **2018**, *9*, 939–949. [CrossRef] [PubMed]

25. Chen, Q.G.; Zhou, W.; Han, T.; Du, S.Q.; Li, Z.H.; Zhang, Z.; Shan, G.Y.; Kong, C.Z. MiR-378 suppresses prostate cancer cell growth through downregulation of MAPK1 in vitro and in vivo. *Tumor Biol.* **2016**, *37*, 2095–2103. [CrossRef] [PubMed]

26. Xu, Z.H.; Yao, T.Z.; Liu, W. miR-378a-3p sensitizes ovarian cancer cells to cisplatin through targeting MAPK1/GRB2. *Biomed. Pharmacother.* **2018**, *107*, 1410–1417. [CrossRef]

27. Liu, H.; Zhu, L.; Liu, B.; Yang, L.; Meng, X.; Zhang, W.; Ma, Y.; Xiao, H. Genome-wide microRNA profiles identify miR-378 as a serum biomarker for early detection of gastric cancer. *Cancer Lett.* **2012**, *316*, 196–203. [CrossRef]

28. Yu, B.L.; Peng, X.H.; Zhao, F.P.; Liu, X.; Lu, J.; Wang, L.; Li, G.; Chen, H.H.; Li, X.P. MicroRNA-378 functions as an onco- miR in nasopharyngeal carcinoma by repressing TOB2 expression. *Int. J. Oncol.* **2014**, *44*, 1215–1222. [CrossRef]

29. Qian, J.; Lin, J.; Qian, W.; Ma, J.C.; Qian, S.X.; Li, Y.; Yang, J.; Li, J.Y.; Wang, C.Z.; Chai, H.Y.; et al. Overexpression of miR-378 is frequent and may affect treatment outcomes in patients with acute myeloid leukemia. *Leuk. Res* **2013**, *37*, 765–768. [CrossRef]

30. Tanaka, H.; Hazama, S.; Iida, M.; Tsunedomi, R.; Takenouchi, H.; Nakajima, M.; Tokumitsu, Y.; Kanekiyo, S.; Shindo, Y.; Tomochika, S.; et al. miR-125b-1 and miR-378a are predictive biomarkers for the efficacy of vaccine treatment against colorectal cancer. *Cancer Sci.* **2017**, *108*, 2229–2238. [CrossRef]

31. Vasilevsky, N.A.; Ruby, C.E.; Hurlin, P.J.; Weinberg, A.D. OX40 engagement stabilizes Mxd4 and Mnt protein levels in antigen-stimulated T cells leading to an increase in cell survival. *Eur. J. Immunol.* **2011**, *41*, 1024–1034. [CrossRef] [PubMed]

32. Campbell, K.J.; Vandenberg, C.J.; Anstee, N.S.; Hurlin, P.J.; Cory, S. Mnt modulates Myc-driven lymphomagenesis. *Cell Death Differ.* **2017**, *24*, 2117–2126. [CrossRef] [PubMed]

33. Yang, G.; Hurlin, P.J. MNT and Emerging Concepts of MNT-MYC Antagonism. *Genes* **2017**, *8*, 83. [CrossRef]

34. Hooker, C.W.; Hurlin, P.J. Of Myc and Mnt. *J. Cell Sci.* **2006**, *119*, 208–216. [CrossRef] [PubMed]

35. Hurlin, P.J.; Zhou, Z.Q.; Toyo-oka, K.; Ota, S.; Walker, W.L.; Hirotsune, S.; Wynshaw-Boris, A. Deletion of Mnt leads to disrupted cell cycle control and tumorigenesis. *EMBO J.* **2003**, *22*, 4584–4596. [CrossRef] [PubMed]

36. Nilsson, J.A.; Maclean, K.H.; Keller, U.B.; Pendeville, H.; Baudino, T.A.; Cleveland, J.L. Mnt loss triggers Myc transcription targets, proliferation, apoptosis, and transformation. *Mol. Cell Biol.* **2004**, *24*, 1560–1569. [CrossRef]

37. Walker, W.; Zhou, Z.Q.; Ota, S.; Wynshaw-Boris, A.; Hurlin, P.J. Mnt-Max to Myc-Max complex switching regulates cell cycle entry. *J. Cell Biol.* **2005**, *169*, 405–413. [CrossRef]

38. Kim, T.W.; Staschke, K.; Bulek, K.; Yao, J.; Peters, K.; Oh, K.H.; Vandenburg, Y.; Xiao, H.; Qian, W.; Hamilton, T.; et al. A critical role for IRAK4 kinase activity in Toll-like receptor-mediated innate immunity. *J. Exp. Med.* **2007**, *204*, 1025–1036. [CrossRef]

39. Kuppers, R. IRAK4 inhibition to shut down TLR signaling in autoimmunity and MyD88-dependent lymphomas. *J. Exp. Med.* **2015**, *212*, 2184. [CrossRef]

40. Akhter, A.; Street, L.; Ghosh, S.; Burns, B.F.; Elyamany, G.; Shabani-Rad, M.T.; Stewart, D.A.; Mansoor, A. Concomitant high expression of Toll-like receptor (TLR) and B-cell receptor (BCR) signalling molecules has clinical implications in mantle cell lymphoma. *Hematol. Oncol.* **2017**, *35*, 79–86. [CrossRef]

41. Akhter, A.; Masir, N.; Elyamany, G.; Phang, K.C.; Mahe, E.; Al-Zahrani, A.M.; Shabani-Rad, M.T.; Stewart, D.A.; Mansoor, A. Differential expression of Toll-like receptor (TLR) and B cell receptor (BCR) signaling molecules in primary diffuse large B-cell lymphoma of the central nervous system. *J. Neuro-Oncol.* **2015**, *121*, 289–296. [CrossRef] [PubMed]

42. Ngo, V.N.; Young, R.M.; Schmitz, R.; Jhavar, S.; Xiao, W.; Lim, K.H.; Kohlhammer, H.; Xu, W.; Yang, Y.; Zhao, H.; et al. Oncogenically active MYD88 mutations in human lymphoma. *Nature* **2011**, *470*, 115–119. [CrossRef] [PubMed]

43. Kelly, P.N.; Romero, D.L.; Yang, Y.B.; Shaffer, A.L.; Chaudhary, D.; Robinson, S.; Miao, W.Y.; Rui, L.X.; Westlin, W.F.; Kapeller, R.; et al. Selective interleukin-1 receptor-associated kinase 4 inhibitors for the treatment of autoimmune disorders and lymphoid malignancy. *J. Exp. Med.* **2015**, *212*, 2189–2201. [CrossRef] [PubMed]

44. Klapproth, K.; Sander, S.; Marinkovic, D.; Baumann, B.; Wirth, T. The IKK2/NF-{kappa}B pathway suppresses MYC-induced lymphomagenesis. *Blood* **2009**, *114*, 2448–2458. [CrossRef] [PubMed]

45. Gehringer, F.; Weissinger, S.E.; Moller, P.; Wirth, T.; Ushmorov, A. Physiological levels of the PTEN-PI3K-AKT axis activity are required for maintenance of Burkitt lymphoma. *Leukemia* **2019**. [CrossRef]

46. Hu, H.; Wang, B.; Borde, M.; Nardone, J.; Maika, S.; Allred, L.; Tucker, P.W.; Rao, A. Foxp1 is an essential transcriptional regulator of B cell development. *Nat. Immunol.* **2006**, *7*, 819–826. [CrossRef]

47. Koon, H.B.; Ippolito, G.C.; Banham, A.H.; Tucker, P.W. FOXP1: A potential therapeutic target in cancer. *Expert Opin. Ther. Targets* **2007**, *11*, 955–965. [CrossRef]

48. Sagardoy, A.; Martinez-Ferrandis, J.I.; Roa, S.; Bunting, K.L.; Aznar, M.A.; Elemento, O.; Shaknovich, R.; Fontan, L.; Fresquet, V.; Perez-Roger, I.; et al. Downregulation of FOXP1 is required during germinal center B-cell function. *Blood* **2013**, *121*, 4311–4320. [CrossRef]

49. van Keimpema, M.; Gruneberg, L.J.; Mokry, M.; van Boxtel, R.; Koster, J.; Coffer, P.J.; Pals, S.T.; Spaargaren, M. FOXP1 directly represses transcription of proapoptotic genes and cooperates with NF-kappaB to promote survival of human B cells. *Blood* **2014**, *124*, 3431–3440. [CrossRef]

50. Tian, D.; Sun, S.; Lee, J.T. The Long Noncoding RNA, Jpx, Is a Molecular Switch for X Chromosome Inactivation. *Cell* **2010**, *143*, 390–403. [CrossRef]

51. Li, J.; Feng, L.; Tian, C.; Tang, Y.L.; Tang, Y.; Hu, F.Q. Long noncoding RNA-JPX predicts the poor prognosis of ovarian cancer patients and promotes tumor cell proliferation, invasion and migration by the PI3K/Akt/mTOR signaling pathway. *Eur. Rev. Med. Pharmacol.* **2018**, *22*, 8135–8144.

52. Jin, M.; Ren, J.; Luo, M.; You, Z.; Fang, Y.; Han, Y.; Li, G.; Liu, H. Long noncoding RNA JPX correlates with poor prognosis and tumor progression in non-small cell lung cancer by interacting with miR-145-5p and CCND2. *Carcinogenesis* **2019**. [CrossRef]

53. Lin, X.Q.; Huang, Z.M.; Chen, X.; Wu, F.; Wu, W. XIST Induced by JPX Suppresses Hepatocellular Carcinoma by Sponging miR-155-5p. *Yonsei Med. J.* **2018**, *59*, 816–826. [CrossRef] [PubMed]

54. Winkle, M.; van den Berg, A.; Tayari, M.; Sietzema, J.; Terpstra, M.; Kortman, G.; de Jong, D.; Visser, L.; Diepstra, A.; Kok, K.; et al. Long noncoding RNAs as a novel component of the Myc transcriptional network. *FASEB J.* **2015**, *29*, 2338–2346. [CrossRef] [PubMed]

55. Yuan, Y.; Kluiver, J.; Koerts, J.; de Jong, D.; Rutgers, B.; Razak, F.R.A.; Terpstra, M.; Plaat, B.E.; Nolte, I.M.; Diepstra, A.; et al. miR-24-3p Is Overexpressed in Hodgkin Lymphoma and Protects Hodgkin and Reed-Sternberg Cells from Apoptosis. *Am. J. Pathol.* **2017**, *187*, 1343–1355. [CrossRef]

56. Friedlander, M.R.; Chen, W.; Adamidi, C.; Maaskola, J.; Einspanier, R.; Knespel, S.; Rajewsky, N. Discovering microRNAs from deep sequencing data using miRDeep. *Nat. Biotechnol.* **2008**, *26*, 407–415. [CrossRef]

57. Griffiths-Jones, S.; Grocock, R.J.; van Dongen, S.; Bateman, A.; Enright, A.J. miRBase: microRNA sequences, targets and gene nomenclature. *Nucleic Acids Res.* **2006**, *34*, D140–D144. [CrossRef]

58. Kluiver, J.; Gibcus, J.H.; Hettinga, C.; Adema, A.; Richter, M.K.; Halsema, N.; Slezak-Prochazka, I.; Ding, Y.; Kroesen, B.J.; van den Berg, A. Rapid generation of microRNA sponges for microRNA inhibition. *PLoS ONE* **2012**, *7*, e29275. [CrossRef]

59. Tan, L.P.; Seinen, E.; Duns, G.; de Jong, D.; Sibon, O.C.M.; Poppema, S.; Kroesen, B.J.; Kok, K.; van den Berg, A. A high throughput experimental approach to identify miRNA targets in human cells. *Nucleic Acids Res.* **2009**, *37*, e137. [CrossRef]
60. Agarwal, V.; Bell, G.W.; Nam, J.W.; Bartel, D.P. Predicting effective microRNA target sites in mammalian mRNAs. *eLife* **2015**, *4*. [CrossRef]

Publisher's Note: MDPI stays neutral with regard to jurisdictional claims in published maps and institutional affiliations.

Communication

Gathering Novel Circulating Exosomal microRNA in Osteosarcoma Cell Lines and Possible Implications for the Disease

Nicola Cuscino [1,†], Lavinia Raimondi [2,†], Angela De Luca [2], Claudia Carcione [3], Giovanna Russelli [1], Laura Conti [4], Jacopo Baldi [5], Pier Giulio Conaldi [1], Gianluca Giavaresi [2] and Alessia Gallo [1,*]

1   IRCCS ISMETT (Istituto Mediterraneo per i Trapianti e Terapie ad alta specializzazione), Research Department, 90127 Palermo, Italy; ncuscino@ismett.edu (N.C.); grusselli@ismett.edu (G.R.); pgconaldi@ismett.edu (P.G.C.)
2   IRCCS ISTITUTO ORTOPEDICO RIZZOLI, 40136 Bologna, Italy; lavinia.raimondi@ior.it (L.R.); angela.deluca@ior.it (A.D.L.); gianluca.giavaresi@ior.it (G.G.)
3   Fondazione Ri.MED, 90133 Palermo, Italy; ccarcione@fondazionerimed.com
4   Body Fluids Biobank, Clinical Pathology, IRCCS Regina Elena Cancer National Institute, 00100 Rome, Italy; laura.conti@ifo.gov.it
5   Dept. of Orthopaedic Oncology, IRCCS Regina Elena National Cancer Institute, 00100 Rome, Italy; jacopo.baldi@ifo.gov.it
*   Correspondence: agallo@ismett.edu; Tel.: +39-9121-92649
†   The authors contributed equally to the study.

Received: 16 October 2019; Accepted: 28 November 2019; Published: 3 December 2019

Abstract: One of the goals of personalized medicine is to understand and treat diseases with greater precision through the molecular profile of the patient. This profiling is becoming a powerful tool for the discovery of novel biomarkers that can guide physicians in assessing, in advance, the disease stage, and monitoring disease progression. Circulating miRNAs and exosomal miRNAs, a group of small non-coding RNAs, are considered the gold standard diagnostic biomarkers for human diseases. We have previously demonstrated that osteosarcoma-derived exosomes are able to influence crucial mechanisms inside tumor niches, inducing osteoclast differentiation, and sustaining bone resorption activity. Here we discovered, through Next-Generation Sequencing (NGS), eight novel microRNAs in three different osteosarcoma cell lines, and assessed the selective packaging into the exosomes released. We then investigated, as proof-of-principle, the presence of the novel microRNAs in osteosarcoma patient samples, and found that 5 of the 8 novel microRNAs were more present in circulating exosomes of osteosarcoma patients compared with the controls. These results raise a question: Could the 8 novel microRNAs play a role for osteosarcoma pathogenesis? Although still premature, the results are encouraging, and further studies with a validation in a larger cohort are needed.

Keywords: microRNAs; exosomes; liquid biopsy

---

1. Introduction

Next-generation sequencing (NGS) and liquid biopsies, also through circulating microRNAs (miRNAs), are both tools that may provide therapeutic strategies to oncologists, which contribute to the development of precision medicine. Several research groups have already investigated circulating miRNAs for their diagnostic and prognostic potential, suggesting them as useful cancer biomarkers [1–4]. In osteosarcoma (OS), some clinical studies have investigated the prognostic and diagnostic potential of microRNAs [5,6]. Moreover, the identification of novel circulating miRNAs, released through exosomes into the blood from malignant cells, can provide novel biomarkers and therapeutic targets for cancer patients [6]. OS is the most common type of primary bone tumor of the skeleton, occurring mainly in children and adolescents, in the metaphyseal region of the long bones, typically in the distal femur, proximal tibia, and humerus [7].

Multidisciplinary treatment, combined with surgery for localized tumors, have led to a five-year survival rate of 60–70% in non-metastatic patients. However, approximately 20% of patients diagnosed with metastatic OS at presentation, primarily in the lungs, are instead characterized by a worse clinical outcome [8]. Among the novel approaches that do not require surgical biopsy, liquid biopsy is useful in improving the prognosis, and monitoring disease course and survival rates of OS patients, offering information on micro-metastasis at diagnosis and minimal residual disease, the latter only partially detectable by conventional diagnostic methods [9].

In our previous study, we highlighted, by RNA sequencing methods, a packaging of specific miRNAs within OS cell-derived exosomes. In detail, we first demonstrated the role of OS cell-derived exosomes, inside the tumor microenvironment, in bone metabolism and tumor angiogenesis; we then focused our attention on some of these specific exosomal miRNAs, and demonstrated their involvement in osteoclast differentiation, bone resorption activity, and angiogenesis [10].

In the present study, in taking advantage of the NGS approach, we discovered and analyzed the expression of eight novel miRNAs in OS cell lines and OS cell-derived exosomes. We also analyzed their expression in a panel of human tissues and in a small group of OS patients. Notably, we found selective packaging of some of these novel miRNAs into the exosomes released by OS cells and into the circulating exosomes from plasma of OS patients. Although, the functional role of exosome-encapsuled miRNAs must be investigated in depth, the results obtained lead us to hypothesize a role for these novel miRNAs as having a circulating biomarker potential in OS, and for enabling personalized treatments in precision medicine.

2. Results

2.1. RNA Profiles of Osteosarcoma Cell Lines and Osteosarcoma Cell-Derived Exosomes

In order to discover novel microRNA specific for OS, we start with three OS cell lines: SAOS-2, MG-63, U-2 OS, and the exosomes released by those cell lines. Exosomes were isolated and characterized as previously described [10]. After assessing the RNA quality, we performed small RNA sequencing on the MG-63 cell line and MG-63 exosomes; the U-2 OS cell line and U-2 OS exosomes; and the SAOS-2 cell line and SAOS-2 exosomes. Due to the NGS approach, we were able to determine, not only the differences in the expression of known miRNAs in the parent cells and the exosomes, but also novel sequences expressed among the different cell lines in relation to their exosome miRNA cargo. Novel miRNA profiles were analyzed in three OS cell lines: MG-63, U-2 OS, and SAOS-2, and their exosomes through miRDeep2. 3116, 5916, and 3381, novel putative miRNA sequences from MG-63, SAOS-2, and U-2 OS, respectively, remained after filtering for known contaminant and highly common sequences, such as rRNA and tRNA, for all known miRNA sequences from miRBase v22 and for exon (https://www.ncbi.nlm.nih.gov/sra/PRJNA575520). We further filtered the miRDeep2 results by score. MiRDeep2 scores ranged from -10 to 10, with a higher number corresponding to increased likelihood that the miRNA is genuine. A cut-off of 0 was used to be included in this study. We chose to validate two miRNAs for MG-63, three for U-2 OS, and three for SAOS-2, with the highest raw counts for each cell line (Table 1). In Supplementary Table S1 and Supplementary Material 2, we report the sequences, the genomic locations and predicted structures of the 8 pre-miRNA candidates.

Table 1. Sequence and location information for the eight candidate miRNAs.

Novel ID	Location (hg19)	Strand	Mature Sequence
Candidate 1	chr9:136204572..136204654	−	CCCCACACUGCUAAAUUUGAC
Candidate 2	chr3:14436198..14436257	+	GGAAUAACGGGUGCUGUAGGCU
Candidate 3	chr9:89037929..89037970	−	CCCCUCACUGCUAAAUUUGAC
Candidate 4	chrX:102411053..102411132	−	CCAUCUGUGGGAUUAUGACUGA
Candidate 5	chr16:33962833..33962878	+	UGCGCAGUGGCAGUAUCGUAGCC
Candidate 6	chr8:56821958..56822010	−	UAUGUGCCUAGUGGCUGCUGUCU
Candidate 7	chr13:27259470..27259536	+	UCUGGGCAACAAGGUGAGACC
Candidate 8	chr9:89037927..89037972	−	AUGGAUUUUUGGAAAUAGGAGA

2.2. Validation of the Candidates' Novel microRNAs

Once we obtained the 8 candidate microRNA sequences (Table 1) with a high score through miRDeep2, we used custom Taqman assays to first validate on the SAOS-2, MG-63, and U-2 OS cell lines and their exosomes (Supplementary Figure S1). We then measured the expression of the candidate microRNAs across a panel of 10 different human tissue RNA (skeletal muscle, stomach, testis, kidney, lung, brain, prostate, liver, spleen, and bone) to assess the presence in different body parts and eventually the tissue specificity. Custom Taqman assays gave reproducible and consistent results, and were able to amplify the target novel miRNAs in most of the tissue types tested. The liver, kidney, and brain RNAs showed the highest candidate microRNAs expression, while skeletal muscle, lung, spleen and bone showed the lowest candidate microRNAs expression (Figure 1).

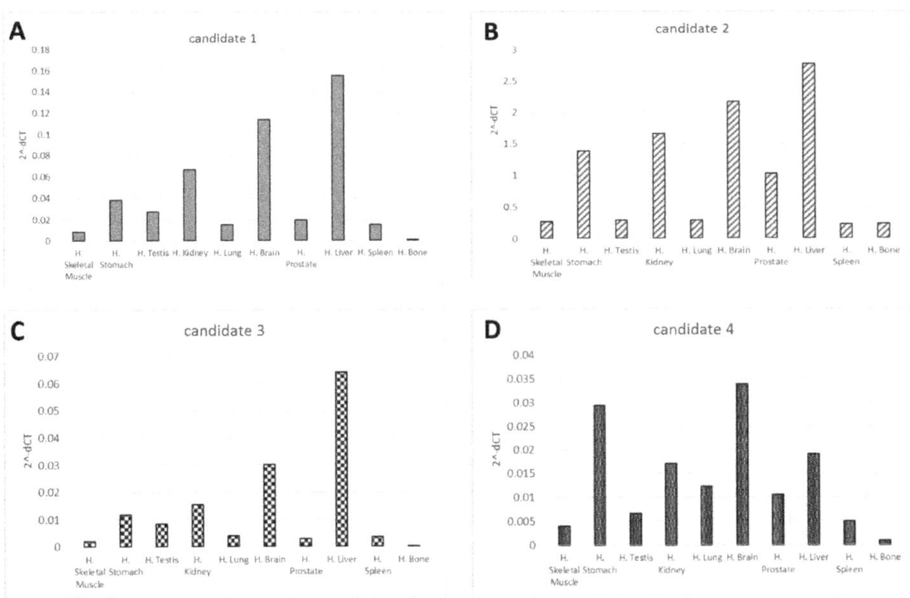

Figure 1. *Cont.*

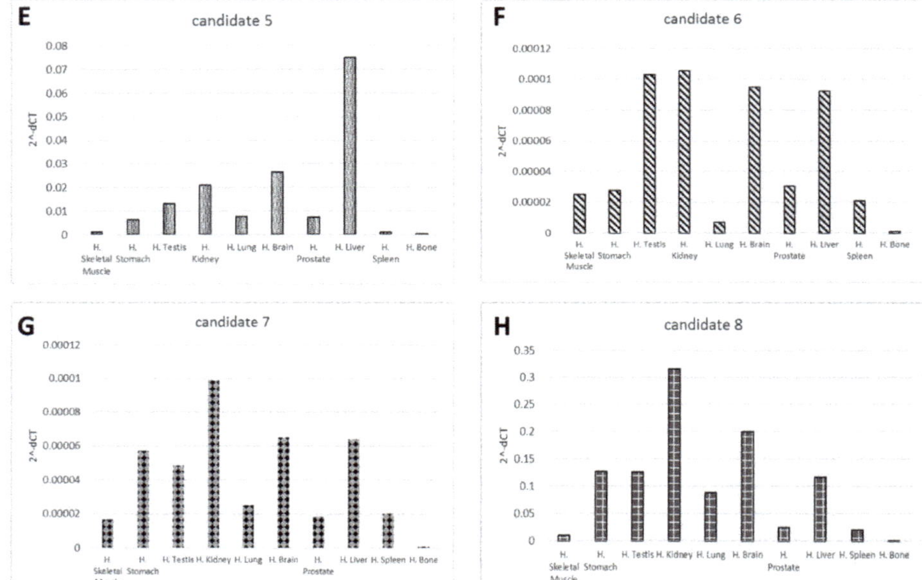

Figure 1. Novel candidate miRNA validation. The graphs (**A–H**) show quantitative RT-PCR results of the 8 novel candidate miRNAs expression, respectively, in 10 different human tissue types. Results are displayed as mean levels to the average expression reported as 2^{-dCT}, and are normalized to U6.

2.3. Candidate Novel microRNA Expression Circulating in Osteosarcoma Samples

Once validated in different human tissues, we wanted to investigate the presence of the candidate microRNAs discovered in OS cell lines, and eventually the differential in samples of OS patients. As proof-of-principle, to test the prognostic potential of the novel candidate microRNAs in liquid biopsies, we analyzed plasma from 5 OS patients whose clinical features are reported in Table 2, and 3 controls by digital PCR.

Table 2. Clinical characteristics of Osteosarcoma patients.

Patient ID	Gender	Age	Site of Origin	Histologic Subtype	metastasis	Treatments
Patient 1	M	14	sx prox Tibia	High Grade Surface OS	No	Chir
Patient 2	M	17	dx distal Femur	OS G3	No	CHT (CDDP/ADM MTX HD × 2)
Patient 3	M	18	dx prox Tibia	OS condrobastic G3	No	CHT (CDDP/ADM MTX HD × 2)
Patient 4	M	19	sx Tibia	OS fibroblastic	Lung	CHT (CDDP/ADM MTX HD × 2)
Patient 5	M	16	dx Pelvis	OS G3	Lung	CHT (CDDP/ADM MTX HD × 2)

We choose this approach because of such advantages as the partitioning of the PCR reaction into thousands of individual reactions; the end-point measurement enables nucleic acid quantitation independent of the reaction efficiency. In this set of samples, 5 novel candidate microRNAs (candidates 2, 4, 5, 6, and 8) out of eight showed a significant differential expression in OS samples compared with the controls (Figure 2). One novel candidate microRNA (candidate 7) did not amplify in any sample;

one novel candidate microRNA (candidate 3) showed no difference between the two groups, and novel candidate 1 showed the highest expression in the control group compared with the OS group.

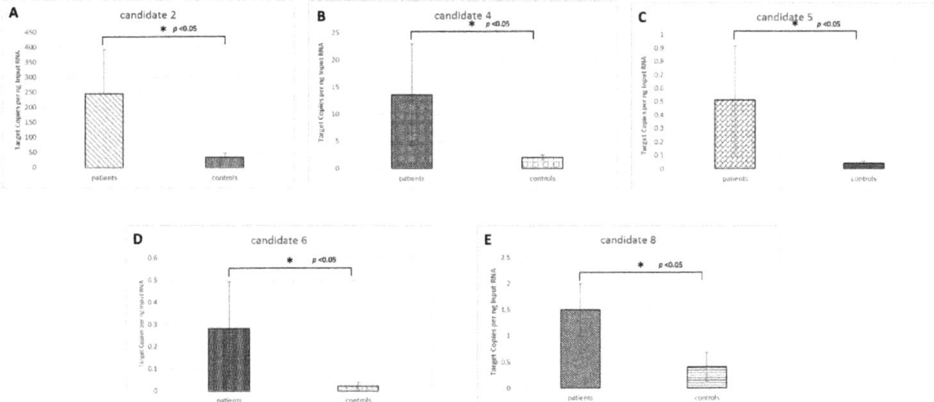

Figure 2. Novel candidate miRNAs expression in osteosarcoma patient samples. The graphs (**A–E**) show quantitative RT-PCR results of the novel candidate miRNAs 2/4/5/6/8 respectively that resulted highly expressed circulating in osteosarcoma sample compared with the controls. Results are displayed as mean levels to the average expression calculated as Target Copies per ng Imput RNA, and are normalized to U6. One star (*) indicates $p < 0.05$, two-tailed Student's test.

We further determined the biologic pathways affected by novel candidate miRNAs using TargetScan web platform able to predict biological targets of microRNAs by searching for the presence of sites that match the seed region of each miRNA. We reported in Table 3 the KEGG biological processes significantly enriched (pathways with $p < 0.05$). All clearly involved with carcinogenesis for each candidate microRNAs determined by Enrichr web platform.

Table 3. Biologic pathways enriched by differentially expressed novel candidate miRNAs.

Candidate 2	KEGG pathway	*p*-value	Genes
	FoxO signaling pathway	6.37×10^{-5}	*SMAD2;SMAD4;CCND2;MAPK1;PIK3R1;IGF1*
	TGF-beta signaling pathway	1.208×10^{-3}	*SMAD2;SMAD4;MAPK1;ACVR1B;BMPR1B;SKP1*
	Hippo signaling pathway	1.29×10^{-3}	*FZD1;SMAD2;PRKCI;SMAD4;CCND2;NF2*
	Proteoglycans in cancer	1.41×10^{-3}	*FZD1;SMAD2;PDCD4;MAPK1;PIK3R1;IGF1*
	Hepatocellular carcinoma	1.75×10^{-3}	*FZD1;SMAD2;SMAD4;CDK6;MAPK1;PIK3R1*
	Glioma	3.09×10^{-3}	*CDK6;MAPK1;PIK3R1;IGF1;CALM1*
	Pancreatic cancer	3.09×10^{-3}	*SMAD2;SMAD4;CDK6;MAPK1;PIK3R1*
	Breast cancer	3.35×10^{-3}	*FZD1;JAG2;CDK6;MAPK1;PIK3R1;IGF1;ESR1*

Candidate 4	KEGG pathway	*p*-value	Genes
	Cell adhesion molecules	3.61×10^{-4}	*NLGN3;CNTNAP1;CD4;NLGN2;PECAM1;NRXN3*
	Focal adhesion	6.73×10^{-4}	*PPP1CB;PPP1CC;RAP1A;COL4A4;ITGB8;MAPK1;CRK*
	TGF-beta signaling pathway	1.08×10^{-3}	*SRC;MAPK1;CRK;CD44;PFN2*
	Renal cell carcinoma	1.12×10^{-3}	*PPP1CB;PPP1CC;FGF7;ITGB8;MAPK1;PIP4K2C;CRK*
	Rap1 signaling pathway	1.32×10^{-3}	*PPP1CB;PPP1CC;PRKAB2;RAPGEF1;MAPK1*
	ErbB signaling pathway	4.50×10^{-3}	*ZFYVE16;SP1;MAPK1;INHBA;RGMB*
	VEGF signaling pathway	8.46×10^{-3}	*PPP1CB;PPP1CC;RAP1A;MAPK1*
	Proteoglycans in cancer	9.37×10^{-3}	*RAP1A;RAPGEF1;MAPK1;CRK*

Candidate 5	KEGG pathway	*p*-value	Genes
	Bladder cancer	1.22×10^{-2}	*VEGFA*
	VEGF signaling pathway	1.75×10^{-2}	*VEGFA*
	Renal cell carcinoma	2.05×10^{-2}	*VEGFA*
	Pancreatic cancer	2.22×10^{-2}	*VEGFA*
	HIF-1 signaling pathway	2.69×10^{-2}	*VEGFA*
	Focal adhesion	2.69×10^{-2}	*VEGFA*
	Proteoglycans in cancer	2.96×10^{-2}	*VEGFA*

Table 3. *Cont.*

Candidate 6	KEGG pathway	p-value	Genes
	Proteoglycans in cancer	2.45×10^{-3}	*PPP1R12A;ARHGEF12;FZD5;ERBB4;KDR;IGF1R*
	MAPK signaling pathway	3.93×10^{-3}	*GABRB2;CACNA1C;GABARAP;TRAK2*
	Focal adhesion	1.13×10^{-2}	*PDGFRA;CACNB4;ERBB4;KDR;CACNA1C;IGF1R*
	Rap1 signaling pathway	5.11×10^{-2}	*PDGFRA;PPP1R12A;KDR;PARVA;IGF1R*
Candidate 8	**KEGG pathway**	***p*-value**	**Genes**
	Hippo signaling pathway	$2,00 \times 10^{-4}$	*PPP1CB;LATS2;DLG3;FZD4;TCF7;YWHAZ;TGFBR1;*
	TGF-beta signaling pathway	2.14×10^{-3}	*CREBBP;TCF7;BAIAP2;TGFBR1;WASF3*
	Breast cancer	5.62×10^{-3}	*CREBBP;RPS6KB1;NEO1;ACVR2B;TGFBR1*
	Colorectal cancer	7.87×10^{-3}	*HDAC4;BECN1;RPS6KB1;ADCY2;CALM2;TGFBR1*
	Prostate cancer	9.07×10^{-3}	*CREBBP;FZD4;TCF7;ADCY2;CALM2*
	Gastric cancer	1.09×10^{-2}	*RPS6KB1;FZD4;NCOA3;TCF7;PGR;LRP6*
	Wnt signaling pathway	2.19×10^{-2}	*NRP1;CREBBP;CDC27;ADCY2;TGFBR1;CREB5*

3. Discussion

In the management of OS, a proper diagnosis and staging of the disease is a major prerequisite for effective surgical and pharmacological treatments. Conventional diagnostic approaches, such as tissue biopsy and imaging, remain the most common diagnostic tests. Nevertheless, tissue biopsy may sometimes be difficult to obtain and not be easily repeatable. In addition, information concerning micro-metastasis or minimal residual disease can sometimes be lost with conventional diagnostic methods [11]. Liquid biopsy may be highly advantageous, given the difficulty of recovering the bone sample for disease staging. In fact, it is a non-invasive and also time-saving approach; moreover, it may offer more precise and accurate information on early detection, therapeutic decisions, and response to therapy in OS [12,13].

In this paper we identified eight novel microRNAs in OS cell lines (MG-63, SAOS-2, and U-2 OS) and their released exosomes. The eight sequences remained after filtering for known contaminant from 3116, 5916, and 3381, novel putative miRNA sequences, obtained by MG-63, SAOS-2, and U-2 OS, respectively. We chose to validate two miRNAs for MG-63, three for U-2 OS, and three for SAOS-2, with the highest raw counts for each cell line. To assess the tissue specificity, we analyzed, by custom TaqMan assays, the expression of the eight candidate microRNAs across a panel of 10 different human tissue RNAs (skeletal muscle, stomach, testis, kidney, lung, brain, prostate, liver, spleen and bone). The candidate microRNAs were found highly expressed in the liver, kidney, and brain, while skeleton muscle, lung, spleen and bone showed the lowest candidate microRNAs expression.

To fill the current gaps and needs advanced by clinical oncologists, we decided to investigate the presence of the novel candidate microRNAs circulating in OS patients, and compare them with controls. Five of eight novel candidate microRNAs were found highly and significantly expressed in OS samples compared with the controls. This fact, together with the low expression of these microRNAs in the healthy bone tissue, make the novel candidates potential disease biomarkers.

Through TargetScan web platform, we identified the putative genes regulated by these five candidate microRNAs. We further determined, through the web platform Enrichr, the biologic pathways predicted and potentially affected. The KEGG biological processes significantly enriched (pathways with $p < 0.05$), reported in Table 3, resulted clearly involved with carcinogenesis for each candidate; Among these, candidate microRNA 5 is predicted to bind and strongly regulate VEGFA which expression is crucial for OS growth and metastasis [14]. Another example is candidate microRNA 2, predicted to bind SMAD2 and SMAD4, genes already related with Osteosarcoma [15], and resulting potentially interesting for further in vitro studies.

Overall, these results allowed us to hypothesize that these 8 novel candidate miRNAs could potentially represent biomarkers for OS. As a result, we plan to carry out future clinical studies on large cohorts of OS patients in order to better analyze and compare the expression of novel miRNAs in patient subgroups, distinguishing metastatic patients from non-metastatic ones. We suppose that such information could correlate the expression of miRNAs with the disease progression. At the same time, functional studies, by in vitro assays, may shed light on the molecular mechanisms beyond the possible pathologic role of these microRNAs.

4. Materials and Methods

4.1. Ethics Approval

The Bioethics Committee of the IFO_AOO - AOO - Istituti Fisioterapici Ospitalieri on February 16, 2017 (resolution number899/17) or osteosarcoma patients approved this study.

The aim of the study, study stages, and sample collection procedures were explained to all subjects. All subjects gave their written informed consent to participate in the study, including for blood sample collection and use of clinical data for research.

4.2. Cell Lines and Reagents

SAOS-2, MG-63, and U-2 OS cell lines were purchased from ECACC (Sigma-Aldrich, Milano, Italy), and grown in DMEM high glucose (Thermo Fisher Scientific, Cambridge, MA, USA) supplemented with 10% fetal bovine serum (FBS, Lonza Group, Basel, Switzerland).

4.3. Exosome Purification

Exosomes released by OS cells (SAOS-2, MG-63, and U-2 OS) during a 48-hour culture period were isolated from conditioned culture medium supplemented with 10% FBS, previously ultra-centrifuged by differential centrifugations as previously described. Exosome protein content was determined by the Bradford assay [16]. Exosomes from plasma of OS patients were isolated with Total Exosome Isolation Kit (Thermo Fisher Scientific, Cambridge, MA, USA), according to the manufacturer's instructions.

4.4. Small RNA Library Construction and Sequencing

To test the quality and assess the quantity of total RNA extracted, we used the RNA ScreenTape assay on a 2200 TapeStation system (Agilent Technologies, Santa Clara, CA, USA). For small RNA-Seq, 1 μg of total RNA per sample was used for library preparation using TruSeq Small RNA Sample Prep Kits (Illumina, San Diego, CA, USA). Size-distribution was measured with the DNA ScreenTape assay on a 2200 TapeStation system (Agilent Technologies, Santa Clara, CA, USA). A total library pool of 4 nM was sequenced using a MiSeq Reagent Kit v3 150 cycle on a MiSeq System (Illumina, San Diego, CA, USA).

4.5. Small RNA-Seq in Silico Analysis

We used the Trimmomatic software [17], v. 0.3633 to remove adaptors, low-quality bases, and reads with less than 16 nucleotides. The parameter "ILLUMINACLIP TruSeq3-SE.fa:2:30:10" was used to remove read adaptors according to Illumina-specific sequences. A sliding window cut was applied to remove bases with average quality below 22 using the parameter "SLIDINGWINDOW:3:22," and reads with less than 16 nucleotides were removed using "MINLEN:16."

Alignment of miRNA sequencing reads to the human reference genome build hg19 was performed using Bowtie v. 1.2.2 [18]. More stringent read length filtering was carried out by miRDeep2 [19] before the identification of novel miRNAs, discarding reads with length less than 18 nucleotides and greater than 23 nucleotides. We used the miRDeep2 v.2.0.1.1 software (Berlin Institute for Medical Systems Biology at Max-Delbrück-Center for Molecular Medicine, Berlin-Buch, Germany) default parameters, using the Fasta format sequences of all mature and hairpin miRNA sequences obtained from miRBASE v.22 database.

4.6. Real-time PCR Validation and Digital PCR of Novel microRNAs

For miRNA validation, total RNA from 10 human tissues (H. Skeletal Muscle, H. Stomach, H. Testis, H. Kidney, H. Lung, H. Brain, H. Prostate, H. Liver, H. spleen from Gentaur S.r.l. Italy and H. Bone from OriGene Technologies GmbH, MD, USA) were reverse transcribed with TaqMan MicroRNA Reverse Transcription Kit, according to the manufacturer's instructions (Thermo Fisher Scientific, Cambridge, MA, USA). Custom TaqMan microRNA Assay are listed in Supplementary Table S2). Novel candidate miRNA expression results are displayed as $2^{\wedge-dCT}$, and are normalized to U6 (Thermo Fisher Scientific, Cambridge, MA, USA). Twenty ng of RNA extracted from plasma of OS patients and controls were reverse transcribed with TaqMan MicroRNA Reverse Transcription, according to the manufacturer's instructions (Thermo Fisher Scientific, Cambridge, MA, USA). The Droplet Digital™ PCR (ddPCR) reactions were prepared in 20 μL total volumes, according to the manufacturer's instructions (Bio-Rad Laboratories, Irvine, CA, USA). The thermal cycling conditions of the droplets generated were as follows: 95 °C for 10 min (1 cycle); then 40 cycles of 94 °C for 30 s, and 60 °C for 1 min, 98 °C for 10 min, and then held at 4 °C. After thermal cycling, droplets were analyzed for positive and negative signals using the QX200 droplet reader (Bio-Rad Laboratories, Irvine, CA, USA). Data analysis was done when the number of droplets produced was more than 20,000. For data analysis, QuantaSoft Version 1.7.4 software (Bio-Rad Laboratories, Irvine, CA, USA) was used to statistically analyze the obtained data. Novel candidate miRNA expression was normalized on U6 (Thermo Fisher Scientific, Cambridge, MA, USA) and the results are displayed as Target Copies per ng Imput RNA [20].

4.7. miRNA Pathway Analysis

To identify potential target genes and pathways of the differentially expressed novel candidate miRNAs found, we conducted in silico analysis using TargetScan web platform [21], in order to identify the putative genes and Enrichr web platform [22] to determine the putative pathways.

5. Conclusions

We have identified eight novel microRNAs, through NGS, in three osteosarcoma cell lines and demonstrated the selective packaging into their released exosomes. Below, the expression of the novel miRNAs was confirmed in circulating exosomes from plasma of osteosarcoma patients. The exact role of the novel miRNAs will be assessed through future clinical and molecular studies. The data obtained suggest these novel miRNAs as having a circulating biomarker potential in osteosarcoma cancer, assuming a role in personalized medicine.

Supplementary Materials: The following are available online at http://www.mdpi.com/2072-6694/11/12/1924/s1, Figure S1: Novel candidate miRNA validation. Table S1: Sequence and location information for the eight pre-miR candidate. Table S2: List of Custom Assay ID for novel candidate miRNA validation. The sequencing raw data are available online at https://www.ncbi.nlm.nih.gov/sra/PRJNA575520.

Author Contributions: Conceptualization: N.C., L.R., and A.G.; methodology, G.R., A.D.L., L.C., and J.B.; software, N.C.; validation, C.C., G.R., and A.D.L.; formal analysis, N.C., L.R., and A.G.; investigation, A.G.; writing—original draft preparation, N.C., L.R., and A.G.; writing—review and editing, A.G., P.G.C., and G.G.; visualization, L.R. and A.G.; supervision, A.G., P.G.C., and G.G.

Funding: This research received no external funding.

Acknowledgments: We would like to thank Warren Blumberg of IRCCS ISMETT's Language Services Department for assistance in editing the manuscript.

Conflicts of Interest: The authors declare no conflict of interest. The funders had no role in the design of the study; in the collection, analyses, or interpretation of data; in the writing of the manuscript, or in the decision to publish the results.

References

1. Mitchell, P.S.; Parkin, R.K.; Kroh, E.M.; Fritz, B.R.; Wyman, S.K.; Pogosova-Agadjanyan, E.L.; Peterson, A.; Noteboom, J.; O'Briant, K.C.; Allen, A.; et al. Circulating Micrornas as Stable Blood-Based Markers for Cancer Detection. *Proc. Natl. Acad. Sci. USA* **2008**, *105*, 10513–10518. [CrossRef] [PubMed]

2. Chen, X.; Ba, Y.; Ma, L.; Cai, X.; Yin, Y.; Wang, K.; Guo, J.; Zhang, Y.; Chen, J.; Guo, X.; et al. Characterization of Micrornas in Serum: A Novel Class of Biomarkers for Diagnosis of Cancer and Other Diseases. *Cell Res.* **2008**, *18*, 997–1006. [CrossRef] [PubMed]

3. Hamam, R.; Hamam, D.; Alsaleh, K.A.; Kassem, M.; Zaher, W.; Alfayez, M.; Aldahmash, A.; Alajez, N.M. Circulating Micrornas in Breast Cancer: Novel Diagnostic and Prognostic Biomarkers. *Cell Death Dis.* **2017**, *8*, e3045. [CrossRef] [PubMed]

4. Sholl, L.M. Next-Generation Sequencing from Liquid Biopsies in Lung Cancer Patients: Advances in Comprehensive Biomarker Testing. *J. Thorac. Oncol.* **2017**, *12*, 1464–1466. [CrossRef] [PubMed]

5. Sarver, A.L.; Phalak, R.; Thayanithy, V.; Subramanian, S. S-Med: Sarcoma Microrna Expression Database. *Lab. Investig.* **2010**, *90*, 753–761. [CrossRef]

6. Brase, J.C.; Wuttig, D.; Kuner, R.; Sultmann, H. Serum Micrornas as Non-Invasive Biomarkers for Cancer. *Mol. Cancer* **2010**, *9*, 306. [CrossRef]

7. Ottaviani, G.; Jaffe, N. The Etiology of Osteosarcoma. *Cancer Treat. Res.* **2009**, *152*, 15–32.

8.   Marina, N.; Gebhardt, M.; Teot, L.; Gorlick, R. Biology and Therapeutic Advances for Pediatric Osteosarcoma. *Oncologist* **2004**, *9*, 422–441. [CrossRef]

9.   Palmirotta, R.; Lovero, D.; Cafforio, P.; Felici, C.; Mannavola, F.; Pelle, E.; Quaresmini, D.; Tucci, M.; Silvestris, F. Liquid Biopsy of Cancer: A Multimodal Diagnostic Tool in Clinical Oncology. *Ther. Adv. Med. Oncol.* **2018**, *10*, 1758835918794630. [CrossRef]

10.   Raimondi, L.; de Luca, A.; Gallo, A.; Costa, V.; Russelli, G.; Cuscino, N.; Manno, M.; Raccosta, S.; Carina, V.; Bellavia, D.; et al. Osteosarcoma Cell-Derived Exosomes Affect Tumor Microenvironment by Specific Packaging of Micrornas. *Carcinogenesis* **2019**. [CrossRef]

11.   Pantel, K.; Alix-Panabieres, C. Real-Time Liquid Biopsy in Cancer Patients: Fact or Fiction? *Cancer Res.* **2013**, *73*, 6384–6388. [CrossRef] [PubMed]

12.   Raimondi, L.; de Luca, A.; Costa, V.; Amodio, N.; Carina, V.; Bellavia, D.; Tassone, P.; Pagani, S.; Fini, M.; Alessandro, R.; et al. Circulating Biomarkers in Osteosarcoma: New Translational Tools for Diagnosis and Treatment. *Oncotarget* **2017**, *8*, 100831–100851. [CrossRef] [PubMed]

13.   Li, X.; Seebacher, N.A.; Hornicek, F.J.; Xiao, T.; Duan, Z. Application of Liquid Biopsy in Bone and Soft Tissue Sarcomas: Present and Future. *Cancer Lett.* **2018**, *439*, 66–77. [CrossRef] [PubMed]

14.   Li, Y.S.; Liu, Q.; Tian, J.; He, H.B.; Luo, W. Angiogenesis Process in Osteosarcoma: An Updated Perspective of Pathophysiology and Therapeutics. *Am. J. Med. Sci.* **2019**, *357*, 280–288. [CrossRef] [PubMed]

15.   Mohseny, A.B.; Cai, Y.; Kuijjer, M.; Xiao, W.; van den Akker, B.; de Andrea, C.E.; Jacobs, R.; Dijke, P.T.; Hogendoorn, P.C.; Cleton-Jansen, A.M. The Activities of Smad and Gli Mediated Signalling Pathways in High-Grade Conventional Osteosarcoma. *Eur. J. Cancer* **2012**, *48*, 3429–3438. [CrossRef]

16.   Thery, C.; Amigorena, S.; Raposo, G.; Clayton, A. Isolation and Characterization of Exosomes from Cell Culture Supernatants and Biological Fluids. *Curr. Protoc. Cell Biol.* **2006**, *30*, 3–22. [CrossRef]

17.   Bolger, A.M.; Lohse, M.; Usadel, B. Trimmomatic: A Flexible Trimmer for Illumina Sequence Data. *Bioinformatics* **2014**, *30*, 2114–2120. [CrossRef]

18.   Langmead, B. Aligning Short Sequencing Reads with Bowtie. *Curr. Protoc. Bioinform.* **2010**, *32*, 11–17. [CrossRef]

19.   Friedlander, M.R.; Mackowiak, S.D.; Li, N.; Chen, W.; Rajewsky, N. Mirdeep2 Accurately Identifies Known and Hundreds of Novel Microrna Genes in Seven Animal Clades. *Nucleic Acids Res.* **2012**, *40*, 37–52. [CrossRef]

20.   Coulter, S.J. Mitigation of the Effect of Variability in Digital Pcr Assays through Use of Duplexed Reference Assays for Normalization. *Biotechniques* **2018**, *65*, 86–91. [CrossRef]

21.   Agarwal, V.; Bell, G.W.; Nam, J.W.; Bartel, D.P. Predicting Effective Microrna Target Sites in Mammalian Mrnas. *Elife* **2015**, *4*, e05005. [CrossRef] [PubMed]

22.   Chen, E.Y.; Tan, C.M.; Kou, Y.; Duan, Q.; Wang, Z.; Meirelles, G.V.; Clark, N.R.; Ma'ayan, A. Enrichr: Interactive and Collaborative Html5 Gene List Enrichment Analysis Tool. *BMC Bioinform.* **2013**, *14*, 128. [CrossRef] [PubMed]

Article

The Emerging Role of miRNAs for the Radiation Treatment of Pancreatic Cancer

Lily Nguyen [1,2], Daniela Schilling [1,2], Sophie Dobiasch [1,2,3], Susanne Raulefs [1,2], Marina Santiago Franco [1], Dominik Buschmann [4], Michael W. Pfaffl [4], Thomas E. Schmid [1,2] and Stephanie E. Combs [1,2,3,*]

1 Institute of Radiation Medicine (IRM), Department of Radiation Sciences (DRS), Helmholtz Zentrum München, 85764 Neuherberg, Germany; lily.nguyen@tum.de (L.N.); daniela.schilling@tum.de (D.S.); sophie.dobiasch@tum.de (S.D.); susanne.raulefs@helmholtz-muenchen.de (S.R.); franco.marinasantiago@gmail.com (M.S.F.); thomas.schmid@helmholtz-muenchen.de (T.E.S.)
2 Department of Radiation Oncology, School of Medicine, Technical University of Munich (TUM), Klinikum rechts der Isar, 81675 Munich, Germany
3 Deutsches Konsortium für Translationale Krebsforschung (DKTK), Partner Site Munich, 81675 Munich, Germany
4 Division of Animal Physiology and Immunology, TUM School of Life Sciences Weihenstephan, Technical University of Munich (TUM), 85354 Freising, Germany; dominik.buschmann@wzw.tum.de (D.B.); michael.pfaffl@wzw.tum.de (M.W.P.)
* Correspondence: stephanie.combs@tum.de; Tel.: +49-89-4140-4501

Received: 21 October 2020; Accepted: 4 December 2020; Published: 9 December 2020

Simple Summary: Pancreatic cancer is an aggressive disease with a high mortality rate. Radiotherapy is one treatment option within a multimodal therapy approach for patients with locally advanced, non-resectable pancreatic tumors. However, radiotherapy is only effective in about one-third of the patients. Therefore, biomarkers that can predict the response to radiotherapy are of utmost importance. Recently, microRNAs, small non-coding RNAs regulating gene expression, have come into focus as there is growing evidence that microRNAs could serve as diagnostic, predictive and prognostic biomarkers in various cancer entities, including pancreatic cancer. Moreover, their high stability in body fluids such as serum and plasma render them attractive candidates for non-invasive biomarkers. This article describes the role of microRNAs as suitable blood biomarkers and outlines an overview of radiation-induced microRNAs changes and the association with radioresistance in pancreatic cancer.

Abstract: Today, pancreatic cancer is the seventh leading cause of cancer-related deaths worldwide with a five-year overall survival rate of less than 7%. Only 15–20% of patients are eligible for curative intent surgery at the time of diagnosis. Therefore, neoadjuvant treatment regimens have been introduced in order to downsize the tumor by chemotherapy and radiotherapy. To further increase the efficacy of radiotherapy, novel molecular biomarkers are urgently needed to define the subgroup of pancreatic cancer patients who would benefit most from radiotherapy. MicroRNAs (miRNAs) could have the potential to serve as novel predictive and prognostic biomarkers in patients with pancreatic cancer. In the present article, the role of miRNAs as blood biomarkers, which are associated with either radioresistance or radiation-induced changes of miRNAs in pancreatic cancer, is discussed. Furthermore, the manuscript provides own data of miRNAs identified in a pancreatic cancer mouse model as well as radiation-induced miRNA changes in the plasma of tumor-bearing mice.

Keywords: pancreatic cancer; miRNA; radiotherapy; radioresistance; personalized medicine; biomarker; target

1. Introduction

Pancreatic cancer is one of the most lethal cancers and could be the second leading cause of cancer-related deaths within the next decade [1]. The most common type of malignant pancreatic neoplasms is of ductal origin and is classified as pancreatic ductal adenocarcinoma (PDAC) [2]. Despite intense research efforts, the prognosis of patients with PDAC still remains very poor, with a five-year overall survival (OS) rate of less than 7%, without any significant improvements over the past years. The majority of patients are diagnosed in locally advanced or metastatic stages because of unspecific symptoms, a lack of early sensitive and specific markers, and difficulties in imaging early-stage tumors [2].

The only potentially curative treatment option for patients with pancreatic cancer is surgical resection. However, only 15–20% of patients are eligible for surgery at the time of diagnosis due to highly aggressive tumor growth with perineural and vascular invasion and early distant metastases [3]. Therefore, the importance of neoadjuvant treatments, including chemotherapy, radiotherapy (RT), or combined chemoradiotherapy (CRT), is evident. In particular, the aim of neoadjuvant strategies is a tumor downsizing enabling a secondary resection to improve a long-term prognosis in patients with borderline and primary non-resectable, locally advanced pancreatic cancer (LAPC) [4].

Clinical trials have shown the efficacy of RT in about 30% of pancreatic cancer patients [5]. However, international standardized therapeutic guidelines are lacking and the role of RT as a treatment option for patients with LAPC is controversially discussed in the literature [6]. Different clinical trials reported conflicting results regarding the benefit of combining RT and chemotherapy. While no benefit of subsequent RT was observed in various clinical trials when compared to chemotherapy alone [7,8], others reported an improvement in OS for LAPC patients treated with CRT [9]. Concurrent chemotherapy agents comprising capecitabine, 5-fluorouracil, or gemcitabine are used in neoadjuvant CRT treatment regimens for patients with LAPC [10]. Addition of 5-fluorouracil to RT significantly improved OS compared to RT alone [11]. Capecitabine-based CRT showed a trend toward preferable progression free survival (PFS) in comparison to gemcitabine-based CRT after induction chemotherapy for LAPC [12]. More prospective studies are needed to elucidate the benefits of associating RT with systemic therapy.

The recent focus of neoadjuvant treatment strategies for LAPC lies in investigating more effective chemotherapy schemes, such as FOLFIRINOX (leucovorin, 5-fluorouracil, irinotecan, and oxaliplatin) [13–17], and subsequent CRT or RT with modern techniques (e.g., stereotactic, intensity modulated, and particle RT) [18]. The high failure rate of RT in PDAC can be attributed to the high intrinsic radioresistance of the tumors [6]. Additionally, pancreatic cancer is characterized by high heterogeneity, genetic diversity, presence of a dense desmoplastic tumor stroma, cancer stem cells, and a complex tumor microenvironment. These factors contribute to a high resistance to conventional treatment options such as RT, chemotherapy, and molecularly targeted therapies [19,20].

Consideration of molecular profiles or tumor subtypes for therapy decisions is not yet implemented in clinical routine. To further increase the efficacy of CRT, novel molecular biomarkers are urgently needed to define the subgroup of pancreatic cancer patients who benefit from CRT more precisely [21].

MicroRNAs (miRNAs) are highly conserved small non-coding RNAs consisting of 20–24 nucleotides and regulate protein output and gene expression at transcriptional or post-transcriptional levels [22]. Previous studies have attributed critical roles to miRNAs in biological processes such as cell proliferation, differentiation, and survival. Furthermore, miRNAs have been implicated as crucial players during development, physiology, homeostasis, and disease, and have been shown to regulate the initiation and progression of many malignancies by controlling oncogenic and tumor-suppressive pathways [23,24]. Therefore, miRNAs could have potential as novel biomarkers for diagnosis, monitoring of recurrence, and predicting prognosis and survival of patients with pancreatic cancer [22]. A recently published review article analyzed the association between miRNAs and response to chemotherapy (gemcitabine and 5-fluorouracil) in pancreatic cancer [25]. Multiple miRNAs were identified that might contribute to chemotherapeutic resistance or sensitivity.

Our article provides an overview of the current literature concerning miRNAs as blood biomarkers and miRNAs associated with radioresistance as well as radiation-induced changes of miRNAs in pancreatic cancer. Additionally, own data identifying miRNAs in a pancreatic cancer mouse model as well as radiation-induced miRNA changes in the plasma of tumor-bearing mice are included.

2. Results

2.1. miRNAs as Biomarker in Pancreatic Cancer

It has been reported that serum and other body fluids contain stable miRNAs signatures. Circulating miRNAs in PDAC blood samples represent a valuable source of information for either defining eligible therapy options, monitoring therapeutic response or predicting prognosis [26]. In a meta-study, Chhatriya et al. published a list of in total 21 miRNAs, which are designated as a "meta-signature" of miRNAs in PDAC altered in both serum and cancer tissue [27].

Ouyang et al. showed that miR-10b levels in PDAC plasma samples are highly increased compared to healthy controls or patients with chronic pancreatitis [28]. The authors claim that miR-10b is not only suitable as a diagnostic marker but could serve as a therapeutic target by interrupting the growth-promoting deleterious EGF-TGF-β interactions and antagonizing the metastatic process. In addition, a study published in 2017 by Qu et al. described that miR-21-5p might be a stable and high-accuracy diagnostic biomarker for PDAC patients [29]. Recent studies confirmed the high discriminative impact of miR-21-5p for PDAC [30] and uncovered an association with a significant unfavorable prognostic outcome [31]. miR-221 seems to be a plasma marker for monitoring the tumor status due to its high preoperative plasma concentration and significantly reduced levels post-surgery [32]. These findings suggest that miR-221 may be released from the tumor into the bloodstream and therefore reflects tumor dynamics, which could be used for monitoring a possible tumor recurrence.

Several altered miRNA expression signatures, e.g., the downregulation of miR-141 and miR-720 activating ZEB-1 and TWIST1 transcription, have been identified to discriminate between PDAC patients with and without nodal metastasis [33]. A very recent study established a 4-miRNA signature (miR-29c, miR-125a, miR-200b, miR-155) that predicts local-regional failure and overall survival of pancreatic cancer patients who underwent tumor resection with and without chemotherapy but did not receive radiotherapy [34]. Therefore, defining the miRNA expression profile of individual tumors could improve the diagnosis, the ability to select better treatment options, e.g., aggressive treatment for patients with lymph node metastasis or adjuvant chemoradiation for patients with local-regional failure and ultimately predict the respective patient's outcome.

The Carbohydrate antigen CA19-9 is a blood antigen, and its increased level is approved as a biomarker for pancreatic cancer [35]. CA19-9 is released from the cell surface of pancreatic cancer cells, and its serum concentration is related to tumor mass and recurrence. Nevertheless, despite its low specificity and sensitivity, it is the most common diagnostic marker in PDAC patients. Recent studies postulate that the combination of CA19-9 together with plasma miRNAs can effectively be used for screening of early tumor stages and prognostic stratification due to improved specificity and sensitivity. This strategy has already been validated by combining CA19-9 with miR-16 and miR-196a [36] or miR-33a-3p and miR-320a [30].

The study from LaConti et al. analyzed serum miRNAs in a transgenic PDAC mouse model to establish novel circulating biomarkers for PDAC progression [37]. This study uncovered that miRNA changes show remarkable similarities between pancreatic cancer in patients and a transgenic pancreatic cancer mouse model. In summary, the authors analyzed eight different miRNAs and identified two miRNAs, miR-10 and miR-155, that were increased in serum of PDAC mice compared to control mice [37].

In our study, we aimed to identify all tumor-specific miRNAs—without pre-selecting miRNAs—that are present in the plasma of mice harboring human MIA PaCa-2 tumors compared to non-tumor-bearing mice (supplementary methods and Table S1). Principal component analysis of miRNA expression distinguished tumor-bearing from non-tumor-bearing mice (Figure 1A). Seven miRNAs that are significantly upregulated in the plasma of tumor-bearing mice were identified (Figure 1B): miR-339-3p, miR-320d, miR-92b-3p, miR-584-5p, miR-197-3p, miR-1307-3p, and miR-1246. The predicted targets of these miRNA are summarized in Table S1.

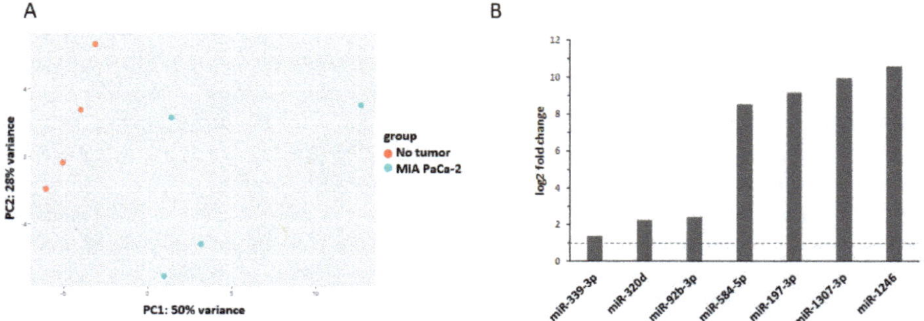

Figure 1. Detection of tumor-specific miRNAs in the plasma. (**A**). Principal component analysis of miRNA expression in plasma derived from non-tumor and tumor-bearing (MIA PaCa-2) mice. Analysis of all miRNAs in the dataset separates plasma samples derived from control mice (no tumor) to tumor-bearing (MIA PaCa-2) mice (PC1) and PC2 shows intra-group variability. (**B**). The upregulated miRNAs in the plasma of tumor-bearing (MIA PaCa-2) mice compared to the plasma of non-tumor-bearing mice are shown. Significant miRNAs were selected based on a log2 fold change $\geq$ |1| and an adjusted *p*-value of $\leq$0.05. Only transcripts with a base mean $\geq$50 were included.

Hereinafter, the miRNAs upregulated in plasma samples of tumor-bearing mice are discussed in the context of the recent literature and summarized in Table 1.

Table 1. Overview of miRNAs found to be deregulated in our analysis of plasma samples of tumor-bearing mice and associated publications.

miRNA	(Tumor) Entity	Expression	Source	Reference
miR-339-3p	Vater's papilla adenocarcinoma	downregulated	tissue	[38]
	colorectal cancer	downregulated	tissue	[39]
miR-320d	chronic pancreatitis	upregulated	tissue	[40]
	colorectal cancer	downregulated	plasma and tissue	[41]
	hepatocellular carcinoma (HCC)	downregulated	serum exosomes	[42]
miR-92b-3p	pancreatic ductal adenocarcinoma (PDAC)	downregulated	tissue	[43]
	gastric cancer	upregulated	serum exosomes	[44]
miR-584-5p	gastric cancer	downregulated	tissue	[45]
	lung cancer	upregulated	plasma	[46]
miR-197-3p	HCC	downregulated	tissue	[47]
miR-1307-3p	breast cancer	upregulated	serum	[48]
miR-1246	breast cancer	upregulated	serum	[48]
	PDAC	upregulated	plasma exosomes	[49]
	PDAC	upregulated	serum	[50]
	PDAC	upregulated	serum exosomes	[51]
	gastric cancer	upregulated	serum exosomes	[52]
	esophageal squamous cell carcinoma (SCC)	upregulated	serum	[53]
	cervical cancer	upregulated	serum	[54]

miR-339-3p was found to be deregulated in Vater's papilla adenocarcinoma [38]. In addition, miR-339-3p was downregulated in colorectal cancer (CRC) and its low-level expression was associated with lymph node metastasis in patients with CRC [39].

In relationship to pancreatic disorders, miR-320d was identified as a potential marker for late chronic pancreatitis [40]. miR-320d was recognized as a promising biomarker for early diagnosis of CRC because miR-320d expression could discriminate adenoma and CRC patients from healthy controls [41]. In serum samples of hepatocellular carcinoma (HCC) patients, the expression level of exosomal serum miR-320d was remarkably reduced compared to the respective controls [42].

A study from Long et al. described reduced miR-92b-3p expression levels in human PDAC tissues and a correlation with advanced tumor/node/metastasis (TNM) stages. Moreover, it was postulated that miR-92b-3p might act as a tumor suppressor in PDAC [43]. In gastric cancer, the serum level of exosomal miR-92b-3p was found to be higher than in healthy individuals, thus serving as a potential non-invasive biomarker for early diagnosis of gastric cancer [44].

miR-584-5p was negatively associated with tumor size in gastric cancer and therefore highlighted as a tumor suppressor and potential therapeutic target [45]. In lung adenocarcinoma, miR-584-5p was one of six upregulated miRNAs which might serve as circulating biomarkers for the early tumor detection [46].

The analysis from Ni et al. showed that miR-197-3p was downregulated in HCC tissues and low levels were correlated with larger tumor size and enhanced invasion, indicating an aggressive subtype [47].

Interestingly, miR-1307-3p and miR-1246, which were also deregulated in our study (Figure 1B), are two of five miRNAs (miR-1246, miR-1307-3p, miR4634, miR-6861-5p, and miR-6875-5p) that are suitable for early detection of breast cancer [48]. In addition, the authors claim that the combination of these five serum miRNAs can be used to differentiate breast cancer from benign diseases of the pancreas, biliary tract, and prostate, or from other cancers.

Additionally, the upregulation of miR-1246 in combination with miR-196a in plasma exosomes was found to be a potential indicator for localized pancreatic cancer. Therefore, these miRNAs might serve as circulating biomarkers for the early detection of this severe disease [49]. The study from Wei et al. showed that the expression level of serum miR-1246 was upregulated in PDAC patients compared to healthy controls and strongly reduced after tumor resection [50]. In addition, miR-1246 was part of a four-miRNA panel that was significantly upregulated in the serum of pancreatic cancer patients [51]. Further studies described elevated serum exosomal miR-1246 as a potential biomarker for the early diagnosis of gastric cancer [52] as well as esophageal squamous cell carcinoma (SCC) [53] and cervical cancer [54].

The listed publications demonstrate that circulating miRNAs represent a rich source of potential biomarkers for the early diagnosis, classification, therapeutic success, and recurrence monitoring in various human cancers, especially PDAC. Therefore, miRNA-based liquid biopsy holds promising impact for further implementation in the clinical routine.

2.2. miRNA Response to Ionizing Radiation

Radiation therapy uses ionizing radiation (IR) to induce double-strand breaks in the genomic DNA of tumor cells. Failure to restore genomic integrity before mitosis can lead to cell death or malignant transformation. Therefore, cells trigger the DNA double-strand repair. In this complex process, first sensor proteins recognize the DNA damage. Subsequently, transducer proteins recruit effector proteins responsible for cell cycle arrest, apoptosis, transcription arrest, and DNA repair [55].

Increasing evidence demonstrates that miRNAs play a critical role in the cellular response to IR, as multiple examples of miRNA expression changes in response to IR have been reported [56,57]. It was shown that miRNAs are responsible for regulating almost every cellular pathway, including the DNA damage response (DDR), after IR [55]. Whereas some miRNAs involved in the DDR are upregulated by IR (e.g., miR-34a, miR-100, and miR-143), others (e.g., miR-15b, miR-222, and miR-181a) are downregulated [55]. Interestingly, several miRNAs participating in the DDR (e.g., let-7 family, miR-15a, miR-16, miR-21, miR-24, miR-155, miR-182, and miR-302 cluster) have been reported to be both down- and upregulated, depending on the cell type, radiation dose, and time point of measurement post-RT [55].

Conversely, different members of the DDR pathway are involved in radiation-induced miRNA regulation. Ataxia telangiectasia mutated (ATM), a DDR pathway member, plays an important role in the regulation of miRNA biogenesis [58]. ATM-dependent activation of KH-type splicing regulatory protein (KSRP) upon DNA damage increases miRNA processing and expression of a specific class of miRNAs. miR-21 and miR-16, both playing a role in DDR, belong to the miRNAs that are upregulated by ATM-KSRP. Activation of the transcription factor p53 by IR leads to enhanced expression of the miR-34 family and also regulates other miRNAs, including let-7a and let-7b [59,60]. IR also affects the activity of other transcription factors (e.g., NF-kB, Myc, and E2F), which modulate the expression of several miRNAs in the DDR [57].

For pancreatic cancer, only few in vitro studies investigated radiation-induced miRNA changes [61]. Radiation significantly reduced the levels of miR-99b, leading to enhanced mTOR expression and radioresistance [61]. Very recently Jiang et al. performed miRNA sequencing and identified miR-196b-5p and miR-194-5p to be upregulated after irradiation in exosomes derived from dying pancreatic cancer cells (SW1990, Panc-1) [62]. Furthermore, irradiation with 10 Gy induced miR-194-5p upregulation in different pancreatic cancer cell lines (SW1990, Panc-1, MIA Paca-2) [62]. miR-194-5p mimics reduced the DNA damage in irradiated pancreatic cancer cells suggesting that miR-194-5p promotes survival and radioresistance of pancreatic cancer cells [62].

A very recent systematic review and meta-analysis summarize the effects of RT on circulating miRNAs in humans, nonhuman primates, and mice [63]. Radiation treatment in the included studies was very heterogeneous, including total body irradiation and local tumor-specific irradiation with different fractionation schemes and irradiation doses. Nevertheless, the meta-analysis identified 28 miRNAs with significant radiation-induced changes (18 miRNAs upregulated and 10 miRNAs downregulated). Whereas in this meta-analysis nine publications analyzed the effect of total body irradiation on the miRNA levels in plasma/serum of healthy mice, only one study used a tumor xenograft model (breast cancer) and investigated the impact of localized radiotherapy on tumor-specific miRNA plasma levels [64]. In this xenograft mammary tumor mouse model, decreased plasma levels of miR-155, miR-10b, and miR-21 have been detected after RT [64]. Specifically, for pancreatic cancer, only one study analyzed the impact of RT on the expression level of one specific miRNA and found enhanced miR-194-5p levels in exosomes derived from the plasma of irradiated PDX (patient derived xenograft) mice [62]. No in vivo studies describing the effect of RT on the expression levels of different miRNAs in pancreatic cancer were found in the literature.

Therefore, we investigated the impact of radiation on circulating miRNA levels in a pancreatic cancer xenograft mouse model (supplementary methods and Table S2). MIA PaCa-2 tumors were irradiated with a single dose of 5 Gy. Sham-irradiated (0 Gy) tumor-bearing mice served as control. Mice were sacrificed 24 h after irradiation and the plasma was collected. Small RNA sequencing revealed 21 miRNAs that were significantly modified (Figure 2). Interestingly, 20 miRNAs were down-regulated after irradiation with 5 Gy, and only one miRNA (miR-184) was upregulated. The predicted targets of these miRNAs are summarized in Table S2.

From an earlier study, it is known that miR-374b functions as an oncogene by targeting PTEN, resulting in the activation of the PI3K/Akt signaling cascade in human gastrointestinal stromal tumor cells [65]. However, there are no reports concerning the role of miR-374b in pancreatic cancer yet. It is known that the miR-15b family is involved in cell cycle regulation, proliferation, and apoptosis [66]. A recent study showed that the expression of SMURF2, a tumor suppressor gene, is inhibited by miR-15b in pancreatic cancer [67]. There is also evidence of the downregulation of miR-652 in various cancerous diseases, but its role in the radiation response is still unknown [68].

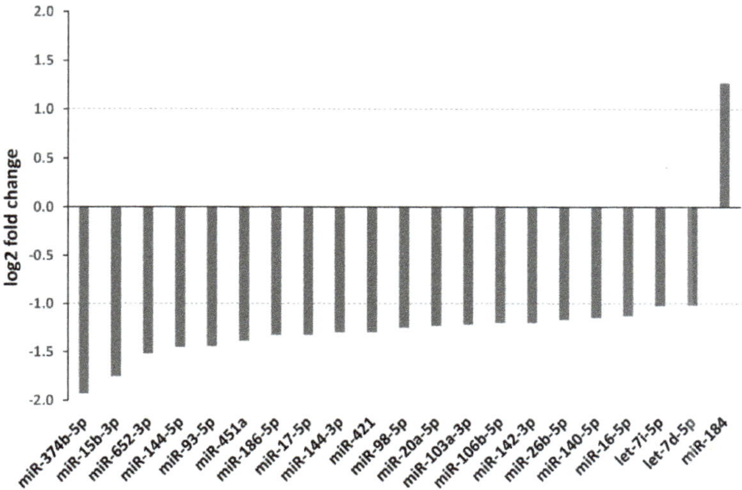

Figure 2. Radiation-induced changes in plasma miRNA levels. The deregulated miRNAs in the plasma of tumor-bearing (MIA PaCa-2) mice after 5 Gy irradiation of the tumor when compared with the plasma of tumor-bearing (MIA PaCa-2) mice receiving sham (0 Gy) irradiation are shown. Positive log2 fold changes indicate miRNAs upregulated in the plasma of irradiated mice compared to plasma of unirradiated mice whereas negative log2 fold changes indicate miRNAs downregulated in plasma of irradiated mice. Significant miRNAs were selected based on a log2 fold change ≥ |1| and an adjusted *p*-value of ≤0.05. Only transcripts with a base mean ≥50 were included.

In PDAC, the overexpression of miR-144-3p reduces cell migration, proliferation, and invasion [69]. Currently, an increasing number of miR-93 and miR-106b target genes have been identified, suggesting miR-93 and miR-106b may differentially affect the behavior of tumors. A study found that the upregulation of miR-93/106b enhances PD-L1 in response to irradiation [70]. It is known that the PD-1/PD-L1 pathway serves as a mechanism for tumors to evade an antigen-specific T cell immunologic response. However, pancreatic cancer patients have been shown to be mostly resistant to PD-1 inhibition [71].

The downregulation of the tumor-suppressive miR-451a leads to enhanced cancer cell migration and invasion in hypopharyngeal SCC [72]. miR-186 overexpression has been observed in PDAC and was shown to contribute to its invasive potential [73]. Different studies revealed upregulation of miR-17 in pancreatic cancer, leading to increased proliferation and invasion [74,75]. Furthermore, inhibition of miR-17 increased the sensitivity of pancreatic cancer cell lines to chemotherapy by upregulation of Bim [76]. miR-421 has been found to be upregulated in pancreatic cancer as an oncogene and potential regulator of DPC4/Smad4 [77].

miR-98, a member of the let-7 family, regulates many cellular biological responses, including cell migration and apoptosis, after irradiation. The radiation-induced inhibition of miR-98 is closely related to both the p53 and ATM signaling pathways [78,79]. In a recent publication, it was shown that downregulation of miR-98-5p and other members of the let-7 family leads to increased PDAC proliferation and metastasis by reversely regulating the MAP4K4 pathway [80]. In the study of Morimura and colleagues, miR-20a levels in the plasma of pancreatic cancer patients were increased compared with healthy donors [81]. It was shown that the expression of miR-20a promotes radioresistance in nasopharyngeal cancer cells [82]. Higher levels of miR-20a activate the PTEN/PI3K/Akt signaling pathway and induce radioresistance of HCC [83]. A recent study identified miR-103 as one of the most important miRNAs in a functional screen for DNA damage disrupting agents. The authors identified Rad51 and Rad51D as key targets of miR-103 in tumor cells [84].

Another in vitro study demonstrated that miR-142 has a regulatory relationship with HIF-1α. miR-142 inhibits pancreatic cancer cell proliferation and invasion, partly by targeting HIF-1α at its binding site [85]. An earlier study analyzed the role of miR-26a in pancreatic tissue by quantifying its expression levels in 106 PDAC tissue samples [86]. The authors found that miR-26a expression was downregulated in PDAC compared to adjacent normal tissue. miR-140 inhibits cell viability, proliferation, and invasion in PDAC [87]. One study showed that miR-16 can enhance radiation sensitivity by regulating the TLR1/NF-κB signaling pathway and act as a potential therapeutic approach to overcome radioresistance for lung cancer treatment [88]. Jiao et al. compared the miRNA expression profile in PDAC with benign cystic tumors to identify miRNAs deregulated during PDAC development [89]. Amongst others, let-7i and let-7d were downregulated in tissue samples of PDAC [89].

The only miRNA that was significantly upregulated after irradiation in our study was miR-184. In recent years, miR-184 has been extensively explored in various cancer types. miR-184 was not only found to be upregulated in pancreatic cancer but can also facilitate the proliferation and invasion of tumor cells while suppressing apoptosis [90].

2.3. miRNAs and Radioresistanc

miRNAs are well known to affect the radiosensitivity by modulating DNA damage repair, cell cycle checkpoints, apoptosis, signal transduction pathways, and the tumor microenvironment [91]. Both DNA double-strand repair pathways, the non-homologous end-joining (NHEJ) and homologous recombination (HR), are targeted by miRNAs. Furthermore, miRNAs can influence signaling pathways mediating radioresistance, such as PI3-K/Akt, NF-κB (nuclear factor-kappa B), MAPK (mitogen-activated protein kinase), and TGFβ (transforming growth factor-β). Prominent examples of miRNA targets affecting radioresistance are H2AX, BRCA1, ATM, DNA-PK, RAD51, Chk1, Cdc25, and p53 [91]. Overexpression of miR-138 and miR-24 was reported to reduce the expression of H2AX and subsequently diminishes DNA repair capacity [92,93]. miR-21 targets the cell cycle checkpoint gene Cdc25 and thereby modulates radiosensitivity [94]. miR-421 and miR-101 suppress ATM expression and sensitize tumor cells to radiation [95,96].

In the following section the role of the miRNAs—that were found to be down- or upregulated by irradiation (Figure 2)—in mediating radioresistance/radiosensitivity was investigated. Whereas five of these miRNAs contribute to increased radioresistance (miR-374b, miR-93, miR-20a, miR-106b, miR-140), nine miRNAs (miR-15b, miR-451, miR-186, miR-421, miR-98, miR-142, miR-26b, miR-16, let-7 family) mediate decreased radioresistance (Table 2). Two miRNAs (mir-144, miR-17) are controversially described in the literature concerning their effect on radioresistance. For miR-652, miR-103a, and miR-184, no data regarding their impact on the radioresponse have been found in the literature.

Table 2. miRNAs from Figure 2 and their association with radioresistance. miRNAs, their effect on radioresistance (increased or decreased), the entity, the identified target genes, and proteins/pathways.

miRNA from Figure 2	miRNA in Reference	Radio-Resistance	Entity (In Vitro/In Vivo) *	Targets #	Proteins/Pathways §	Ref.
miR-374b-5p	miR-374b-5p	increased	Canine oral melanoma (in vitro)	PTEN [1]		[97]
miR-15b-3p	miR-15b	decreased	Breast cancer (in vitro)	Chk1, Wee1 [1]		[98]
miR-144-5p/ miR-144-3p	miR-144-5p	decreased	NSCLC (in vitro/in vivo)	ATF2 [1]		[99]
	miR-144-3p	decreased	Glioblastoma (in vitro)	c-MET [1]	Phosphorylation of STAT3, ERK1/2, AKT, mTOR (all down) [4]	[100]
	miR-144	increased	Breast cancer (in vitro)		PTEN (down) [4] AKT, Snail, N-cadherin, Vimentin (all up) [4]	[101]
	miR-144	decreased	Prostate cancer (in vitro)	PIM1 [1]		[102]

Table 2. *Cont.*

miRNA from Figure 2	miRNA in Reference	Radio-Resistance	Entity (In Vitro/In Vivo) [*]	Targets [#]	Proteins/Pathways [§]	Ref.
miR-93-5p	miR-93-5p	increased	Colorectal cancer (in vitro)	FOXA1 [1]	TGFB3 (up) [4]	[103]
	miR-93	increased	Esophageal squamous carcinoma (in vitro)	BTG3 [1]		[104]
miR-451a	miR-451a	decreased	Mouse colorectal cancer (in vitro)	CAB39, EMSY, MEX3C, EREG [2]		[105]
	miR-451	decreased	NSCLC (in vitro)		PTEN (up) [4]	[106]
	miR-451	decreased	Lung adenocarcinoma (in vitro)	c-MYC [1]	Survivin, rad-51 (both down) [4]	[107]
	miR-451	decreased	Nasopharyngeal carcinoma (in vitro)	RAB14 [1]		[108]
miR-186-5p	miR-186	decreased	Nasopharyngeal carcinoma (in vitro)	FOXD1 [1]		[109]
miR-17-5p	miR-17-5p	decreased	Esophageal adenocarcinoma (in vitro)	PRKACB, C6orf120 [3]	PRKACB, C6orf120 (both down) [5]	[110]
	miR-17-5p	increased	Oral squamous cell carcinoma (in vitro/in vivo)		p21, p-p53, TNF RI, FADD (all down) [4] cIAP1, HIF-1α, TRAIL R1 (all up) [4]	[111]
miR-421	miR-421	decreased	Glioma (in vitro)	MEF2D [1]		[112]
	miR-421	decreased	Cervix carcinoma, NSCLC and SCCHN (in vitro)	ATM [1]		[113]
miR-98-5p	miR-98	decreased	Esophageal squamous cell carcinoma (in vitro)	BCL2 [1]		[114]
miR-20a-5p	miR-20a-5p	increased	Nasopharyngeal cancer (in vitro)	NPAS2 [1]	Notch pathway (down) [6]	[82]
miR-106b-5p	miR-106b	increased	Prostate cancer (in vitro)		p21 (down) [4]	[115]
miR-142-3p	miR-142-3p	decreased	Breast cancer (in vitro)		β-catenin (down) [4]	[116]
	miR-142-3p	decreased	Umbilical cord blood mononuclear cells (in vitro)	CD133 [1]		[117]
miR-26b-5p	miR-26b-5p	decreased	Hepatocellular carcinoma (in vitro)	EphA2 [1]		[118]
miR-140-5p	miR-140	increased	Lung fibroblasts (in vitro/in vivo)			[119]
miR-16-5p	miR-16-5p	decreased	Prostate cancer (in vitro)	Cyclin D1, Cyclin E1 [1]	pRb, E2F1 (both down) [4]	[120]
let 7i-5p let 7d-5p	let-7 family (let-7e)	decreased	Colorectal cancer (in vitro)		IGF-1R (down) [4]	[121]

[*] If not otherwise indicated, human. [#] Targets: all down-regulated by respective miRNA, [1] identified by luciferase assay, [2] identified by miR-TRAP assay, [3] predicted. [§] Proteins/pathways regulated by respective miRNA, [4] demonstrated by Western Blot, [5] demonstrated by RT-PCR, [6] demonstrated by signaling pathway assay.

Direct targets of the miRNAs that mediate radioresistance comprise PTEN, FOXA1, BTG3, and NPAS2. miRNAs that evoke enhanced radiosensitivity target Chk1, Wee1, ATF2, c-MET, PIM1, c-MYC, RAB14, FOXD1, MEF2D, ATM, BCL2, CD133, EphA2, Cyclin D1, and Cyclin E1. Several of these targets (Chk1, c-Myc, ATM, Bcl-2, Cyclins) are prominent key players in the DNA damage response, cell cycle control, and apoptosis [91]. Consequently, the down-regulation of these targets by the respective miRNA (miR-15b, miR-451, miR421, miR-98, miR-16) leads to increased radiosensitivity. However, for other miRNAs the interrelation between target gene expression and radioresistance/radiosensitivity is not evident. Therefore, more research in this field is essential to understand the function and regulation of these miRNAs and their targets in mediating radioresistance, and finally to implement the data in clinical routine to improve the effectiveness of RT.

2.4. miRNAs and Radioresistance in Pancreatic Cancer

For pancreatic cancer specifically, there are only a few publications about the interplay between miRNAs and radioresistance. The details of miRNA expression and radioresistance in pancreatic cancer are described below and summarized in Table 3.

Table 3. miRNAs associated with radioresistance in pancreatic cancer. miRNAs, their effect on radioresistance (reduced or increased), the used cell lines, the identified target genes, and the intervention are listed.

miRNA	Radioresistance	Cell Lines	Targets	Intervention	Ref.
miR-502	reduced	Mia PaCa-2, PaTuT, PaTu02	Ku70, XLF	miR-502 overexpression	[122]
miR-23b	reduced	Panc-1, BxPc3	ATG12	miR-23b mimic/inhibitor	[123]
miR-216a	reduced	Panc-1, BxPc3	beclin-1	miR-216a mimic	[124]
miR-99b	reduced	Panc-1, BxPc3, Capan-2	mTOR	miR-99b precursor/inhibitor	[61]
miR-181b	reduced	Panc-1, MIA PaCa-2	ETS (c-Met)	miR-181b precursor	[125]
miR-34	reduced	BxPc3, MIA PaCa-2	Bcl-2, Notch1-2	miR-34 mimic	[126]
miR-193a	increased	Panc-1, SW1990, AsPc-1	TGF-β2/TGF-βRIII, E2F6	miR-193a antagonist	[127]
miR-620	increased	MIA PaCa-2	HPGD	miR-620 mimic	[128]
Let-7a	reduced	AsPc-1	K-Ras	Lin28 siRNA (repressor of let-7a)	[129]
miR-374	unchanged/reduced	Panc-1, MIA PaCa-2	-	miR-374 overexpression	[130]

A recent paper showed that miR-502 overexpression increased the radiosensitivity of pancreatic cancer cell lines by targeting two proteins of the classical NHEJ repair pathway, Ku70 and XLF [122]. Mechanistically, miR-502 directly inhibits the DNA double-strand repair and also attenuates the cell cycle response upon radiation.

In radioresistant pancreatic cancer cell lines that were established by repeated exposure to radiation, miR-23b expression was reduced compared to the parental cell lines [123]. Overexpression of miR-23b rendered the radioresistant cells more sensitive to radiation. Furthermore, the expression of ATG12 (Ubiquitin-like protein), a target of miR-23b, was increased in the radioresistant cells. As ATG12 is involved in vesicle formation during autophagy, enhanced ATG12 expression increased autophagy and, subsequently, radioresistance. Furthermore, an association between decreased ATG12 expression and elevated miR-23b levels was observed in human pancreatic cancer tissue, indicating that miR-23b might affect autophagy activity in pancreatic cancer cells [123]. Another study showed that miR-216a can render radioresistant pancreatic cancer cells radiosensitive by inhibiting beclin-1-mediated autophagy [124]. Additionally, in a xenograft mouse model, miR-216a expression reduced the tumor growth after irradiation.

Tomihara et al. observed that high c-Met expression is associated with lower PFS and OS in pancreatic cancer patients receiving preoperative radiochemotherapy [125]. Further investigations revealed that miR-181b is downregulated in radioresistant pancreatic cancer cell lines leading to the upregulation of the transcription factor ETS1 and the c-Met pathway [125].

mTOR expression and subsequent mTOR signaling pathways have been upregulated after RT in human pancreatic cancer biopsies [61]. In vitro, mTOR was also upregulated upon radiation and mTOR expression has been shown to contribute to radioresistance. As mTOR is a miR-99b target, down-regulation of miR-99b enhanced mTOR expression and radioresistance in vitro [61].

Increased miR-193a levels have been found in irradiated dying pancreatic cancer cells, leading to elevated miR-193a levels in surviving cells [127]. Furthermore, miR-193a promotes proliferation and repopulation of the surviving cells via TGF-β2/TGF-βRIII signaling [127]. In patient-derived xenograft mouse models, radiation in combination with miR-193 antagonist inhibited cell repopulation and metastasis, and improved the survival. These data suggest that inhibition of miR-193 might increase radiosensitivity.

Transfection with a miR-620 mimic increased the radioresistance of MIA PaCa-2 cells and also of breast and prostate cancer cell lines [128]. The tumor suppressor gene HPGD (15-hydroxyprostaglandin dehydrogenase) was identified as a target of miR-620. miR-620 overexpression leads to degradation of HPGD, enhanced prostaglandin E2 (PGE2) levels and signaling through the EP2 receptor promoting the survival and radioresistance of tumor cells.

Pancreatic cancer cell lines have reduced miR-34a expression compared to normal pancreatic ductal epithelial cells [131]. p53 has been shown to directly regulate miR-34 family members, and subsequently, the expression of miR-34 is strongly reduced in p53-mutant cancer cells [132]. miR-34 restoration in p53-mutant pancreatic cancer cells, which have initially low miR-34 levels leads to the downregulation of anti-apoptotic Bcl-2 and the Notch signaling pathway, which is involved in the maintenance of cancer stem cells [126,132,133]. Furthermore, miR-34 restoration reduces clonogenic cell growth and enhances radio- and chemosensitivity [126]. Therefore, miR-34 may restore the tumor-suppressing function of p53 in p53-deficient human pancreatic cancer cells and might constitute a new approach to treating p53-mutated pancreatic cancer.

Inhibition of Lin28 by siRNA abrogates posttranscriptional degradation of let-7a and increases radiosensitivity of AsPc-1 cells presumably through down-regulation of Kras expression [129].

Baek et al. investigated both the influence of miRNA expression on the radiosensitivity to gamma-rays and to carbon ion beam RT [130]. Overexpression of miR-374 did not affect the sensitivity to conventional gamma-ray radiation, but the sensitivity to carbon ion beam radiation was enhanced. These data suggest that miR-374 could be a potential radiosensitizer for carbon ion RT.

In summary, only a few miRNAs associated with radioresistance in pancreatic cancer have been identified so far. Therefore, comprehensive studies analyzing miRNA expression profiles are necessary to identify miRNAs, which can predict the individual radiosensitivity or constitute targets for radiosensitization.

3. Discussion

Nowadays, early diagnosis is essential for the successful treatment of pancreatic cancer, and surgical resection is the only potentially curative therapy [3]. Since pancreatic cancer is a locally invasive as well as a systemic disease, the disease recurs in most patients, which leads to a five-year survival rate of only 30% after complete resection [134]. miRNAs could help to detect pancreatic cancer in an earlier stage, leading to increased curative treatments. In addition, more research into neoadjuvant treatment options, including RT or combined CRT, is urgently needed. The goal of RT is mainly to downsize the tumor to enable a secondary resection. Therefore, neoadjuvant RT may lead to improved long-term prognosis in patients with borderline and primary non-resectable, locally advanced pancreatic cancer [4]. While RT was historically a central treatment component, its role has been called into question based on the publication of clinical trials with conflicting results and due to the effectiveness of combined chemotherapeutic regimes (e.g., FOLFIRINOX) [135].

Novel molecular biomarkers are required to increase the efficacy of RT and to define which patients benefit most from it [21]. Especially, miRNAs could have the potential as novel biomarkers for diagnosis, monitoring of recurrence, and predicting prognosis and survival of patients with pancreatic cancer [22,136]. Our present study provides an identification of miRNAs in an in vivo pancreatic cancer mouse model. A panel of seven miRNAs was significantly upregulated in plasma of tumor-bearing mice (Figure 1). In the next step, this biomarker panel should be validated in plasma of pancreatic cancer patients. Especially, circulating miRNAs represent a valuable source and are easily accessible [26].

Several miRNAs have been shown to play a role in radioresistance in different tumor entities [56,91]. Furthermore, significant changes in the expression levels of miRNAs can be observed after exposure to IR. It is known that miRNAs are responsible for regulating almost every cellular pathway, including the DNA damage response after IR [57]. Understanding the regulation and function of miRNAs is essential to improve the effectiveness of RT. To determine radiation-induced miRNA changes in the plasma of tumor-bearing mice, local irradiation of subcutaneous MIA PaCa-2 tumors was performed. Small RNA sequencing revealed 21 miRNAs that were significantly regulated (Figure 2). Interestingly, 20 out of these 21 miRNAs were downregulated. Although a mechanistic basis is not yet available, the reduced levels can be assumed to promote translation of specific miRNA target proteins [137]. A current literature search reveals an association between the expression of these miRNAs, their targets (e.g., Chk1, c-Myc, ATM and Cyclins) and either radioresistance or radiosensitivity (Table 2) [98,107,113,120]. Most of these miRNAs (miR-15b, miR-451, miR-186, miR-421, miR-98, miR-142, miR-26b, miR-16, let-7 family) mediate radiosensitivity [98,105–109,112–114,116–118,120,121] suggesting that the radiation-induced downregulation of these miRNAs might enhance radioresistance.

Specifically, for pancreatic cancer, only very few publications about the interplay between miRNAs and radioresistance exist, which were reviewed in this manuscript (Table 3).

Although radioresistance is rather complex, emerging evidence has demonstrated that epigenetic alterations, including miRNA changes, play important roles in resistance to RT. Therefore, more comprehensive preclinical and clinical studies analyzing miRNA expression profiles in context with radioresistance are urgently needed to identify miRNAs that can predict the individual radiosensitivity. Moreover, further studies are required to confirm the molecular mechanisms of the deregulated circulating miRNAs as diagnostic biomarkers for pancreatic cancer. In addition, miRNAs can help to develop innovative cancer therapies, which is of great significance for improving the life expectancy of pancreatic cancer patients [138].

The possibility to detect miRNAs in blood samples makes them a promising tool for rapid PDAC diagnosis, as well as patient stratification, to allow personalized therapy [139]. However, miRNAs are not yet routinely used for cancer management, and we should consider the challenges behind this strategy. One of them is the need to standardize protocols [140]. From sample collection and preparation to miRNA detection, inter-laboratory reproducibility must be ensured. In this regard, sampling tubes, patient-related factors, sample storage, and sample processing can affect RNA yield and purity [141]. The high heterogeneity of PDAC should also be regarded as an additional challenge, as it makes it harder to select reliable biomarkers [139]. Concerning miRNAs applications in RT, it should also be considered that there is still insufficient knowledge about their functions in different RT concepts such as dose fractionation. Information is also lacking regarding the use of alternative radiation qualities such as protons or carbon ions and the tumor microenvironment such as hypoxia. Additionally, the translation of preclinical data from in vitro or xenograft models into clinical routine using patient materials is still poorly realized [137]. Preclinical miRNA candidates will have to be investigated in large prospective trials of different PDAC patient populations to gain confidence in the selected biomarkers [142]. Future trials should be designed to evaluate miRNA biomarkers in a temporal manner, by collecting samples before and after RT, leading us towards personalized RT for PDAC [142].

4. Conclusions

In the last decade, miRNAs have become promising tools as prognostic and diagnostic biomarkers as well as therapeutic targets for innovative and personalized cancer treatment. Several miRNAs have been found differentially expressed and to be predictive for the treatment outcome in multiple cancer entities. With a five-year survival rate of less than 7%, pancreatic cancer presents an urgent need to identify the subset of patients which could benefit from treatment intensification and to establish novel individualized CRT treatment options. In this article, we focused on the role of miRNAs in pancreatic cancer, radioresistance, and radiation-induced changes, which could lead the way to personalized treatment in the future.

Supplementary Materials: The following are available online at http://www.mdpi.com/2072-6694/12/12/3703/s1, Supplementary Methods, Table S1: miRNAs significantly upregulated in the plasma of tumor-bearing mice, Table S2: miRNAs significantly down- or upregulated in plasma of tumor-bearing mice after irradiation with 5 Gy.

Author Contributions: Conceptualization, project administration, and funding acquisition S.E.C.; Writing—original draft preparation: L.N., D.S., S.D., S.R., M.S.F., T.E.S., Methodology and investigation: L.N., D.B., M.W.P. All authors have read and agreed to the published version of the manuscript.

Funding: This work was funded in part by the Deutsche Forschungsgemeinschaft (DFG, German Research Foundation) Projektnummer 329628492—SFB 1321 Project 15 and by "Translational & Clinical Projects", Helmholtz Zentrum München, the German Consortium for Translational Cancer Research, Munich/TUM site and the Medical Faculty of TUM. S.D. acknowledges the "Hans und Klementia Langmatz Stiftung" and the KKF, Medical Faculty of the Technical University of Munich (TUM), for research support.

Acknowledgments: The graphical abstract was created with BioRender.com.

Conflicts of Interest: The authors declare no conflict of interest.

References

1. Rahib, L.; Smith, B.D.; Aizenberg, R.; Rosenzweig, A.B.; Fleshman, J.M.; Matrisian, L.M. Projecting Cancer Incidence and Deaths to 2030: The Unexpected Burden of Thyroid, Liver, and Pancreas Cancers in the United States. *Cancer Res.* **2014**, *74*, 2913–2921. [CrossRef] [PubMed]
2. Kleeff, J.; Korc, M.; Apte, M.; La Vecchia, C.; Johnson, C.D.; Biankin, A.V.; Neale, R.E.; Tempero, M.; Tuveson, D.A.; Hruban, R.H.; et al. Pancreatic cancer. *Nat. Rev. Dis. Prim.* **2016**, *2*, 16022. [CrossRef] [PubMed]
3. Kandel, P.; Wallace, M.B.; Stauffer, J.; Bolan, C.; Raimondo, M.; Woodward, T.A.; Gomez, V.; Ritter, A.W.; Asbun, H.; Mody, K. Survival of Patients with Oligometastatic Pancreatic Ductal Adenocarcinoma Treated with Combined Modality Treatment Including Surgical Resection: A Pilot Study. *J. Pancreat. Cancer* **2018**, *4*, 88–94. [CrossRef] [PubMed]
4. Combs, S.E.; Habermehl, D.; Kessel, K.A.; Bergmann, F.; Werner, J.; Naumann, P.; Jäger, D.; Büchler, M.W.; Debus, J. Prognostic Impact of CA 19-9 on Outcome after Neoadjuvant Chemoradiation in Patients with Locally Advanced Pancreatic Cancer. *Ann. Surg. Oncol.* **2014**, *21*, 2801–2807. [CrossRef]
5. Gillen, S.; Schuster, T.; Büschenfelde, C.M.Z.; Friess, H.; Kleeff, J. Preoperative/Neoadjuvant Therapy in Pancreatic Cancer: A Systematic Review and Meta-analysis of Response and Resection Percentages. *PLoS Med.* **2010**, *7*, e1000267. [CrossRef]
6. Dobiasch, S.; Fietkau, R.; Goerig, N.L.; Combs, S.E. Essential role of radiation therapy for the treatment of pancreatic cancer. *Strahlenther. Onkol.* **2017**, *194*, 185–195. [CrossRef]
7. Chauffert, B.; Mornex, F.; Bonnetain, F.; Rougier, P.; Mariette, C.; Bouché, O.; Bosset, J.F.; Aparicio, T.; Mineur, L.; Azzedine, A.; et al. Phase III trial comparing intensive induction chemoradiotherapy (60 Gy, infusional 5-FU and intermittent cisplatin) followed by maintenance gemcitabine with gemcitabine alone for locally advanced unresectable pancreatic cancer. Definitive results of the 2000–01 FFCD/SFRO study. *Ann. Oncol.* **2008**, *19*, 1592–1599. [CrossRef]
8. Hammel, P.; Huguet, F.F.; Van Laethem, J.-L.; Goldstein, D.D.; Glimelius, B.; Artru, P.P.; Borbath, I.; Bouché, O.; Shannon, J.J.; André, T.; et al. Effect of Chemoradiotherapy vs Chemotherapy on Survival in Patients With Locally Advanced Pancreatic Cancer Controlled after 4 Months of Gemcitabine with or without Erlotinib. *JAMA* **2016**, *315*, 1844–1853. [CrossRef]
9. Loehrer, P.J., Sr.; Feng, Y.; Cardenes, H.; Wagner, L.; Brell, J.M.; Cella, D.; Flynn, P.; Ramanathan, R.K.; Crane, C.H.; Alberts, S.R.; et al. Gemcitabine Alone Versus Gemcitabine Plus Radiotherapy in Patients With Locally Advanced Pancreatic Cancer: An Eastern Cooperative Oncology Group Trial. *J. Clin. Oncol.* **2011**, *29*, 4105–4112. [CrossRef]
10. Brunner, M.; Wu, Z.; Krautz, C.; Pilarsky, C.; Grützmann, R.; Weber, G.F. Current Clinical Strategies of Pancreatic Cancer Treatment and Open Molecular Questions. *Int. J. Mol. Sci.* **2019**, *20*, 4543. [CrossRef]
11. Moertel, C.G.; Frytak, S.; Hahn, R.G.; O'Connell, M.J.; Reitemeier, R.J.; Rubin, J.; Schutt, A.J.; Weiland, L.H.; Childs, D.S.; Holbrook, M.A.; et al. Therapy of locally unresectable pancreatic carcinoma: A randomized comparison of high dose (6000 rads) radiation alone, moderate dose radiation (4000 rads + 5-fluorouracil), and high dose radiation + 5-fluorouracil. The gastrointestinal tumor study group. *Cancer* **1981**, *48*, 1705–1710. [CrossRef]
12. Mukherjee, S.; Hurt, C.N.; Bridgewater, J.; Falk, S.; Cummins, S.; Wasan, H.; Crosby, T.; Jephcott, C.; Roy, R.; Radhakrishna, G.; et al. Gemcitabine-based or capecitabine-based chemoradiotherapy for locally advanced pancreatic cancer (SCALOP): A multicentre, randomised, phase 2 trial. *Lancet Oncol.* **2013**, *14*, 317–326. [CrossRef]
13. Suker, M.; Nuyttens, J.J.; Koerkamp, B.G.; Eskens, F.A.L.M.; Van Eijck, C.H. FOLFIRINOX and radiotherapy for locally advanced pancreatic cancer: A cohort study. *J. Surg. Oncol.* **2018**, *118*, 1021–1026. [CrossRef] [PubMed]
14. Mancini, B.R.; Stein, S.; Lloyd, S.; Rutter, C.E.; James, E.; Chang, B.W.; Lacy, J.; Johung, K.L. Chemoradiation after FOLFIRINOX for borderline resectable or locally advanced pancreatic cancer. *J. Gastrointest. Oncol.* **2018**, *9*, 982–988. [CrossRef] [PubMed]
15. Murphy, J.E.; Wo, J.Y.; Ryan, D.P.; Jiang, W.; Yeap, B.Y.; Drapek, L.C.; Blaszkowsky, L.S.; Kwak, E.L.; Allen, J.N.; Clark, J.W.; et al. Total Neoadjuvant Therapy with FOLFIRINOX Followed by Individualized Chemoradiotherapy for Borderline Resectable Pancreatic Adenocarcinoma. *JAMA Oncol.* **2018**, *4*, 963–969. [CrossRef]

16. Tran, N.H.; Sahai, V.; Griffith, K.A.; Nathan, H.; Kaza, R.; Cuneo, K.C.; Shi, J.; Kim, E.; Sonnenday, C.J.; Cho, C.S.; et al. Phase 2 Trial of Neoadjuvant FOLFIRINOX and Intensity Modulated Radiation Therapy Concurrent With Fixed-Dose Rate-Gemcitabine in Patients With Borderline Resectable Pancreatic Cancer. *Int. J. Radiat. Oncol.* **2020**, *106*, 124–133. [CrossRef]

17. Katz, M.H.G.; Ou, F.S.; Herman, J.M.; Ahmad, S.A.; Wolpin, B.; Marsh, R.; Behr, S.; Shi, Q.; Chuong, M.; Schwartz, L.H.; et al. Alliance for clinical trials in oncology (ALLIANCE) trial A021501: Preoperative extended chemotherapy vs. chemotherapy plus hypofractionated radiation therapy for borderline resectable adenocarcinoma of the head of the pancreas. *BMC Cancer* **2017**, *17*, 505. [CrossRef]

18. Ng, S.P.; Koay, E.J. Current and emerging radiotherapy strategies for pancreatic adenocarcinoma: Stereotactic, intensity modulated and particle radiotherapy. *Ann. Pancreat. Cancer* **2018**, *1*, 22. [CrossRef]

19. Neesse, A.; Bauer, C.A.; Öhlund, D.; Lauth, M.; Buchholz, M.; Michl, P.; Tuveson, D.; Gress, T.M. Stromal biology and therapy in pancreatic cancer: Ready for clinical translation? *Gut* **2019**, *68*, 159–171. [CrossRef]

20. Wu, X.; Tang, W.; Marquez, R.T.; Li, K.; Highfill, C.A.; He, F.; Lian, J.; Lin, J.; Fuchs, J.R.; Ji, M.; et al. Overcoming chemo/radio-resistance of pancreatic cancer by inhibiting STAT3 signaling. *Oncotarget* **2016**, *7*, 11708–11723. [CrossRef]

21. Wang, F.; Xia, X.; Yang, C.; Shen, J.; Mai, J.; Kim, H.-C.; Kirui, D.; Kang, Y.; Fleming, J.B.; Koay, E.J.; et al. SMAD4Gene Mutation Renders Pancreatic Cancer Resistance to Radiotherapy through Promotion of Autophagy. *Clin. Cancer Res.* **2018**, *24*, 3176–3185. [CrossRef] [PubMed]

22. Su, Q.; Zhu, E.C.; Qu, Y.-L.; Wang, D.-Y.; Qu, W.-W.; Zhang, C.; Wu, T.; Gao, Z.-H. Serum level of co-expressed hub miRNAs as diagnostic and prognostic biomarkers for pancreatic ductal adenocarcinoma. *J. Cancer* **2018**, *9*, 3991–3999. [CrossRef] [PubMed]

23. Quattrochi, B.; Gulvady, A.; Driscoll, D.R.; Sano, M.; Klimstra, D.S.; Turner, C.E.; Lewis, B.C. MicroRNAs of the mir-17~92 cluster regulate multiple aspects of pancreatic tumor development and progression. *Oncotarget* **2017**, *8*, 35902–35918. [CrossRef] [PubMed]

24. Vidigal, J.A.; Ventura, A. The biological functions of miRNAs: Lessons from in vivo studies. *Trends Cell Biol.* **2015**, *25*, 137–147. [CrossRef]

25. Royam, M.M.; Ramesh, N.; Shanker, R.; Sabarimurugan, S.; Kumarasamy, C.; Muthukaliannan, G.K.; Baxi, S.; Gupta, A.; Krishnan, S.; Jayaraj, R. miRNA Predictors of Pancreatic Cancer Chemotherapeutic Response: A Systematic Review and Meta-Analysis. *Cancers* **2019**, *11*, 900. [CrossRef]

26. Fesler, A.; Ju, J. Development of microRNA-based therapy for pancreatic cancer. *J. Pancreatol.* **2019**, *2*, 147–151. [CrossRef]

27. Chhatriya, B.; Mukherjee, M.; Ray, S.K.; Sarkar, P.; Chatterjee, S.; Nath, D.; Das, K.; Goswami, S. Comparison of tumour and serum specific microRNA changes dissecting their role in pancreatic ductal adenocarcinoma: A meta-analysis. *BMC Cancer* **2019**, *19*, 1175. [CrossRef]

28. Ouyang, H.; Gore, J.; Deitz, S.; Korc, M. microRNA-10b enhances pancreatic cancer cell invasion by suppressing TIP30 expression and promoting EGF and TGF-β actions. *Oncogene* **2013**, *33*, 4664–4674. [CrossRef]

29. Qu, K.; Zhang, X.; Lin, T.; Liu, T.; Wang, Z.; Liu, S.; Zhou, L.; Wei, J.; Chang, H.; Li, K.; et al. Circulating miRNA-21-5p as a diagnostic biomarker for pancreatic cancer: Evidence from comprehensive miRNA expression profiling analysis and clinical validation. *Sci. Rep.* **2017**, *7*, 1692. [CrossRef]

30. Vila-Navarro, E.; Duran-Sanchon, S.; Vila-Casadesús, M.; Moreira, L.; Ginès, À.; Cuatrecasas, M.; Lozano, J.J.; Bujanda, L.; Castells, A.; Gironella, M. Novel Circulating miRNA Signatures for Early Detection of Pancreatic Neoplasia. *Clin. Transl. Gastroenterol.* **2019**, *10*, e00029. [CrossRef]

31. Karasek, P.; Gablo, N.; Hlavsa, J.; Kiss, I.; Vychytilova-Faltejskova, P.; Hermanova, M.; Kala, Z.; Slaby, O.; Prochazka, V. Pre-operative Plasma miR-21-5p Is a Sensitive Biomarker and Independent Prognostic Factor in Patients with Pancreatic Ductal Adenocarcinoma Undergoing Surgical Resection. *Cancer Genom. Proteom.* **2018**, *15*, 321–327. [CrossRef] [PubMed]

32. Kawaguchi, T.; Komatsu, S.; Ichikawa, D.; Morimura, R.; Tsujiura, M.; Konishi, H.; Takeshita, H.; Nagata, H.; Arita, T.; Hirajima, S.; et al. Clinical impact of circulating miR-221 in plasma of patients with pancreatic cancer. *Br. J. Cancer* **2013**, *108*, 361–369. [CrossRef] [PubMed]

33. Lemberger, M.; Loewenstein, S.; Lubezky, N.; Nizri, E.; Pasmanik-Chor, M.; Barazovsky, E.; Klausner, J.M.; Lahat, G. MicroRNA profiling of pancreatic ductal adenocarcinoma (PDAC) reveals signature expression related to lymph node metastasis. *Oncotarget* **2019**, *10*, 2644–2656. [CrossRef] [PubMed]

34. Wolfe, A.R.; Wald, P.; Webb, A.; Sebastian, N.; Walston, S.; Robb, R.; Chen, W.; Vedaie, M.; Dillhoff, M.; Frankel, W.L.; et al. A microRNA-based signature predicts local-regional failure and overall survival after pancreatic cancer resection. *Oncotarget* **2020**, *11*, 913–923. [CrossRef]

35. Schmiegel, W.-H.; Kreiker, C.; Eberl, W.; Arndt, R.; Classen, M.; Greten, H.; Jessen, K.; Kalthoff, H.; Soehendra, N.; Thiele, H.-G. Monoclonal antibody defines CA 19-9 in pancreatic juices and sera. *Gut* **1985**, *26*, 456–460. [CrossRef]

36. Liu, J.; Gao, J.; Du, Y.; Li, Z.; Ren, Y.; Gu, J.; Wang, X.; Gong, Y.; Wang, W.; Kong, X. Combination of plasma microRNAs with serum CA19-9 for early detection of pancreatic cancer. *Int. J. Cancer* **2012**, *131*, 683–691. [CrossRef]

37. LaConti, J.J.; Shivapurkar, N.; Preet, A.; Mays, A.D.; Peran, I.; Kim, S.E.; Marshall, J.L.; Riegel, A.T.; Wellstein, A. Tissue and Serum microRNAs in the KrasG12D Transgenic Animal Model and in Patients with Pancreatic Cancer. *PLoS ONE* **2011**, *6*, e20687. [CrossRef]

38. Mazza, T.; Copetti, M.; Capocefalo, D.; Fusilli, C.; Biagini, T.; Carella, M.; De Bonis, A.; Mastrodonato, N.; Piepoli, A.; Pazienza, V.; et al. MicroRNA co-expression networks exhibit increased complexity in pancreatic ductal compared to Vater's papilla adenocarcinoma. *Oncotarget* **2017**, *8*, 105320–105339. [CrossRef]

39. Zhou, C.; Lu, Y.; Li, X. miR-339-3p inhibits proliferation and metastasis of colorectal cancer. *Oncol. Lett.* **2015**, *10*, 2842–2848. [CrossRef]

40. Xin, L.; Gao, J.; Wang, D.; Lin, J.-H.; Liao, Z.; Ji, J.-T.; Du, T.-T.; Jiang, F.; Hu, L.-H.; Li, Z. Novel blood-based microRNA biomarker panel for early diagnosis of chronic pancreatitis. *Sci. Rep.* **2017**, *7*, 40019. [CrossRef]

41. Liu, X.; Xu, X.; Pan, B.; He, B.; Chen, X.; Zeng, K.; Xu, M.; Pan, Y.; Sun, H.; Xu, T.; et al. Circulating miR-1290 and miR-320d as Novel Diagnostic Biomarkers of Human Colorectal Cancer. *J. Cancer* **2019**, *10*, 43–50. [CrossRef] [PubMed]

42. Li, W.; Ding, X.; Wang, S.; Xu, L.; Yin, T.; Han, S.; Geng, J.; Sun, W. Downregulation of serum exosomal miR-320d predicts poor prognosis in hepatocellular carcinoma. *J. Clin. Lab. Anal.* **2020**, *34*, e23239. [CrossRef] [PubMed]

43. Long, M.; Zhan, M.; Xu, S.; Yang, R.; Chen, W.; Zhang, S.; Shi, Y.; Yongheng, S.; Mohan, M.; Liu, Q.; et al. miR-92b-3p acts as a tumor suppressor by targeting Gabra3 in pancreatic cancer. *Mol. Cancer* **2017**, *16*, 167. [CrossRef] [PubMed]

44. Tang, S.; Cheng, J.; Yao, Y.; Lou, C.; Wang, L.; Huang, X.; Zhang, Y. Combination of Four Serum Exosomal MiRNAs as Novel Diagnostic Biomarkers for Early-Stage Gastric Cancer. *Front. Genet.* **2020**, *11*, 237. [CrossRef]

45. Li, Q.; Li, Z.; Wei, S.; Wang, W.; Chen, Z.; Zhang, L.; Chen, L.; Li, B.; Sun, G.; Xu, J.; et al. Overexpression of miR-584-5p inhibits proliferation and induces apoptosis by targeting WW domain-containing E3 ubiquitin protein ligase 1 in gastric cancer. *J. Exp. Clin. Cancer Res.* **2017**, *36*, 59. [CrossRef]

46. Zhou, X.; Wen, W.; Shan, X.; Zhu, W.; Xu, J.; Guo, R.; Cheng, W.; Wang, F.; Qi, L.-W.; Chen, Y.; et al. A six-microRNA panel in plasma was identified as a potential biomarker for lung adenocarcinoma diagnosis. *Oncotarget* **2017**, *8*, 6513–6525. [CrossRef]

47. Ni, J.; Zheng, H.; Huang, Z.; Hong, Y.; Ou, Y.; Tao, Y.; Wang, M.; Wang, Z.; Yang, Y.; Zhou, W. MicroRNA-197-3p acts as a prognostic marker and inhibits cell invasion in hepatocellular carcinoma. *Oncol. Lett.* **2018**, *17*, 2317–2327. [CrossRef]

48. Shimomura, A.; Shiino, S.; Kawauchi, J.; Takizawa, S.; Sakamoto, H.; Matsuzaki, J.; Ono, M.; Takeshita, F.; Niida, S.; Shimizu, C.; et al. Novel combination of serum microRNA for detecting breast cancer in the early stage. *Cancer Sci.* **2016**, *107*, 326–334. [CrossRef]

49. Xu, Y.-F.; Hannafon, B.N.; Zhao, Y.D.; Postier, R.G.; Ding, W.-Q. Plasma exosome miR-196a and miR-1246 are potential indicators of localized pancreatic cancer. *Oncotarget* **2017**, *8*, 77028–77040. [CrossRef]

50. Wei, J.; Yang, L.; Wu, Y.-N.; Xu, J. Serum miR-1290 and miR-1246 as Potential Diagnostic Biomarkers of Human Pancreatic Cancer. *J. Cancer* **2020**, *11*, 1325–1333. [CrossRef]

51. Madhavan, B.; Yue, S.; Galli, U.; Rana, S.; Gross, W.; Müller, M.; Giese, N.A.; Kalthoff, H.; Becker, T.; Büchler, M.W.; et al. Combined evaluation of a panel of protein and miRNA serum-exosome biomarkers for pancreatic cancer diagnosis increases sensitivity and specificity. *Int. J. Cancer* **2015**, *136*, 2616–2627. [CrossRef] [PubMed]

52. Shi, Y.; Wang, Z.; Zhu, X.; Chen, L.; Ma, Y.; Wang, J.; Yang, X.-Z.; Liu, Z. Exosomal miR-1246 in serum as a potential biomarker for early diagnosis of gastric cancer. *Int. J. Clin. Oncol.* **2019**, *25*, 89–99. [CrossRef]

53. Takeshita, N.; Hoshino, I.; Mori, M.; Akutsu, Y.; Hanari, N.; Yoneyama, Y.; Ikeda, N.; Isozaki, Y.; Maruyama, T.; Akanuma, N.; et al. Serum microRNA expression profile: miR-1246 as a novel diagnostic and prognostic biomarker for oesophageal squamous cell carcinoma. *Br. J. Cancer* **2013**, *108*, 644–652. [CrossRef] [PubMed]

54. Nagamitsu, Y.; Nishi, H.; Sasaki, T.; Takaesu, Y.; Terauchi, F.; Isaka, K. Profiling analysis of circulating microRNA expression in cervical cancer. *Mol. Clin. Oncol.* **2016**, *5*, 189–194. [CrossRef] [PubMed]

55. Czochor, J.R.; Glazer, P.M. microRNAs in Cancer Cell Response to Ionizing Radiation. *Antioxid. Redox Signal.* **2014**, *21*, 293–312. [CrossRef]

56. Chaudhry, M.A. Radiation-induced microRNA: Discovery, functional analysis, and cancer radiotherapy. *J. Cell. Biochem.* **2014**, *115*, 436–449. [CrossRef] [PubMed]

57. Mao, A.; Liu, Y.; Zhang, H.; Di, C.; Sun, C. microRNA Expression and Biogenesis in Cellular Response to Ionizing Radiation. *DNA Cell Biol.* **2014**, *33*, 667–679. [CrossRef]

58. Zhang, X.; Wan, G.; Berger, F.G.; He, X.; Lu, X. The ATM Kinase Induces MicroRNA Biogenesis in the DNA Damage Response. *Mol. Cell* **2011**, *41*, 371–383. [CrossRef]

59. He, L.; He, X.; Lim, L.P.; De Stanchina, E.; Xuan, Z.; Liang, Y.; Xue, W.; Zender, L.; Magnus, J.F.; Ridzon, D.; et al. A microRNA component of the p53 tumour suppressor network. *Nat. Cell Biol.* **2007**, *447*, 1130–1134. [CrossRef]

60. Saleh, A.D.; Savage, J.E.; Cao, L.; Soule, B.P.; Ly, D.; DeGraff, W.; Harris, C.C.; Mitchell, J.B.; Simone, N.L. Cellular Stress Induced Alterations in MicroRNA let-7a and let-7b Expression Are Dependent on p53. *PLoS ONE* **2011**, *6*, e24429. [CrossRef]

61. Wei, F.; Liu, Y.; Guo, Y.; Xiang, A.; Wang, G.-Y.; Xue, X.; Lu, Z. miR-99b-targeted mTOR induction contributes to irradiation resistance in pancreatic cancer. *Mol. Cancer* **2013**, *12*, 81. [CrossRef] [PubMed]

62. Jiang, M.-J.; Chen, Y.-Y.; Dai, J.-J.; Gu, D.-N.; Mei, Z.; Liu, F.-R.; Huang, Q.; Tian, L. Dying tumor cell-derived exosomal miR-194-5p potentiates survival and repopulation of tumor repopulating cells upon radiotherapy in pancreatic cancer. *Mol. Cancer* **2020**, *19*, 1–15. [CrossRef] [PubMed]

63. Małachowska, B.; Tomasik, B.; Stawiski, K.; Kulkarni, S.; Guha, C.; Chowdhury, D.; Fendler, W. Circulating microRNAs as Biomarkers of Radiation Exposure: A Systematic Review and Meta-Analysis. *Int. J. Radiat. Oncol.* **2020**, *106*, 390–402. [CrossRef]

64. Farsinejad, S.; Rahaie, M.; Alizadeh, A.M.; Mir-Derikvand, M.; Gheisary, Z.; Nosrati, H.; Khalighfard, S. Expression of the circulating and the tissue microRNAs after surgery, chemotherapy, and radiotherapy in mice mammary tumor. *Tumor Biol.* **2016**, *37*, 14225–14234. [CrossRef]

65. Long, Z.-W.; Wu, J.-H.; Hong, C.; Wang, Y.-N.; Zhou, Y. MiR-374b Promotes Proliferation and Inhibits Apoptosis of Human GIST Cells by Inhibiting PTEN through Activation of the PI3K/Akt Pathway. *Mol. Cells* **2018**, *41*, 532–544.

66. Sun, W.; Lan, J.; Chen, L.; Qiu, J.; Luo, Z.; Li, M.; Wang, J.; Zhao, J.; Zhang, T.; Long, X.; et al. A mutation in porcine pre-miR-15b alters the biogenesis of MiR-15b\16-1 cluster and strand selection of MiR-15b. *PLoS ONE* **2017**, *12*, e0178045. [CrossRef]

67. Zhang, W.-L.; Zhang, J.; Wu, X.-Z.; Yan, T.; Lv, W. miR-15b promotes epithelial-mesenchymal transition by inhibiting SMURF2 in pancreatic cancer. *Int. J. Oncol.* **2015**, *47*, 1043–1053. [CrossRef]

68. Yang, W.; Zhou, C.; Luo, M.; Shi, X.; Li, Y.; Sun, Z.; Zhou, F.; Chen, Z.; He, J. MiR-652-3p is upregulated in non-small cell lung cancer and promotes proliferation and metastasis by directly targeting Lgl1. *Oncotarget* **2016**, *7*, 16703–16715. [CrossRef]

69. Liu, S.; Luan, J.; Ding, Y. miR-144-3p Targets FosB Proto-oncogene, AP-1 Transcription Factor Subunit (FOSB) to Suppress Proliferation, Migration, and Invasion of PANC-1 Pancreatic Cancer Cells. *Oncol. Res. Featur. Preclin. Clin. Cancer Ther.* **2018**, *26*, 683–690. [CrossRef]

70. Cioffi, M.; Trabulo, S.M.; Vallespinos, M.; Raj, D.; Kheir, T.B.; Lin, M.-L.; Begum, J.; Baker, A.-M.; Amgheib, A.; Saif, J.; et al. The miR-25-93-106b cluster regulates tumor metastasis and immune evasion via modulation of CXCL12 and PD-L1. *Oncotarget* **2017**, *8*, 21609–21625. [CrossRef]

71. Brahmer, J.R.; Tykodi, S.S.; Chow, L.Q.M.; Hwu, W.-J.; Topalian, S.L.; Hwu, P.; Drake, C.G.; Camacho, L.H.; Kauh, J.; Odunsi, K.; et al. Safety and Activity of Anti–PD-L1 Antibody in Patients with Advanced Cancer. *N. Engl. J. Med.* **2012**, *366*, 2455–2465. [CrossRef] [PubMed]

72. Fukumoto, I.; Kinoshita, T.; Hanazawa, T.; Kikkawa, N.; Chiyomaru, T.; Enokida, H.; Yamamoto, N.; Goto, Y.; Nishikawa, R.; Nakagawa, M.; et al. Identification of tumour suppressive microRNA-451a in hypopharyngeal squamous cell carcinoma based on microRNA expression signature. *Br. J. Cancer* **2014**, *111*, 386–394. [CrossRef] [PubMed]

73. Zhang, Z.-L.; Bai, Z.-H.; Wang, X.-B.; Bai, L.; Miao, F.; Pei, H.-H. miR-186 and 326 Predict the Prognosis of Pancreatic Ductal Adenocarcinoma and Affect the Proliferation and Migration of Cancer Cells. *PLoS ONE* **2015**, *10*, e0118814. [CrossRef]

74. Yu, J.; Ohuchida, K.; Mizumoto, K.; Fujita, H.; Nakata, K.; Tanaka, M. MicroRNAmiR-17-5pis overexpressed in pancreatic cancer, associated with a poor prognosis, and involved in cancer cell proliferation and invasion. *Cancer Biol. Ther.* **2010**, *10*, 748–757. [CrossRef]

75. Zhu, Y.; Gu, J.; Li, Y.; Peng, C.; Shi, M.; Wang, X.; Wei, G.; Ge, O.; Wang, D.; Zhang, B.; et al. MiR-17-5p enhances pancreatic cancer proliferation by altering cell cycle profiles via disruption of RBL2/E2F4-repressing complexes. *Cancer Lett.* **2018**, *412*, 59–68. [CrossRef]

76. Yan, H.-J.; Liu, W.-S.; Sun, W.-H.; Wu, J.; Ji, M.; Wang, Q.; Zheng, X.; Jiang, J.; Wu, C. miR-17-5p Inhibitor Enhances Chemosensitivity to Gemcitabine Via Upregulating Bim Expression in Pancreatic Cancer Cells. *Dig. Dis. Sci.* **2012**, *57*, 3160–3167. [CrossRef]

77. Hao, J.; Zhang, S.; Zhou, Y.; Liu, C.; Hu, X.-G.; Shao, C. MicroRNA 421 suppresses DPC4/Smad4 in pancreatic cancer. *Biochem. Biophys. Res. Commun.* **2011**, *406*, 552–557. [CrossRef]

78. Siragam, V.; Rutnam, Z.J.; Yang, W.; Fang, L.; Luo, L.; Yang, X.; Li, M.; Deng, Z.; Qian, J.; Peng, C.; et al. MicroRNA miR-98 inhibits tumor angiogenesis and invasion by targeting activin receptor-like kinase-4 and matrix metalloproteinase-11. *Oncotarget* **2012**, *3*, 1370–1385. [CrossRef]

79. Wang, L.; Yuan, C.; Lv, K.; Xie, S.; Fu, P.; Liu, X.; Chen, Y.; Qin, C.; Deng, W.; Hu, W. Lin28 Mediates Radiation Resistance of Breast Cancer Cells via Regulation of Caspase, H2A.X and Let-7 Signaling. *PLoS ONE* **2013**, *8*, e67373. [CrossRef]

80. Fu, Y.; Liu, X.; Chen, Q.; Liu, T.; Lu, C.; Yu, J.; Miao, Y.; Wei, J. Downregulated miR-98-5p promotes PDAC proliferation and metastasis by reversely regulating MAP4K4. *J. Exp. Clin. Cancer Res.* **2018**, *37*, 130. [CrossRef]

81. Morimura, R.; Komatsu, S.; Ichikawa, D.; Takeshita, H.; Tsujiura, M.; Nagata, H.; Konishi, H.; Shiozaki, A.; Ikoma, H.; Okamoto, K.; et al. Novel diagnostic value of circulating miR-18a in plasma of patients with pancreatic cancer. *Br. J. Cancer* **2011**, *105*, 1733–1740. [CrossRef]

82. Zhao, F.; Pu, Y.; Qian, L.; Zang, C.; Tao, Z.; Gao, J. MiR-20a-5p promotes radio-resistance by targeting NPAS2 in nasopharyngeal cancer cells. *Oncotarget* **2017**, *8*, 105873–105881. [CrossRef]

83. Zhang, Y.; Zheng, L.; Ding, Y.; Li, Q.; Wang, R.; Liu, T.; Sun, Q.; Yang, H.; Peng, S.; Wang, W.; et al. MiR-20a Induces Cell Radioresistance by Activating the PTEN/PI3K/Akt Signaling Pathway in Hepatocellular Carcinoma. *Int. J. Radiat. Oncol.* **2015**, *92*, 1132–1140. [CrossRef]

84. Huang, J.-W.; Wang, Y.; Dhillon, K.K.; Calses, P.; Villegas, E.; Mitchell, P.S.; Tewari, M.; Kemp, C.J.; Taniguchi, T. Systematic Screen Identifies miRNAs That Target RAD51 and RAD51D to Enhance Chemosensitivity. *Mol. Cancer Res.* **2013**, *11*, 1564–1573. [CrossRef] [PubMed]

85. Lu, Y.; Ji, N.; Wei, W.; Sun, W.; Gong, X.; Wang, X. MiR-142 modulates human pancreatic cancer proliferation and invasion by targeting hypoxia-inducible factor 1 (HIF-1α) in the tumor microenvironments. *Biol. Open* **2017**, *6*, 252–259. [CrossRef] [PubMed]

86. Deng, J.; He, M.; Chen, L.; Chen, C.; Zheng, J.; Cai, Z. The Loss of miR-26a-Mediated Post-Transcriptional Regulation of Cyclin E2 in Pancreatic Cancer Cell Proliferation and Decreased Patient Survival. *PLoS ONE* **2013**, *8*, e76450. [CrossRef] [PubMed]

87. Liang, S.; Gong, X.; Zhang, G.; Huang, G.; Lu, Y.; Li, Y.-X. MicroRNA-140 regulates cell growth and invasion in pancreatic duct adenocarcinoma by targeting iASPP. *Acta Biochim. Biophys. Sin. (Shanghai)* **2016**, *48*, 174–181. [CrossRef]

88. Lan, F.; Yue, X.; Ren, G.; Li, H.; Ping, L.; Wang, Y.; Xia, T. miR-15a/16 Enhances Radiation Sensitivity of Non-Small Cell Lung Cancer Cells by Targeting the TLR1/NF-κB Signaling Pathway. *Int. J. Radiat. Oncol.* **2015**, *91*, 73–81. [CrossRef] [PubMed]

89. Jiao, L.R.; Frampton, A.E.; Jacob, J.; Pellegrino, L.; Krell, J.; Giamas, G.; Tsim, N.; Vlavianos, P.; Cohen, P.; Ahmad, R.; et al. MicroRNAs Targeting Oncogenes Are Down-Regulated in Pancreatic Malignant Transformation from Benign Tumors. *PLoS ONE* **2012**, *7*, e32068. [CrossRef] [PubMed]

90. Li, H.; Xiang, H.; Ge, W.; Wang, H.; Wang, T.; Xiong, M. Expression and functional perspectives of miR-184 in pancreatic ductal adenocarcinoma. *Int. J. Clin. Exp. Pathol.* **2015**, *8*, 12313–12318.

91. Zhao, L.; Bode, A.M.; Cao, Y.; Dong, Z. Regulatory mechanisms and clinical perspectives of miRNA in tumor radiosensitivity. *Carcinogenesis* **2012**, *33*, 2220–2227. [CrossRef]

92. Lal, A.; Pan, Y.; Navarro, F.; Dykxhoorn, D.M.; Moreau, L.; Meire, E.; Bentwich, Z.; Lieberman, J.; Chowdhury, D. miR-24–mediated downregulation of H2AX suppresses DNA repair in terminally differentiated blood cells. *Nat. Struct. Mol. Biol.* **2009**, *16*, 492–498. [CrossRef] [PubMed]

93. Yang, H.; Luo, J.; Liu, Z.; Zhou, R.; Luo, H. MicroRNA-138 Regulates DNA Damage Response in Small Cell Lung Cancer Cells by Directly Targeting H2AX. *Cancer Investig.* **2015**, *33*, 126–136. [CrossRef] [PubMed]

94. Wang, P.; Zou, F.; Zhang, X.; Li, H.; Dulak, A.; Tomko, R.J.; Lazo, J.S.; Wang, Z.; Zhang, L.; Yu, J.; et al. microRNA-21 negatively regulates Cdc25A and cell cycle progression in colon cancer cells. *Cancer Res.* **2009**, *69*, 8157–8165. [CrossRef]

95. Yan, D.; Ng, W.L.; Zhang, X.; Wang, P.; Zhang, Z.; Mo, Y.-Y.; Mao, H.; Hao, C.; Olson, J.J.; Curran, W.J.; et al. Targeting DNA-PKcs and ATM with miR-101 Sensitizes Tumors to Radiation. *PLoS ONE* **2010**, *5*, e11397. [CrossRef]

96. Hu, H.; Du, L.; Nagabayashi, G.; Seeger, R.C.; Gatti, R.A. ATM is down-regulated by N-Myc-regulated microRNA-421. *Proc. Natl. Acad. Sci. USA* **2010**, *107*, 1506–1511. [CrossRef]

97. Noguchi, S.; Ogusu, R.; Wada, Y.; Matsuyama, S.; Mori, T. PTEN, A Target of Microrna-374b, Contributes to the Radiosensitivity of Canine Oral Melanoma Cells. *Int. J. Mol. Sci.* **2019**, *20*, 4631. [CrossRef]

98. Mei, Z.; Su, T.; Ye, J.; Yang, C.; Zhang, S.; Xie, C. The miR-15 Family Enhances the Radiosensitivity of Breast Cancer Cells by Targeting G2Checkpoints. *Radiat. Res.* **2015**, *183*, 196–207. [CrossRef] [PubMed]

99. Song, L.; Peng, L.; Hua, S.; Li, X.; Ma, L.; Jie, J.; Chen, D.; Wang, Y.; Li, D. miR-144-5p Enhances the Radiosensitivity of Non-Small-Cell Lung Cancer Cells via Targeting ATF2. *BioMed Res. Int.* **2018**, *2018*, 5109497. [CrossRef]

100. Lan, F.; Yu, H.; Hu, M.; Xia, T.; Yue, X. miR-144-3p exerts anti-tumor effects in glioblastoma by targeting c-Met. *J. Neurochem.* **2015**, *135*, 274–286. [CrossRef]

101. Yu, L.; Yang, Y.; Hou, J.; Zhai, C.; Song, Y.; Zhang, Z.; Qiu, L.; Jia, X. MicroRNA-144 affects radiotherapy sensitivity by promoting proliferation, migration and invasion of breast cancer cells. *Oncol. Rep.* **2015**, *34*, 1845–1852. [CrossRef] [PubMed]

102. Gu, H.; Liu, M.; Ding, C.; Wang, X.; Wang, R.; Wu, X.; Fan, R. Hypoxia-responsive miR-124 and miR-144 reduce hypoxia-induced autophagy and enhance radiosensitivity of prostate cancer cells via suppressing PIM 1. *Cancer Med.* **2016**, *5*, 1174–1182. [CrossRef] [PubMed]

103. Chen, X.; Liu, J.; Zhang, Q.; Liu, B.; Cheng, Y.; Zhang, Y.; Sun, Y.; Ge, H.; Liu, Y. Exosome-mediated transfer of miR-93-5p from cancer-associated fibroblasts confer radioresistance in colorectal cancer cells by downregulating FOXA1 and upregulating TGFB3. *J. Exp. Clin. Cancer Res.* **2020**, *39*, 65. [CrossRef]

104. Cui, H.; Zhang, S.; Zhou, H.; Guo, L. Direct Downregulation of B-Cell Translocation Gene 3 by microRNA-93 Is Required for Desensitizing Esophageal Cancer to Radiotherapy. *Dig. Dis. Sci.* **2017**, *62*, 1995–2003. [CrossRef] [PubMed]

105. Ruhl, R.; Rana, S.; Kelley, K.; Espinosa-Diez, C.; Hudson, C.; Lanciault, C.; Thomas, C.R., Jr.; Liana Tsikitis, V.; Anand, S. microRNA-451a regulates colorectal cancer proliferation in response to radiation. *BMC Cancer* **2018**, *18*, 517. [CrossRef] [PubMed]

106. Tian, F.; Han, Y.; Yan, X.; Zhong, D.; Yang, G.; Lei, J.; Li, X.; Wang, X. Upregulation of microrna-451 increases the sensitivity of A 549 cells to radiotherapy through enhancement of apoptosis. *Thorac. Cancer* **2015**, *7*, 226–231. [CrossRef] [PubMed]

107. Wang, R.; Chen, D.-Q.; Huang, J.-Y.; Zhang, K.; Feng, B.; Pan, B.-Z.; Chen, J.; De, W.; Chen, L.-B. Acquisition of radioresistance in docetaxel-resistant human lung adenocarcinoma cells is linked with dysregulation of miR-451/c-Myc-survivin/rad-51 signaling. *Oncotarget* **2014**, *5*, 6113–6129. [CrossRef]

108. Zhang, T.; Sun, Q.; Liu, T.; Chen, J.; Du, S.; Ren, C.; Liao, G.; Yuan, Y. MiR-451 increases radiosensitivity of nasopharyngeal carcinoma cells by targeting ras-related protein 14 (RAB14). *Tumor Biol.* **2014**, *35*, 12593–12599. [CrossRef]

109. Zhang, Y.; Zhang, W. FOXD1, negatively regulated by miR-186, promotes the proliferation, metastasis and radioresistance of nasopharyngeal carcinoma cells. *Cancer Biomark.* **2020**, *28*, 511–521. [CrossRef]

110. Lynam-Lennon, N.; Heavey, S.; Sommerville, G.; Bibby, B.A.; Ffrench, B.; Quinn, J.; Gasch, C.; O'Leary, J.J.; Gallagher, M.F.; Reynolds, J.V.; et al. MicroRNA-17 is downregulated in esophageal adenocarcinoma cancer stem-like cells and promotes a radioresistant phenotype. *Oncotarget* **2016**, *8*, 11400–11413. [CrossRef]

111. Wu, S.-Y.; Wu, A.T.; Liu, S.-H. MicroRNA-17-5p regulated apoptosis-related protein expression and radiosensitivity in oral squamous cell carcinoma caused by betel nut chewing. *Oncotarget* **2016**, *7*, 51482–51493. [CrossRef] [PubMed]

112. Liu, L.; Cui, S.; Zhang, R.; Shi, Y.; Luo, L. MiR-421 inhibits the malignant phenotype in glioma by directly targeting MEF2D. *Am. J. Cancer Res.* **2017**, *7*, 857–868. [PubMed]

113. Mansour, W.Y.; Bogdanova, N.V.; Kasten-Pisula, U.; Rieckmann, T.; Köcher, S.; Borgmann, K.; Baumann, M.; Krause, M.; Petersen, C.; Hu, H.; et al. Aberrant overexpression of miR-421 downregulates ATM and leads to a pronounced DSB repair defect and clinical hypersensitivity in SKX squamous cell carcinoma. *Radiother. Oncol.* **2013**, *106*, 147–154. [CrossRef] [PubMed]

114. Jin, Y.-Y.; Chen, Q.-J.; Wei, Y.; Wang, Y.-L.; Wang, Z.-W.; Xu, K.; He, Y.; Ma, H. Upregulation of microRNA-98 increases radiosensitivity in esophageal squamous cell carcinoma. *J. Radiat. Res.* **2016**, *57*, 468–476. [CrossRef] [PubMed]

115. Li, B.; Shi, X.-B.; Nori, D.; Chao, C.K.; Chen, A.M.; Valicenti, R.; White, R.D. Down-regulation of microRNA 106b is involved in p21-mediated cell cycle arrest in response to radiation in prostate cancer cells. *Prostate* **2011**, *71*, 567–574. [CrossRef] [PubMed]

116. Troschel, F.M.; Böhly, N.; Borrmann, K.; Braun, T.; Schwickert, A.; Kiesel, L.; Eich, H.T.; Götte, M.; Greve, B. miR-142-3p attenuates breast cancer stem cell characteristics and decreases radioresistance in vitro. *Tumor Biol.* **2018**, *40*. [CrossRef]

117. Yuan, F.; Liu, L.; Lei, Y.; Hu, Y. MiRNA-142-3p increases radiosensitivity in human umbilical cord blood mononuclear cells by inhibiting the expression of CD133. *Sci. Rep.* **2018**, *8*, 5674. [CrossRef]

118. Jin, Q.; Li, X.J.; Cao, P.G. MicroRNA-26b Enhances the Radiosensitivity of Hepatocellular Carcinoma Cells by Targeting EphA2. *Tohoku J. Exp. Med.* **2016**, *238*, 143–151. [CrossRef]

119. Duru, N.; Gernapudi, R.; Zhang, Y.; Yao, Y.; Lo, P.-K.; Wolfson, B.; Zhou, Q. NRF2/miR-140 signaling confers radioprotection to human lung fibroblasts. *Cancer Lett.* **2015**, *369*, 184–191. [CrossRef]

120. Wang, F.; Mao, A.; Tang, J.; Zhang, Q.; Yan, J.; Wang, Y.; Di, C.; Gan, L.; Sun, C.; Zhang, H. microRNA-16-5p enhances radiosensitivity through modulating Cyclin D1/E1–pRb–E2F1 pathway in prostate cancer cells. *J. Cell. Physiol.* **2019**, *234*, 13182–13190. [CrossRef]

121. Samadi, P.; Afshar, S.; Amini, R.; Najafi, R.; Mahdavinezhad, A.; Pashaki, A.S.; Gholami, M.H.; Saidijam, M. Let-7e enhances the radiosensitivity of colorectal cancer cells by directly targeting insulin-like growth factor 1 receptor. *J. Cell. Physiol.* **2019**, *234*, 10718–10725. [CrossRef] [PubMed]

122. Smolinska, A.; Swoboda, J.; Fendler, W.; Lerch, M.M.; Sendler, M.; Moskwa, P. MiR-502 is the first reported miRNA simultaneously targeting two components of the classical non-homologous end joining (C-NHEJ) in pancreatic cell lines. *Heliyon* **2020**, *6*, e03187. [CrossRef] [PubMed]

123. Wang, P.; Zhang, J.; Zhang, L.; Zhu, Z.; Fan, J.; Chen, L.; Zhuang, L.; Luo, J.; Chen, H.; Liu, L.; et al. MicroRNA 23b Regulates Autophagy Associated With Radioresistance of Pancreatic Cancer Cells. *Gastroenterology* **2013**, *145*, 1133–1143.e12. [CrossRef] [PubMed]

124. Zhang, X.; Shi, H.; Lin, S.; Ba, M.-C.; Cui, S. MicroRNA-216a enhances the radiosensitivity of pancreatic cancer cells by inhibiting beclin-1-mediated autophagy. *Oncol. Rep.* **2015**, *34*, 1557–1564. [CrossRef]

125. Tomihara, H.; Yamada, D.; Eguchi, H.; Iwagami, Y.; Noda, T.; Asaoka, T.; Wada, H.; Kawamoto, K.; Gotoh, K.; Takeda, Y.; et al. MicroRNA-181b-5p, ETS1, and the c-Met pathway exacerbate the prognosis of pancreatic ductal adenocarcinoma after radiation therapy. *Cancer Sci.* **2017**, *108*, 398–407. [CrossRef]

126. Ji, Q.; Hao, X.; Zhang, M.; Tang, W.; Yang, M.; Li, L.; Xiang, D.; DeSano, J.T.; Bommer, G.T.; Fan, D.; et al. MicroRNA miR-34 Inhibits Human Pancreatic Cancer Tumor-Initiating Cells. *PLoS ONE* **2009**, *4*, e6816. [CrossRef]

127. Fang, C.; Dai, C.-Y.; Mei, Z.; Jiang, M.-J.; Gu, D.-N.; Huang, Q.; Tian, L. microRNA-193a stimulates pancreatic cancer cell repopulation and metastasis through modulating TGF-β2/TGF-βRIII signalings. *J. Exp. Clin. Cancer Res.* **2018**, *37*, 25. [CrossRef]

128. Huang, X.; Taeb, S.; Jahangiri, S.; Korpela, E.; Cadonic, I.; Yu, N.; Krylov, S.N.; Fokas, E.; Boutros, P.C.; Liu, S.K. miR-620 promotes tumor radioresistance by targeting 15-hydroxyprostaglandin dehydrogenase (HPGD). *Oncotarget* **2015**, *6*, 22439–22451. [CrossRef]

129. Oh, J.-S.; Kim, J.-J.; Byun, J.-Y.; Kim, I.A. Lin28-let7 Modulates Radiosensitivity of Human Cancer Cells with Activation of K-Ras. *Int. J. Radiat. Oncol.* **2010**, *76*, 5–8. [CrossRef]

130. Baek, S.-J.; Azuma, R.; Hayashi, K.; Ishii, H.; Sato, K.; Nishida, N.; Koseki, J.; Kawamoto, K.; Konno, M.; Satoh, T.; et al. MicroRNA miR-374, a potential radiosensitizer for carbon ion beam radiotherapy. *Oncol. Rep.* **2016**, *36*, 2946–2950. [CrossRef]

131. Chang, T.-C.; Wentzel, E.A.; Kent, O.A.; Ramachandran, K.; Mullendore, M.; Lee, K.H.; Feldmann, G.; Yamakuchi, M.; Ferlito, M.; Lowenstein, C.J.; et al. Transactivation of miR-34a by p53 Broadly Influences Gene Expression and Promotes Apoptosis. *Mol. Cell* **2007**, *26*, 745–752. [CrossRef] [PubMed]

132. Ji, Q.; Hao, X.; Meng, Y.; Zhang, M.; DeSano, J.; Fan, D.; Xu, L. Restoration of tumor suppressor miR-34 inhibits human p53-mutant gastric cancer tumorspheres. *BMC Cancer* **2008**, *8*, 266. [CrossRef] [PubMed]

133. Fan, X.; Matsui, W.; Khaki, L.; Stearns, D.; Chun, J.; Li, Y.-M.; Eberhart, C.G. Notch Pathway Inhibition Depletes Stem-like Cells and Blocks Engraftment in Embryonal Brain Tumors. *Cancer Res.* **2006**, *66*, 7445–7452. [CrossRef] [PubMed]

134. Raufi, A.G.; Manji, G.A.; Chabot, J.A.; Bates, S.E. Neoadjuvant Treatment for Pancreatic Cancer. *Semin. Oncol.* **2019**, *46*, 19–27. [CrossRef] [PubMed]

135. Hall, W.A.; Goodman, K.A. Radiation therapy for pancreatic adenocarcinoma, a treatment option that must be considered in the management of a devastating malignancy. *Radiat. Oncol.* **2019**, *14*, 114. [CrossRef]

136. Guo, S.; Fesler, A.; Hwang, G.-R.; Ju, J. microRNA based prognostic biomarkers in pancreatic Cancer. *Biomark. Res.* **2018**, *6*, 18. [CrossRef]

137. Moertl, S.; Mutschelknaus, L.; Heider, T.; Atkinson, M.J. MicroRNAs as novel elements in personalized radiotherapy. *Transl. Cancer Res.* **2016**, *5*, S1262–S1269. [CrossRef]

138. Li, X.; Gao, P.; Wang, Y.; Wang, X. Blood-Derived microRNAs for Pancreatic Cancer Diagnosis: A Narrative Review and Meta-Analysis. *Front. Physiol.* **2018**, *9*, 685. [CrossRef]

139. Buscail, E.; Maulat, C.; Muscari, F.; Chiche, L.; Cordelier, P.; Dabernat, S.; Alix-Panabières, C.; Buscail, L. Liquid Biopsy Approach for Pancreatic Ductal Adenocarcinoma. *Cancers* **2019**, *11*, 852. [CrossRef]

140. Buschmann, D.; Haberberger, A.; Kirchner, B.; Spornraft, M.; Riedmaier, I.; Schelling, G.; Pfaffl, M.W. Toward reliable biomarker signatures in the age of liquid biopsies—How to standardize the small RNA-Seq workflow. *Nucleic Acids Res.* **2016**, *44*, 5995–6018. [CrossRef]

141. Cacheux, J.; Bancaud, A.; Leïchlé, T.; Cordelier, P. Technological Challenges and Future Issues for the Detection of Circulating MicroRNAs in Patients with Cancer. *Front. Chem.* **2019**, *7*, 815. [CrossRef] [PubMed]

142. Korpela, E.; Vesprini, D.; Liu, S.K. MicroRNA in radiotherapy: miRage or miRador? *Br. J. Cancer* **2015**, *112*, 777–782. [CrossRef] [PubMed]

Publisher's Note: MDPI stays neutral with regard to jurisdictional claims in published maps and institutional affiliations.

 *cancers* 

Review

MicroRNA-361: A Multifaceted Player Regulating Tumor Aggressiveness and Tumor Microenvironment Formation

Daozhi Xu [1,†] , Peixin Dong [1,*,†] , Ying Xiong [2,†], Junming Yue [3,4], Kei Ihira [1], Yosuke Konno [1], Noriko Kobayashi [1], Yukiharu Todo [5] and Hidemichi Watari [1,*]

1   Department of Obstetrics and Gynecology, Hokkaido University School of Medicine, Hokkaido University, Sapporo 060-8638, Japan
2   Department of Gynecology, State Key Laboratory of Oncology in South China, Sun Yat-sen University Cancer Center, Guangzhou 510060, China
3   Department of Pathology and Laboratory Medicine, University of Tennessee Health Science Center, Memphis, TN 38163, USA
4   Center for Cancer Research, University of Tennessee Health Science Center, Memphis, TN 38163, USA
5   Division of Gynecologic Oncology, National Hospital Organization, Hokkaido Cancer Center, Sapporo 003-0804, Japan
*   Correspondence: dpx1cn@gmail.com (P.D.); watarih@med.hokudai.ac.jp (H.W.); Tel.: +81-11-706-5941 (P.D. & H.W.)
†   These authors contributed equaly.

Received: 8 July 2019; Accepted: 1 August 2019; Published: 7 August 2019

Abstract: MicroRNA-361-5p (miR-361) expression frequently decreases or is lost in different types of cancers, and contributes to tumor suppression by repressing the expression of its target genes implicated in tumor growth, epithelial-to-mesenchymal transition (EMT), metastasis, drug resistance, glycolysis, angiogenesis, and inflammation. Here, we review the expression pattern of miR-361 in human tumors, describe the mechanisms responsible for its dysregulation, and discuss how miR-361 modulates the aggressive properties of tumor cells and alter the tumor microenvironment by acting as a novel tumor suppressor. Furthermore, we describe its potentials as a promising diagnostic or prognostic biomarker for cancers and a promising target for therapeutic development.

Keywords: microRNA-361; EMT; angiogenesis; tumor microenvironment; cancer diagnosis; cancer treatment

1. Introduction

Large-scale transcriptional analysis reveals that more than 80% of the human genome is transcribed into RNA, whereas less than 2% of the human genome is used for protein translation [1], suggesting that the vast majority of the human transcriptome is composed of non-coding RNAs (ncRNAs). MicroRNAs (miRNAs) are a class of endogenous regulatory noncoding RNAs, typically 20–23 nucleotides in length, thereby exerting essential roles in a wide range of physiological processes [2]. Although some miRNAs can bind to the 5′-untranslated regions (5′-UTRs) or the coding regions of target messenger RNAs (mRNAs) [3,4], miRNAs primarily suppress the expression of their target genes by targeting the 3′-UTRs of target mRNAs for mRNA degradation or translation inhibition [2]. A single miRNA can target many genes, and multiple miRNAs can regulate a single gene. Previous studies indicated that miRNAs may regulate as many as one third of human genes [2]. miRNAs are evolutionary highly conserved ncRNAs and are expressed in a tissue-specific or developmental stage-specific manner, thereby contributing to cell or tissue-specific protein expression profiles [5–7].

In human cancer cells, miRNAs exert either pro- or anti-tumorigenic effects through tissue-dependent mechanisms. Some miRNAs that are amplified or overexpressed in cancer could act as oncogenes to either directly or indirectly downregulate the expression of tumor suppressors [8,9]. On the other hand, some miRNAs, such as miR-361-5p (miR-361), can target mRNAs encoding oncogenic proteins and therefore be categorized as tumor suppressors [8–10].

Dysregulation of miRNA expression was reported in most cancer types [8,9]. miRNA expression profiles can distinguish cancer tissues from normal tissues and separate different cancers subtypes [11]. miRNAs have already been described as non-invasive biomarkers useful for cancer diagnosis, patient stratification, and the prediction of patient prognosis, and treatment efficacy [12–14].

miRNAs mediate tumor initiation and progression by regulating a variety of biological processes, including cell proliferation, migration, invasion, metastasis, glycolysis, apoptosis, cancer stem cell (CSC)-like phenotype, chemoresistance, and epithelial-to-mesenchymal transition (EMT) [11,15]. The components of the tumor microenvironment, which includes the extracellular matrix (ECM), fibroblasts, immune cells, inflammatory cells, endothelial cells, lymphatic endothelial cells, growth factors, and cytokines, play an important role during tumor progression and metastasis [16]. Recent works demonstrated the importance of miRNAs in regulating complex signaling networks involved in multiple aspects of the microenvironment remodeling, including the hypoxic response, angiogenesis, anti-tumor immune response, inflammation, and ECM organization [16].

In this review, we discuss the expression pattern of miR-361 in human tumors and the mechanisms responsible for its dysregulation. Furthermore, we elucidate the diverse mechanisms by which downregulation of miR-361 expression confers the aggressive properties of tumor cells and alters the tumor microenvironment. Finally, we describe its potential as a promising biomarker for cancer diagnosis and prediction of prognosis in patients with cancers.

2. Evidence Acquisition

PubMed and Google Scholar were used to search for articles published up to April 2019 using the following keywords: miR-361-5p, microRNA-361-5p, tumor, cancer, and carcinoma. All recognized studies were assessed for relevance by two authors by checking the title and abstract. All irrelevant articles, studies without access to the full text of the publication, case reports, letters, expert opinions, meeting proceedings, review articles, non-English articles, and articles whose methods do not contain biomedical experimental validation were excluded. After this, the full text of any selected article was reviewed independently by two authors. A weakness of our study relates to the lack of access to some relevant research works that may contain information on miR-361 and its target genes. We also searched the reference lists of the reviewed articles to identify additional relevant articles. A flow diagram of the study selection process is shown in Figure 1.

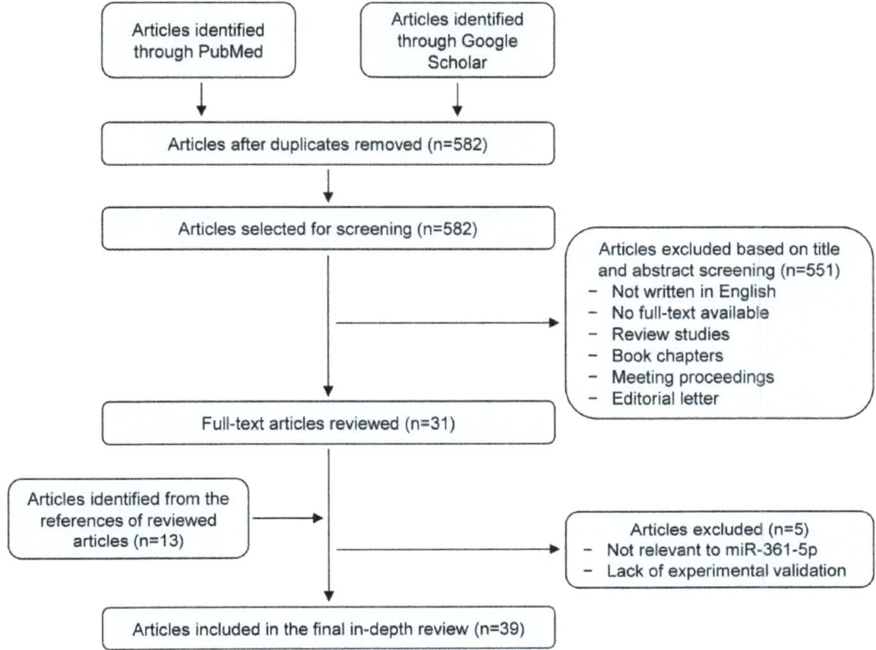

Figure 1. Summary of literature search, screening and selection.

3. Dysregulation of miR-361 in Tumor

Previous studies of solid tumors showed that miR-361 is frequently downregulated in various tumor tissues, including cutaneous squamous cell carcinoma [17], osteosarcoma [18], breast cancer [19–22], glioma [23,24], papillary thyroid carcinoma [25], lung cancer [26–29], gastric cancer [30–32], hepatocellular carcinoma [33], colorectal cancer [32], ovarian cancer [34], endometrial cancer [10], cervical cancer [35], and prostate cancer [36–38]. However, increased expression of miR-361 was detected in acute myeloid leukemia [39], indicating the possibility that miR-361 dysregulation might be required to impair differentiation programs in leukemia, and miR-361 may regulate the expression of the hematopoietic differentiation-specific genes, which have a weak importance in solid tumors. Several works demonstrated that low levels of miR-361 were associated with shorter overall survival in patients with breast cancer [20,21], gastric cancer [32], and colorectal cancer [32].

4. Mechanisms of miR-361 Regulation in Tumor

Large-scale profiling studies have revealed that dysregulation of miRNA is a common event during cancer carcinogenesis and metastasis [11,12]. The molecular mechanisms regulating miRNA expression include genomic amplification or deletion of miRNA genes, abnormal transcriptional control of miRNAs, epigenetic silencing, and defects in miRNA biogenesis and processing machinery [11,40]. The downregulation of miR-361 in tumor tissues could be caused by several mechanisms (Figure 2A).

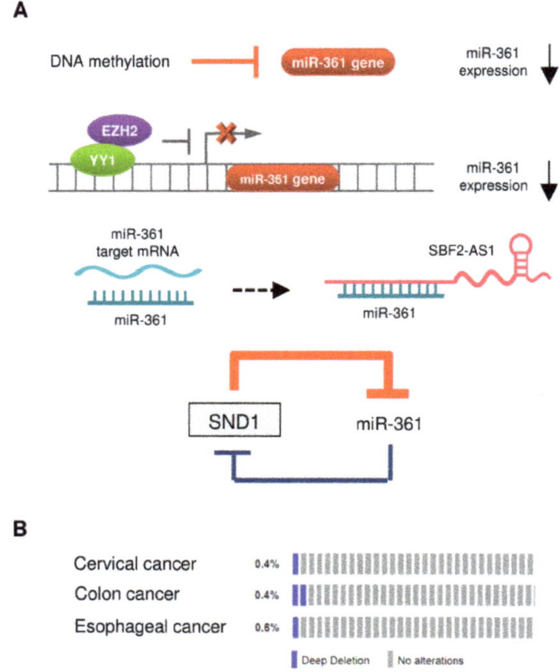

Figure 2. Mechanisms of miR-361 dysregulation in tumors. (**A**) Reported mechanisms responsible for miR-361 downregulation in tumors. (**B**) OncoPrint of cBioPortal showing the genetic alterations of *miR-361* (deep deletion) in tumor samples obtained from The Cancer Genome Atlas (TCGA)-cervical cancer, TCGA-colon cancer, and TCGA-esophageal cancer datasets. Each bar indicates the individual cases and % on the left indicates the percentage of cases altered in the human *miR-361* gene.

4.1. DNA Hypermethylation

Hypermethylation of tumor suppressor gene promoter regions leads to silencing of those genes. A previous study reported that the expression of miR-361 in hepatocellular carcinoma cell lines was restored upon treatment with an inhibitor of DNA methylation 5-azacytidine (5-AZA) [33]. Similar results were obtained from another study, where 5-AZA treatment significantly upregulated miR-361 expression in endometrial cancer cells [10].

4.2. Transcriptional Control of miR-361 Expression

We previously reported that enhancer of zeste homolog 2 (EZH2), which represses gene expression by methylation of histone H3 on lysine 27, acted as a co-suppressor of transcription factor YY1 to epigenetically suppress the transcription of miR-361 [10]. The use of GSK343 (a specific EZH2 inhibitor) was found to increase the levels of miR-361 in endometrial cancer cells [10].

4.3. Long Non-Coding RNA (lncRNA) SBF2-AS1 Acts as a Sponge for miR-361

LncRNAs are non-protein coding transcripts longer than 200 nucleotides that could serve as miRNA sponges to inhibit the interaction between miRNA-target mRNAs or suppressing the levels of miRNAs [41]. For instance, in mouse cardiomyocytes, mitochondrial dynamic related lncRNA (MDRL) directly binds to miR-361 and acts as its sponge to promote the processing of pri-miR-484 [42]. Another lncRNA, Maternally Expressed Gene 3 (MEG3), was shown to facilitate cardiac hypertrophy by sponging miR-361 [43]. In cervical cancer cells, lncRNA SBF2-AS1 (SBF2 Antisense RNA 1) was shown to function as an endogenous RNA sponge that interacted with miR-361 and suppressed its expression [35].

4.4. The SND1/miR-361 Feedback Loop Controls miR-361 Expression

miR-361 directly targeted SND1 (Staphylococcal nuclease and tudor domain containing 1), and SND1 conversely suppressed the expression of miR-361 by binding to pre-miR-361, thus creating a double-negative feedback loop, in which miR-361 and SND1 repress expression of each other in gastric and colon cancer cells [24].

4.5. Deletion of the Human miR-361 Gene

The reduced miRNA expression in tumor cells could arise from copy-number alterations and chromosomal aberrations (such as amplification, deletion, or translocation) [11]. However, whether genomic alterations of the human miR-361 gene can lead to decreased expression of miR-361 in cancer is poorly understood. We investigated copy-number alterations and nucleotide changes of the miR-361 gene in human cervical, colon and esophageal cancer samples from the cBioPortal database. As shown in Figure 2B, gene deletion is the most frequent alteration type in these cancers, supporting the notion that the loss of miR-361 expression plays an important role in the development of colon and cervical cancers [32,35].

5. The Impact of miR-361 on the Aggressive Properties of Tumor Cells and Tumor Microenvironment Remodeling

MiR-361 has been shown to act as a novel tumor suppressor that represses a large number of downstream target transcripts implicated in cellular proliferation, glycolysis, migration, invasion, EMT, chemoresistance, cancer stemness, angiogenesis, and inflammation (Figure 3).

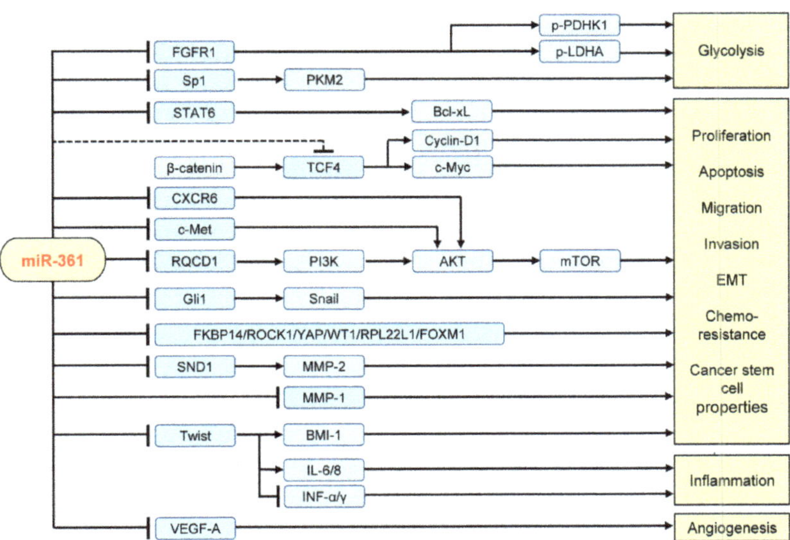

Figure 3. Validated targets and signaling pathways regulated by miR-361 in human tumor cells. FGFR1: Fibroblast growth factor receptor 1; Sp1: Transcription factor; STAT6: Signal transducer and activator of transcription 6; PKM2: pyruvate kinase M2; PDHK1: pyruvate dehydrogenase kinase 1; LDHA: lactate dehydrogenase A; TCF4: Transcription factor 4; CXCR6: C-X-C chemokine receptor type 6; RQCD1: CCR4-NOT transcription complex subunit 9; PI3K: phosphoinositide 3-kinase; mTOR: mammalian target of rapamycin; SND1: Staphylococcal nuclease and tudor domain containing 1; MMP: Matrix metallopeptidase; IL: interleukin; INF-α/γ: interferon α/γ; VEGF-A: vascular endothelial growth factor A.

5.1. Inhibiting Tumor Growth, Invasion, EMT, Metastasis, and Glycolysis

Restoration of miR-361 expression by transfection with miR-361 mimics inhibited the proliferation of osteosarcoma, breast cancer, thyroid papillary carcinoma, lung cancer, gastric cancer, colorectal cancer, hepatocellular carcinoma, cervical cancer, and prostate cancer cells [18,19,25–33,35,36]. Moreover, over-expression of miR-361 attenuated cell migration and invasion in endometrial cancer, breast cancer, glioma, papillary thyroid carcinoma, lung cancer, gastric cancer, ovarian cancer, and prostate cancer [10,19,23–26,28,30,32,34,36]. Conversely, knocking down miR-361 expression using anti-miR-361 inhibitor promoted cell migration and invasion in endometrial cancer, breast cancer, glioma, papillary thyroid carcinoma, lung cancer, gastric cancer, and ovarian cancer [10,19,22–26,28,30,32,34,36]. Mechanically, miR-361 impairs tumor cell proliferation, migration, and invasion by directly targeting and downregulating the expression of FKBP14 [18], MMP-1 [19], SND1 [24], ROCK1 [25], YAP [27], WT1 [28], RPL22L1 [34] and STAT6 [29,37].

The phosphoinositide 3-kinase (PI3K)/AKT pathway is activated in a wide range of cancers, and associated with cell growth, proliferation, survival, motility, tumor progression and resistance to cancer therapies. CCR4-NOT transcription complex subunit 9 (CNOT9/RQCD1) has been identified as a key activator of the PI3K/AKT pathway [44], and upregulation of miR-361 in breast cancer cells can suppress its direct target RQCD1, leading to the downregulation of PI3K, AKT, and MMP-9 [22]. The C-X-C motif chemokine receptor 6 (CXCR6), when bound with its ligand CXCL16, induced the activation of the PI3K/AKT signaling in cancer cells [45]. A previous study demonstrated that miR-361 inhibited the proliferation of hepatocellular carcinoma cells by directly targeting CXCR6 [33]. These results provided examples of miR-361-mediated repression of the PI3K/AKT signaling at different levels and illustrated the importance of miR-361 regulation in carcinogenesis and tumor progression. EMT encompasses a series of phenotypic and biochemical changes that enable epithelial cells to acquire a mesenchymal cell phenotype, which includes enhanced migration, invasion, metastasis, CSC-like features, resistance to conventional chemotherapy, radiotherapy, and small-molecule drugs [46–48].

EMT is mediated by a core set of key transcription factors, including Twist, Zinc finger E-box binding homeobox 1 (ZEB1), ZEB2, Snail and Slug, and the expression of these transcription factors are finely mediated at the transcriptional, translational, and post-translational levels [46–48].

In accordance with its reported anti-tumor functions, ectopic expression of miR-361 was found to cause dramatic suppression of EMT process in various cancer cells. For example, experiments show that enforced overexpression of miR-361 greatly suppressed EMT, invasion and metastasis in many tumors, including endometrial cancer, glioma, lung cancer, gastric cancer, colorectal cancer, ovarian cancer, and prostate cancer [10,23,26,32,34,36]. By inhibiting the expression of Twist, miR-361 played a crucial role in suppressing EMT characteristics and cancer stem cell (CSC)-like properties of endometrial cancer cells [10].

In addition to targeting EMT-promoting transcription factors directly, miR-361 also modulated the expression of key mediators of the EMT program. For example, the loss of miR-361 expression activated Gli1 and its downstream effector Snail to promote EMT and prostate cancer cell invasion [36]. In ovarian cancer cells, miR-361 targeted and reduced the levels of RPL22L1 and another target gene *c-Met* [34], which could serve as an upstream stimulator of the PI3K/AKT signaling and EMT-associated signaling pathways [49]. Additionally, miR-361 attenuated EMT and chemoresistance in cancer cells by suppressing the expression of FOXM1 [26,31], an oncogenic transcription factor required for EMT and metastasis [50].

Activation of the Wingless-type MMTV integration site family (Wnt)/β-catenin signaling in cancer cells is responsible for EMT induction, metastasis, CSC self-renewal, increased resistance to chemotherapy or radiotherapy and immunosuppression [51]. Although the introduction of miR-361 into gastric cancer cells downregulated the expression of Wnt/β-catenin pathway-related proteins (TCF4, Cyclin-D1 and c-Myc) [30], it remains unknown whether these genes are direct targets of miR-361.

Cancer cells are known to consume more glucose to produce lactate by glycolysis rather than oxidative phosphorylation, even in oxygen-rich conditions [52]. Recent data suggested that miR-361 directly targeted the 3'-UTR of *FGFR1*, which promotes glycolysis through activation of two critical glycolytic enzymes lactate dehydrogenase A (LDHA) and pyruvate dehydrogenase kinase 1 (PDK1/PDHK1), thereby suppressing glucose consumption and lactate production of breast cancer cells [19]. The glycolytic enzyme pyruvate kinase M2 (PKM2) is often highly expressed in cancer cells but is present at a very low level in normal cells [52]. PKM2 catalyzes the rate-limiting ATP-generating step of glycolysis, controlling the conversion of phosphoenolpyruvate and ADP to pyruvate and ATP, respectively [52]. Transfection of miR-361 mimic was shown to inhibit glucose metabolism by targeting Sp1 and subsequently downregulating the expression of PKM2 in prostate cancer [38]. The role of miR-361 in the regulation of glucose metabolism in human cancers has not yet been fully investigated.

5.2. Suppressing Angiogenesis and Inflammation

Accumulated evidence showed that miRNAs participate in the remodeling of tumor microenvironments through several mechanisms, including the regulation of the expression of cell membrane proteins, secretion of cytokines, as well as transmission of mature miRNAs between different cell types via exosomes [53,54]. It has become apparent that miR-361 is able to regulate cancer progression through modulating tumor microenvironments (Figure 3).

Angiogenesis, an important hallmark of cancer, plays an essential role in providing tumor cells with oxygen and nutrients. Some miRNAs modulate the expression of regulatory molecules driving angiogenesis, including vascular endothelial growth factors (such as vascular endothelial growth factor A (VEGF-A)), cytokines, metalloproteinases, and growth factors [55]. MiR-361 targeted the 3'-UTR of *VEGF-A* to repress its expression in skin squamous cell carcinoma [17]. Moreover, overexpression of miR-361 was shown to indirectly reduce the expression of VEGF-A through inhibiting the Wnt/β-catenin pathway in gastric cancer cells [30]. Consistent with these data, transient transfection with miR-361 mimic significantly downregulated the expression of VEGF-A, whereas the silencing of miR-361 with miRNA inhibitor enhanced the levels of VEGF-A in endometrial cancer cells [10]. Collectively, these results suggest that reduced levels of miR-361 could be an important driving mechanism for the formation of a pro-angiogenic tumor microenvironment.

Numerous studies have indicated that chronic inflammation actively promotes tumor initiation, progression, and metastasis via multiple mechanisms, including generation of an immunosuppressive tumor microenvironment [56]. Tumor cells undergoing EMT could modulate the surrounding microenvironment via enhanced secretion of inflammatory cytokines (including IL-6 and IL-8) [57,58]. We reported that miR-361 could downregulate the mRNA expression of IL-6 and IL-8 in endometrial cancer cells through targeting Twist [10]. Additionally, the activation of signal transducer and activator of transcription (STAT) family members (for example STAT6) is closely linked to tumor-promoting inflammation and the suppression of anti-tumor immunity in multiple cancer tissues [59,60]. MiR-361 was shown to directly inhibit the expression of STAT6 by binding to its 3'-UTR region [29,37]. These data support a novel function of miR-361 in exerting anti-angiogenesis and anti-inflammatory effects, at least by regulating the EMT-associated signaling and the production of pro-inflammatory cytokines.

6. Diagnostic and Prognostic Value of miR-361 in Tumor

Early studies showed that miRNA expression signatures can be useful in distinguishing cancer tissues from normal tissues, categorizing cancer subtypes, and predicting the progression, prognosis, and treatment response in many cancer types [61–65]. miRNAs are more stable than mRNA in the peripheral blood, serum, and formalin-fixed tissues [61,62] and often exhibit tumor-specific and tissue-specific expression profiles, making them attractive candidates for diagnostic and prognostic applications.

More specifically, downregulation of miR-361 was implicated in the progression of many tumor types, including breast cancer [19,21], glioma [24], papillary thyroid carcinoma [25], and lung cancer [66]. Lower miR-361 levels in patients with breast cancer [19,21], colon cancer [32], and lung cancer [66] were associated with shorter overall survival, suggesting that reduced miR-361 expression serves as a potential biomarker that predicts poor clinical outcomes in cancer patients.

Circulating miRNAs escape degradation by residing within microvesicles, exosomes, and apoptotic bodies, and dysregulated miRNAs have been detected in the blood, plasma, and serum of cancer patients [67]. The levels of circulating miRNAs (such as let-7, miR-155 and miR-195) were able to distinguish those patients with breast cancer from healthy controls [68]. Another study showed that several circulating miRNAs were detected in stage I/II breast cancer patients' plasma at a significantly higher level compared to healthy controls, suggesting that these miRNAs might be used for early cancer diagnosis [69]. Changes in circulating miRNA expression were linked to lymph node metastasis in breast cancer patients [70]. Furthermore, the diagnostic value of a panel of cancer-associated miRNAs was verified in patients with various cancer types [71]. To date, only two studies described miR-361 signatures in plasma of patients with cancers [39,72]. Quantitative PCR analysis of a group of miRNAs in acute myeloid leukemia (AML) indicated that miR-361 was significantly increased in plasma of newly diagnosed AML patients at diagnosis compared to healthy controls and decreased after chemotherapy [39]. In addition, deep sequencing of circulating miRNAs in plasma of lung cancer patients demonstrated that the levels of miR-361 were upregulated in cancer patients compared with healthy controls, and were relatively higher in patients with adenocarcinoma than in squamous cell carcinoma [72], suggesting that circulating miR-361 may be used for the differentiation of different cancer subtypes.

7. Treating Cancer with miR-361 Replacement Therapy

Although many miRNA-based therapeutics are processed in the preclinical stage, only one miRNA therapeutic, the compound SPC3649 (miravirsen, an inhibitor of miR-122) has undergone successful phase II clinical trials for the treatment of hepatitis C virus. Therapeutically restoring the expression of tumor suppressor miRNAs using synthetic miRNA mimics or miRNA expression plasmids has been developed for the clinical modulation of miRNAs [73]. Numerous studies showed that reintroduction of miR-361 exhibited significant anti-tumor activities in experimental xenograft models of breast cancer, thyroid papillary carcinoma, lung cancer, gastric cancer, colorectal cancer, hepatocellular and prostate cancer [19,25,26,30,32,33,37], highlighting its potential as a therapeutic target for treatment of these cancers. In endometrial cancer cells, we identified that EZH2 was a key upstream suppressor of miR-361, and showed that EZH2 blockade using GSK343, a specific EZH2 inhibitor that showed effective anti-cancer effects and minimal toxicity against normal cells, led to the reactivation of miR-361 and the suppression of endometrial cancer progression [10]. These findings provided an insightful cancer therapeutic strategy to indirectly restore miR-361 function via targeting EZH2.

8. Future Perspectives

Although it is becoming clear that miR-361 exerts a tumor-suppressive function in most solid tumors via reducing cancer aggressiveness and producing a suppressive tumor microenvironment, the multifaceted biological roles of miR-361 are yet to be fully characterized.

Using the TargetScan, miRDB, and miRSystem online analysis tools, we explored the potential target genes of miR-361. All these tools identified over 200 overlapping target genes for miR-361. However, only a small proportion of these miR-361 target genes (around 8%) have been experimentally validated in tumor cells (Figure 3). As shown in Figure 4, we identified six unreported miR-361 targets (*ARF4, DEPDC1, EPHA4, PHACTR4,* and *BSG*) and two previously reported miR-361 targets (*Twist* and *VEGF-A*) [10,17]. We investigated the expression of these genes in human ovarian cancer tissues and normal tissues using the Oncomine database (https://www.oncomine.org/resource/login.html). The expression levels of these genes were significantly increased in ovarian cancer tissues (Figure 4), indicating that these genes might be important components of miR-361-mediated gene networks.

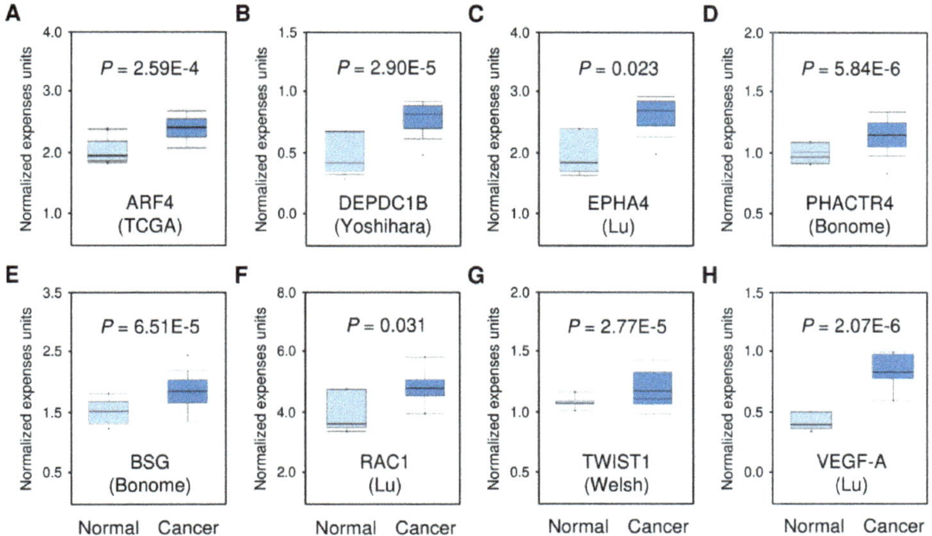

Figure 4. Oncomine analysis indicates higher expression of the predicted miR-361 targets in ovarian cancer tissues. The box plots revealed the expression levels of *ARF4* (**A**, The Cancer Genome Atlas (TCGA)), *DEPDC1B* (**B**, Yoshihara), *EPHA4* (**C**, Lu), *PHACTR4* (**D**, Bonome), *BSG* (**E**, Bonome), *RAC1* (**F**, Lu), *Twist* (**G**, Welsh) and *VEGF-A* (**H**, Lu) in ovarian cancer tissues with respect to normal tissues. *P*-values were calculated using the Oncomine database through two-sided Student's *t*-test. A value of $p < 0.05$ was considered as statistically significant.

Despite the recent progress that has been made towards the identification of the molecular mechanisms causing dysregulation of miR-361, there are currently many unclear points. Given that lncRNAs interact with miRNAs to form the intertwined and regulatory networks that control cancer development and progression [74], detailed analysis of the interactions between lncRNAs and miR-361 may partly explain the frequent downregulation of miR-361 observed in cancers. With the perspective of therapeutic miR-361 restoration, the existence of upstream suppressors (such as EZH2) should be taken into consideration.

The most common types of genetic variations in the human genome are single nucleotide polymorphisms (SNPs), which are the results of point mutations that produce single base-pair differences among chromosome sequences [75]. SNPs are located in different regions of genes (such as promoters, exons, introns, 5′-UTRs, and 3′-UTRs) and alter gene expression through complex mechanisms [76]. The occurrence of SNPs may affect cancer susceptibility and represent genetic markers for cancer risk [77,78]. The function of miRNAs may be influenced through SNPs in their own sequences and in their target gene sequences [79]. Some SNPs were shown to interfere with the function of certain miRNAs and affect the expression of the miRNA targets [80]. A study confirmed that a functional SNP in *CD80* 3′-UTR disrupted the inhibitory effect of miR-361 on CD80 expression in gastric cancer cells [81]. Further characterization of SNPs in the miR-361 gene and its potential targets in cancer cells would shed light on the molecular mechanisms responsible for miR-361-mediated carcinogenesis and metastasis.

9. Conclusions

Emerging works on miR-361 demonstrated its importance in controlling multiple malignant features of tumor cells and regulating critical aspects of the tumor microenvironment. MiR-361 has great potential to be used as a promising diagnostic, prognostic, and predictive biomarker for cancers and has therapeutic potential to improve cancer treatment. Additional works will continue to elucidate how miR-361 exerts significant effects on tumor progression and will offer crucial therapeutic opportunities for cancer patients.

Author Contributions: Writing, review, and editing: D.X., P.D., Y.X., J.Y., K.I., Y.K., N.K., Y.T., and H.W.

Funding: This work was supported by a grant from JSPS Grant-in-Aid for Scientific Research (C) (18K09278 and 19K09769), the Science and Technology Planning Project of Guangdong Province, China (2014A020212124) and an NIH/NCI grant 1R21CA216585-01A1 to J. Yue.

Acknowledgments: We thank Zhujie Xu for her full contribution in preparing the figures.

Conflicts of Interest: The authors declare no conflict of interest.

References

1. ENCODE Project Consortium. An integrated encyclopedia of DNA elements in the human genome. *Nature* **2012**, *489*, 57–74. [CrossRef] [PubMed]

2. Shivdasani, R.A. MicroRNAs: Regulators of gene expression and cell differentiation. *Blood* **2006**, *108*, 3646–3653. [CrossRef] [PubMed]

3. Da Sacco, L.; Masotti, A. Recent Insights and Novel Bioinformatics Tools to Understand the Role of MicroRNAs Binding to 5′ Untranslated Region. *Int. J. Mol. Sci.* **2012**, *14*, 480–495. [CrossRef] [PubMed]

4. O'Brien, J.; Hayder, H.; Zayed, Y.; Peng, C. Overview of MicroRNA Biogenesis, Mechanisms of Actions, and Circulation. *Front. Endocrinol.* **2018**, *9*, 402. [CrossRef] [PubMed]

5. Pasquinelli, A.E.; Reinhart, B.J.; Slack, F.; Martindale, M.Q.; Kuroda, M.I.; Maller, B.; Hayward, D.C.; Ball, E.E.; Degnan, B.; Müller, P.; et al. Conservation of the sequence and temporal expression of let-7 heterochronic regulatory RNA. *Nature* **2000**, *408*, 86–89. [CrossRef] [PubMed]

6. Ruvkun, G. Glimpses of a Tiny RNA World. *Science* **2001**, *294*, 797–799. [CrossRef] [PubMed]

7. Lee, R.C. An Extensive Class of Small RNAs in Caenorhabditis elegans. *Science* **2001**, *294*, 862–864. [CrossRef] [PubMed]

8. Ebert, M.S.; Sharp, P.A. Roles for microRNAs in conferring robustness to biological processes. *Cell* **2012**, *149*, 515–524. [CrossRef] [PubMed]

9. Król, J.; Loedige, I.; Filipowicz, W. The widespread regulation of microRNA biogenesis, function and decay. *Nat. Rev. Genet.* **2010**, *11*, 597–610. [CrossRef]

10. Ihira, K.; Dong, P.; Xiong, Y.; Watari, H.; Konno, Y.; Hanley, S.J.; Noguchi, M.; Hirata, N.; Suizu, F.; Yamada, T.; et al. EZH2 inhibition suppresses endometrial cancer progression via miR-361/Twist axis. *Oncotarget* **2017**, *8*, 13509–13520. [CrossRef] [PubMed]

11. Peng, Y.; Croce, C.M. The role of MicroRNAs in human cancer. *Signal. Transduct. Target. Ther.* **2016**, *1*, 15004. [CrossRef] [PubMed]

12. Hayes, J.; Peruzzi, P.P.; Lawler, S. MicroRNAs in cancer: Biomarkers, functions and therapy. *Trends Mol. Med.* **2014**, *20*, 460–469. [CrossRef] [PubMed]

13. Schwarzenbach, H.; Nishida, N.; Calin, G.A.; Pantel, K. Clinical relevance of circulating cell-free microRNAs in cancer. *Nat. Rev. Clin. Oncol.* **2014**, *11*, 145–156. [CrossRef] [PubMed]

14. Bouchie, A. First microRNA mimic enters clinic. *Nat. Biotechnol.* **2013**, *31*, 577. [CrossRef] [PubMed]

15. Dong, P.; Xiong, Y.; Yu, J.; Chen, L.; Tao, T.; Yi, S.; Hanley, S.J.B.; Yue, J.; Watari, H.; Sakuragi, N. Control of PD-L1 expression by miR-140/142/340/383 and oncogenic activation of the OCT4–miR-18a pathway in cervical cancer. *Oncogene* **2018**, *37*, 5257–5268. [CrossRef] [PubMed]

16. Rupaimoole, R.; Calin, G.A.; Lopez-Berestein, G.; Sood, A.K. MicroRNA deregulation in cancer cells and the tumor microenvironment. *Cancer Discov.* **2016**, *6*, 235–246. [CrossRef] [PubMed]

17. Kanitz, A.; Imig, J.; Dziunycz, P.J.; Primorac, A.; Galgano, A.; Hofbauer, G.F.L.; Gerber, A.P.; Detmar, M. The Expression Levels of MicroRNA-361-5p and Its Target VEGFA Are Inversely Correlated in Human Cutaneous Squamous Cell Carcinoma. *PLoS ONE* **2012**, *7*, e49568. [CrossRef] [PubMed]

18. Wang, K.; Qi, X.; Liu, H.; Su, H. MiR-361 inhibits osteosarcoma cell lines invasion and proliferation by targeting FKBP14. *Eur. Rev. Med. Pharmacol. Sci.* **2018**, *22*, 79–86. [CrossRef] [PubMed]

19. Ma, F.; Zhang, L.; Ma, L.; Zhang, Y.; Zhang, J.; Guo, B. MiR-361-5p inhibits glycolytic metabolism, proliferation and invasion of breast cancer by targeting FGFR1 and MMP-1. *J. Exp. Clin. Cancer Res.* **2017**, *36*, 158. [CrossRef]

20. Chang, J.T.H.; Wang, F.; Chapin, W.; Huang, R.S. Identification of MicroRNAs as Breast Cancer Prognosis Markers through the Cancer Genome Atlas. *PLoS ONE* **2016**, *11*, e0168284. [CrossRef] [PubMed]

21. Cao, Z.G.; Huang, Y.N.; Yao, L.; Liu, Y.R.; Hu, X.; Hou, Y.F.; Shao, Z.M. Positive expression of miR-361-5p indicates better prognosis for breast cancer patients. *J. Thorac. Dis.* **2016**, *8*, 1772–1779. [CrossRef] [PubMed]

22. Han, J.; Yu, J.; Dai, Y.; Li, J.; Guo, M.; Song, J.; Zhou, X. Overexpression of miR-361-5p in triple-negative breast cancer (TNBC) inhibits migration and invasion by targeting RQCD1 and inhibiting the EGFR/PI3K/Akt pathway. *Bosn. J. Basic Med. Sci.* **2019**, *19*, 52–59. [CrossRef] [PubMed]

23. Wei, C.; Qu, J.; Zhang, X.; Li, J.; Liu, J. MicroRNA-361-5p inhibits epithelial-to-mesenchymal transition of glioma cells through targeting Twist1. *Oncol. Rep.* **2017**, *37*, 1849–1856.

24. Liu, J.; Yang, J.; Yu, L.; Rao, C.; Wang, Q.; Sun, C.; Shi, C.; Hua, D.; Zhou, X.; Luo, W.; et al. miR-361-5p inhibits glioma migration and invasion by targeting SND1. *OncoTargets Ther.* **2018**, *11*, 5239–5252. [CrossRef] [PubMed]

25. Li, R.; Dong, B.; Wang, Z.; Jiang, T.; Chen, G. MicroRNA-361-5p inhibits papillary thyroid carcinoma progression by targeting ROCK1. *Biomed. Pharmacother.* **2018**, *102*, 988–995. [CrossRef] [PubMed]

26. Hou, X.W.; Sun, X.; Yu, Y.; Zhao, H.M.; Yang, Z.J.; Wang, X.; Cao, X.C. miR-361-5p suppresses lung cancer cell lines progression by targeting FOXM1. *Neoplasma* **2017**, *64*, 526–534. [CrossRef] [PubMed]

27. Zhang, S.; Liu, Z.; Wu, L.; Wang, Y. MiR-361 targets Yes-associated protein (YAP) mRNA to suppress cell proliferation in lung cancer. *Biochem. Biophys. Res. Commun.* **2017**, *492*, 468–473. [CrossRef] [PubMed]

28. Wang, J.; Shang, J.; Yang, S.; Zhang, Y.; Zhao, X. microRNA-361 targets Wilms' tumor 1 to inhibit the growth, migration and invasion of non-small-cell lung cancer cells. *Mol. Med. Rep.* **2016**, *14*, 5415–5421.

29. Ma, Y.; Bao, C.; Kong, R.; Xing, X.; Zhang, Y.; Li, S.; Zhang, W.; Jiang, J.; Zhang, J.; Qiao, Z.; et al. MicroRNA-361-5p suppresses cancer progression by targeting signal transducer and activator of transcription 6 in non-small cell lung cancer. *Mol. Med. Rep.* **2015**, *12*, 7367–7373. [CrossRef]

30. Tian, L.; Zhao, Z.; Xie, L.; Zhu, J. MiR-361-5p inhibits the mobility of gastric cancer cells through suppressing epithelial-mesenchymal transition via the Wnt/β-catenin pathway. *Gene* **2018**, *675*, 102–109. [CrossRef]

31. Tian, L.; Zhao, Z.; Xie, L.; Zhu, J. MiR-361-5p suppresses chemoresistance of gastric cancer cells by targeting FOXM1 via the PI3K/Akt/mTOR pathway. *Oncotarget* **2017**, *9*, 4886–4896. [CrossRef] [PubMed]

32. Ma, F.; Song, H.; Guo, B.; Zhang, Y.; Zheng, Y.; Lin, C.; Wu, Y.; Guan, G.; Sha, R.; Zhou, Q.; et al. MiR-361-5p inhibits colorectal and gastric cancer growth and metastasis by targeting staphylococcal nuclease domain containing-1. *Oncotarget* **2015**, *6*, 17404–17416. [CrossRef] [PubMed]

33. Sun, J.J.; Chen, G.Y.; Xie, Z.T. MicroRNA-361-5p Inhibits Cancer Cell Growth by Targeting CXCR6 in Hepatocellular Carcinoma. *Cell. Physiol. Biochem.* **2016**, *38*, 777–785. [CrossRef] [PubMed]

34. Ma, J.; Jing, X.; Chen, Z.; Duan, Z.; Zhang, Y. MiR-361-5p decreases the tumorigenicity of epithelial ovarian cancer cells by targeting at RPL22L1 and c-Met signaling. *Int. J. Clin. Exp. Pathol.* **2018**, *11*, 2588–2596.

35. Gao, F.; Feng, J.; Yao, H.; Li, Y.; Xi, J.; Yang, J. LncRNA SBF2-AS1 promotes the progression of cervical cancer by regulating miR-361-5p/FOXM1 axis. *Artif. Cells Nanomed. Biotechnol.* **2019**, *47*, 776–782. [CrossRef] [PubMed]

36. Chen, S.; Zhang, G.; Yu, Q.; Zhang, X.; Han, G. MicroRNA-361 inhibited prostate carcinoma cell invasion by targeting Gli1. *Int. J. Clin. Exp. Pathol.* **2017**, *10*, 6108–6116.

37. Liu, D.; Tao, T.; Xu, B.; Chen, S.; Liu, C.; Zhang, L.; Lu, K.; Huang, Y.; Jiang, L.; Zhang, X.; et al. MiR-361-5p acts as a tumor suppressor in prostate cancer by targeting signal transducer and activator of transcription-6(STAT6). *Biochem. Biophys. Res. Commun.* **2014**, *445*, 151–156. [CrossRef]

38. Ling, Z.; Liu, D.; Zhang, G.; Liang, Q.; Xiang, P.; Xu, Y.; Han, C.; Tao, T. miR-361-5p modulates metabolism and autophagy via the Sp1-mediated regulation of PKM2 in prostate cancer. *Oncol. Rep.* **2017**, *38*, 1621–1628. [CrossRef]

39. Koutova, L.; Sterbova, M.; Pazourková, E.; Pospisilova, S.; Svobodová, I.; Horinek, A.; Lysak, D.; Korabečná, M. The impact of standard chemotherapy on miRNA signature in plasma in AML patients. *Leuk. Res.* **2015**, *39*, 1389–1395. [CrossRef]

40. Ha, M.; Kim, V.N. Regulation of microRNA biogenesis. *Nat. Rev. Mol. Cell Biol.* **2014**, *15*, 509–524. [CrossRef]

41. Dong, P.; Xiong, Y.; Yue, J.; Hanley, S.J.B.; Kobayashi, N.; Todo, Y.; Watari, H. Exploring lncRNA-Mediated Regulatory Networks in Endometrial Cancer Cells and the Tumor Microenvironment: Advances and Challenges. *Cancers* **2019**, *11*, 234. [CrossRef]

42. Wang, K.; Sun, T.; Li, N.; Wang, Y.; Wang, J.X.; Zhou, L.Y.; Long, B.; Liu, C.Y.; Liu, F.; Li, P.F. MDRL lncRNA Regulates the Processing of miR-484 Primary Transcript by Targeting miR-361. *PLoS Genet.* **2014**, *10*, e1004467. [CrossRef]

43. Zhang, J.; Liang, Y.; Huang, X.; Guo, X.; Liu, Y.; Zhong, J.; Yuan, J. STAT3-induced upregulation of lncRNA MEG3 regulates the growth of cardiac hypertrophy through miR-361-5p/HDAC9 axis. *Sci. Rep.* **2019**, *9*, 460. [CrossRef]

44. Ajiro, M.; Katagiri, T.; Ueda, K.; Nakagawa, H.; Fukukawa, C.; Lin, M.L.; Park, J.H.; Nishidate, T.; Daigo, Y.; Nakamura, Y. Involvement of RQCD1 overexpression, a novel cancer-testis antigen, in the Akt pathway in breast cancer cells. *Int. J. Oncol.* **2009**, *35*, 673–681.

45. Wang, J.; Lu, Y.; Wang, J.; Koch, A.E.; Zhang, J.; Taichman, R.S. CXCR6 Induces Prostate Cancer Progression by the AKT/Mammalian Target of Rapamycin Signaling Pathway. *Cancer Res.* **2008**, *68*, 10367–10376. [CrossRef]

46. Nieto, M.A.; Huang, R.Y.J.; Jackson, R.A.; Thiery, J.P. EMT: 2016. *Cell* **2016**, *166*, 21–45. [CrossRef]

47. Huo, W.; Zhao, G.; Yin, J.; Ouyang, X.; Wang, Y.; Yang, C.; Wang, B.; Dong, P.; Wang, Z.; Watari, H.; et al. Lentiviral CRISPR/Cas9 vector mediated miR-21 gene editing inhibits the epithelial to mesenchymal transition in ovarian cancer cells. *J. Cancer* **2017**, *8*, 57–64. [CrossRef]

48. Dong, P.; Xiong, Y.; Watari, H.; Hanley, S.J.; Konno, Y.; Ihira, K.; Suzuki, F.; Yamada, T.; Kudo, M.; Yue, J.; et al. Suppression of iASPP-dependent aggressiveness in cervical cancer through reversal of methylation silencing of microRNA. *Sci. Rep.* **2016**, *6*, 35480. [CrossRef]

49. Zhang, Y.; Xia, M.; Jin, K.; Wang, S.; Wei, H.; Fan, C.; Wu, Y.; Li, X.; Li, X.; Li, G.; et al. Function of the c-Met receptor tyrosine kinase in carcinogenesis and associated therapeutic opportunities. *Mol. Cancer* **2018**, *17*, 45. [CrossRef]

50. Liao, G.B.; Li, X.Z.; Zeng, S.; Liu, C.; Yang, S.M.; Yang, L.; Hu, C.J.; Bai, J.Y. Regulation of the master regulator FOXM1 in cancer. *Cell Commun. Signal.* **2018**, *16*, 57. [CrossRef]

51. Zhan, T.; Rindtorff, N.; Boutros, M. Wnt signaling in cancer. *Oncogene* **2017**, *36*, 1461–1473. [CrossRef]

52. Jang, M.; Kim, S.S.; Lee, J. Cancer cell metabolism: Implications for therapeutic targets. *Exp. Mol. Med.* **2013**, *45*, e45. [CrossRef]

53. Zhang, Y.; Yang, P.; Wang, X.-F. Microenvironmental regulation of cancer metastasis by miRNAs. *Trends Cell Biol.* **2014**, *24*, 153–160. [CrossRef]

54. Yang, N.; Zhu, S.; Lv, X.; Qiao, Y.; Liu, Y.J.; Chen, J. MicroRNAs: Pleiotropic Regulators in the Tumor Microenvironment. *Front. Immunol.* **2018**, *9*, 2491. [CrossRef]

55. Salinas-Vera, Y.M.; Marchat, L.A.; Gallardo-Rincón, D.; Ruiz-García, E.; Astudillo-De La Vega, H.; Echavarría-Zepeda, R.; López-Camarillo, C. AngiomiRs: MicroRNAs driving angiogenesis in cancer (Review). *Int. J. Mol. Med.* **2019**, *43*, 657–670. [CrossRef]

56. Wang, D.; Dubois, R.N. Immunosuppression associated with chronic inflammation in the tumor microenvironment. *Carcinogenesis* **2015**, *36*, 1085–1093. [CrossRef]

57. Palena, C.; Hamilton, D.H.; Fernando, I.R. Influence of IL-8 on the epithelial–mesenchymal transition and the tumor microenvironment. *Futur. Oncol.* **2012**, *8*, 713–722. [CrossRef]

58. Landskron, G.; De La Fuente, M.; Thuwajit, P.; Thuwajit, C.; Hermoso, M.A. Chronic Inflammation and Cytokines in the Tumor Microenvironment. *J. Immunol. Res.* **2014**, *2014*, 1–19. [CrossRef]

59. Yu, H.; Pardoll, D.; Jove, R. STATs in cancer inflammation and immunity: A leading role for STAT3. *Nat. Rev. Cancer* **2009**, *9*, 798–809. [CrossRef]

60. León-Cabrera, S.; Molina-Guzman, E.; Delgado-Ramírez, Y.; Vazquez-Sandoval, A.; Ledesma-Soto, Y.; Perez-Plasencia, C.; Chirino, I.Y.; Delgado-Buenrostro, N.L.; Rodriguez-Sosa, M.; Vaca-Paniagua, F.; et al. Lack of STAT6 Attenuates Inflammation and Drives Protection against Early Steps of Colitis-Associated Colon Cancer. *Cancer Immunol. Res.* **2017**, *5*, 385–396. [CrossRef]

61. Calin, G.A.; Croce, C.M. MicroRNA signatures in human cancers. *Nat. Rev. Cancer* **2006**, *6*, 857–866. [CrossRef]

62. Lu, J.; Getz, G.; Miska, E.A.; Alvarez-Saavedra, E.; Lamb, J.; Peck, D.; Sweet-Cordero, A.; Ebert, B.L.; Mak, R.H.; Ferrando, A.A.; et al. MicroRNA expression profiles classify human cancers. *Nature* **2005**, *435*, 834–838. [CrossRef]

63. Scholl, V.; Hassan, R.; Zalcberg, I.R. miRNA-451: A putative predictor marker of Imatinib therapy response in chronic myeloid leukemia. *Leuk. Res.* **2012**, *36*, 119–121. [CrossRef]

64. Sebio, A.; Paré, L.; Páez, D.; Salazar, J.; González, A.; Sala, N.; del Río, E.; Martín-Richard, M.; Tobeña, M.; Barnadas, A. The LCS6 polymorphism in the binding site of let-7 microRNA to the KRAS 3-untranslated region: Its role in the efficacy of anti-EGFR-based therapy in metastatic colorectal cancer patients. *Pharm. Genom.* **2013**, *23*, 142–147. [CrossRef]

65. Pardini, B.; Rosa, F.; Barone, E.; Di Gaetano, C.; Slyskova, J.; Novotny, J.; Levy, M.; Garritano, S.; Vodickova, L.; Buchler, T.; et al. Variation within 3-UTRs of Base Excision Repair Genes and Response to Therapy in Colorectal Cancer Patients: A Potential Modulation of microRNAs Binding. *Clin. Cancer Res.* **2013**, *19*, 6044–6056. [CrossRef]

66. Zhuang, Z.L.; Tian, F.M.; Sun, C.L. Downregulation of miR-361-5p associates with aggressive clinicopathological features and unfavorable prognosis in non-small cell lung cancer. *Eur. Rev. Med. Pharmacol. Sci.* **2016**, *20*, 5132–5136.

67. Mitchell, P.S.; Parkin, R.K.; Kroh, E.M.; Fritz, B.R.; Wyman, S.K.; Pogosova-Agadjanyan, E.L.; Peterson, A.; Noteboom, J.; O'Briant, K.C.; Allen, A.; et al. Circulating microRNAs as stable blood-based markers for cancer detection. *Proc. Natl. Acad. Sci. USA* **2008**, *105*, 10513–10518. [CrossRef]

68. Heneghan, H.M.; Miller, N.; Kelly, R.; Newell, J.; Kerin, M.J. Systemic miRNA-195 Differentiates Breast Cancer from Other Malignancies and Is a Potential Biomarker for Detecting Noninvasive and Early Stage Disease. *Oncologist* **2010**, *15*, 673–682. [CrossRef]

69. Cuk, K.; Zucknick, M.; Madhavan, D.; Schott, S.; Golatta, M.; Heil, J.; Marmé, F.; Turchinovich, A.; Sinn, P.; Sohn, C.; et al. Plasma MicroRNA Panel for Minimally Invasive Detection of Breast Cancer. *PLoS ONE* **2013**, *8*, e76729. [CrossRef]

70. Inns, J.; James, V. Circulating microRNAs for the prediction of metastasis in breast cancer patients diagnosed with early stage disease. *Breast* **2015**, *24*, 364–369. [CrossRef]

71. Wang, H.; Peng, R.; Wang, J.; Qin, Z.; Xue, L. Circulating microRNAs as potential cancer biomarkers: The advantage and disadvantage. *Clin. Epigenetics* **2018**, *10*, 59. [CrossRef]

72. Jin, X.; Chen, Y.; Chen, H.; Fei, S.; Chen, D.; Cai, X.; Liu, L.; Lin, B.; Su, H.; Zhao, L.; et al. Evaluation of Tumor-Derived Exosomal miRNA as Potential Diagnostic Biomarkers for Early-Stage Non–Small Cell Lung Cancer Using Next-Generation Sequencing. *Clin. Cancer Res.* **2017**, *23*, 5311–5319. [CrossRef] [PubMed]

73. Rupaimoole, R.; Slack, F.J. MicroRNA therapeutics: Towards a new era for the management of cancer and other diseases. *Nat. Rev. Drug Discov.* **2017**, *16*, 203–222. [CrossRef] [PubMed]

74. Anastasiadou, E.; Jacob, L.S.; Slack, F.J. Non-coding RNA networks in cancer. *Nat. Rev. Cancer* **2018**, *18*, 5–18. [CrossRef] [PubMed]

75. Erichsen, H.C.; Chanock, S.J. SNPs in cancer research and treatment. *Br. J. Cancer* **2004**, *90*, 747–751. [CrossRef] [PubMed]

76. Shukla, A.; Alsarraj, J.; Hunter, K. Understanding susceptibility to breast cancer metastasis: The genetic approach. *Breast Cancer Manag.* **2014**, *3*, 165–172. [CrossRef]

77. Deng, N.; Zhou, H.; Fan, H.; Yuan, Y. Single nucleotide polymorphisms and cancer susceptibility. *Oncotarget* **2017**, *8*, 110635–110649. [CrossRef] [PubMed]

78. Ding, C.; Li, C.; Wang, H.; Li, B.; Guo, Z. A miR-SNP of the XPO5 gene is associated with advanced non-small-cell lung cancer. *OncoTargets Ther.* **2013**, *6*, 877–881.

79. Ziebarth, J.D.; Bhattacharya, A.; Cui, Y. Integrative Analysis of Somatic Mutations Altering MicroRNA Targeting in Cancer Genomes. *PLoS ONE* **2012**, *7*, e47137. [CrossRef]

80. Wu, Y.; Xiao, Y.; Ding, X.; Zhuo, Y.; Ren, P.; Zhou, C.; Zhou, J. A miR-200b/200c/429-Binding Site Polymorphism in the 3′ Untranslated Region of the AP-2α Gene Is Associated with Cisplatin Resistance. *PLoS ONE* **2011**, *6*, e29043. [CrossRef]

81. Wu, R.; Li, F.; Zhu, J.; Tang, R.; Qi, Q.; Zhou, X.; Li, R.; Wang, W.; Hua, D.; Chen, W. A functional variant at miR-132-3p, miR-212-3p, and miR-361-5p binding site in CD80 gene alters susceptibility to gastric cancer in a Chinese Han population. *Med. Oncol.* **2014**, *31*, 60. [CrossRef]

Review

MicroRNAs and Metastasis

Carla Solé [1] and Charles H. Lawrie [1,2,3,*]

[1] Molecular Oncology Group, Biodonostia Research Institute, 20014 San Sebastián, Spain; carla.sole@biodonostia.org

[2] IKERBASQUE, Basque Foundation for Science, 48013 Bilbao, Spain

[3] Radcliffe Department of Medicine, University of Oxford, Oxford OX3 9DU, UK

* Correspondence: charles.lawrie@biodonostia.org or charles.lawrie@ndcls.ox.ac.uk; Tel.: +34-943-006138

Received: 27 November 2019; Accepted: 27 December 2019; Published: 30 December 2019 

Abstract: Metastasis, the development of secondary malignant growths at a distance from the primary site of a cancer, is associated with almost 90% of all cancer deaths, and half of all cancer patients present with some form of metastasis at the time of diagnosis. Consequently, there is a clear clinical need for a better understanding of metastasis. The role of miRNAs in the metastatic process is beginning to be explored. However, much is still to be understood. In this review, we present the accumulating evidence for the importance of miRNAs in metastasis as key regulators of this hallmark of cancer.

Keywords: miRNA; metastasis; cancer; liquid biopsies

---

1. Introduction

Nearly half of all patients with cancer present with some form of metastasis at time of diagnosis [1]. Unfortunately, with very few exceptions, metastatic disease remains essentially incurable and almost 90% of all cancer deaths are associated with metastasis [2,3]. Consequently, there is a clear clinical need for a better understanding of metastasis and the development of novel therapeutics targeting this process.

In essence, metastasis, the development of secondary malignant growths at a distance from the primary site of a cancer, is a multiphase process that requires tumor cells to detach from the primary tumor mass, enter and travel through the blood or lymph system, to leave circulation, and to form a new tumor in other organs or tissues of the body. The process of metastasis is very inefficient, with the survival rate of circulating tumor cells (CTCs) being as low as 0.2%, and then only those survivors can successfully metastasize target organs only when optimal conditions occur [4].

MicroRNAs (miRNAs) are small (19–25 nt) non-coding single-stranded RNAs that regulate gene expression through imperfect binding to the 3′ untranslated region (UTR) of target genes [5] (Figure 1). Because a single miRNA can target several hundred genes, and a single target gene often contains multiple miRNA binding sites, it is believed that more than 60% of all human genes are a direct target for miRNA regulation [6]. Consequently, miRNAs have been shown to play key regulatory roles in virtually every aspect of biology including both physiological and pathological processes, most notably in cancer. The importance of miRNAs in controlling cancer development and progression is well established [7]. In this review, we will consider the identity and role of miRNAs in the metastatic process.

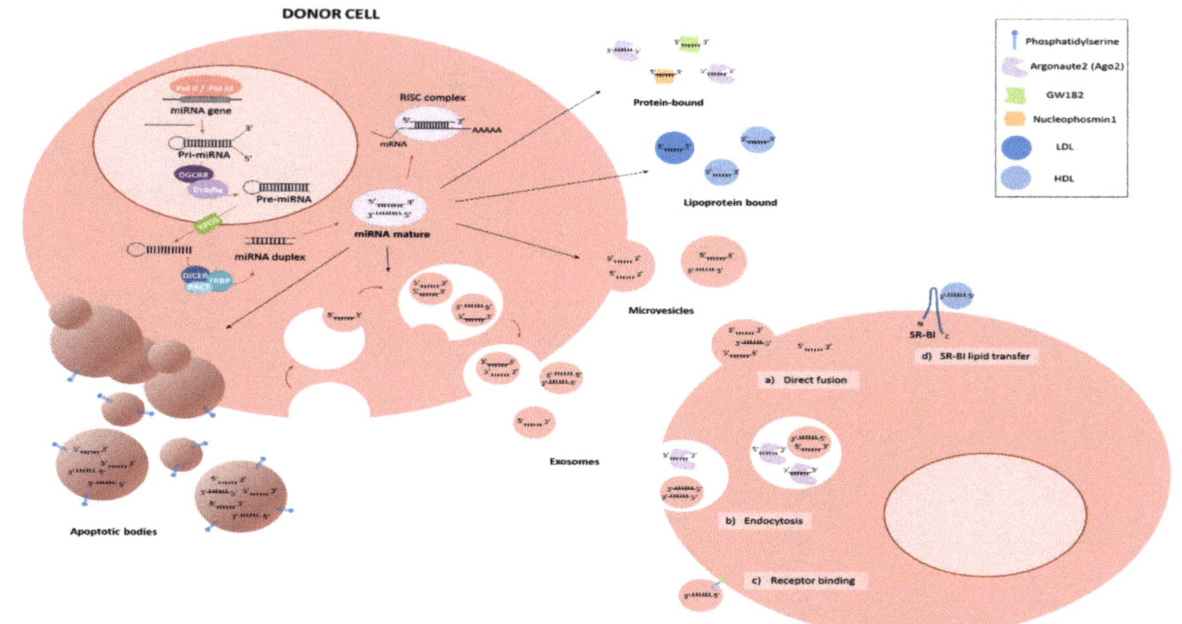

Figure 1. The release of microRNAs (miRNAs) in the extracellular environment. MiRNAs could enter the RNA induced silencing complex (RISC) to regulate mRNA expression and/or translation or can be released outside cell to reach biological fluids. Circulating miRNAs could be found in two major forms, vesicle associated and non-vesicle associated. When miRNAs are released by donor cells through vesicles, these could be microvesicles by outward budding (100–1000 nm), exosomes (50–100 nm) or apoptotic bodies as a result of apoptosis (1–5 μm). On the other hand, cell-free miRNAs could be found bound to proteins, such as Argonaute2 (AGO2), GW182 and Nucleophosmin1 (NPM1), or bound to lipoproteins, high-density lipoproteins (HDLs) or low-density lipoproteins (LDLs) [8]. Circulating miRNAs uptake by receptor cells occurred by membrane direct fusion (**a**), endocytosis (**b**) and receptor binding (**c**), which could trigger a downstream cascade or could produce internalization of the vesicle [9,10]. Moreover, it has been demonstrated that delivery of HDL miRNAs is dependent on scavenger receptor class B type I (SR-BI) [11,12].

The first reports associating miRNAs with metastasis came in 2007, with the demonstration that *miR-10b* was induced by Twist1 binding and could promote metastasis in breast cancer in vitro and in vivo through targeting of Homeobox D10 (HOXD10) [13]. In the same year, also in breast cancer, *let-7* was identified as a suppressor of metastasis acting to target the GTPase H-RAS and High Mobility Group AT-Hook 2 (*HMGA2*) gene in tumor-initiating cells, resulting in reduced proliferation and mammosphere formation in vitro and decreased metastasis in a NOD/SCID murine model [14]. In subsequent studies, breast cancer has remained the main focus of research investigating miRNAs in metastasis, and many studies have shown that miRNAs can act as both promoters or inhibitors of metastasis in cancer and modulate many steps of the metastatic pathway, including migration, invasion, adhesion, the epithelial–mesenchymal transition (EMT), niche conditioning and proliferation in secondary site (Table 1) [15,16].

2. Metastasis—Promoting miRNAs

In common with *miR-10b*, many of the metastasis-promoting miRNAs that have been characterized have been identified in breast cancer. For example, *miR-105* has been identified as being up-regulated in tumor cells and exosomes derived from breast cancer cells were demonstrated to breakdown vascular endothelial barriers and induce vascular permeability, thereby promoting metastasis by targeting of *ZO-1*, a component of cell–cell adhesion complexes in endothelial and epithelial cells [17]. Furthermore, these authors used exosomes to reduce tight junction formation in endothelial monolayers and induce vascular permeability and metastasis in vivo. *MiR-181b-3p* was demonstrated to promote EMT in vitro, with its inhibition reducing the expression of mesenchymal markers, migration and invasion in highly metastatic cell lines [18]. *YWHAG* was identified as a direct target of *miR-181b-3p*, which in turn led to protein stabilization of the EMT regulator, *Snail*. The expression of *miR-374a* was found to be up-regulated in patients with distant metastases and poor prognostic outcome. *MiR-374a* A was also demonstrated to promote EMT and metastasis in vitro and in vivo by activation of the Wnt/β-catenin pathway by targeting *WIF1*, *PTEN*, and *WNT5A* which inhibit this cascade [19]. Similarly, *miR-135a* is also highly expressed in metastatic breast cancer and has been demonstrated to promote migration and invasion mediated by targeting *HOXA10* and *APC* [20,21]. The same miRNA is up-regulated in hepatocellular carcinoma (HCC) patients and was found to promote migration and invasion through targeting of *FOXO1* [22]. *MiR-96* is also highly expressed in metastatic breast cancer and was demonstrated to promote cell proliferation, migration and invasion in vitro and enhanced tumor growth in vivo via targeting of *PTPN9* [23]. This miRNA was also observed to be highly expressed in HCC tissue, where it was demonstrated to increase proliferation and migration in vitro through inhibition of *ephrinA5* expression [23]. The hypoxia-induced miRNA, *miR-210*, was found to be up-regulated in breast cancer stem cells (BCSCs) and the expression of this miRNA was shown to promote migration and invasion of these cells mediated by direct targeting of *E-Cadherin* and its transcription repressor, *Snail* [24]. BCSCs were also shown to express high levels of *miR-29a*, which increased levels of migration, invasion and EMT through targeting of the methyltransferase *SUV420H2* [25]. Further, in breast cancer, *miR-130* was found to target *FOSL1* and suppress the inhibition of *ZO-1*, thereby promoting cell migration and invasion [26]. Both *miR-8084* and *miR-1204* were found to enhance migration and invasion through EMT induction via targeting of *ING2* and *VDR* respectively [27,28].

Outside of breast cancer, several other miRNAs have been described to promote EMT. For example, *miR-93* was demonstrated to target *FOXA1* in endometrial carcinoma [29] and *miR-197*, highly up-regulated in metastatic HCC, was found to target *NKD1* and *DKK2*, leading to inhibition of Wnt/β-catenin signaling [30]. In gastric cancer (GC), TGF-β was shown to up-regulate *miR-577* expression, which in turn targets *SDPR*, leading to EMT induction [31]; and *miR-520c*, which was up-regulated in both cell lines and tissues and shown to increase proliferation, migration and invasion of cancer cells through *IRF2* targeting [32]. Similarly, in colorectal cancer (CRC), *miR-1269a* has been found to enhance TGF-β signaling through targeting of *Smad7* and *HOXD10* [33].

Esophageal carcinomas have been observed to over-express both *miR-20b* and *miR-9*, which were demonstrated to promote tumorigenic processes including metastasis [34,35]. In particular, the over-expression of *miR-20b* was shown to promote migration and invasion in vitro by targeting and regulation of *PTEN* [34]. In contrast, *miR-9* was shown to stimulate metastasis through promotion of cell migration and induction of the EMT pathway by inhibiting E-Cadherin expression which in turn was demonstrated to induce c-myc and CD44 expression [35]. In non-small-cell lung cancer (NSCLC) *miR-574-5p* was found to promote metastasis in vitro and in vivo through targeting of *TPRU* [36]. In ovarian cancer, miR-194 was demonstrated to increase the growth, migration, and invasion of cells in vitro through targeting of the *PTPN12* gene [37].

3. Metastasis—Suppressing miRNAs

In addition to miRNAs that promote metastasis, other miRNAs negatively regulate this process and are consequently found to be down-regulated in cancer tissues and/or cell lines. For example, members of the *miR-200* family (i.e., *miR-200a*, *miR-200b*, *miR-200c*, *miR-141* and *miR-429*) as well as *miR-205* have been shown to inhibit the expression of transcription repressors ZEB1 and ZEB2 to enhance E-Cadherin expression, thereby inhibiting EMT in breast cancer [38–40]. In a study that looked at miRNA expression in 59 of the NCI60 cell lines that had a E-Cadherin high and vimentin low (EMT inhibitory) phenotype, they observed a strong negative correlation with *miR-200* expression, suggesting that this miRNA is a universal regulator of metastasis in many cancer types including lung, kidney, colon and ovarian cancer [41]. Specifically, *miR-200b* was shown to be down-regulated in triple negative breast cancer (TNBC) as a result of the recruitment of DNMT3A by MYC, which in turn binds to the *miR-200b* promoter region, resulting in promoter methylation and silencing, thereby inhibiting migration, invasion and mammosphere formation in TNBC cells [42]. Part of the *miR-200* family, *miR-141* is also down-regulated in prostate cancer (PC), and its ectopic expression was shown to inhibit invasion and metastasis and to convey a strong epithelial phenotype with a partial mesenchymal phenotype [43].

Similar to the *miR-200* family, the *miR-29* family (*miR-29a/b/c*) is another group of negative regulators of metastasis. In head and neck squamous cell carcinoma (HNSCC), these miRNAs were found to inhibit migration and invasion through targeting of the focal adhesion laminin–integrin pathway (LAMC2, ITGA6 and LOXL2) [44,45]. LOXL2 is an enzyme that facilitates metastasis by changing cell morphology and is also regulated by the miR-29 family in clear cell renal cell carcinoma (ccRCC) [46]. LOXL2 is also regulated by *miR-26a/b* and *miR-218* in HNSCC [45], and by *miR-504* in NSCLC [47]. Moreover, one member of the *miR-29* family, *miR-29c*, has been found to be down-regulated in metastatic lung cancer and to reduce adhesion, invasion, migration and metastasis in vitro and in vivo through targeting integrin β1 and metalloproteinase 2 (MMP2), which is involved in extracellular matrix breakdown [48].

miR-203 has also been found to act as a negative regulator of metastasis in several different cancers. In HNSCC, it was shown to inhibit factors involved in cytoskeletal (LASP1), extracellular matrix (SPARC) and metabolic genes (NUAK1) [49]. In breast cancer, *miR-203* regulates EMT via a double-negative feedback loop formed by targeting *TGF-β* and *Slug* [50]. In melanoma, low levels of *miR-203* were associated with poorer overall survival in metastatic patients and its expression in vivo was demonstrated to inhibit metastasis via regulation of Slug [51]. In ovarian cancer cell lines, *miR-203* was also found to be down-regulated and was shown to inhibit the EMT pathway by targeting *BIRC5* and, thereby, attenuating TGFβ activity [52]. In gastric cancer, *miR-203* has been demonstrated to inhibit invasion and EMT through targeting of Annexin A4 [53]. In CRC, *miR-203* expression was linked with clinical stage, lymph node metastasis and poor survival and was demonstrated to regulate cell proliferation, migration and invasion by targeting EIF5A2 expression [54].

Similarly, *miR-124* has also been shown to be down-regulated in metastatic CRC when compared with healthy individuals and non-metastatic CRC patients [55]. This miRNA was found to regulate cell proliferation and invasive properties in cell lines through targeting of *ROCK1* expression. *miR-135a* is another down-regulated miRNA that was shown to target ROCK1, thereby inhibiting EMT, invasion and migration of GC cell lines [56]. This miRNA was also observed to be down-regulated in GC patients and its expression was associated with more advance-stage disease and higher rate of lymph node metastases [56]. *ROCK1* expression has also been described to be modulated by *miR-381* in breast cancer, which also regulates other molecules of the Wnt signaling pathway [57].

Several miRNAs have been described as metastasis suppressors in breast cancer such as *miR-7*, which was found to regulate FAK and its levels were positively correlated with E-Cadherin expression and negatively correlated with Vimentin and Fibronectin expression [58]. Loss of *miR-31* was associated with invasion and metastasis in breast cancer by regulating genes involved in invasion and metastasis including multiple α subunit partners of $\beta 1$ and $\beta 3$ integrins and WAVE3, [59,60]. *MiR-154* is also down-regulated in breast cancer tissues and cell lines where it was shown to inhibit proliferation, migration and invasion by targeting E2F5 [61]. In TNBC, *miR-150* was found to be down-regulated in tumor tissues, and to regulate HMGA2, leading to suppression of migration in vitro [62]. In breast cancer, *miR-124* was observed to be significantly down-regulated in metastatic bone tissues [63]. Recently, nanoparticle delivery of *miR-708*, another inhibitor of metastasis, was shown to reduce lung metastasis in breast cancer in vivo [64]. In another study, *miR-33b* expression was shown to be inversely correlated with the presence of lymph node metastases in breast cancer patients and to inhibit stemness, migration and invasion potential in vitro by targeting HMGA2, SALL4 and Twist1 [65]. Similarly, *miR-34c* was demonstrated to regulate migration and invasion of tumor cells in vitro through targeting of GIT1 [66], a protein whose expression has been linked to the presence of lymph node metastases in breast cancer patients [67].

MiR-34c and other family members (i.e., *miR-34a/b/c*) have been shown to be induced by activation of p53 and to target Snail, Slug, CD44 and ZEB1 [68]. ZEB1 and ZEB2, and E-Cadherin inhibitors have been demonstrated to be negatively regulated by several miRNAs, including *miR-101* in ovarian carcinoma [69], *miR-139-5p* in glioblastoma multiforme [70], *miR-215* in NSCLC [71] and *miR-132* in CRC and NSCLC [72,73].

In addition to direct regulation of metastasis by specific miRNAs, indirect regulation of metastasis can occur by regulation of components of the miRNA biosynthetic machinery. Two of these components, namely Dicer and Drosha, have been shown to be down-regulated in many cancer types [74,75]. It has been found that hypoxia can down-regulate Dicer expression through epigenetic silencing mediated by oxygen-sensitive H3K27me3 demethylases KDM6A and KDM6AB [76]. The authors demonstrated that this global reduction in miRNA expression resulted in down-regulation of *miR-200* which in turn increased levels of ZEB1 regulating metastasis. It has been noted that Dicer is down-regulated in metastatic human tumors deficient in TAp63, which can bind to the Dicer promoter and activates its expression. Deletion of TAp63 in mice reduced Dicer levels in tumors and increased the frequency of metastases [77]. Similarly, another important component of the miRNA biogenesis pathway, AGO2, has been found to be phosphorylated under hypoxic conditions by EGFR in breast cancer cells where it was shown to mediate EGFR-associated tumor cell invasiveness [78].

MiRNAs themselves can also directly target biosynthetic components—such is the case for *miR-103*, *miR-107* and *miR-630*. Hypoxia was shown to up-regulate *miR-630* expression, leading to targeting of Dicer [79]. Using an orthotopic murine model of ovarian cancer, the authors demonstrated that delivery of *miR-630* resulted in increased tumor growth and metastasis, along with decreased Dicer expression. In breast cancer, high levels of *miR-103/107* were associated with the presence of metastasis and poor clinical outcome and were demonstrated to directly target Dicer as well as increase the migratory properties of cells in vitro and metastasis in vivo [80].

In addition to whole tumors, several studies have looked specifically at the role of miRNAs in cancer stem cells (CSCs) [81,82] which play key roles in metastasis and resistance to therapies [83–86]. For example, breast CSCs were found to express lower levels of *miR-7* and higher levels of *KLF4*, an essential gene for induced pluripotent stem cells, and the expression of this miRNA was shown to down-regulate metastasis in vitro and in vivo [87,88]. Further identified as being down-regulated in breast CSCs, *miR-4319* was shown to inhibit tumor initiation and metastasis in vivo by targeting E2F2 [89]. In contrast, *miR-31* and *miR-29a* have been found to be up-regulated in breast CSCs, and inhibition of these miRNAs reduced the number and tumor-initiating ability of CSCs along with their metastatic ability 25 [90]. In prostate CSCs, *miR-34a* was found to be down-regulated and restoration of levels of this miRNA inhibited self-renewal capabilities and metastasis through targeting of CD44 [91]. In another study, the ectopic expression of down-regulated *miR-141* in prostate CSCs were demonstrated to inhibit EMT, spheroid formation, invasion and metastatic capabilities via targeting multiple pro-metastatic genes, such as *EZH2*, *CD44* and Rho GTPases [43]. In gastric CSCs, up-regulation of *miR-106b* was shown to enhance self-renewal, invasion and EMT, through activation of the TGF-β/Smad signaling pathway [92].

4. Metastasis and Circulating miRNAs

Unlike other RNA types—the vast majority of which are degraded by high levels of RNases found in the blood [93]—miRNAs are stable in the blood and are surprisingly resistant to fragmentation by either chemical or enzymatic agents [94]. Consequently, there has been a great deal of interest in circulating miRNAs in recent years [95]. Although the majority of studies relate to the biomarker potential of circulating miRNAs, they have also been demonstrated to act functionally with the ability to regulate spatially separated cells, a characteristic that lends itself to metastatic regulation [96–98]. Indeed, it has been described that cancer cells interact with other cells in the metastatic site to promote their own survival [99–101]. MiRNAs can exist in a circulating form either cell-free bound to proteins such as Argonaute2 (Ago2) [102,103], to lipids such as HDLs or LDLs [104], or they can be present inside extracellular vesicles such as exosomes [105,106]. They can act in an autocrine, paracrine and endocrine manner [96]. Several studies have reported higher levels of circulating miRNAs in metastatic patients. For example, *miR-141* levels in serum from prostate cancer patients [94], and levels of *miR-200c* and *miR-141* in breast cancer patients [107].

Circulating miRNAs have also been found to be present in tumor-secreted extracellular vesicles (EVs), mostly exosomes, which are known to participate in the metastatic process (Figure 2) [108–110]. For example, *miR-25-3p*, present in exosomes derived from CRC cells, were demonstrated to enter surrounding epithelial cells and to promote liver and lung metastasis in vivo [111]. The exosome-delivered *miR-25-3p* was shown disrupt the integrity of junctions in epithelial cells and to induce angiogenesis. Furthermore, this effect was mediated through targeting of KLF2, an inhibitor of VEGFR2, thereby decreasing the integrity of the endothelial barrier and targeting related molecule KLF4, leading to the decreased expression of Occludin, Claudin5 and ZO-1—all molecules implicated in maintenance of the cell–cell junction. Similarly, in breast cancer, *miR-105* present in exosomes was demonstrated to target ZO-1, resulting in destruction of vascular structures and enhancing vascular permeability [17]. The authors demonstrated that in vivo exosomal *miR-105* resulted in increased lung and brain metastasis. Furthermore, they observed that serum levels of *miR-105* were higher in patients with distant or lymph node metastasis. In another study, *miR-181c* derived from brain metastasis breast cancer cells could induce abnormal localization of claudin-5, Occludin, ZO-1, N-Cadherin and Actin through transfer of *miR-181c* into blood–brain barrier endothelial cells, resulting in destruction of cell–cell contact [112]. Similarly, levels of exosome-associated miR-181c from breast cancer patient serum were also observed to be significantly higher in patients that suffered brain metastasis. In HCC, when exosomal *miR-103a-3p* was delivered into endothelial cells, the miRNA was shown to abrogate junction integrity and promoted tumor metastasis through targeting of VE-Cad, p120 and ZO-1 [113]. Again, levels of *miR-103a-3p* in serum from HCC patients were associated with higher metastasis potential, higher TNM and higher recurrent risk.

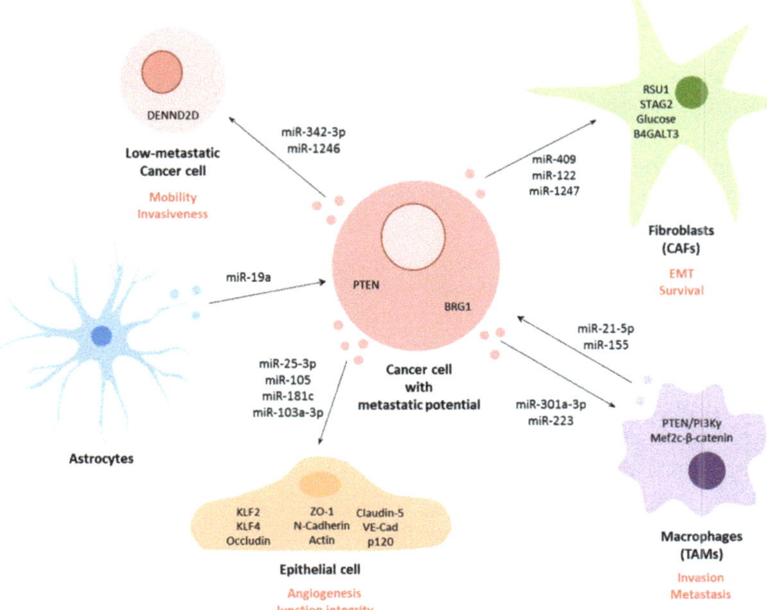

Figure 2. Cancer cell communication through extracellular vesicles (EVs). Cancer cells can communicate with surrounding cells or distant cells via miRNAs contained inside the EVs. Non-tumor cells are usually epithelial cells, macrophages or fibroblasts, although, communication between cancer cells with low-metastatic potential with astrocytes has also been described. Schematic representation of miRNAs involved in each communication and targets described in receptor cells.

MiRNAs contained in exosomes have also been shown to influence non-tumor cells in the tumor microenvironment such as tumor-associated macrophages (TAMs) that promote invasion and metastasis in cancer. For example, pancreatic cancer cells under hypoxic conditions were shown to release exosomes that contained *miR-301a-3p*, which was demonstrated to induce TAM polarization resulting in increased pancreatic cell migration and EMT in vitro and lung metastasis in vivo [114]. This polarization was induced by activation of the PTEN/PI3Kγ signaling pathway. In contrast, TAMs themselves have also been shown to secrete exosomes containing functional miRNAs that can promote metastasis. For example, exosomal *miR-223* derived from TAMs of breast cancer patients were demonstrated to promote tumor cell invasion through targeting of the Mef2c-β-catenin pathway [115,116]. In colon cancer, activated TAMs were shown to release exosomes containing *miR-21-5p* and *miR-155-5p*, which were demonstrated to regulate migration and invasion of colorectal cancer cells through targeting of BRG1 [116]. In addition to TAMs, cancer-associated fibroblasts (CAFs), which initiate remodeling of the extracellular matrix, thereby facilitating metastasis, can also release and respond to miRNA-containing exosomes [117]. This is the case for example in prostate cancer, where EV-associated *miR-409* was demonstrated to promote EMT both in vitro and in vivo through down-regulation of RSU1 and STAG2 [118]. In breast cancer, tumor cells were demonstrated to secrete exosomes containing *miR-122* that could induce glucose reallocation in pre-metastatic sites in fibroblast and astrocyte populations, thereby making sites more conducive to metastasizing cancer cells [119]. In liver cancer, tumor-derived exosomal *miR-1247-3p* was shown to promote the activation of fibroblasts to form CAFs through the down-regulation of B4GALT3, leading to activation of the β1-integrin-NF-κB signaling pathway, thereby promoting stemness, EMT, spheroid formation, mobility and chemoresistance in vitro and increasing lung metastasis in vivo [120]. Moreover, higher levels of *miR-1247-3p* were detected in serum from liver cancer patients with lung metastases [120].

Other tumor microenvironment cells have also been shown to be able to communicate with tumor cells as a result of exosome-associated miRNAs. For example, astrocytes in breast cancer patients were found to release exosomes containing *miR-19a* which was demonstrated to regulate PTEN in tumor cells and to promote brain metastasis after tumor extravasation [121]. In oral cancer, highly metastatic tumor cells were demonstrated to release exosomes containing *miR-342-3p* and *miR-1246*, which could be taken up by less metastatic tumor cells, resulting in an increase in their mobility and invasiveness through regulation of DENND2D [122].

5. The Metastatic Targetome

As shown above, many miRNAs have now been identified that are associated with the regulation of cancer metastasis. However, the functional significance of such deregulation is poorly understood, as the target genes (the targetome) of miRNAs are notoriously difficult to predict computationally, and moreover differ according to the cellular context [123,124]. An alternative approach is to directly sample the targetome in situ using cross-linking immunoprecipitation (CLIP) techniques coupled with high-throughput sequencing. This technology has evolved through the development of several variations, with arguably the most promising being Photoactivatable-Ribonucleoside-Enhanced CLIP (PAR-CLIP), which has a far better signal-to-noise ratio than other CLIP-based technologies [125]. For example, *miR-200*, a major regulator of cellular migration and invasion, was demonstrated to target *WIPF1*, *CFL2* and *MPRIP* by HITS-CLIP in breast cancer—all genes which promote invadopodia and invasion of cells [126]. In addition, PAR-CLIP was used with prostate cancer cells to show that *miR-148a* reduced migration and invasion by direct interaction with *CENPF* 3′UTR [127]. Similarly, *miR-141* was shown to mediate cell invasion by directly targeting *RAC1*, *CDC42*, and i [43]. Targets of *miR-346* were also identified by CLIP including the oncogene *YWHAZ* that modulates cell invasion and levels were correlated with Gleason grade, biochemical recurrence, non-organ-confined disease and lymph node metastases in patients [128].

6. Concluding Remarks and Future Perspectives

As is clear from the evidence presented above (Table 1), many miRNAs play crucial roles in cancer metastasis and, as a result, the therapeutic targeting of these miRNAs has generated a lot of interest in recent years [129–135]. In general terms, there are two strategies to modulate miRNA expression—either the restoration of down-regulated tumor suppressor miRNAs, or the inhibition of pro-metastatic miRNAs. The former group, requiring the expression of a specific miRNA using techniques such as direct delivery, viral or other vector formats, is much less challenging from a practical point of view than the inhibition of a specific miRNA whose over-expression could be very localized to a few cells or may require complete inhibition to be effective. The latter approach is generally achieved using some type of specific inhibitor such as antagomiRs or miRNA sponges [136,137]. All of these approaches, whether expression or inhibition, face common challenges including poor delivery, low cellular efficiency, endosomal escape and off-target effects, amongst others. Perhaps most attention has been focused on improving delivery systems for miRNAs and/or miRNA inhibitors. Two general approaches have been taken to deliver miRNAs in vivo; viral vectors and non-viral delivery systems. Viral vectors such as lentivirus, adenovirus or adeno-associated viruses [138,139] have been demonstrated to be able to deliver miRNAs with high efficiency in vivo. However, these vectors can trigger an immune response in patients [140]. Consequently, many studies have chosen to use non-viral vectors, in particular nanoparticles such as lipid and polymeric nanoparticles that can protect miRNA from degradation in vivo and, thereby, increase their half-life in circulation but do, however, have much lower transfection efficiencies than viral vectors [141–144]. For example, lipid-derived nanoparticles carrying *miR-34a* were demonstrated to reduce metastasis and increase survival in an orthotopic model of prostate cancer [91]. Lipid nanoparticles were also used for the systemic delivery of *miR-200* to reduce angiogenesis and metastasis in murine models of ovarian, lung, renal and breast cancer, through regulation of Interleukin 8 and CXCL1 [145]. In NSCLC, cationic liposomes were used to deliver *miR-143* in mice and were demonstrated to inhibit metastasis and prolong survival [146]. CRISPR/Cas9 technology has been used as an alternative to inhibitor sequences in several cancer types [147–149]. For example, lentivirus-mediated disruption of *miR-21* by CRISPR-Cas9 technology was shown to inhibit EMT in ovarian cancer [150], and, in glioblastoma, lentivirus-mediated *miR-10b* CRISPR/Cas9 inhibition was found to be lethal for GMB cells and GBM-initiating stem cells both in vitro and in orthotopic mice [151]. In addition to addressing the problem of general delivery, several approaches have been made to improve specific delivery by targeting technologies. For example, *let-7g* was conjugated with an aptamer that binds and antagonizes the oncogenic receptor tyrosine kinase Axl (GL21.T) in lung cell line (A549—Axl+) and breast cancer cell line (MCF7—Axl−) [152]. These constructs were shown to retain cell and tissue specificity in vivo and produce a reduction of tumor volume. Specific cell-targeted aptamers have also been used including delivery of anti-*miR-155* using poly lactic-co-glycolic acid (PLGA) nanoparticles and a peptide with a low pH-induced transmembrane structure (pHLIP) that facilitated the delivery of the inhibitor across the plasma membrane under acidic conditions, such as those found within tumors [153]. This strategy was used in a mouse lymphoma model that led to a reduction of tumor growth without any discernable toxicity. Disulfide-cross-linked polyethylenimine (PEI-SS) was employed along with conjugated folic acid to target in breast cancer [143]. EVs themselves have been used as a vehicle delivery system for metastatic miRNAs. For example, EVs produced by B-cells that contained a *miR-335* synthetic mimic were demonstrated to inhibit *SOX4* expression and to reduce tumor growth in vivo in a breast cancer model [154].

In addition to direct modulation of metastasis-associated miRNAs, these delivery systems have also been used to modulate treatment response in a metastatic context. For example, liposomal nanoparticle delivery of anti-*miR-155* was used to reverse cisplatin chemoresistance in a murine lung cancer model of metastasis resulting in reduced proliferation and angiogenesis [155]. In glioblastoma, antagomirs directed against *miR-21* and *miR-10b* were incorporated within nanoparticles (cRGD-tagged PEG-PGLA) and shown to have high levels of uptake by cells and to increase chemosensitivity to Temozolomide in vivo [156]. The same antagomirs were also used for TNBC [157]. Liposomal nanoparticles loaded with *miR-200c* were demonstrated to sensitize metastatic lung cancer cells to radiotherapy in vivo [158]. In addition, multiple studies have been used combining miRNA modulators (mimics or anti-miRNAs) along with chemotherapy—most commonly, doxorubicin with miRNAs such as *let-7a* and *miR-21* in breast cancer [159,160], *miR-31* in cervical cancer [161] and *miR-34a* in prostate cancer [162]. Co-polymer nano-assemblies (PEG5K-VE4-DET20) were co-loaded with *let-7b* mimic and paclitaxel in NSCLC and demonstrated to potentiate the cytotoxicity of paclitaxel and induced apoptosis and inhibition of invasiveness in vivo [163]. Researchers developed a polymeric dual delivery nanoscale device (DDND) to delivery *miR-345* mimic and gemcitabine for metastatic pancreatic cancer [164]. This system was used to reduce tumor growth and decrease metastasis in a murine xenograft model. Gemcitabine was co-delivered with *miR-203a* using an EGFR-targeted cationic polymeric misted micelle system and shown to reduce tumor growth, increase apoptosis and inhibit EMT in an orthotopic pancreatic tumor model [165]. In breast cancer, *miR-34a* and doxorubicin were co-delivered using multi-functional nano-micellar carriers, resulting in reduced tumor formation and metastasis in vivo [166].

In addition to these pre-clinical studies, there are a number of clinical trials targeting metastasis-associated miRNAs. For example, *miR-34* mimics encapsulated in liposomal carriers have been intravenously administered to patients with metastatic primary liver cancer, small-cell lung cancer (SCLC), lymphoma, melanoma, multiple myeloma, renal cell carcinoma and NSCLC during a phase 1 trial (MRX34, miRNA Therapeutics Inc.) [167]. However, this trial was terminated before completion in 2016 after serious adverse immune-related effects were developed by some patients (ClinicalTrials.gov identifiers: NCT01829971, NCT02862145). MiRNA Therapeutics have several other miRNA-based clinical trials underway including a phase I trial to deliver a *miR-155* antagomir (MRG-106 or Cobomarsen™), which is currently recruiting patients with lymphoma or leukemia (NCT02580552), and a phase II trial, which is currently recruiting cutaneous T-cell lymphoma patients to compare with

Vorinostat treatment (NCT03713320) and a separate follow-up trial (NCT03837457). The MesomiR-1 phase I clinical trial used TargomiR delivery vehicles (bacterially derived minicells containing a targeting antibody and miRNA mimic) to deliver *miR-16* to 26 metastatic pleural mesothelioma patients using an anti-EGFR targeting antibody (ClinicalTrials.gov identifier: NCT02369198) [168,169]. The trial closed in 2017, with a reported objective response of 5% with a duration of 32 weeks.

In summary, it can be seen from the breadth of evidence presented above that miRNAs (Table 1) represent a key regulatory control of metastasis in multiple cancers and, as a result, are promising targets for novel therapeutic approaches, although it is equally clear that much more research is still needed to translate this knowledge into the clinic.

Table 1. List of miRNAs associated with metastasis.

miRNA	Cancer	Express.	Sample	Target	Ref
Let-7	Breast	Low	Cells & Tissue	*RAS & HMGA2*	[14]
miR-7	Breast	Low	Cells & Tissue	*FAK*	[58]
miR-10b	Breast	High	Cells & Tissue	*HOXD10*	[13]
miR-26a/b	HNSCC	Low	Cells & Tissue	*LOXL2*	[45]
miR-29 family	HNSCC	Low	Cells	*LAMC2 & ITGA6*	[44]
miR-29 family	HNSCC	Low	Cells & Tissue	*LOXL2*	[45]
miR-29 family	ccRCC	Low	Cells & Tissue	*LOXL2*	[46]
miR-29a	Breast	High	Cells & Tissue	*SUV420H2*	[25]
miR-29c	Lung	Low	Cells	*Integrin β1 & MMP2*	[48]
miR-31	Breast	Low	Cells	*Integrin α subunits*	[60]
miR-31	Breast	Low	Cells & Tissue	*WAVE3*	[59]
miR-33b	Breast	Low	Cells & Tissue	*HMGA2, SALL4 & Twist1*	[65]
miR-34	CRC	Low	Cells	*ZEB1 & Slug*	[68]
miR-34a/b/c	CRC	Low	Cells	*Snail*	[170]
miR-34a	Breast	Low	Cells & Tissue	*CXCL10*	[171]
miR-34c	Breast	Low	Cells & Tissue	*GIT1*	[66]
miR-93	Endometrial	High	Cells & Tissue	*FOXA1*	[29]
miR-96	Breast	High	Cells & Tissue	*PTPN9*	[23]
miR-96	HCC	High	Cells & Tissue	*ephrinA5*	[172]
miR-101	Ovarian	Low	Cells	*ZEB1 & ZEB2*	[69]
miR-101	NSCLC	Low	Cells & Tissue	*ZEB1*	[173]
miR-105	Breast	High	Cells, Tissue & Serum	*ZO-1*	[17]
miR-124	Breast	Low	Cells & Tissue	*ZEB2*	[174]
miR-124	Breast	Low	Cells & Tissue	*IL-11*	[63]
miR-124	CRC	Low	Cells & Tissue	*ROCK1*	[55]
miR-128-3p	ESCC	Low	Cells & Tissue	*ZEB1*	[175]
miR-130a	Breast	Low	Cells	*FOSL1*	[26]
miR-132	NSCLC	Low	Cells & Tissue	*ZEB2*	[73]
miR-132	CRC	Low	Cells & Tissue	*ZEB2*	[72]
miR-135a	Breast	High	Cells & Tissue	*HOXA10*	[20]
miR-135a	HCC	High	Cells & Tissue	*FOXO1*	[22]
miR-135a	Gastric	Low	Cells & Tissue	*ROCK1*	[56]
miR-135b	Breast	High	Cells & Tissue	*APC*	[21]
miR-138	Breast	Low	Cells & Tissue	*ROCK1*	[57]
miR-139-5p	Glioblastoma	Low	Cells & Tissue	*ZEB1 & ZEB2*	[70]
miR-141	Breast	Low	Cells & Tissue	*HIPK1*	[176]
miR-141	Prostate	Low	Cells & Tissue	*Rho GTases, CD44 & EZH2*	[43]
miR-150	Breast	Low	Cells & Tissue	*HMGA2*	[62]
miR-154	Breast	Low	Cells & Tissue	*E2F5*	[61]
miR-181b	Breast	High	Cells & Tissue	*YWHAG*	[18]
miR-182	HCC	High	Cells & Tissue	*ephrinA5*	[172]
miR-186-5p	CRC	Low	Cells	*ZEB1*	[177]
miR-190	Breast	Low	Cells	*STC2*	[178]
miR-190	Breast	Low	Cells & Tissue	*SMAD2*	[179]
miR-194	Ovarian	High	Cells & Tissue	*PTPN12*	[37]

Table 1. *Cont.*

miRNA	Cancer	Express.	Sample	Target	Ref
miR-197	HCC	High	Cells & Tissue	*Axin-2, NKD1 & DKK2*	[30]
miR-200	Breast	Low	Cells & Tissue	*ZEB1 & ZEB2*	[39]
miR-200a/b	Breast	Low	Cells	*ZEB1 & ZEB2*	[38]
miR-200c	Breast	Low	Cells	*CRKL*	[180]
miR-200c	Breast	Low	Cells	*ZEB1*	[181]
miR-200c	Breast	Low	Cells & Tissue	*HIPK1*	[176]
miR-203	HNSCC	Low	Cells & Tissue	*LASP1, SPARC & NUAK*	[49]
miR-203	Breast	Low	Cells	*Slug*	[50]
miR-203	CRC	Low	Cells & Tissue	*EIF5A2*	[54]
miR-203	Melanoma	Low	Cells & Tissue	*Slug*	[51]
miR-203	Ovarian	Low	Cells & Tissue	*BIRC5*	[52]
miR-203	Gastric	Low	Cells & Tissue	*Annexin A4*	[53]
miR-205	Breast	Low	Cells	*ZEB1*	[182]
miR-205	Breast	Low	Cells & Tissue	*ZEB1 & ZEB2*	[39]
miR-210	Breast	High	Cells & Tissue	*E-Cadherin & Snail*	[24]
miR-215	NSCLC	Low	Cells & Tissue	*ZEB2*	[71]
miR-218	HNSCC	Low	Cells & Tissue	*LOXL2*	[45]
miR-374a	Breast	High	Cells & Tissue	*WIF1, PTEN & WNT5A*	[19]
miR-409-3p	Osteosarcoma	Low	Cells & Tissue	*ZEB1*	[183]
miR-429	Breast	Low	Cells	*ZEB1 & ZEB2*	[38]
miR-504	NSCLC	Low	Cells & Tissue	*LOXL2*	[47]
miR-508-3p	Breast	Low	Cells & Tissue	*ZEB1*	[184]
miR-520c	Gastic	High	Cells & Tissue	*IRF2*	[32]
miR-574-5p	NSCLC	High	Cells, Tissue & Serum	*PTPRU*	[36]
miR-577	Gastric	High	Cells & Tissue	*SDPR*	[31]
miR-641	Cervical	Low	Cells & Tissue	*ZEB1*	[185]
miR-708-3p	Breast	Low	Cells & Tissue	*ZEB1, CDH2 & Vimentin*	[186]
miR-1204	Breast	High	Cells & Tissue	*VDR*	[28]
miR-1269a	Colorectal	High	Cells & Tissue	*Smad7 & HOXD10*	[33]
miR-8084	Breast	High	Cells, Tissue & Serum	*ING2*	[27]

HNSCC; head and neck squamous cell carcinoma; ccRCC, clear cell renal cell carcinoma; CRC, colorectal cancer; HCC, hepatocellular carcinoma; NSCLC, non-small-cell lung carcinoma; ESCC, esophageal cancer.

Author Contributions: C.S. and C.H.L. conceived, drafted, edited and wrote the final manuscript. All authors have read and agreed to the published version of the manuscript.

Funding: This research received no external funding.

Acknowledgments: C.H.L. and his research are supported by grants from the IKERBASQUE foundation for science, the Starmer-Smith memorial fund, Ministerio de Economía y Competitividad (MINECO) of the Spanish Central Government and FEDER funds (PI12/00663, PIE13/00048, DTS14/00109, PI15/00275, PI18/01710), Departamento de Desarrollo Económico y Competitividad y Departamento de Sanidad of the Basque government, Asociación Española Contra el Cancer (AECC), and the Diputación Foral de Guipuzcoa (DFG).

Conflicts of Interest: The authors declare no conflict of interest.

References

1. Winter, M.; Gibson, R.; Ruszkiewicz, A.; Thompson, S.K.; Thierry, B. Beyond conventional pathology: Towards preoperative and intraoperative lymph node staging. *Int. J. Cancer* **2015**, *136*, 743. [CrossRef] [PubMed]
2. Mehlen, P.; Puisieux, A. Metastasis: A question of life or death. *Nat. Rev. Cancer* **2006**, *6*, 449. [CrossRef] [PubMed]
3. Monteiro, J.; Fodde, R. Cancer stemness and metastasis: Therapeutic consequences and perspectives. *Eur. J. Cancer* **2010**, *46*, 1198. [CrossRef] [PubMed]
4. Chambers, A.F.; Groom, A.C.; MacDonald, I.C. Dissemination and growth of cancer cells in metastatic sites. *Nat. Rev. Cancer* **2002**, *2*, 563. [CrossRef] [PubMed]

5. Bartel, D.P. MicroRNAs: Target recognition and regulatory functions. *Cell* **2009**, *136*, 215. [CrossRef] [PubMed]

6. Friedman, R.C.; Farh, K.K.; Burge, C.B.; Bartel, D.P. Most mammalian mRNAs are conserved targets of microRNAs. *Genome Res.* **2009**, *19*, 92. [CrossRef]

7. Lawrie, C.H. *MicroRNAs in Medicine*; Wiley: New York, NY, USA, 2014.

8. Cui, M.; Wang, H.; Yao, X.; Zhang, D.; Xie, Y.; Cui, R.; Zhang, X. Circulating MicroRNAs in Cancer: Potential and Challenge. *Front. Genet.* **2019**, *10*, 626. [CrossRef]

9. Mulcahy, L.A.; Pink, R.C.; Carter, D.R. Routes and mechanisms of extracellular vesicle uptake. *J. Extracell. Vesicles* **2014**, *3*, 24641. [CrossRef]

10. Gangoda, L.; Boukouris, S.; Liem, M.; Kalra, H.; Mathivanan, S. Extracellular vesicles including exosomes are mediators of signal transduction: Are they protective or pathogenic? *Proteomics* **2015**, *15*, 260. [CrossRef]

11. Shen, W.J.; Azhar, S.; Kraemer, F.B. SR-B1: A Unique Multifunctional Receptor for Cholesterol Influx and Efflux. *Annu. Rev. Physiol.* **2018**, *80*, 95. [CrossRef]

12. Vickers, K.C.; Palmisano, B.T.; Shoucri, B.M.; Shamburek, R.D.; Remaley, A.T. MicroRNAs are transported in plasma and delivered to recipient cells by high-density lipoproteins. *Nat. Cell Biol.* **2011**, *13*, 423. [CrossRef] [PubMed]

13. Ma, L.; Teruya-Feldstein, J.; Weinberg, R.A. Tumour invasion and metastasis initiated by microRNA-10b in breast cancer. *Nature* **2007**, *449*, 682. [CrossRef] [PubMed]

14. Yu, F.; Yao, H.; Zhu, P.; Zhang, X.; Pan, Q.; Gong, C.; Huang, Y.; Hu, X.; Su, F.; Lieberman, J.; et al. let-7 regulates self renewal and tumorigenicity of breast cancer cells. *Cell* **2007**, *131*, 1109. [CrossRef] [PubMed]

15. Nicoloso, M.S.; Spizzo, R.; Shimizu, M.; Rossi, S.; Calin, G.A. MicroRNAs—The micro steering wheel of tumour metastases. *Nat. Rev. Cancer* **2009**, *9*, 293. [CrossRef]

16. Ma, L.; Weinberg, R.A. Micromanagers of malignancy: Role of microRNAs in regulating metastasis. *Trends Genet. Genet.* **2008**, *24*, 448. [CrossRef]

17. Zhou, W.; Fong, M.Y.; Min, Y.; Somlo, G.; Liu, L.; Palomares, M.R.; Yu, Y.; Chow, A.; O'Connor, S.T.; Chin, A.R.; et al. Cancer-secreted miR-105 destroys vascular endothelial barriers to promote metastasis. *Cancer Cell* **2014**, *25*, 501. [CrossRef]

18. Yoo, J.O.; Kwak, S.Y.; An, H.J.; Bae, I.H.; Park, M.J.; Han, Y.H. miR-181b-3p promotes epithelial-mesenchymal transition in breast cancer cells through Snail stabilization by directly targeting YWHAG. *Biochim. Biophys. Acta* **2016**, *1863 Pt 7 A*, 1601. [CrossRef]

19. Cai, J.; Guan, H.; Fang, L.; Yang, Y.; Zhu, X.; Yuan, J.; Wu, J.; Li, M. MicroRNA-374a activates Wnt/beta-catenin signaling to promote breast cancer metastasis. *J. Clin. Investig.* **2013**, *123*, 566.

20. Chen, Y.; Zhang, J.; Wang, H.; Zhao, J.; Xu, C.; Du, Y.; Luo, X.; Zheng, F.; Liu, R.; Zhang, H.; et al. miRNA-135a promotes breast cancer cell migration and invasion by targeting HOXA10. *BMC Cancer* **2012**, *12*, 111. [CrossRef]

21. Lv, Z.D.; Xin, H.N.; Yang, Z.C.; Wang, W.J.; Dong, J.J.; Jin, L.Y.; Li, F.N. miR-135b promotes proliferation and metastasis by targeting APC in triple-negative breast cancer. *J. Cell. Physiol.* **2019**, *234*, 10819. [CrossRef]

22. Zeng, Y.B.; Liang, X.H.; Zhang, G.X.; Jiang, N.; Zhang, T.; Huang, J.Y.; Zhang, L.; Zeng, X.C. miRNA-135a promotes hepatocellular carcinoma cell migration and invasion by targeting forkhead box O1. *Cancer Cell Int.* **2016**, *16*, 63. [CrossRef] [PubMed]

23. Hong, Y.; Liang, H.; Uzair Ur, R.; Wang, Y.; Zhang, W.; Zhou, Y.; Chen, S.; Yu, M.; Cui, S.; Liu, M.; et al. miR-96 promotes cell proliferation, migration and invasion by targeting PTPN9 in breast cancer. *Sci. Rep.* **2016**, *6*, 37421. [CrossRef] [PubMed]

24. Tang, T.; Yang, Z.; Zhu, Q.; Wu, Y.; Sun, K.; Alahdal, M.; Zhang, Y.; Xing, Y.; Shen, Y.; Xia, T.; et al. Up-regulation of miR-210 induced by a hypoxic microenvironment promotes breast cancer stem cells metastasis, proliferation, and self-renewal by targeting E-cadherin. *FASEB J.* **2018**, *32*, 6965–6981. [CrossRef] [PubMed]

25. Wu, Y.; Shi, W.; Tang, T.; Wang, Y.; Yin, X.; Chen, Y.; Zhang, Y.; Xing, Y.; Shen, Y.; Xia, T.; et al. miR-29a contributes to breast cancer cells epithelial-mesenchymal transition, migration, and invasion via down-regulating histone H4K20 trimethylation through directly targeting SUV420H2. *Cell Death Dis.* **2019**, *10*, 176. [CrossRef]

26. Chen, X.; Zhao, M.; Huang, J.; Li, Y.; Wang, S.; Harrington, C.A.; Qian, D.Z.; Sun, X.X.; Dai, M.S. microRNA-130a suppresses breast cancer cell migration and invasion by targeting FOSL1 and upregulating ZO-1. *J. Cell. Biochem.* **2018**, *119*, 4945. [CrossRef]

27. Gao, Y.; Ma, H.; Gao, C.; Lv, Y.; Chen, X.; Xu, R.; Sun, M.; Liu, X.; Lu, X.; Pei, X.; et al. Tumor-promoting properties of miR-8084 in breast cancer through enhancing proliferation, suppressing apoptosis and inducing epithelial-mesenchymal transition. *J. Transl. Med.* **2018**, *16*, 38. [CrossRef]

28. Liu, X.; Bi, L.; Wang, Q.; Wen, M.; Li, C.; Ren, Y.; Jiao, Q.; Mao, J.H.; Wang, C.; Wei, G.; et al. miR-1204 targets VDR to promotes epithelial-mesenchymal transition and metastasis in breast cancer. *Oncogene* **2018**, *37*, 3426. [CrossRef]

29. Chen, S.; Chen, X.; Sun, K.X.; Xiu, Y.L.; Liu, B.L.; Feng, M.X.; Sang, X.B.; Zhao, Y. MicroRNA-93 Promotes Epithelial-Mesenchymal Transition of Endometrial Carcinoma Cells. *PLoS ONE* **2016**, *11*, e0165776. [CrossRef]

30. Hu, Z.; Wang, P.; Lin, J.; Zheng, X.; Yang, F.; Zhang, G.; Chen, D.; Xie, J.; Gao, Z.; Peng, L.; et al. MicroRNA-197 Promotes Metastasis of Hepatocellular Carcinoma by Activating Wnt/beta-Catenin Signaling. *Cell. Physiol. Biochem. Int. J. Exp. Cell. Physiol. Biochem. Pharm.* **2018**, *51*, 470. [CrossRef]

31. Luo, Y.; Wu, J.; Wu, Q.; Li, X.; Wu, J.; Zhang, J.; Rong, X.; Rao, J.; Liao, Y.; Bin, J.; et al. miR-577 Regulates TGF-beta Induced Cancer Progression through a SDPR-Modulated Positive-Feedback Loop with ERK-NF-kappaB in Gastric Cancer. *Mol. Ther. J. Am. Soc. Gene Ther.* **2019**. [CrossRef]

32. Li, Y.R.; Wen, L.Q.; Wang, Y.; Zhou, T.C.; Ma, N.; Hou, Z.H.; Jiang, Z.P. MicroRNA-520c enhances cell proliferation, migration, and invasion by suppressing IRF2 in gastric cancer. *FEBS Openbio* **2016**, *6*, 1257. [CrossRef]

33. Bu, P.; Wang, L.; Chen, K.Y.; Rakhilin, N.; Sun, J.; Closa, A.; Tung, K.L.; King, S.; Kristine Varanko, A.; Xu, Y.; et al. miR-1269 promotes metastasis and forms a positive feedback loop with TGF-beta. *Nat. Commun.* **2015**, *6*, 6879. [CrossRef]

34. Wang, B.; Yang, J.; Xiao, B. MicroRNA-20b (miR-20b) Promotes the Proliferation, Migration, Invasion, and Tumorigenicity in Esophageal Cancer Cells via the Regulation of Phosphatase and Tensin Homologue Expression. *PLoS ONE* **2016**, *11*, e0164105. [CrossRef] [PubMed]

35. Song, Y.; Li, J.; Zhu, Y.; Dai, Y.; Zeng, T.; Liu, L.; Li, J.; Wang, H.; Qin, Y.; Zeng, M.; et al. MicroRNA-9 promotes tumor metastasis via repressing E-cadherin in esophageal squamous cell carcinoma. *Oncotarget* **2014**, *5*, 11669. [CrossRef] [PubMed]

36. Zhou, R.; Zhou, X.; Yin, Z.; Guo, J.; Hu, T.; Jiang, S.; Liu, L.; Dong, X.; Zhang, S.; Wu, G. MicroRNA-574-5p promotes metastasis of non-small cell lung cancer by targeting PTPRU. *Sci. Rep.* **2016**, *6*, 35714. [CrossRef] [PubMed]

37. Liang, T.; Li, L.; Cheng, Y.; Ren, C.; Zhang, G. MicroRNA-194 promotes the growth, migration, and invasion of ovarian carcinoma cells by targeting protein tyrosine phosphatase nonreceptor type 12. *Onco Targets Ther.* **2016**, *9*, 4307. [CrossRef] [PubMed]

38. Bracken, C.P.; Gregory, P.A.; Kolesnikoff, N.; Bert, A.G.; Wang, J.; Shannon, M.F.; Goodall, G.J. A double-negative feedback loop between ZEB1-SIP1 and the microRNA-200 family regulates epithelial-mesenchymal transition. *Cancer Res.* **2008**, *68*, 7846. [CrossRef]

39. Gregory, P.A.; Bert, A.G.; Paterson, E.L.; Barry, S.C.; Tsykin, A.; Farshid, G.; Vadas, M.A.; Khew-Goodall, Y.; Goodall, G.J. The miR-200 family and miR-205 regulate epithelial to mesenchymal transition by targeting ZEB1 and SIP1. *Nat. Cell Biol.* **2008**, *10*, 593. [CrossRef]

40. Korpal, M.; Lee, E.S.; Hu, G.; Kang, Y. The miR-200 family inhibits epithelial-mesenchymal transition and cancer cell migration by direct targeting of E-cadherin transcriptional repressors ZEB1 and ZEB2. *J. Biol. Chem.* **2008**, *283*, 14910. [CrossRef]

41. Park, S.M.; Gaur, A.B.; Lengyel, E.; Peter, M.E. The miR-200 family determines the epithelial phenotype of cancer cells by targeting the E-cadherin repressors ZEB1 and ZEB2. *Genes Dev.* **2008**, *22*, 894. [CrossRef]

42. Pang, Y.; Liu, J.; Li, X.; Xiao, G.; Wang, H.; Yang, G.; Li, Y.; Tang, S.C.; Qin, S.; Du, N.; et al. MYC and DNMT3A-mediated DNA methylation represses microRNA-200b in triple negative breast cancer. *J. Cell. Mol. Med.* **2018**, *22*, 6262. [CrossRef] [PubMed]

43. Liu, C.; Liu, R.; Zhang, D.; Deng, Q.; Liu, B.; Chao, H.P.; Rycaj, K.; Takata, Y.; Lin, K.; Lu, Y.; et al. MicroRNA-141 suppresses prostate cancer stem cells and metastasis by targeting a cohort of pro-metastasis genes. *Nat. Commun.* **2017**, *8*, 14270. [CrossRef] [PubMed]

44. Kinoshita, T.; Nohata, N.; Hanazawa, T.; Kikkawa, N.; Yamamoto, N.; Yoshino, H.; Itesako, T.; Enokida, H.; Nakagawa, M.; Okamoto, Y.; et al. Tumour-suppressive microRNA-29s inhibit cancer cell migration and invasion by targeting laminin-integrin signalling in head and neck squamous cell carcinoma. *Br. J. Cancer* **2013**, *109*, 2636. [CrossRef] [PubMed]

45. Fukumoto, I.; Kikkawa, N.; Matsushita, R.; Kato, M.; Kurozumi, A.; Nishikawa, R.; Goto, Y.; Koshizuka, K.; Hanazawa, T.; Enokida, H.; et al. Tumor-suppressive microRNAs (miR-26a/b, miR-29a/b/c and miR-218) concertedly suppressed metastasis-promoting LOXL2 in head and neck squamous cell carcinoma. *J. Hum. Genet.* **2016**, *61*, 109. [CrossRef]

46. Nishikawa, R.; Chiyomaru, T.; Enokida, H.; Inoguchi, S.; Ishihara, T.; Matsushita, R.; Goto, Y.; Fukumoto, I.; Nakagawa, M.; Seki, N. Tumour-suppressive microRNA-29s directly regulate LOXL2 expression and inhibit cancer cell migration and invasion in renal cell carcinoma. *Febs Lett.* **2015**, *589*, 2136. [CrossRef]

47. Ye, M.F.; Zhang, J.G.; Guo, T.X.; Pan, X.J. MiR-504 inhibits cell proliferation and invasion by targeting LOXL2 in non small cell lung cancer. *Biomed. Pharm.* **2018**, *97*, 1289. [CrossRef]

48. Wang, H.; Zhu, Y.; Zhao, M.; Wu, C.; Zhang, P.; Tang, L.; Zhang, H.; Chen, X.; Yang, Y.; Liu, G. miRNA-29c suppresses lung cancer cell adhesion to extracellular matrix and metastasis by targeting integrin beta1 and matrix metalloproteinase2 (MMP2). *PLoS ONE* **2013**, *8*, e70192.

49. Benaich, N.; Woodhouse, S.; Goldie, S.J.; Mishra, A.; Quist, S.R.; Watt, F.M. Rewiring of an epithelial differentiation factor, miR-203, to inhibit human squamous cell carcinoma metastasis. *Cell Rep.* **2014**, *9*, 104. [CrossRef]

50. Ding, X.; Park, S.I.; McCauley, L.K.; Wang, C.Y. Signaling between transforming growth factor beta (TGF-beta) and transcription factor SNAI2 represses expression of microRNA miR-203 to promote epithelial-mesenchymal transition and tumor metastasis. *J. Biol. Chem.* **2013**, *288*, 10241. [CrossRef]

51. Lohcharoenkal, W.; Das Mahapatra, K.; Pasquali, L.; Crudden, C.; Kular, L.; Akkaya Ulum, Y.Z.; Zhang, L.; Xu Landen, N.; Girnita, L.; Jagodic, M.; et al. Genome-Wide Screen for MicroRNAs Reveals a Role for miR-203 in Melanoma Metastasis. *J. Investig. Dermatol.* **2018**, *138*, 882. [CrossRef]

52. Wang, B.; Li, X.; Zhao, G.; Yan, H.; Dong, P.; Watari, H.; Sims, M.; Li, W.; Pfeffer, L.M.; Guo, Y.; et al. miR-203 inhibits ovarian tumor metastasis by targeting BIRC5 and attenuating the TGFbeta pathway. *J. Exp. Clin. Cancer Res.* **2018**, *37*, 235. [CrossRef]

53. Li, J.; Zhang, B.; Cui, J.; Liang, Z.; Liu, K. miR-203 inhibits the invasion and EMT of gastric cancer cells by directly targeting Annexin A4. *Oncol. Res.* **2019**. [CrossRef]

54. Deng, B.; Wang, B.; Fang, J.; Zhu, X.; Cao, Z.; Lin, Q.; Zhou, L.; Sun, X. MiRNA-203 suppresses cell proliferation, migration and invasion in colorectal cancer via targeting of EIF5A2. *Sci. Rep.* **2016**, *6*, 28301. [CrossRef]

55. Xi, Z.W.; Xin, S.Y.; Zhou, L.Q.; Yuan, H.X.; Wang, Q.; Chen, K.X. Downregulation of rho-associated protein kinase 1 by miR-124 in colorectal cancer. *World J. Gastroenterol.* **2015**, *21*, 5454. [CrossRef]

56. Shin, J.Y.; Kim, Y.I.; Cho, S.J.; Lee, M.K.; Kook, M.C.; Lee, J.H.; Lee, S.S.; Ashktorab, H.; Smoot, D.T.; Ryu, K.W.; et al. MicroRNA 135a suppresses lymph node metastasis through down-regulation of ROCK1 in early gastric cancer. *PLoS ONE* **2014**, *9*, e85205. [CrossRef]

57. Mohammadi-Yeganeh, S.; Hosseini, V.; Paryan, M. Wnt pathway targeting reduces triple-negative breast cancer aggressiveness through miRNA regulation in vitro and in vivo. *J. Cell. Physiol.* **2019**. [CrossRef]

58. Kong, X.; Li, G.; Yuan, Y.; He, Y.; Wu, X.; Zhang, W.; Wu, Z.; Chen, T.; Wu, W.; Lobie, P.E.; et al. MicroRNA-7 inhibits epithelial-to-mesenchymal transition and metastasis of breast cancer cells via targeting FAK expression. *PLoS ONE* **2012**, *7*, e41523. [CrossRef]

59. Sossey-Alaoui, K.; Downs-Kelly, E.; Das, M.; Izem, L.; Tubbs, R.; Plow, E.F. WAVE3, an actin remodeling protein, is regulated by the metastasis suppressor microRNA, miR-31, during the invasion-metastasis cascade. *Int. J. Cancer* **2011**, *129*, 1331. [CrossRef]

60. Augoff, K.; Das, M.; Bialkowska, K.; McCue, B.; Plow, E.F.; Sossey-Alaoui, K. miR-31 is a broad regulator of beta1-integrin expression and function in cancer cells. *Mol. Cancer Res.* **2011**, *9*, 1500. [CrossRef]

61. Xu, H.; Fei, D.; Zong, S.; Fan, Z. MicroRNA-154 inhibits growth and invasion of breast cancer cells through targeting E2F5. *Am. J. Transl. Res.* **2016**, *8*, 2620.

62. Tang, W.; Xu, P.; Wang, H.; Niu, Z.; Zhu, D.; Lin, Q.; Tang, L.; Ren, L. MicroRNA-150 suppresses triple-negative breast cancer metastasis through targeting HMGA2. *Onco Targets Ther.* **2018**, *11*, 2319. [CrossRef] [PubMed]

63. Cai, W.L.; Huang, W.D.; Li, B.; Chen, T.R.; Li, Z.X.; Zhao, C.L.; Li, H.Y.; Wu, Y.M.; Yan, W.J.; Xiao, J.R. microRNA-124 inhibits bone metastasis of breast cancer by repressing Interleukin-11. *Mol. Cancer* **2018**, *17*, 9. [CrossRef] [PubMed]

64. Ramchandani, D.; Lee, S.K.; Yomtoubian, S.; Han, M.S.; Tung, C.H.; Mittal, V. Nanoparticle Delivery of miR-708 Mimetic Impairs Breast Cancer Metastasis. *Mol. Cancer Ther.* **2019**, *18*, 579. [CrossRef] [PubMed]

65. Lin, Y.; Liu, A.Y.; Fan, C.; Zheng, H.; Li, Y.; Zhang, C.; Wu, S.; Yu, D.; Huang, Z.; Liu, F.; et al. MicroRNA-33b Inhibits Breast Cancer Metastasis by Targeting HMGA2, SALL4 and Twist1. *Sci. Rep.* **2015**, *5*, 9995. [CrossRef]

66. Tao, W.Y.; Wang, C.Y.; Sun, Y.H.; Su, Y.H.; Pang, D.; Zhang, G.Q. MicroRNA-34c Suppresses Breast Cancer Migration and Invasion by Targeting GIT1. *J. Cancer* **2016**, *7*, 1653. [CrossRef]

67. Goicoechea, I.; Rezola, R.; Arestin, M.; Caffarel, M.; Cortazar, A.R.; Manterola, L.; Fernandez-Mercado, M.; Armesto, M.; Sole, C.; Larrea, E.; et al. Spatial intratumoural heterogeneity in the expression of GIT1 is associated with poor prognostic outcome in oestrogen receptor positive breast cancer patients with synchronous lymph node metastases. *F1000Research* **2017**, *6*, 1606. [CrossRef]

68. Siemens, H.; Jackstadt, R.; Hunten, S.; Kaller, M.; Menssen, A.; Gotz, U.; Hermeking, H. miR-34 and SNAIL form a double-negative feedback loop to regulate epithelial-mesenchymal transitions. *Cell Cycle* **2011**, *10*, 4256. [CrossRef]

69. Guo, F.; Cogdell, D.; Hu, L.; Yang, D.; Sood, A.K.; Xue, F.; Zhang, W. MiR-101 suppresses the epithelial-to-mesenchymal transition by targeting ZEB1 and ZEB2 in ovarian carcinoma. *Oncol. Rep.* **2014**, *31*, 2021. [CrossRef]

70. Yue, S.; Wang, L.; Zhang, H.; Min, Y.; Lou, Y.; Sun, H.; Jiang, Y.; Zhang, W.; Liang, A.; Guo, Y.; et al. miR-139-5p suppresses cancer cell migration and invasion through targeting ZEB1 and ZEB2 in GBM. *Tumour Biol. J. Int. Soc. Oncodevelopmental Biol. Med.* **2015**, *36*, 6741. [CrossRef]

71. Hou, Y.; Zhen, J.; Xu, X.; Zhen, K.; Zhu, B.; Pan, R.; Zhao, C. miR-215 functions as a tumor suppressor and directly targets ZEB2 in human non-small cell lung cancer. *Oncol. Lett.* **2015**, *10*, 1985. [CrossRef]

72. Zheng, Y.B.; Luo, H.P.; Shi, Q.; Hao, Z.N.; Ding, Y.; Wang, Q.S.; Li, S.B.; Xiao, G.C.; Tong, S.L. miR-132 inhibits colorectal cancer invasion and metastasis via directly targeting ZEB2. *World J. Gastroenterol.* **2014**, *20*, 6515. [CrossRef] [PubMed]

73. You, J.; Li, Y.; Fang, N.; Liu, B.; Zu, L.; Chang, R.; Li, X.; Zhou, Q. MiR-132 suppresses the migration and invasion of lung cancer cells via targeting the EMT regulator ZEB2. *PLoS ONE* **2014**, *9*, e91827. [CrossRef] [PubMed]

74. Merritt, W.M.; Lin, Y.G.; Han, L.Y.; Kamat, A.A.; Spannuth, W.A.; Schmandt, R.; Urbauer, D.; Pennacchio, L.A.; Cheng, J.F.; Nick, A.M.; et al. Dicer, Drosha, and outcomes in patients with ovarian cancer. *N. Engl. J. Med.* **2008**, *359*, 2641. [CrossRef] [PubMed]

75. Lawrie, C.H.; Cooper, C.D.; Ballabio, E.; Chi, J.; Tramonti, D.; Hatton, C.S. Aberrant expression of microRNA biosynthetic pathway components is a common feature of haematological malignancy. *Br. J. Haematol.* **2009**, *145*, 545. [CrossRef] [PubMed]

76. Van den Beucken, T.; Koch, E.; Chu, K.; Rupaimoole, R.; Prickaerts, P.; Adriaens, M.; Voncken, J.W.; Harris, A.L.; Buffa, F.M.; Haider, S.; et al. Hypoxia promotes stem cell phenotypes and poor prognosis through epigenetic regulation of DICER. *Nat. Commun.* **2014**, *5*, 5203. [CrossRef] [PubMed]

77. Su, X.; Chakravarti, D.; Cho, M.S.; Liu, L.; Gi, Y.J.; Lin, Y.L.; Leung, M.L.; El-Naggar, A.; Creighton, C.J.; Suraokar, M.B.; et al. TAp63 suppresses metastasis through coordinate regulation of Dicer and miRNAs. *Nature* **2010**, *467*, 986. [CrossRef]

78. Shen, J.; Xia, W.; Khotskaya, Y.B.; Huo, L.; Nakanishi, K.; Lim, S.O.; Du, Y.; Wang, Y.; Chang, W.C.; Chen, C.H.; et al. EGFR modulates microRNA maturation in response to hypoxia through phosphorylation of AGO2. *Nature* **2013**, *497*, 383. [CrossRef]

79. Rupaimoole, R.; Ivan, C.; Yang, D.; Gharpure, K.M.; Wu, S.Y.; Pecot, C.V.; Previs, R.A.; Nagaraja, A.S.; Armaiz-Pena, G.N.; McGuire, M.; et al. Hypoxia-upregulated microRNA-630 targets Dicer, leading to increased tumor progression. *Oncogene* **2016**, *35*, 4312. [CrossRef]

80. Martello, G.; Rosato, A.; Ferrari, F.; Manfrin, A.; Cordenonsi, M.; Dupont, S.; Enzo, E.; Guzzardo, V.; Rondina, M.; Spruce, T.; et al. A MicroRNA targeting dicer for metastasis control. *Cell* **2010**, *141*, 1195. [CrossRef]

81. Garg, M. Emerging role of microRNAs in cancer stem cells: Implications in cancer therapy. *World J. Stem Cells* **2015**, *7*, 1078. [CrossRef]

82. Asadzadeh, Z.; Mansoori, B.; Mohammadi, A.; Aghajani, M.; Haji-Asgarzadeh, K.; Safarzadeh, E.; Mokhtarzadeh, A.; Duijf, P.H.G.; Baradaran, B. microRNAs in cancer stem cells: Biology, pathways, and therapeutic opportunities. *J. Cell. Physiol.* **2019**, *234*, 10002. [CrossRef] [PubMed]

83. Clevers, H. The cancer stem cell: Premises, promises and challenges. *Nat. Med.* **2011**, *17*, 313. [CrossRef] [PubMed]

84. Yu, Z.; Pestell, T.G.; Lisanti, M.P.; Pestell, R.G. Cancer stem cells. *Int. J. Biochem. Cell Biol.* **2012**, *44*, 2144. [CrossRef] [PubMed]

85. Seton-Rogers, S. Cancer stem cells: Easily moulded. *Nat. Rev. Cancer* **2013**, *13*, 519. [CrossRef] [PubMed]

86. Prasetyanti, P.R.; Medema, J.P. Intra-tumor heterogeneity from a cancer stem cell perspective. *Mol. Cancer* **2017**, *16*, 41. [CrossRef]

87. Okuda, H.; Xing, F.; Pandey, P.R.; Sharma, S.; Watabe, M.; Pai, S.K.; Mo, Y.Y.; Iiizumi-Gairani, M.; Hirota, S.; Liu, Y.; et al. miR-7 suppresses brain metastasis of breast cancer stem-like cells by modulating KLF4. *Cancer Res.* **2013**, *73*, 1434. [CrossRef]

88. Zhang, H.; Cai, K.; Wang, J.; Wang, X.; Cheng, K.; Shi, F.; Jiang, L.; Zhang, Y.; Dou, J. MiR-7, inhibited indirectly by lincRNA HOTAIR, directly inhibits SETDB1 and reverses the EMT of breast cancer stem cells by downregulating the STAT3 pathway. *Stem Cells* **2014**, *32*, 2858. [CrossRef]

89. Chu, J.; Li, Y.; Fan, X.; Ma, J.; Li, J.; Lu, G.; Zhang, Y.; Huang, Y.; Li, W.; Huang, X.; et al. MiR-4319 Suppress the Malignancy of Triple-Negative Breast Cancer by Regulating Self-Renewal and Tumorigenesis of Stem Cells. *Cell. Physiol. Biochem. Int. J. Exp. Cell. Physiol. Biochem. Pharmacol.* **2018**, *48*, 593. [CrossRef]

90. Lv, C.; Li, F.; Li, X.; Tian, Y.; Zhang, Y.; Sheng, X.; Song, Y.; Meng, Q.; Yuan, S.; Luan, L.; et al. MiR-31 promotes mammary stem cell expansion and breast tumorigenesis by suppressing Wnt signaling antagonists. *Nat. Commun.* **2017**, *8*, 1036. [CrossRef]

91. Liu, C.; Kelnar, K.; Liu, B.; Chen, X.; Calhoun-Davis, T.; Li, H.; Patrawala, L.; Yan, H.; Jeter, C.; Honorio, S.; et al. The microRNA miR-34a inhibits prostate cancer stem cells and metastasis by directly repressing CD44. *Nat. Med.* **2011**, *17*, 211. [CrossRef]

92. Yu, D.; Shin, H.S.; Lee, Y.S.; Lee, Y.C. miR-106b modulates cancer stem cell characteristics through TGF-beta/Smad signaling in CD44-positive gastric cancer cells. *Lab. Investig. A J. Tech. Methods Pathol.* **2014**, *94*, 1370. [CrossRef] [PubMed]

93. Duttagupta, R.; Jiang, R.; Gollub, J.; Getts, R.C.; Jones, K.W. Impact of Cellular miRNAs on Circulating miRNA Biomarker Signatures. *PLoS ONE* **2011**, *6*, e20769. [CrossRef] [PubMed]

94. Mitchell, P.S.; Parkin, R.K.; Kroh, E.M.; Fritz, B.R.; Wyman, S.K.; Pogosova-Agadjanyan, E.L.; Peterson, A.; Noteboom, J.; O'Briant, K.C.; Allen, A.; et al. Circulating microRNAs as stable blood-based markers for cancer detection. *Proc. Natl. Acad. Sci. USA* **2008**, *105*, 10513. [CrossRef] [PubMed]

95. Fernandez-Mercado, M.; Manterola, L.; Larrea, E.; Goicoechea, I.; Arestin, M.; Armesto, M.; Otaegui, D.; Lawrie, C.H. The circulating transcriptome as a source of non-invasive cancer biomarkers: Concepts and controversies of non-coding and coding RNA in body fluids. *J. Cell. Mol. Med.* **2015**, *19*, 2307. [CrossRef] [PubMed]

96. Cortez, M.A.; Bueso-Ramos, C.; Ferdin, J.; Lopez-Berestein, G.; Sood, A.K.; Calin, G.A. MicroRNAs in body fluids—The mix of hormones and biomarkers. *Nat. Rev. Clin. Oncol.* **2011**, *8*, 467. [CrossRef]

97. Weber, J.A.; Baxter, D.H.; Zhang, S.; Huang, D.Y.; Huang, K.H.; Lee, M.J.; Galas, D.J.; Wang, K. The microRNA spectrum in 12 body fluids. *Clin. Chem.* **2010**, *56*, 1733. [CrossRef]

98. Sole, C.; Arnaiz, E.; Manterola, L.; Otaegui, D.; Lawrie, C.H. The circulating transcriptome as a source of cancer liquid biopsy biomarkers. *Semin. Cancer Biol.* **2019**, *58*, 100. [CrossRef]

99. Aguirre-Ghiso, J.A. On the theory of tumor self-seeding: Implications for metastasis progression in humans. *Breast Cancer Res.* **2010**, *12*, 304. [CrossRef]

100. Fidler, I.J. The pathogenesis of cancer metastasis: The 'seed and soil' hypothesis revisited. *Nat. Rev. Cancer* **2003**, *3*, 453. [CrossRef]

101. Aguirre-Ghiso, J.A. How dormant cancer persists and reawakens. *Science* **2018**, *361*, 1314. [CrossRef]

102. Arroyo, J.D.; Chevillet, J.R.; Kroh, E.M.; Ruf, I.K.; Pritchard, C.C.; Gibson, D.F.; Mitchell, P.S.; Bennett, C.F.; Pogosova-Agadjanyan, E.L.; Stirewalt, D.L.; et al. Argonaute2 complexes carry a population of circulating microRNAs independent of vesicles in human plasma. *Proc. Natl. Acad. Sci. USA* **2011**, *108*, 5003. [CrossRef] [PubMed]

103. Turchinovich, A.; Weiz, L.; Langheinz, A.; Burwinkel, B. Characterization of extracellular circulating microRNA. *Nucleic Acids Res.* **2011**, *39*, 7223. [CrossRef]

104. Vickers, K.C.; Remaley, A.T. Lipid-based carriers of microRNAs and intercellular communication. *Curr. Opin. Lipidol.* **2012**, *23*, 91. [CrossRef] [PubMed]

105. Stoorvogel, W. Functional transfer of microRNA by exosomes. *Blood* **2012**, *119*, 646. [CrossRef] [PubMed]

106. Caby, M.P.; Lankar, D.; Vincendeau-Scherrer, C.; Raposo, G.; Bonnerot, C. Exosomal-like vesicles are present in human blood plasma. *Int. Immunol.* **2005**, *17*, 879. [CrossRef] [PubMed]

107. Zhang, G.; Zhang, W.; Li, B.; Stringer-Reasor, E.; Chu, C.; Sun, L.; Bae, S.; Chen, D.; Wei, S.; Jiao, K.; et al. MicroRNA-200c and microRNA- 141 are regulated by a FOXP3-KAT2B axis and associated with tumor metastasis in breast cancer. *Breast Cancer Res.* **2017**, *19*, 73. [CrossRef] [PubMed]

108. Tominaga, N.; Katsuda, T.; Ochiya, T. Micromanaging of tumor metastasis by extracellular vesicles. *Semin. Cell Dev. Biol.* **2015**, *40*, 52. [CrossRef] [PubMed]

109. Hoshino, A.; Costa-Silva, B.; Shen, T.L.; Rodrigues, G.; Hashimoto, A.; Tesic Mark, M.; Molina, H.; Kohsaka, S.; Di Giannatale, A.; Ceder, S.; et al. Tumour exosome integrins determine organotropic metastasis. *Nature* **2015**, *527*, 329. [CrossRef]

110. Peinado, H.; Lavotshkin, S.; Lyden, D. The secreted factors responsible for pre-metastatic niche formation: Old sayings and new thoughts. *Semin. Cancer Biol.* **2011**, *21*, 139. [CrossRef]

111. Zeng, Z.; Li, Y.; Pan, Y.; Lan, X.; Song, F.; Sun, J.; Zhou, K.; Liu, X.; Ren, X.; Wang, F.; et al. Cancer-derived exosomal miR-25-3p promotes pre-metastatic niche formation by inducing vascular permeability and angiogenesis. *Nat. Commun.* **2018**, *9*, 5395. [CrossRef]

112. Tominaga, N.; Kosaka, N.; Ono, M.; Katsuda, T.; Yoshioka, Y.; Tamura, K.; Lotvall, J.; Nakagama, H.; Ochiya, T. Brain metastatic cancer cells release microRNA-181c-containing extracellular vesicles capable of destructing blood-brain barrier. *Nat. Commun.* **2015**, *6*, 6716. [CrossRef] [PubMed]

113. Fang, J.H.; Zhang, Z.J.; Shang, L.R.; Luo, Y.W.; Lin, Y.F.; Yuan, Y.; Zhuang, S.M. Hepatoma cell-secreted exosomal microRNA-103 increases vascular permeability and promotes metastasis by targeting junction proteins. *Hepatology* **2018**, *68*, 1459. [CrossRef] [PubMed]

114. Wang, X.; Luo, G.; Zhang, K.; Cao, J.; Huang, C.; Jiang, T.; Liu, B.; Su, L.; Qiu, Z. Hypoxic Tumor-Derived Exosomal miR-301a Mediates M2 Macrophage Polarization via PTEN/PI3Kgamma to Promote Pancreatic Cancer Metastasis. *Cancer Res.* **2018**, *78*, 4586. [CrossRef] [PubMed]

115. Yang, M.; Chen, J.; Su, F.; Yu, B.; Su, F.; Lin, L.; Liu, Y.; Huang, J.D.; Song, E. Microvesicles secreted by macrophages shuttle invasion-potentiating microRNAs into breast cancer cells. *Mol. Cancer* **2011**, *10*, 117. [CrossRef]

116. Lan, J.; Sun, L.; Xu, F.; Liu, L.; Hu, F.; Song, D.; Hou, Z.; Wu, W.; Luo, X.; Wang, J.; et al. M2 Macrophage-Derived Exosomes Promote Cell Migration and Invasion in Colon Cancer. *Cancer Res.* **2019**, *79*, 146. [CrossRef]

117. Erez, N.; Truitt, M.; Olson, P.; Arron, S.T.; Hanahan, D. Cancer-Associated Fibroblasts Are Activated in Incipient Neoplasia to Orchestrate Tumor-Promoting Inflammation in an NF-kappaB-Dependent Manner. *Cancer Cell* **2010**, *17*, 135. [CrossRef]

118. Josson, S.; Gururajan, M.; Sung, S.Y.; Hu, P.; Shao, C.; Zhau, H.E.; Liu, C.; Lichterman, J.; Duan, P.; Li, Q.; et al. Stromal fibroblast-derived miR-409 promotes epithelial-to-mesenchymal transition and prostate tumorigenesis. *Oncogene* **2015**, *34*, 2690. [CrossRef]

119. Fong, M.Y.; Zhou, W.; Liu, L.; Alontaga, A.Y.; Chandra, M.; Ashby, J.; Chow, A.; O'Connor, S.T.; Li, S.; Chin, A.R.; et al. Breast-cancer-secreted miR-122 reprograms glucose metabolism in premetastatic niche to promote metastasis. *Nat. Cell Biol.* **2015**, *17*, 183. [CrossRef]

120. Fang, T.; Lv, H.; Lv, G.; Li, T.; Wang, C.; Han, Q.; Yu, L.; Su, B.; Guo, L.; Huang, S.; et al. Tumor-derived exosomal miR-1247-3p induces cancer-associated fibroblast activation to foster lung metastasis of liver cancer. *Nat. Commun.* **2018**, *9*, 191. [CrossRef]

121. Zhang, L.; Zhang, S.; Yao, J.; Lowery, F.J.; Zhang, Q.; Huang, W.C.; Li, P.; Li, M.; Wang, X.; Zhang, C.; et al. Microenvironment-induced PTEN loss by exosomal microRNA primes brain metastasis outgrowth. *Nature* **2015**, *527*, 100. [CrossRef]

122. Sakha, S.; Muramatsu, T.; Ueda, K.; Inazawa, J. Exosomal microRNA miR-1246 induces cell motility and invasion through the regulation of DENND2D in oral squamous cell carcinoma. *Sci. Rep.* **2016**, *6*, 38750. [CrossRef] [PubMed]

123. Nam, J.W.; Rissland, O.S.; Koppstein, D.; Abreu-Goodger, C.; Jan, C.H.; Agarwal, V.; Yildirim, M.A.; Rodriguez, A.; Bartel, D.P. Global analyses of the effect of different cellular contexts on microRNA targeting. *Mol. Cell* **2014**, *53*, 1031. [CrossRef] [PubMed]

124. Clark, P.M.; Loher, P.; Quann, K.; Brody, J.; Londin, E.R.; Rigoutsos, I. Argonaute CLIP-Seq reveals miRNA targetome diversity across tissue types. *Sci. Rep.* **2014**, *4*, 5947. [CrossRef] [PubMed]

125. Ascano, M.; Hafner, M.; Cekan, P.; Gerstberger, S.; Tuschl, T. Identification of RNA-protein interaction networks using PAR-CLIP. *Wiley Interdiscip. Rev. RNA* **2012**, *3*, 159. [CrossRef]

126. Bracken, C.P.; Li, X.; Wright, J.A.; Lawrence, D.M.; Pillman, K.A.; Salmanidis, M.; Anderson, M.A.; Dredge, B.K.; Gregory, P.A.; Tsykin, A.; et al. Genome-wide identification of miR-200 targets reveals a regulatory network controlling cell invasion. *EMBO J.* **2014**, *33*, 2040. [CrossRef]

127. Hamilton, M.P.; Rajapakshe, K.I.; Bader, D.A.; Cerne, J.Z.; Smith, E.A.; Coarfa, C.; Hartig, S.M.; McGuire, S.E. The Landscape of microRNA Targeting in Prostate Cancer Defined by AGO-PAR-CLIP. *Neoplasia* **2016**, *18*, 356. [CrossRef]

128. Fletcher, C.E.; Sulpice, E.; Combe, S.; Shibakawa, A.; Leach, D.A.; Hamilton, M.P.; Chrysostomou, S.L.; Sharp, A.; Welti, J.; Yuan, W.; et al. Androgen receptor-modulatory microRNAs provide insight into therapy resistance and therapeutic targets in advanced prostate cancer. *Oncogene* **2019**, *38*, 5700. [CrossRef]

129. He, L.; Thomson, J.M.; Hemann, M.T.; Hernando-Monge, E.; Mu, D.; Goodson, S.; Powers, S.; Cordon-Cardo, C.; Lowe, S.W.; Hannon, G.J.; et al. A microRNA polycistron as a potential human oncogene. *Nature* **2005**, *435*, 828. [CrossRef]

130. Akao, Y.; Nakagawa, Y.; Naoe, T. let-7 microRNA functions as a potential growth suppressor in human colon cancer cells. *Biol. Pharm. Bull.* **2006**, *29*, 903. [CrossRef]

131. Bonci, D.; Coppola, V.; Musumeci, M.; Addario, A.; Giuffrida, R.; Memeo, L.; D'Urso, L.; Pagliuca, A.; Biffoni, M.; Labbaye, C.; et al. The miR-15a-miR-16-1 cluster controls prostate cancer by targeting multiple oncogenic activities. *Nat. Med.* **2008**, *14*, 1271. [CrossRef]

132. Garzon, R.; Liu, S.; Fabbri, M.; Liu, Z.; Heaphy, C.E.; Callegari, E.; Schwind, S.; Pang, J.; Yu, J.; Muthusamy, N.; et al. MicroRNA-29b induces global DNA hypomethylation and tumor suppressor gene reexpression in acute myeloid leukemia by targeting directly DNMT3A and 3B and indirectly DNMT1. *Blood* **2009**, *113*, 6411. [CrossRef] [PubMed]

133. Trang, P.; Medina, P.P.; Wiggins, J.F.; Ruffino, L.; Kelnar, K.; Omotola, M.; Homer, R.; Brown, D.; Bader, A.G.; Weidhaas, J.B.; et al. Regression of murine lung tumors by the let-7 microRNA. *Oncogene* **2010**, *29*, 1580. [CrossRef] [PubMed]

134. Klein, U.; Lia, M.; Crespo, M.; Siegel, R.; Shen, Q.; Mo, T.; Ambesi-Impiombato, A.; Califano, A.; Migliazza, A.; Bhagat, G.; et al. The DLEU2/miR-15a/16-1 cluster controls B cell proliferation and its deletion leads to chronic lymphocytic leukemia. *Cancer Cell* **2010**, *17*, 28. [CrossRef] [PubMed]

135. Zhao, J.J.; Lin, J.; Lwin, T.; Yang, H.; Guo, J.; Kong, W.; Dessureault, S.; Moscinski, L.C.; Rezania, D.; Dalton, W.S.; et al. microRNA expression profile and identification of miR-29 as a prognostic marker and pathogenetic factor by targeting CDK6 in mantle cell lymphoma. *Blood* **2010**, *115*, 2630. [CrossRef]

136. Rupaimoole, R.; Slack, F.J. MicroRNA therapeutics: Towards a new era for the management of cancer and other diseases. *Nat. Rev. Drug Discov.* **2017**, *16*, 203. [CrossRef]

137. Hanna, J.; Hossain, G.S.; Kocerha, J. The Potential for microRNA Therapeutics and Clinical Research. *Front. Genet.* **2019**, *10*, 478. [CrossRef]

138. Liu, Y.P.; Vink, M.A.; Westerink, J.T.; Ramirez de Arellano, E.; Konstantinova, P.; Ter Brake, O.; Berkhout, B. Titers of lentiviral vectors encoding shRNAs and miRNAs are reduced by different mechanisms that require distinct repair strategies. *RNA* **2010**, *16*, 1328. [CrossRef]

139. Lou, W.; Chen, Q.; Ma, L.; Liu, J.; Yang, Z.; Shen, J.; Cui, Y.; Bian, X.W.; Qian, C. Oncolytic adenovirus co-expressing miRNA-34a and IL-24 induces superior antitumor activity in experimental tumor model. *J. Mol. Med.* **2013**, *91*, 715. [CrossRef]

140. Collins, S.A.; Guinn, B.A.; Harrison, P.T.; Scallan, M.F.; O'Sullivan, G.C.; Tangney, M. Viral vectors in cancer immunotherapy: Which vector for which strategy? *Curr. Gene Ther.* **2008**, *8*, 66. [CrossRef]

141. Campani, V.; Salzano, G.; Lusa, S.; De Rosa, G. Lipid Nanovectors to Deliver RNA Oligonucleotides in Cancer. *Nanomaterials* **2016**, *6*, 131. [CrossRef]

142. Ando, H.; Okamoto, A.; Yokota, M.; Asai, T.; Dewa, T.; Oku, N. Polycation liposomes as a vector for potential intracellular delivery of microRNA. *J. Gene Med.* **2013**, *15*, 375. [CrossRef] [PubMed]

143. Dai, Y.; Zhang, X. MicroRNA Delivery with Bioreducible Polyethylenimine as a Non-Viral Vector for Breast Cancer Gene Therapy. *Macromol. Biosci.* **2019**, *19*, e1800445. [CrossRef] [PubMed]

144. Li, Y.; Dai, Y.; Zhang, X.; Chen, J. Three-layered polyplex as a microRNA targeted delivery system for breast cancer gene therapy. *Nanotechnology* **2017**, *28*, 285101. [CrossRef] [PubMed]

145. Pecot, C.V.; Rupaimoole, R.; Yang, D.; Akbani, R.; Ivan, C.; Lu, C.; Wu, S.; Han, H.D.; Shah, M.Y.; Rodriguez-Aguayo, C.; et al. Tumour angiogenesis regulation by the miR-200 family. *Nat. Commun.* **2013**, *4*, 2427. [CrossRef] [PubMed]

146. Jiang, Q.; Yuan, Y.; Gong, Y.; Luo, X.; Su, X.; Hu, X.; Zhu, W. Therapeutic delivery of microRNA-143 by cationic lipoplexes for non-small cell lung cancer treatment in vivo. *J. Cancer Res. Clin. Oncol.* **2019**, *145*, 2951. [CrossRef] [PubMed]

147. Chang, H.; Yi, B.; Ma, R.; Zhang, X.; Zhao, H.; Xi, Y. CRISPR/cas9, a novel genomic tool to knock down microRNA in vitro and in vivo. *Sci. Rep.* **2016**, *6*, 22312. [CrossRef]

148. Aquino-Jarquin, G. Emerging Role of CRISPR/Cas9 Technology for MicroRNAs Editing in Cancer Research. *Cancer Res.* **2017**, *77*, 6812. [CrossRef]

149. Yang, J.; Meng, X.; Pan, J.; Jiang, N.; Zhou, C.; Wu, Z.; Gong, Z. CRISPR/Cas9-mediated noncoding RNA editing in human cancers. *RNA Biol.* **2018**, *15*, 35. [CrossRef]

150. Huo, W.; Zhao, G.; Yin, J.; Ouyang, X.; Wang, Y.; Yang, C.; Wang, B.; Dong, P.; Wang, Z.; Watari, H.; et al. Lentiviral CRISPR/Cas9 vector mediated miR-21 gene editing inhibits the epithelial to mesenchymal transition in ovarian cancer cells. *J. Cancer* **2017**, *8*, 57. [CrossRef]

151. El Fatimy, R.; Subramanian, S.; Uhlmann, E.J.; Krichevsky, A.M. Genome Editing Reveals Glioblastoma Addiction to MicroRNA-10b. *Mol. Ther. J. Am. Soc. Gene Ther.* **2017**, *25*, 368. [CrossRef]

152. Esposito, C.L.; Cerchia, L.; Catuogno, S.; De Vita, G.; Dassie, J.P.; Santamaria, G.; Swiderski, P.; Condorelli, G.; Giangrande, P.H.; de Franciscis, V. Multifunctional aptamer-miRNA conjugates for targeted cancer therapy. *Mol. Ther. J. Am. Soc. Gene Ther.* **2014**, *22*, 1151. [CrossRef] [PubMed]

153. Cheng, C.J.; Bahal, R.; Babar, I.A.; Pincus, Z.; Barrera, F.; Liu, C.; Svoronos, A.; Braddock, D.T.; Glazer, P.M.; Engelman, D.M.; et al. MicroRNA silencing for cancer therapy targeted to the tumor microenvironment. *Nature* **2015**, *518*, 107–110. [CrossRef] [PubMed]

154. Almanza, G.; Rodvold, J.J.; Tsui, B.; Jepsen, K.; Carter, H.; Zanetti, M. Extracellular vesicles produced in B cells deliver tumor suppressor miR-335 to breast cancer cells disrupting oncogenic programming in vitro and in vivo. *Sci. Rep.* **2018**, *8*, 17581. [CrossRef] [PubMed]

155. Van Roosbroeck, K.; Fanini, F.; Setoyama, T.; Ivan, C.; Rodriguez-Aguayo, C.; Fuentes-Mattei, E.; Xiao, L.; Vannini, I.; Redis, R.S.; D'Abundo, L.; et al. Combining Anti-Mir-155 with Chemotherapy for the Treatment of Lung Cancers. *Clin. Cancer Res.* **2017**, *23*, 2891. [CrossRef] [PubMed]

156. Malhotra, M.; Sekar, T.V.; Ananta, J.S.; Devulapally, R.; Afjei, R.; Babikir, H.A.; Paulmurugan, R.; Massoud, T.F. Targeted nanoparticle delivery of therapeutic antisense microRNAs presensitizes glioblastoma cells to lower effective doses of temozolomide in vitro and in a mouse model. *Oncotarget* **2018**, *9*, 21478. [CrossRef]

157. Devulapally, R.; Sekar, N.M.; Sekar, T.V.; Foygel, K.; Massoud, T.F.; Willmann, J.K.; Paulmurugan, R. Polymer nanoparticles mediated codelivery of antimiR-10b and antimiR-21 for achieving triple negative breast cancer therapy. *ACS Nano* **2015**, *9*, 2290. [CrossRef]

158. Cortez, M.A.; Valdecanas, D.; Zhang, X.; Zhan, Y.; Bhardwaj, V.; Calin, G.A.; Komaki, R.; Giri, D.K.; Quini, C.C.; Wolfe, T.; et al. Therapeutic delivery of miR-200c enhances radiosensitivity in lung cancer. *Mol. Ther. J. Am. Soc. Gene Ther.* **2014**, *22*, 1494. [CrossRef]

159. Yin, P.T.; Pongkulapa, T.; Cho, H.Y.; Han, J.; Pasquale, N.J.; Rabie, H.; Kim, J.H.; Choi, J.W.; Lee, K.B. Overcoming Chemoresistance in Cancer via Combined MicroRNA Therapeutics with Anticancer Drugs Using Multifunctional Magnetic Core-Shell Nanoparticles. *ACS Appl. Mater. Interfaces* **2018**, *10*, 26954. [CrossRef]

160. Rui, M.; Qu, Y.; Gao, T.; Ge, Y.; Feng, C.; Xu, X. Simultaneous delivery of anti-miR21 with doxorubicin prodrug by mimetic lipoprotein nanoparticles for synergistic effect against drug resistance in cancer cells. *Int. J. Nanomed.* **2017**, *12*, 217. [CrossRef]

161. Wang, F.; Zhang, L.; Bai, X.; Cao, X.; Jiao, X.; Huang, Y.; Li, Y.; Qin, Y.; Wen, Y. Stimuli-Responsive Nanocarrier for Co-delivery of MiR-31 and Doxorubicin To Suppress High MtEF4 Cancer. *ACS Appl. Mater. Interfaces* **2018**, *10*, 22767. [CrossRef]

162. Yao, C.; Liu, J.; Wu, X.; Tai, Z.; Gao, Y.; Zhu, Q.; Li, J.; Zhang, L.; Hu, C.; Gu, F.; et al. Reducible self-assembling cationic polypeptide-based micelles mediate co-delivery of doxorubicin and microRNA-34a for androgen-independent prostate cancer therapy. *J. Control. Release* **2016**, *232*, 203. [CrossRef] [PubMed]

163. Dai, X.; Fan, W.; Wang, Y.; Huang, L.; Jiang, Y.; Shi, L.; McKinley, D.; Tan, W.; Tan, C. Combined Delivery of Let-7b MicroRNA and Paclitaxel via Biodegradable Nanoassemblies for the Treatment of KRAS Mutant Cancer. *Mol. Pharm.* **2016**, *13*, 520. [CrossRef] [PubMed]

164. Uz, M.; Kalaga, M.; Pothuraju, R.; Ju, J.; Junker, W.M.; Batra, S.K.; Mallapragada, S.; Rachagani, S. Dual delivery nanoscale device for miR-345 and gemcitabine co-delivery to treat pancreatic cancer. *J. Control. Release* **2019**, *294*, 237. [CrossRef] [PubMed]

165. Mondal, G.; Almawash, S.; Chaudhary, A.K.; Mahato, R.I. EGFR-Targeted Cationic Polymeric Mixed Micelles for Codelivery of Gemcitabine and miR-205 for Treating Advanced Pancreatic Cancer. *Mol. Pharm.* **2017**, *14*, 3121. [CrossRef]

166. Xu, J.; Sun, J.; Ho, P.Y.; Luo, Z.; Ma, W.; Zhao, W.; Rathod, S.B.; Fernandez, C.A.; Venkataramanan, R.; Xie, W.; et al. Creatine based polymer for codelivery of bioengineered MicroRNA and chemodrugs against breast cancer lung metastasis. *Biomaterials* **2019**, *210*, 25. [CrossRef]

167. Bader, A.G. miR-34—A microRNA replacement therapy is headed to the clinic. *Front. Genet.* **2012**, *3*, 120. [CrossRef]

168. Reid, G.; Kao, S.C.; Pavlakis, N.; Brahmbhatt, H.; MacDiarmid, J.; Clarke, S.; Boyer, M.; van Zandwijk, N. Clinical development of TargomiRs, a miRNA mimic-based treatment for patients with recurrent thoracic cancer. *Epigenomics* **2016**, *8*, 1079. [CrossRef]

169. Van Zandwijk, N.; Pavlakis, N.; Kao, S.C.; Linton, A.; Boyer, M.J.; Clarke, S.; Huynh, Y.; Chrzanowska, A.; Fulham, M.J.; Bailey, D.L.; et al. Safety and activity of microRNA-loaded minicells in patients with recurrent malignant pleural mesothelioma: A first-in-man, phase 1, open-label, dose-escalation study. *Lancet Oncol.* **2017**, *18*, 1386. [CrossRef]

170. Haghi, M.; Taha, M.F.; Javeri, A. Suppressive effect of exogenous miR-16 and miR-34a on tumorigenesis of breast cancer cells. *J. Cell. Biochem.* **2019**. [CrossRef]

171. Xu, M.; Li, D.; Yang, C.; Ji, J.S. MicroRNA-34a Inhibition of the TLR Signaling Pathway Via CXCL10 Suppresses Breast Cancer Cell Invasion and Migration. *Cell. Physiol. Biochem. Int. J. Exp. Cell. Physiol. Biochem. Pharmacol.* **2018**, *46*, 1286. [CrossRef]

172. Wang, T.H.; Yeh, C.T.; Ho, J.Y.; Ng, K.F.; Chen, T.C. OncomiR miR-96 and miR-182 promote cell proliferation and invasion through targeting ephrinA5 in hepatocellular carcinoma. *Mol. Carcinog.* **2016**, *55*, 366. [CrossRef]

173. Han, L.; Chen, W.; Xia, Y.; Song, Y.; Zhao, Z.; Cheng, H.; Jiang, T. MiR-101 inhibits the proliferation and metastasis of lung cancer by targeting zinc finger E-box binding homeobox 1. *Am. J. Transl. Res.* **2018**, *10*, 1172.

174. Ji, H.; Sang, M.; Liu, F.; Ai, N.; Geng, C. miR-124 regulates EMT based on ZEB2 target to inhibit invasion and metastasis in triple-negative breast cancer. *Pathol. Res. Pract.* **2019**, *215*, 697. [CrossRef] [PubMed]

175. Zhao, L.; Li, R.; Xu, S.; Li, Y.; Zhao, P.; Dong, W.; Liu, Z.; Zhao, Q.; Tan, B. Tumor suppressor miR-128-3p inhibits metastasis and epithelial-mesenchymal transition by targeting ZEB1 in esophageal squamous-cell cancer. *Acta Biochim. Et Biophys. Sin.* **2018**, *50*, 171. [CrossRef] [PubMed]

176. Liu, B.; Du, R.; Zhou, L.; Xu, J.; Chen, S.; Chen, J.; Yang, X.; Liu, D.X.; Shao, Z.M.; Zhang, L.; et al. miR-200c/141 Regulates Breast Cancer Stem Cell Heterogeneity via Targeting HIPK1/beta-Catenin Axis. *Theranostics* **2018**, *8*, 5801. [CrossRef] [PubMed]

177. Li, J.; Xia, L.; Zhou, Z.; Zuo, Z.; Xu, C.; Song, H.; Cai, J. MiR-186-5p upregulation inhibits proliferation, metastasis and epithelial-to-mesenchymal transition of colorectal cancer cell by targeting ZEB1. *Arch. Biochem. Biophys.* **2018**, *640*, 53. [CrossRef]

178. Sun, G.; Liu, M.; Han, H. Overexpression of microRNA-190 inhibits migration, invasion, epithelial-mesenchymal transition, and angiogenesis through suppression of protein kinase B-extracellular signal-regulated kinase signaling pathway via binding to stanniocalcin 2 in breast cancer. *J. Cell. Physiol.* **2019**, *10*, 17824.

179. Yu, Y.; Luo, W.; Yang, Z.J.; Chi, J.R.; Li, Y.R.; Ding, Y.; Ge, J.; Wang, X.; Cao, X.C. miR-190 suppresses breast cancer metastasis by regulation of TGF-beta-induced epithelial-mesenchymal transition. *Mol. Cancer* **2018**, *17*, 70. [CrossRef]

180. Bian, X.; Liang, Z.; Feng, A.; Salgado, E.; Shim, H. HDAC inhibitor suppresses proliferation and invasion of breast cancer cells through regulation of miR-200c targeting CRKL. *Biochem. Pharmacol.* **2018**, *147*, 30. [CrossRef]

181. Liu, J.; He, D.; Zhang, G.J. Bioluminescence Imaging for Monitoring miR-200c Expression in Breast Cancer Cells and its Effects on Epithelial-Mesenchymal Transition Progress in Living Animals. *Mol. Imaging Biol. Mib Off. Publ. Acad. Mol. Imaging* **2018**, *20*, 761. [CrossRef]

182. Seo, S.; Moon, Y.; Choi, J.; Yoon, S.; Jung, K.H.; Cheon, J.; Kim, W.; Kim, D.; Lee, C.H.; Kim, S.W.; et al. The GTP binding activity of transglutaminase 2 promotes bone metastasis of breast cancer cells by downregulating microRNA-205. *Am. J. Cancer Res.* **2019**, *9*, 597. [PubMed]

183. Wu, L.; Zhang, Y.; Huang, Z.; Gu, H.; Zhou, K.; Yin, X.; Xu, J. MiR-409-3p Inhibits Cell Proliferation and Invasion of Osteosarcoma by Targeting Zinc-Finger E-Box-Binding Homeobox-1. *Front. Pharmacol.* **2019**, *10*, 137. [CrossRef] [PubMed]

184. Guo, S.J.; Zeng, H.X.; Huang, P.; Wang, S.; Xie, C.H.; Li, S.J. MiR-508-3p inhibits cell invasion and epithelial-mesenchymal transition by targeting ZEB1 in triple-negative breast cancer. *Eur. Rev. Med. Pharmacol. Sci.* **2018**, *22*, 6379. [PubMed]

185. Yao, R.; Zheng, H.; Wu, L.; Cai, P. miRNA-641 inhibits the proliferation, migration, and invasion and induces apoptosis of cervical cancer cells by directly targeting ZEB1. *Onco Targets Ther.* **2018**, *11*, 8965. [CrossRef]

186. Lee, J.W.; Guan, W.; Han, S.; Hong, D.K.; Kim, L.S.; Kim, H. MicroRNA-708-3p mediates metastasis and chemoresistance through inhibition of epithelial-to-mesenchymal transition in breast cancer. *Cancer Sci.* **2018**, *109*, 1404. [CrossRef]

Review

Interplay among SNAIL Transcription Factor, MicroRNAs, Long Non-Coding RNAs, and Circular RNAs in the Regulation of Tumor Growth and Metastasis

Klaudia Skrzypek * and **Marcin Majka ***

Jagiellonian University Medical College, Faculty of Medicine, Institute of Pediatrics, Department of Transplantation, Wielicka 265, 30-663 Cracow, Poland
* Correspondence: klaudia.skrzypek@uj.edu.pl (K.S.); mmajka@cm-uj.krakow.pl (M.M);
  Tel.: +48-12-659-15-93 (K.S. & M.M.)

Received: 20 November 2019; Accepted: 9 January 2020; Published: 14 January 2020

Abstract: SNAIL (SNAI1) is a zinc finger transcription factor that binds to E-box sequences and regulates the expression of genes. It usually acts as a gene repressor, but it may also activate the expression of genes. SNAIL plays a key role in the regulation of epithelial to mesenchymal transition, which is the main mechanism responsible for the progression and metastasis of epithelial tumors. Nevertheless, it also regulates different processes that are responsible for tumor growth, such as the activity of cancer stem cells, the control of cell metabolism, and the regulation of differentiation. Different proteins and microRNAs may regulate the SNAIL level, and SNAIL may be an important regulator of microRNA expression as well. The interplay among SNAIL, microRNAs, long non-coding RNAs, and circular RNAs is a key event in the regulation of tumor growth and metastasis. This review for the first time discusses different types of regulation between SNAIL and non-coding RNAs with a focus on feedback loops and the role of competitive RNA. Understanding these mechanisms may help develop novel therapeutic strategies against cancer based on microRNAs.

Keywords: tumor; metastasis; microRNA; SNAIL (SNAI1) transcription factor; epithelial to mesenchymal transition (EMT); long non-coding RNAs (lncRNAs); circular RNAs

---

1. Introduction: Background of SNAIL Transcription Factor

SNAIL is a member of the group of conservative zinc finger transcription factors. It was first described in *Drosophila melanogaster* as an essential factor for the mesoderm formation [1]. Subsequently, its homologues have been described in many species, including humans. The SNAIL family consists of three members: SNAIL (SNAI1), SLUG (SNAI2), and SMUG (SNAI3) [2]. The SNAIL protein contains C-terminal zinc finger domains that are responsible for DNA binding, the N-terminal SNAG domain responsible for interaction with several co-repressors or epigenetic remodeling complexes, the serine-rich domain (SRD) regulating ubiquitination and proteasome degradation, and the nuclear export sequence (NES) that controls the protein stability and subcellular localization [3].

1.1. SNAIL Expression and Regulation

SNAIL expression may be regulated by many signaling pathways. At the transcriptional level, SNAIL is regulated by multiple growth factors and signaling molecules that are responsible for the subsequent regulation of the SNAIL promoter, including transforming growth factor β (TGF-β), fibroblast growth factor 2 (FGF2), epidermal growth factor (EGF), Harvey rat sarcoma viral oncogene homolog (H-ras), Akt kinase-transforming protein (v-Akt), and nuclear factor kappa-light-chain-enhancer of activated B cells/protein 65 (NF-κB/p65) [4,5]. Post-translational modifications, such as phosphorylation, ubiquitination, and lysine oxidation also regulate SNAIL level. Glycogen synthase kinase 3 beta (GSK-3β) phosphorylates SNAIL at two consensus motifs. Phosphorylation of the first motif regulates ubiquitination and degradation in the proteasome, whereas phosphorylation of the second motif regulates its subcellular localization [6]. Lysyl oxidase-like 2 (LOXL2) enzyme interaction regulates SNAIL stability [7] by interfering with FBXL14 binding SNAIL. FBXL14 (F-box and leucine-rich repeat protein 14) is a ubiquitin ligase that targets both phosphorylated and unphosphorylated SNAIL for proteasome degradation [8]. SNAIL can also be stabilized by hyperglycemia-regulated O-linked β-N-acetylglucosamine (O-GlcNAc) modification of serine [9]. Moreover, SNAIL can be stabilized by NF-κB, which induces COP9 signalosome 2 (CSN2), which, in turn, blocks the ubiquitination and degradation of SNAIL [10]. The phosphorylation of SNAIL may result in an increased retention of the protein in the nucleus. That mechanism of action was described for p21-activated kinase (PAK1), which phosphorylates SNAIL at Ser 246 [11].

1.2. Different Pathways Regulated by SNAIL

SNAIL plays an important role in the regulation of epithelial to mesenchymal transition in embryo development: gastrulation and mesoderm formation [2]. However, molecular mechanisms of certain pathological stages resemble those observed in physiological process. One of them is epithelial to mesenchymal transition (EMT) during cancer progression. It is the main mechanism responsible for the invasiveness and metastasis of neoplasm at the advanced stages [12]. SNAIL exerts its effects by decreasing the expression of E-cadherin by binding to its promoter [13]. Nevertheless, SNAIL is a transcriptional repressor, which binds to regulatory regions and promoters containing sequences called E-boxes, and thereby it regulates the expression of many different genes and in this way, it may also regulate EMT. The SNAIL family contains a highly conserved region of four to six zinc fingers that allows them to interact with those E-box sequences (CANNTG). Since these sequences are also recognized by transcription factors from the basic helix-loop-helix (bHLH) family, the role of SNAIL factors is mainly focused on transcription repression by excluding these proteins from their binding sites [2]. SNAIL is capable of interacting with HDAC1/2 histone deacetylase, which causes a local modification of the chromatin structure and blocks the expression of E-cadherin, the loss of which is a marker of epithelial–mesenchymal transition (EMT) [13]. As E-box sequences are present in the promoters of many different genes, in the literature, SNAIL is described as a regulator of many genes important in tumorigenesis, such as cyclin D2, proliferating cell nuclear antigen (PCNA), prostaglandin dehydrogenase, ATPase1, etc. [12]. SNAIL turned out to be also a direct regulator of not only EMT in tumor progression, but also of myogenic differentiation. The binding of SNAIL to E-box sequences in the myogenic factor 5 (MYF5) promoter and recruiting histone deacetylases (HDACs) was described in the regulation of rhabdomyosarcoma development [14]. Another example of the non-canonical actions of SNAIL is the regulation of myoblast determination protein 1 (MyoD) function in myogenic differentiation by the competitive binding of SNAIL to its regulatory sequences [15]. Nevertheless, SNAIL is not only described as a transcriptional repressor, but also as the transcriptional activator. For example, SNAIL induces the expression of mesenchymal genes, such as vimentin, fibronectin, matrix metalloproteinases MMP-2, and MMP-9. In that way, it further facilitates the increased motility of cells [16].

What is more, the recent data demonstrated the mechanism of self-regulation by members of the SNAIL family: the SNAIL-binding site is present in the *SNAIL* promoter (negative feedback) [17], and avian Slug can self-activate during the neural crest development [18]. Moreover, in ovarian cancer cells, SNAIL binds to two E-box sequences in *SLUG* promoter and represses SLUG, which is predominantly mediated through the recruitment of the HDACs [19].

SNAIL plays a role in many physiological and pathological processes, such as chronic inflammation, fibrosis, EMT induction, the regulation of cancer stem cells, the control of cell metabolism, the suppression of estrogen receptor signaling, and in particular the development and metastasis of tumors [3]. Currently, many research papers focus not only on interaction between SNAIL and different genes, but also on the interplay between SNAIL and non-coding RNAs, such as microRNAs, long non-coding RNAs, and circular RNAs [20]. In this review, we discuss recent advances in those fields. We present bidirectional crosstalk between SNAIL and non-coding RNAs with implications of these new findings on tumor progression, which may help develop novel therapeutic strategies in future.

2. Non-Coding RNAs as Regulators of Tumor Progression

Non-coding RNAs (ncRNAs) are a class of RNA transcripts that do not encode proteins, but they may play a role in the regulation of gene expression at transcriptional, translational, and post-translational levels. Among regulatory ncRNAs, long non-coding RNAs, small RNAs, and circular RNAs may be distinguished [21] (Figure 1), and they are described in this review.

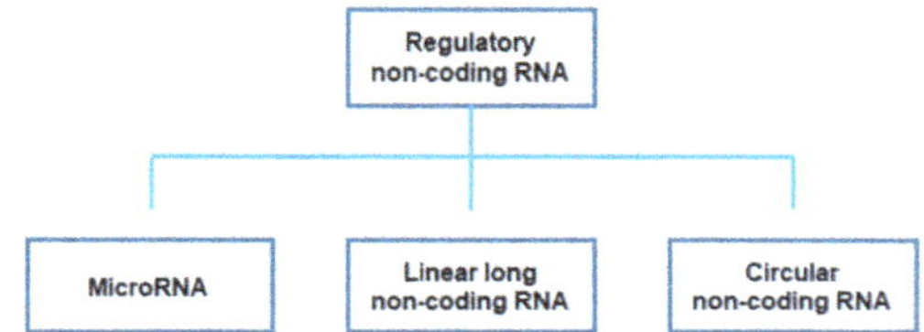

Figure 1. Scheme presenting the selected regulatory non-coding RNAs.

Long non-coding RNAs (lncRNAs) are RNA transcripts with a length greater than 200 nucleotides. They can regulate gene expressions and functions. Therefore, they are involved in the pathogenesis of many diseases, including cancer. Nevertheless, there are papers revealing that some lncRNAs contain cryptic open reading frames (ORFs), which may blur the distinction between protein-coding and non-coding transcripts [22]. lncRNAs can originate from their own promoters or from the promoters shared with other coding or non-coding genes, or from enhancer sequences. lncRNAs are usually transcribed by RNA polymerase II or RNA polymerase III. They are often 5'-capped, spliced, and polyadenylated, but they are usually shorter than mRNAs [23,24]. lncRNAs may be co-regulated with mRNAs in expression networks. lncRNAs may also be generated from the divergent transcription from shared protein-coding gene promoters. Divergent transcription generates the sense (mRNA) and anti-sense RNAs [24,25]. lncRNA promoters are usually evolutionarily conserved and tightly regulated, and they are prone to epigenetic modification [23]. lncRNAs may also be processed in different ways than mRNAs, such as RNase P-processed 3' maturation, which was shown for MALAT1 (metastasis associated lung adenocarcinoma transcript 1) [24]. DICER1 endonuclease is an important factor in both the biogenesis of miRNAs that may also act as a downstream activator of many lncRNAs [26]. What is also interesting is that few miRNAs are derived from lncRNA exons [27]. lncRNAs participate in and modulate the various cellular processes, such as cellular transcription, the modulation of chromatin structure, DNA methylation, or histone modification. They may act as a sponge for microRNAs and as a competing endogenous RNAs (ceRNAs) [28].

Circular RNA (circRNA) is a type of single-stranded RNA that forms a covalently closed continuous loop that is insensitive to ribonucleases. circRNAs are formed by exon skipping or back-splicing events. circRNAs are produced by nonsequential exon-exon back-splicing, which results in a chemically circularized transcript in which 3' sequences are spliced upstream of 5' sequences, and they have special 5' and 3'-end processing [24]. Alternative splicing factor quaking is a regulator of that circularization during EMT [29]. There is also a class of circular intronic lncRNA (ciRNAs) that are generated from stabilized introns after canonical splicing. They display regulatory functions, mostly at their transcription sites [30]. There are also exon–intron circRNAs (elciRNAs) that represent a class of circular RNAs that retain unspliced introns. Their role involves induction of the transcription of their parental genes via interaction with polymerase II and U1 snRNP (small nuclear ribonucleoprotein) [31].

circRNAs are closely associated with tumor metastasis and patient prognosis, because they are differentially expressed in different tumor types. They may act as a microRNA sponge and interact with proteins [32]. Nevertheless, recent research papers provide initial evidence for certain endogenous circRNAs coding for proteins [33].

MicroRNAs (miRNAs) are a class of approximately 22 nucleotides small non-coding RNAs. They can regulate the expression of genes and translation of proteins by interfering with ribosomal machinery. They commonly target the 3′ untranslated regions (3′ UTRs) of mRNAs and in that way decrease their stability and suppress translation. Nevertheless, they can also activate other genes [34,35]. Genes highly and constitutively expressed usually display shorter 3′ UTR sites and in consequence only a few binding sites for miRNAs. Accordingly, genes potently regulated during development display multiple binding sites for miRNAs [36].

miRNAs can be expressed at high levels (even up to tens of thousands of copies per cell), and they act as important regulatory factors, controlling hundreds of mRNA targets [37]. Animal miRNAs target the 3′ UTRs of different mRNAs by seed sequence complementarity. They usually repress translation more often than they cleave mRNA [35,38,39].

miRNAs are located in introns of coding genes, in exons, or in non-protein coding DNA regions. miRNAs have their own promoters, and they are independently expressed. Some of them are also organized in clusters sharing the same transcriptional regulation. miRNAs can arise from spliced introns, which are often termed miRtrons, or their own promoter, driving the expression of a single miRNA or polycistron yielding multiple pre-miRNA stem loops [40]. Nevertheless, miRNA transcription may also be dependent on the host gene. Intronic miRNAs can be expressed together with their host gene mRNA, and they can be derived from a common transcript [41]. Many non-canonical miRNA biogenesis pathways have also been characterized [42].

miRNAs are transcribed by polymerase II, sometimes as polycistronic transcripts. miRNA stem loops are excised from the primary transcripts (pri-miRNA) in the nucleus by endoribonuclease Drosha, acting together with DGCR8. Then, the excised 70–100 nt hairpin called pre-miRNA is actively transported from the nucleus to cytoplasm in a GTP (guanosine-5′-triphosphate)-dependent manner. The export is mediated by exportin 5 and Ran GTPase. Subsequently in the cytoplasm, the pre-miRNA is cleaved by Dicer endonuclease, giving the mature miRNA—a base-paired double-stranded processing intermediate with a 2 nt 3′ overhang. Two strands are generated. Then, one strand of the duplex is incorporated into RNA-induced silencing complexes (RISC) with the Argonaute protein, which is capable of endonucleolytic cleavage [42,43]. The translational repression is characterized by low miRNA–target complementarity, whereas mRNA degradation requires a high miRNA–target complementarity [44].

Alterations of miRNAs expression in various cancers have been described in the literature. Firstly, in 2002, they were shown in the most common form of adult leukemia, B cell chronic lymphocytic leukemia [45] and then in 2003 in colorectal cancers [46]. It soon turned out that miRNAs can be differentially expressed in different tumor types as either benign or malignant, and they can also act as biomarkers [47].

Global miRNA downregulation is a common trait of many tumors [48,49]. Accordingly, the diminished expression of miRNA processing factors is also associated with the poor prognosis of different cancer types [50].

What is more, some miRNAs' loci often display genomic instability in cancer, and they are located in cancer-associated genomic regions or in fragile sites. It was also demonstrated that several miRNAs located in deleted regions are expressed at low levels in cancer [51].

Cancer cells can also escape from miRNA regulation by the production of mRNAs with shortened 3′UTR and fewer miRNA target sites. This global switch of the use of miRNA-mediated gene regulation is associated with an increased proliferation or cellular transformation [50]. These findings are consistent with the widespread decrease of miRNAs in cancer [48,49].

Some miRNAs can behave as oncogenes favoring tumorigenesis. They are called oncomirs. They can reduce the levels of proteins blocking proliferation and migration and activating apoptosis. Many miRNAs were identified as oncomirs in different types of tumors. For example, members encoded by the miR-17-92 cluster were previously associated with carcinogenesis and usually display increased expression in tumors, including lung cancer [52,53].

On the other hand, tumor-suppressive miRNAs can inhibit cancer development. Their inactivation in tumors is followed by the accumulation of proteins stimulating proliferation and migration and decreasing apoptosis. For example, miR-181a and miR-181b were described to act as tumor suppressors in glioma [54] and miR-181a in non-small cell lung carcinoma [55]. Interestingly, plenty of miRNAs may behave oppositely in different types of tumors. For example, miR-34c can exert tumor-suppressive functions in prostate cancer [56], but in lung adenocarcinoma with different oncogenic mutations, it was reported to be upregulated [57].

miRNAs can affect tumor progression also by modulation of the development of new blood vessels. miRNAs promoting angiogenesis are called angiomirs, and they can target genes that are important in angiogenic processes [58].

Currently, miRNAs' role in the regulation of epithelial to mesenchymal transition has been widely described in the literature [59]. Since SNAIL is one of the crucial factors regulating EMT, the interplay between SNAIL and miRNAs may be a key factor in the regulation of tumor progression.

3. MicroRNAs Regulating SNAIL

3.1. MicroRNAs Directly Targeting SNAIL

MiRNAs can act as regulators of SNAIL expression by binding to the 3′UTR of *SNAIL*. Bioinformatical analysis using TargetScanHuman 7.1 [60] revealed several binding sites for different miRNAs in this region in human cells (Figure 2), and most of them have been already verified in the literature. For example, the *SNAIL* 3′UTR was shown to function as a sponge for multiple migration and invasion-related miRNA candidates including miR-153, miR-199a-5p, miR-203, miR-204, miR-22, miR-34a and miR-34c [61].

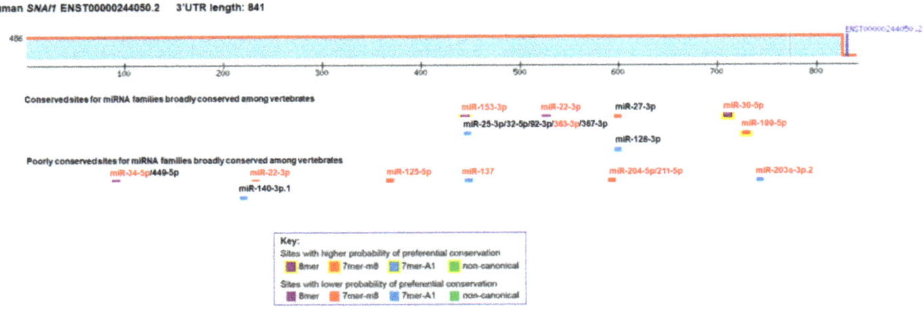

Figure 2. MicroRNAs targeting the 3′ untranslated (3′UTR) region of *SNAIL* from bioinformatical analysis using TargetScanHuman 7.1 (access: 22 October 2019). Experimental evidence for direct binding to *SNAIL* 3′UTR was shown in the literature for miR-153, miR-22, miR-30, miR-363, miR-199, miR-34, miR-22, miR-137, miR-203, miR-125, miR-211, and miR-203 (marked in red), which is described in the text below.

Several miRNAs were experimentally validated to target *SNAIL* 3′UTR, and subsequently, their role was described in different tumor types. One of the crucial regulators of SNAIL expression widely described in the literature is the miR-30 family. Members of this family target the 3′UTR of *SNAIL* mRNA in non-small cell lung carcinoma [62], breast cancer [63], pancreatic cancer stem cells [64], melanoma [65], esophageal squamous cell carcinoma [66], rhabdomyosarcoma [14], or in hepatocytes [67,68]. This inhibition usually regulates EMT in epithelial tumor types, but in mesenchymal tumors, such as rhabdomyosarcoma, it may be responsible for non-canonical SNAIL action [14]; it might also be important in different processes, such as atherosclerosis [69]. Moreover, miR-30a was also shown to regulate not only SNAIL but also SLUG in breast cancer to suppress EMT and metastasis [70].

SNAIL-dependent EMT in cancer has also been demonstrated to be regulated by p53 and miR-34 axis. In the absence of wild-type p53 function, SNAIL-dependent EMT is activated in colon, breast, lung carcinoma cells [71], and ovarian cancer [72] as a consequence of a decrease in miR-34 levels. A conserved miR-34a/b/c seed-matching sequence was detected in the *SNAIL* 3′-UTR. Moreover, there is a double-negative feedback loop in the regulation of EMT formed by miR-34 and SNAIL [73]. Luciferase reporter assays revealed that in pancreatic cancer, miR-34a targets both *SNAIL* and *NOTCH1* to inhibit pancreatic cancer progression through the regulation of EMT and NOTCH signaling pathways [74].

Another example of miRNA that is described as a direct regulator of SNAIL expression in plenty tumor types is miR-153. The downregulation of SNAIL by miR-153 suppresses human laryngeal squamous cell carcinoma migration and invasion [75], melanoma cells proliferation and invasion [76], esophageal squamous cell carcinoma progression [77], and gastric cancer metastasis [78]; regulates EMT in hepatocellular carcinoma [79]; and diminishes pancreatic ductal adenocarcinoma migration and invasion with miR-153 serving as a prognostic marker [80].

MiR-22 was demonstrated to target *SNAIL* and thereby inhibit tumor cell EMT and invasion in lung [81] and bladder cancer [82], in melanoma [83] and gastric cancer [84]. In bladder cancer, it inhibits both SNAIL and MAPK1 (mitogen-activated protein kinase 1) /SLUG/vimentin feedback loop [82], whereas in melanoma and gastric cancer it acts as a tumor suppressor by targeting both SNAIL and MMP14 [83,84].

SNAIL was found to be a target of multiple miRNAs in different tumor types. *SNAIL* was targeted in breast cancer by miR-125b [85], miR-203 [86], miR-410-3p [87], and miR-182 [88]; in gastric cancer by miR-491-5p [89] and miR-204 [90]; in lung cancer by miR-199a [91] and miR-940 [92]; in papillary thyroid carcinoma by miR-199a [93]; in ovarian cancer by miR-137 [72] and miR-363 [94]; in hepatocellular carcinoma by miR-122 [95] and miR-502-5p [96]; in prostate cancer by miR-486-5p [97]; and in renal cancer by miR-211-5p [98]. What is more, besides tumorigenesis, SNAIL is also regulated in different processes by miRNAs. For example, miR-133 promotes cardiac reprogramming by the direct repression of SNAIL and silencing fibroblast signatures [99], whereas miR-130b directly targets SNAIL in the regulation of diabetic nephropathy [100]. The results described above are summarized in Table 1.

Table 1. MicroRNAs regulating SNAIL.

MicroRNA	Cancer/Cell Type	References
miR-22	lung cancer	[81]
	bladder cancer	[82]
	melanoma	[83]
	gastric cancer	[84]
miR-30 family	non-small cell lung carcinoma	[62]
	breast cancer	[63]
	pancreatic cancer	[64]
	melanoma	[65]
	esophageal squamous cell carcinoma	[66]
	rhabdomyosarcoma	[14]
	hepatocytes	[67,68]

Table 1. *Cont.*

MicroRNA	Cancer/Cell Type	References
miR-34	colon carcinoma	[71]
	breast carcinoma	[71]
	lung carcinoma	[71]
	ovarian cancer	[72]
	pancreatic cancer	[74]
miR-122	hepatocellular carcinoma	[95]
miR-125b	breast cancer	[85]
miR-130b	diabetic nephropathy	[100]
miR-133	fibroblasts	[99]
miR-137	ovarian cancer	[72]
miR-153	laryngeal squamous cell carcinoma	[75]
	melanoma	[76]
	esophageal squamous cell carcinoma	[77]
	gastric cancer	[78]
	hepatocellular carcinoma	[79]
	pancreatic ductal adenocarcinoma	[80]
miR-182	breast cancer	[88]
miR-199a	lung cancer	[91]
	papillary thyroid carcinoma	[93]
miR-203	breast cancer	[86]
miR-204	gastric cancer	[90]
miR-211-5p	renal cancer	[98]
miR-363	ovarian cancer	[94]
miR-410-3p	breast cancer	[87]
miR-486-5p	prostate cancer	[97]
miR-491-5p	gastric cancer	[89]
miR-502-5p	hepatocellular carcinoma	[96]
miR-940	lung cancer	[92]

3.2. Other Examples of SNAIL Regulation by MicroRNAs

The indirect regulation of SNAIL involves several different mechanisms. One of the examples is inhibition of the GSK-3β (glycogen synthase kinase 3 beta) pathway. miR-148a binds to the 3′-UTR region of *MET*, which results in the attenuation of its downstream signaling, inhibition of AKT-Ser473 and GSK-3β phosphorylation, and in consequence reduced accumulation of SNAIL in the nucleus, the inhibition of EMT, and the metastasis of hepatoma cells [101]. In lung cancer cells, miR-126 affects the PI3K/AKT/SNAIL (phosphatidylinositol 3-kinase/protein kinase B/SNAIL) signaling pathway to regulate EMT [102]. A similar mechanism was described for miR-215 in papillary thyroid cancer [103]. In thyroid carcinoma, miR-101 targets the CXCL12 (C-X-C motif chemokine ligand 12, stromal cell-derived factor 1)-mediated AKT and SNAIL signaling pathways to inhibit invasion and the EMT-associated signaling pathways [104]. On the other hand, in hepatocellular carcinoma, miR-1306-3p targets FBXL5 to suppress SNAIL degradation and promote metastasis [105]. The SNAIL level can also be stabilized by miRNAs. miR-181b-3p promotes EMT in breast cancer cells through SNAIL stabilization by directly targeting the YWHAG protein [106]. In breast cancer cells, miR-5003-3p promotes EMT also through SNAIL stabilization via MDM2 and the direct targeting of E-cadherin [107]. In melanoma growth and metastasis, miR-9 is described as a downregulator of NF-κB1-SNAIL pathway [108]. The results described above are summarized in Table 2.

Table 2. Signaling pathways involving microRNAs that regulate SNAIL.

MicroRNA	Regulated Pathway and Genes	Mechanism of SNAIL Regulation	Cancer/Cell Type	References
miR-9	NF-κB1	SNAIL expression	melanoma	[108]
miR-101	CXCL12-mediated AKT	SNAIL localization	thyroid carcinoma	[104]
miR-126	PI3K-AKT	SNAIL localization	lung cancer	[102]
miR-148a	MET/AKT/GSK-3β	SNAIL localization and degradation	hepatoma cells	[101]
miR-181b-3p	YWHAG protein	SNAIL stabilization	breast cancer	[106]
miR-215	PI3K-AKT	SNAIL localization	papillary thyroid cancer	[103]
miR-1306-3p	FBXL5	Suppression of SNAIL degradation	hepatocellular carcinoma	[105]
miR-5003-3p	MDM2, E-cadherin	SNAIL stabilization	breast cancer	[107]

Sometimes, the research data demonstrate the regulation of SNAIL expression by miRNAs, but it is not described if the regulation is direct or indirect. There are also several other examples of miRNAs regulating the SNAIL level. In ovarian cancer, miR-16 is associated with the downregulation of mesenchymal markers, such as SNAIL, SLUG, and vimentin [109]. In Wilms' tumor cells, miR-483-3p regulates EMT by the modulation of E-cadherin, N-cadherin, SNAIL, and vimentin expression [110]. In osteosarcoma, the downregulation of miR-145 promotes EMT by regulation of the SNAIL level [111]. In rhabdomyosarcoma, miR-410-3p inhibits tumor growth and progression by inhibition of the expression of SNAIL, SLUG, N-cadherin, and Bcl-2 [112]. However, miR-410-3p was shown previously in different tumor types to directly target SNAIL [87].

The miRNAs–SNAIL axis may regulate not only EMT, but also the activity of cancer stem cells. miR-210 induced by a hypoxic microenvironment favored breast cancer stem cells' metastasis, proliferation, and self-renewal by targeting E-cadherin and the upregulation of SNAIL [113]. Another example is miR-146a, which directs the symmetric division of SNAIL-dominant colorectal cancer stem cells [114].

3.3. Regulation of SLUG Expression by MicroRNAs

MiRNAs can regulate not only SNAIL, but also SLUG, which is another important factor from the SNAIL family. Some miRNAs can regulate both factors. Among them are miR-30a [70], miR-122 [95], miR-182 [115], and miR-203 [115] and miR-204 [116]. *SLUG* is targeted in in oral squamous cell carcinoma by miR-204 [116]; glioblastoma by miR-203 [117]; in lung cancer by miR-1 [118]; in breast cancer by miR-124 [119,120], miR-30a [70], miR-497 [121], miR-1271 [122], and miR-203 [123,124]; in gastric cancer by miR-33a [125]; in lung cancer by miR-218 [126]; in clear cell renal cell carcinoma by miR-1 [127]; in osteosarcoma by miR-124 [128]; and in gingival fibroblasts by miR-200b [129]. Similarly to SNAIL, miRNAs–SLUG action regulates EMT in cancer progression, as well as different processes, such as the modulation of cancer stem cells' activity. miR-204 binds to the 3′UTR regions of both *SLUG* and *SOX4* to suppress osteosarcoma cancer stem cells [117], whereas the loss of miR-124 enhances the stem-like traits of glioma cells [130]. The miRNAs–SLUG axis is also important in other biological processes, such as for example in traumatic heterotopic ossification. miR-630 inhibits endothelial–mesenchymal transition by targeting SLUG [131]. The regulation of SLUG expression by miRNAs is summarized in Table 3.

Table 3. MicroRNAs regulating SLUG.

MicroRNA	Cancer/Cell Type	References
miR-1	lung cancer	[118]
miR-30a	breast cancer	[70]
miR-33a	gastric cancer	[125]
miR-124	breast cancer	[119,120]
	osteosarcoma	[128]
	glioma	[130]
miR-200b	gingival fibroblasts	[129]
miR-203	glioblastoma	[117]
	breast cancer	[123,124]
miR-204	oral squamous cell carcinoma	[116]
miR-218	lung cancer	[126]
miR-497	breast cancer	[121]
miR-630	dermal microvascular endothelial cells	[131]
miR-1271	breast cancer	[122]

4. LncRNA, CircRNAs, and their Relationship to SNAIL and Targeting MicroRNAs

Besides miRNAs, an interesting mechanism of action in the regulation of SNAIL or SLUG expression is also described for long non-coding RNAs (lncRNA). The may act as sponges for miRNAs targeting SNAIL. LncRNA MALAT1 (metastasis associated lung adenocarcinoma transcript 1) acts as a competing endogenous RNA (ceRNA) by sponging miR-22 to promote melanoma growth and metastasis [83]. MALAT1 turned out to be a regulator of not only miR-22, but also miR-1-3p expression. In that way, it inhibits migration, invasion, and EMT, which leads to the increased expression of E-cadherin and decreased expression of vimentin, SLUG, and SNAIL [132]. Another interesting feature of MALAT1 is the modulation of cancer stem cells' (CSC) activity by regulation of the miR-1/SLUG axis in nasopharyngeal carcinoma [133]. In gastric cancer, miR-22 is also regulated by lncRNA H19 with effects on metastasis via the miR-22-3p/SNAIL axis [134]. Another example in gastric cancer is lncRNA SNHG7 (small nucleolar RNA host gene 7), which directly binds to miR-34a and suppresses the miR-34a–SNAIL–EMT axis, which regulates gastric cancer cell migration and invasion [135].

SLUG level can also be regulated by other lncRNAs. For example, lncRNA GAPLINC (gastric adenocarcinoma associated) promotes the invasion of colorectal cancer by binding to PSF/NONO (probable DNA replication complex GINS protein PSF/non-POU domain-containing octamer-binding protein) and partly by stimulating the expression of SLUG [136]. lncRNA CAR10 directly binds two miRNAs: miR-30 and miR-203 and hence regulates the expression of both SNAIL and SLUG. In that way, it induces EMT and promotes lung adenocarcinoma metastasis [137]. In that cancer type, another example is lncRNA HCP5 acting as a sponge for miR-203 [138]. miR-203 interacts also with lncRNA UCA1 in hepatocellular carcinoma, and in that way, SLUG expression is regulated in tumor progression [139]. In that cancer type, lncRNA–AB209371 binds to hsa-miR199a-5p and weakens the inhibitory effect of hsa-miR199a-5p on SNAIL expression to promote EMT [140]. In breast cancer, lncRNA TINCR (terminal differentiation-induced ncRNA) targets miR-125b, and in that way regulates SNAIL and EMT [85].

LncRNAs may regulate the SNAIL level not only by miRNAs, but also epigenetically. LncRNA SATB2-AS1 (the antisense transcript of *SAT2B*—special AT-rich sequence-binding protein 2) mediates the epigenetic regulation of SNAIL expression in colorectal cancer progression. SATB2-AS1 recruits p300, whose acetylation of H3K27 and H3K9 at the *SATB2* promoter and subsequently the elevated SATB2 recruits HDAC1 to the *SNAIL* promoter to repress its transcription [141].

The interaction of lncRNAs with SNAIL is also possible. lncRNA NEAT1 (nuclear enriched abundant transcript 1) epigenetically suppresses E-cadherin expression in osteosarcoma cells by association with the G9a–DNMT1 (DNA methyltransferase 1)—SNAIL complex [142].

lncRNAs may also regulate the level of transcription factor by increasing their stability. For example, lncRNA SNHG15 impedes SLUG ubiquitination and its proteasomal degradation by interaction with the zinc finger domain of SLUG [143].

Besides lncRNAs, circular RNAs (circRNAs) were also described as SNAIL regulators. In hepatocellular carcinoma, circ-ZNF652 could physically interact with miR-203 and miR-502-5p to increase the expression of SNAIL. circ-ZNF652 was identified as a novel driver of EMT [96]. Similarly, in melanoma, circRNA_0084043 promotes progression via the miR-153-3p/SNAIL axis [144]. In urothelial carcinoma, circRNA PRMT5 acts as a sponge for miR-30c, which affects the SNAIL/E-cadherin pathway and thereby induces EMT [145]. circRNAs may be also implicated in the regulation of SLUG level. For example, circRNA-000284 can positively regulate the SLUG level in cervical cancer by sponging miR-506, which directly binds to *SLUG* 3′UTR [146].

The indirect regulation of SNAIL level by several mediators is also possible. Circular RNA hsa_circ_0008305 (circPTK2) inhibits TGF-β-induced EMT in non-small cell lung cancer by direct binding to miR-429/miR-200b-3p, which act as direct regulators of TIF1γ (transcriptional intermediary factor 1 γ), resulting in diminished SNAIL expression [147]. CircPIP5K1A induces non-small cell lung cancer progression by the regulation of miR-600/HIF-1α (hypoxia-inducible factor 1-alpha), which results in the upregulation of EMT-related factors, such as SNAIL [148]. Circ_0026344 promotes colorectal carcinoma invasion by targeting miR-183, which increases EMT and upregulates mesenchymal markers and SNAIL [149].

To summarize, SNAIL is regulated by signaling networks involving plenty of miRNAs, long non-coding RNAs, and circular RNAs (Table 4). lncRNAs and circRNAs usually act as sponges for miRNAs targeting SNAIL (Figure 3). This mechanism may be responsible for the regulation of tumor progression.

Table 4. Long non-coding RNAs regulating SNAIL and SLUG.

LncRNA/CircRNA	Regulated MicroRNAs	Regulated Factors	Cancer	References
lncRNA MALAT1	miR-22	SNAIL	melanoma	[83]
	miR-22 and miR-1-3p	E-cadherin, vimentin, SLUG and SNAIL	prostate cancer	[132]
	miR-1	SLUG	nasopharyngeal carcinoma	[133]
lncRNA H19	miR-22-3p	SNAIL	gastric cancer	[134]
lncRNA SNHG7	miR-34a	SNAIL	gastric cancer	[135]
lncRNA CAR10	miR-30 and miR-203	SNAIL and SLUG	lung adenocarcinoma	[137]
lncRNA HCP5	miR-203	SNAIL	lung adenocarcinoma	[138]
lncRNA UCA1	miR-203	SLUG	hepatocellular carcinoma	[139]
lncRNA AB209371	miR199a-5p	SNAIL	hepatocellular carcinoma	[140]
lncRNA TINCR	miR-125b	SNAIL	breast cancer	[85]
lncRNA SATB2-AS1	-	SNAIL (epigenetic regulation involving SATB2)	colorectal cancer	[141]
lncRNA NEAT1	-	E-cadherin by association with G9a-DNMT1-SNAIL complex	osteosarcoma cells	[142]
lncRNA SNHG15	-	SNAIL (ubiquitination by interaction with zinc finger domain)	colon cancer	[143]
lncRNA GAPLINC	-	SLUG (by binding to PSF/NONO)	colorectal cancer	[136]
circ-ZNF652	miR-203 and miR-502-5p	SNAIL	hepatocellular carcinoma	[96]
circRNA_0084043	miR-153-3p	SNAIL	melanoma	[144]
circRNA PRMT5	miR-30c	SNAIL	urothelial carcinoma	[145]
circRNA-000284	miR-506	SLUG	cervical cancer	[146]
hsa_circ_0008305 (circPTK2)	miR-429 and miR-200b-3p	SNAIL (indirectly by TIF1γ)	non-small cell lung cancer	[147]
circPIP5K1A	miR-600	SNAIL (indirectly by HIF-1α)	non-small cell lung cancer	[148]
circ_0026344	miR-183	SNAIL (indirectly)	colorectal cancer	[149]

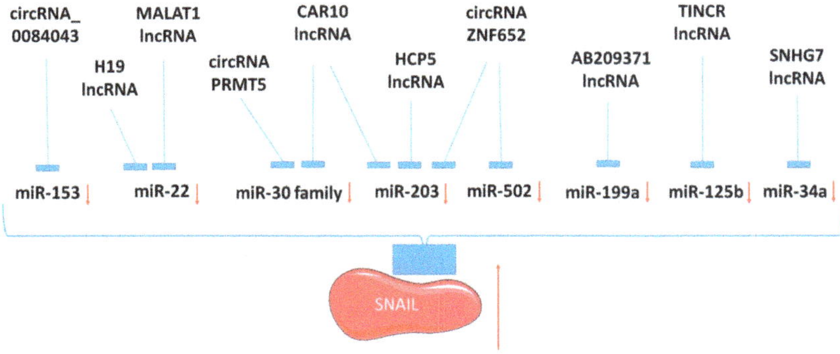

Figure 3. Role of long non-coding RNAs and circular RNAs as sponges for microRNAs in the regulation of SNAIL expression in tumors.

5. SNAIL Regulation of Non-Coding RNAs

MiRNAs were presented as regulators of SNAIL expression. On the other hand, there are several cases describing SNAIL as a regulator of miRNA level with implications to epithelial tumor progression and the role of EMT in this process. MiRNAs may be regulated either indirectly or by the direct binding of SNAIL to E-box sequences in miRNA promoters or regulatory sequences.

For example, in breast cancer cells, SNAIL directly suppresses miR-182 [88] and miR-203 [86]. In head and neck cancers, SNAIL binds to the miR-493 promoter [150]. SNAIL also significantly represses the miR-145 promoter. miR-145 plays a role in antagonizing SNAIL-mediated stemness in colorectal cancer [151]. In gastric cancer, SNAIL binds to the putative promoter of miR-375 [152]. SNAIL directly activates the transcription of miR-21 to produce exosomes abundant in miR-21, which promotes the M2-like polarization of tumor-associated macrophages [153].

In non-epithelial tumor types, such as glioma, SNAIL suppresses miR-128b expression by direct binding to the miR-128b-specific promoter motif; then, miR-128 and SP1 regulate tumor progression [154]. A similar direct mechanism was demonstrated for miR-128-2 in mammary epithelial cells. The loss of SNAIL-regulated miR-128-2 targets multiple stem cell factors to promote the oncogenic transformation of mammary epithelial cells [155]. The SNAIL/miR-128 axis regulates the growth, invasion, metastasis, and EMT of gastric cancer. miR-128 targets directly Bmi11, and it can reverse EMT induced by Bmi-1 via the PI3K/AKT pathway, whereas SNAIL curbs the expression of miR-128, and then down-regulated miR-128 promotes the expression of Bmi-1 [156]. The loss of SNAIL was also shown to inhibit cellular growth and metabolism through the miR-128-mediated signaling pathway in prostate cancer cells [157].

Interestingly, SNAIL may also exert its effects by epigenetic modifications. SNAIL is involved in CpG DNA methylation of the miR-200f loci, which is essential for maintenance of the mesenchymal phenotype. In the MDCK (Madin-Darby canine kidney) epithelial kidney cells model, it has been shown that ZEB1 and SNAIL engage miR-200f transcriptional and epigenetic regulation during EMT [158]. Regulation of the miR-200 family by SNAIL also plays a role in vasculogenesis and may be significant both in malignant cancer and in early developing embryos [159].

SNAIL overexpression increases the level of miR-125b through the SNAIL-activated Wnt/β-catenin/TCF4 (transcription factor 4) axis. This mechanism was described for SNAIL-induced stem cell propagation [160]. Another example of SNAIL action in cancer stem cells is signaling axis involving SNAIL, miR-146a, and Numb in regulation of the switch between symmetric and asymmetric cell division in colorectal cancer stem cells [161].

As indicated previously, SNAIL is a regulator of not only EMT and cancer stem cells, but also of myogenic differentiation. In rhabdomyosarcoma, SNAIL regulates the expression of myogenic-associated miRNAs, such as miR-1, miR-206, and miR-378 [14]. What is more, the

SNAIL/miR-199a-5p axis promotes the differentiation of fibroblasts into myofibroblasts by the induction of endothelial–mesenchymal transition [162].

There are also examples of interaction among lncRNAs, miRNAs, and SNAIL. SNAIL binds to the promoter of lncRNA PCA3 and activates its expression. Then, lncRNA PCA3 inhibits the translation of PRKD3 (serine/threonine-protein kinase D3) protein via competitive miR-1261 sponging and in that way promotes the invasion of prostate cancer cells [163].

SNAIL's role has been also described in controlling telomere transcription and integrity, which may be significant features of cancer stem cells, since telomere maintenance is essential for stemness. SNAIL turned out to be a negative regulator of lncRNA that controls telomere integrity, which is called telomeric repeat-containing RNA (TERRA). What is more, TERRA can also affect the transcription of some genes induced during EMT [164].

SNAIL may also not only regulate the level of lncRNAs, but it may also interact with them to modify the chromatin. lncRNA HOTAIR (HOX Transcript Antisense Intergenic RNA) mediates a physical interaction between SNAIL and EZH2 (enhancer of zeste homolog 2), which is an enzymatic subunit of the polycomb-repressive complex 2. In that way, SNAIL recruits EZH2 to specific genomic sites during EMT [165].

SNAIL may also regulate circRNAs. For example, SNAIL targets the E-box motif on the promoter of circ-ZNF652 to increase its expression [96].

Besides SNAIL, similar mechanisms of binding to miRNA promoters were also described for SLUG. In colorectal cancer, SLUG binds to miR-145 promoter and represses it to modulate 5-fluorouracil sensitivity [166]. In lung cancer cells, SLUG binds directly to the E-box in the promoter of miR-137 and acts as an activator, which promotes cancer invasion and progression by directly suppressing TFAP2C (transcription factor AP-2 gamma) [167]. In prostate cancer, SLUG is a direct repressor of miR-1 and miR-200 transcription [168]. In breast cancer cells, SLUG directly binds to miR-203 promoter, downregulating its expression [124]. SLUG-upregulated miR-221 promotes breast cancer progression through suppressing E-cadherin expression, which indicates that miR-221 is an additional blocker of E-cadherin besides SNAIL and SLUG [169].

Sometimes, both SNAIL and SLUG collaborate on EMT and tumor metastasis through miRNAs. In oral tongue squamous cell carcinoma, those transcription factors act through the miR-101-mediated EZH2 axis [170]. miR-101 functions as a tumor suppressor by directly targeting *ZEB1* (zinc finger E-Box binding homeobox 1) in various cancers, including colorectal cancer [171].

To summarize, SNAIL and SLUG may be direct or indirect regulators of miRNAs, lncRNAs, and circRNAs (Table 5). There are several examples of direct binding SNAIL to promoters or regulatory sequences of non-coding RNAs (Figure 4). Subsequently, those RNAs target plenty of genes to regulate tumor progression.

Table 5. Non-coding RNAs regulated by SNAIL and SLUG.

Non-Coding RNA	Mechanism	Cancer/Cell Type	References
miR-1	SLUG binding to promoter	prostate cancer	[168]
	regulation by SNAIL (unknown mechanism)	rhabdomyosarcoma	[14]
miR-21	SNAIL binding to promoter	head and neck cancer	[153]
miR-101	transcriptional control by SNAIL and SLUG	squamous cell carcinoma	[170]
miR-125b	SNAIL-activated Wnt/β-catenin/TCF4 axis	breast cancer stem cells	[160]

Table 5. *Cont.*

Non-Coding RNA	Mechanism	Cancer/Cell Type	References
miR-128	SNAIL binding to promoter	glioma	[154]
		prostate cancer	[157]
		gastric cancer	[156]
miR-137	SLUG binding to promoter	lung cancer	[167]
miR-145	SNAIL binding to promoter	colorectal cancer	[151]
	SLUG binding to promoter	colorectal cancer	[166]
miR-146a	SNAIL-induced β-catenin-TCF4 complex	colorectal cancer stem cells	[161]
miR-182	SNAIL binding to promoter	breast cancer	[88]
miR-200	SNAIL involved in CpG DNA methylation	human kidney cells	[158]
	SLUG binding to promoter	prostate cancer	[168]
miR-203	SNAIL binding to promoter	breast cancer	[86]
	SLUG binding to promoter	breast cancer	[124]
miR-206	regulation by SNAIL (unknown mechanism)	rhabdomyosarcoma	[14]
miR-221	transcriptional control by SLUG	breast cancer	[169]
miR-375	SNAIL binding to promoter	gastric cancer	[152]
miR-378	regulation by SNAIL (unknown mechanism)	rhabdomyosarcoma	[14]
miR-493	SNAIL binding to promoter	head and neck cancer	[150]
lncRNA PCA3	SNAIL binding to promoter	prostate cancer	[163]
lncRNA TERRA	transcriptional control by SNAIL	mesenchymal stem cells and mammary cells	[164]
lncRNA HOTAIR	interaction of SNAIL with HOTAIR and EZH2	hepatocytes	[165]
circ-ZNF652	SNAIL binding to promoter	hepatocellular carcinoma	[96]

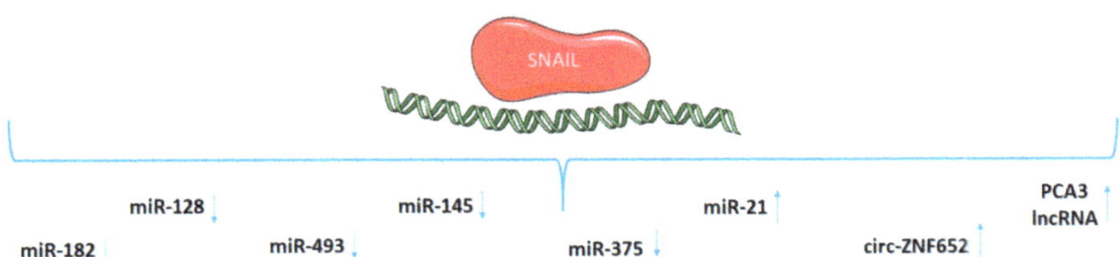

Figure 4. MiRNAs, long non-coding RNAs (lncRNAs), and circular RNAs regulated directly by SNAIL transcription factor.

6. Multi-Component Feedback Loops and Multi-Component Signaling Networks

The literature also describes several examples of multi-component feedback loops and multi-component signaling networks involving the SNAIL transcription factor and non-coding RNAs.

Selected different multi-component feedback loops and multi-component signaling networks are presented in Figure 5.

An interesting example is miR-182, which is directly suppressed by SNAIL in breast cancer cells, which can also target its suppressor (Figure 5A). This mechanism regulates an epithelial-like phenotype in vitro and enhances macrometastases in vivo [88].

Similarly in breast cancer, miR-203 forms also a double-negative miR-203/SNAIL feedback loop, as SNAIL reduces the activity of the miR-203 promoter (Figure 5B) [86].

Moreover, miR-34 and SNAIL form a double-negative feedback loop (Figure 5C) [73] that may feed-forward regulate ZNF281/ZBP99 to promote EMT, which has implications for human colon and breast cancer [172]. The expression of ZNF281 (zinc finger protein 281) is induced by SNAIL and inhibited by miR-34a, which mediates the repression of ZNF281 by the p53 tumor suppressor. The deregulation of this circuitry by mutational and epigenetic alterations in the p53/miR-34a axis promotes colorectal cancer metastasis [173].

In head and neck cancers, SNAIL binds to miR-493 promoter to repress it, and subsequently, miR-493 forms a negative feedback loop with the insulin-like growth factor 1 receptor pathway to block tumorigenesis (Figure 5D) [150].

Besides miRNAs, SNAIL may also form feedback loops with circular RNAs. SNAIL upregulates circ-ZNF652 by binding to the E-box motif on the promoter. Subsequently, circ-ZNF652 acts a sponge for miR-203 and miR-502-5p, which target *SNAIL* 3'UTR (Figure 5E) [96].

In cancer stem cells, SNAIL forms a feedback circuit to maintain Wnt activity. SNAIL induces miR-146a expression through the β-catenin-TCF4 complex, and subsequently, miR-146a targets Numb to stabilize β-catenin (Figure 5F) [161].

An interesting example is also SNAIL action in ZEB1 circuit in melanoma cells. SNAIL is considered as an external signal that transcriptionally regulates the ZEB1/miR-200a/cicrZEB1 axis. circZEB1, generated from the ZEB1 gene, contains a binding site for mir200a, which is a post-transcriptional regulator of ZEB1 (Figure 5G) [174].

SLUG and microRNAs may also form regulatory loops. In breast cancer cells, SLUG and miR-203 form a double-negative feedback loop and SLUG directly binds to miR-203 promoter, downregulating its expression in metastatic breast cancer cells (Figure 5H) [124]. Furthermore, SLUG and miR-1/miR-200 act in a self-reinforcing regulatory loop, which results in EMT amplification (Figure 5I) [168].

What is also interesting is that sometimes, gene transcripts may also act as a competitive endogenous RNA (ceRNA) to regulate biological processes. *FN1* (fibronectin 1) acts as a ceRNA for miR-200c in the canonical SNAIL-ZEB-miR200 pathway in breast cancer cells (Figure 5J), whereas *TGFBI* (transforming growth factor-beta-induced) is a transcript that is highly induced during EMT in lung cancer cells, which acts as the ceRNA for miR-21 to modulate EMT [175].

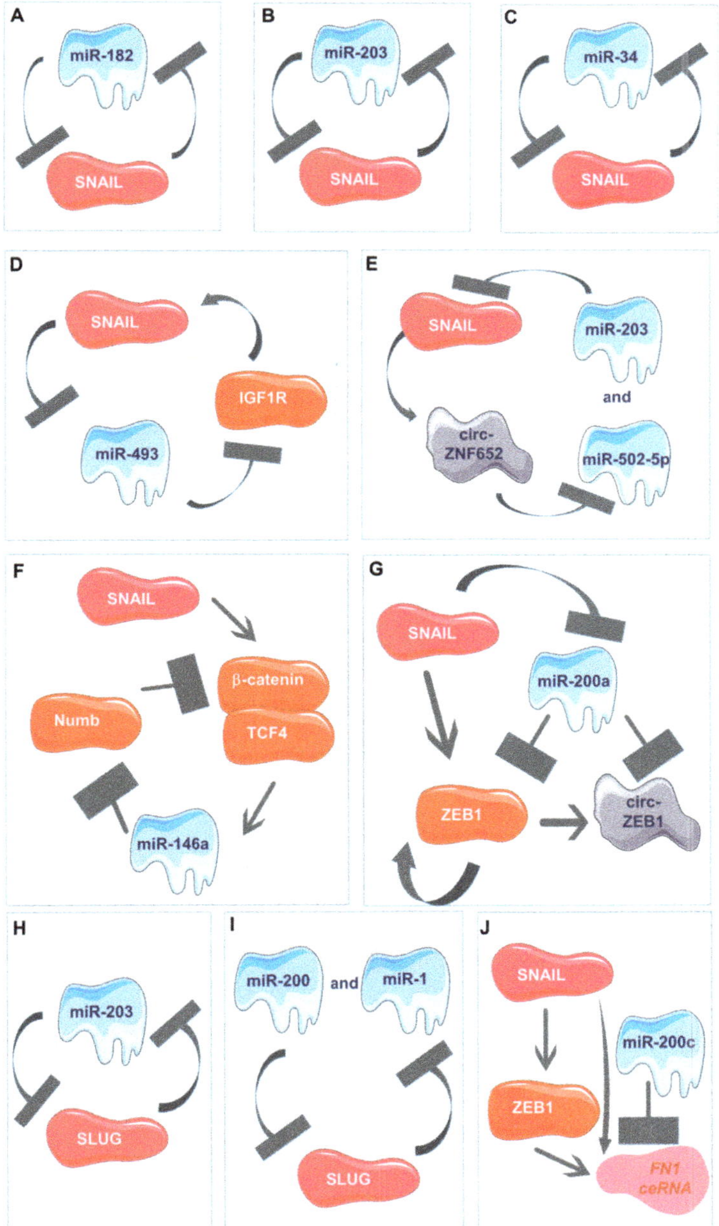

Figure 5. Multi-component feedback loops and multi-component signaling networks involving SNAIL and non-coding RNAs. (**A**) Negative regulation between SNAIL and miR-182. (**B**) Negative regulation between SNAIL and miR-203. (**C**) Negative regulation between SNAIL and miR-34. (**D**) Feedback loop between SNAIL, miR-493, and IGF1R. (**E**) Feedback loop between SNAIL, circ-ZNF652, miR-203, and miR-502-5p. (**F**) Signaling pathway involving SNAIL, β-catenin, miR-146a, and Numb. (**G**) Signaling pathway involving SNAIL, ZEB1, miR-200a, and circ-ZEB1. (**H**) Negative regulation between SLUG and miR-203. (**I**) Negative regulation among SLUG, miR-1, and miR-200. (**J**) Signaling pathway involving SNAIL, ZEB1, miR-200c, and *FN1* ceRNA.

7. Conclusions

SNAIL participates in many physiological and pathological processes, including embryonic development and cancer metastasis. Therefore, the identification of its crosstalk with non-coding RNAs can help in understanding the complex signaling networks that drive tumor progression. Unraveling these signaling networks may help generate new types of cancer therapeutics. miRNAs and other non-coding RNAs play key roles in tumor progression or suppression. One miRNA may target multiple genes besides *SNAIL*. Therapies targeting miRNA may enable the regulation of more than one signaling pathway. An interesting example of miRNA (described in this review) therapeutics is a drug based on miR-34a mimics, which has been already enrolled in clinical trials [176]. The identification of miRNA downstream and upstream of SNAIL may create novel possibilities for biomarker determination during cancer progression, which may lead to improvements in prognosis and therapy. As those miRNAs usually regulate epithelial to mesenchymal transition, their identification may help to distinguish different stages of tumor development, as well as benign and malignant tumors. For the identification of novel biomarkers, the next step is verification of whether miRNA candidates can be secreted from tumor to blood vessels.

Funding: This research was funded by the National Science Centre in Poland to K.S.: 2015/17/D/NZ5/02202 and to M.M.: 2013/09/B/NZ5/00769.

Acknowledgments: K.S. was a recipient of the Foundation for Polish Science's START scholarship for outstanding young scientists and the Polish Ministry of Science and Higher Education's scholarship for outstanding young scientists.

Conflicts of Interest: The authors declare no conflict of interest.

References

1. Alberga, A.; Boulay, J.L.; Kempe, E.; Dennefeld, C.; Haenlin, M. The snail gene required for mesoderm formation in Drosophila is expressed dynamically in derivatives of all three germ layers. *Development* **1991**, *111*, 983–992. [PubMed]
2. Nieto, M.A. The snail superfamily of zinc-finger transcription factors. *Nat. Rev. Mol. Cell Biol.* **2002**, *3*, 155–166. [CrossRef] [PubMed]
3. Wang, Y.; Shi, J.; Chai, K.; Ying, X.; Zhou, B.P. The Role of Snail in EMT and Tumorigenesis. *Curr. Cancer Drug Targets* **2013**, *13*, 963–972. [CrossRef] [PubMed]
4. Barberà, M.J.; Puig, I.; Domínguez, D.; Julien-Grille, S.; Guaita-Esteruelas, S.; Peiró, S.; Baulida, J.; Francí, C.; Dedhar, S.; Larue, L.; et al. Regulation of Snail transcription during epithelial to mesenchymal transition of tumor cells. *Oncogene* **2004**, *23*, 7345–7354. [CrossRef] [PubMed]
5. De Craene, B.; van Roy, F.; Berx, G. Unraveling signalling cascades for the Snail family of transcription factors. *Cell. Signal.* **2005**, *17*, 535–547. [CrossRef] [PubMed]
6. Zhou, B.P.; Deng, J.; Xia, W.; Xu, J.; Li, Y.M.; Gunduz, M.; Hung, M.-C. Dual regulation of Snail by GSK-3β-mediated phosphorylation in control of epithelial–mesenchymal transition. *Nat. Cell Biol.* **2004**, *6*, 931–940. [CrossRef]
7. Peinado, H.; del Carmen Iglesias-de la Cruz, M.; Olmeda, D.; Csiszar, K.; Fong, K.S.K.; Vega, S.; Nieto, M.A.; Cano, A.; Portillo, F. A molecular role for lysyl oxidase-like 2 enzyme in Snail regulation and tumor progression. *EMBO J.* **2005**, *24*, 3446–3458. [CrossRef]
8. Viñas-Castells, R.; Beltran, M.; Valls, G.; Gómez, I.; García, J.M.; Montserrat-Sentís, B.; Baulida, J.; Bonilla, F.; de Herreros, A.G.; Díaz, V.M. The Hypoxia-controlled FBXL14 Ubiquitin Ligase Targets SNAIL1 for Proteasome Degradation. *J. Biol. Chem.* **2010**, *285*, 3794–3805. [CrossRef]
9. Park, S.Y.; Kim, H.S.; Kim, N.H.; Ji, S.; Cha, S.Y.; Kang, J.G.; Ota, I.; Shimada, K.; Konishi, N.; Nam, H.W.; et al. Snail1 is stabilized by O-GlcNAc modification in hyperglycaemic condition. *EMBO J.* **2010**, *29*, 3787–3796. [CrossRef]
10. Wu, Y.; Deng, J.; Rychahou, P.G.; Qiu, S.; Evers, B.M.; Zhou, B.P. Stabilization of Snail by NF-κB Is Required for Inflammation-Induced Cell Migration and Invasion. *Cancer Cell* **2009**, *15*, 416–428. [CrossRef]

11. Yang, Z.; Rayala, S.; Nguyen, D.; Vadlamudi, R.K.; Chen, S.; Kumar, R. Pak1 Phosphorylation of Snail, a Master Regulator of Epithelial-to-Mesenchyme Transition, Modulates Snail's Subcellular Localization and Functions. *Cancer Res.* **2005**, *65*, 3179–3184. [CrossRef] [PubMed]

12. Peinado, H.; Olmeda, D.; Cano, A. Snail, Zeb and bHLH factors in tumour progression: An alliance against the epithelial phenotype? *Nat. Rev. Cancer* **2007**, *7*, 415–428. [CrossRef] [PubMed]

13. Peinado, H.; Ballestar, E.; Esteller, M.; Cano, A. Snail mediates E-cadherin repression by the recruitment of the Sin3A/histone deacetylase 1 (HDAC1)/HDAC2 complex. *Mol. Cell. Biol.* **2004**, *24*, 306–319. [CrossRef] [PubMed]

14. Skrzypek, K.; Kusienicka, A.; Trzyna, E.; Szewczyk, B.; Ulman, A.; Konieczny, P.; Adamus, T.; Badyra, B.; Kortylewski, M.; Majka, M. SNAIL is a key regulator of alveolar rhabdomyosarcoma tumor growth and differentiation through repression of MYF5 and MYOD function. *Cell Death Dis.* **2018**, *9*, 643. [CrossRef] [PubMed]

15. Soleimani, V.D.; Yin, H.; Jahani-Asl, A.; Ming, H.; Kockx, C.E.M.; van Ijcken, W.F.J.; Grosveld, F.; Rudnicki, M.A. Snail regulates MyoD binding-site occupancy to direct enhancer switching and differentiation-specific transcription in myogenesis. *Mol. Cell* **2012**, *47*, 457–468. [CrossRef] [PubMed]

16. Barrallo-Gimeno, A.; Nieto, M.A. The Snail genes as inducers of cell movement and survival: Implications in development and cancer. *Development* **2005**, *132*, 3151–3161. [CrossRef] [PubMed]

17. Peiro, S.; Escrivà, M.; Puig, I.; Barberà, M.J.; Dave, N.; Herranz, N.; Larriba, M.J.; Takkunen, M.; Francí, C.; Muñoz, A.; et al. Snail1 transcriptional repressor binds to its own promoter and controls its expression. *Nucleic Acids Res.* **2006**, *34*, 2077–2084. [CrossRef]

18. Sakai, D.; Suzuki, T.; Osumi, N.; Wakamatsu, Y. Cooperative action of Sox9, Snail2 and PKA signaling in early neural crest development. *Development* **2006**, *133*, 1323–1333. [CrossRef]

19. Sundararajan, V.; Tan, M.; Tan, T.Z.; Ye, J.; Thiery, J.P.; Huang, R.Y.-J. SNAI1 recruits HDAC1 to suppress SNAI2 transcription during epithelial to mesenchymal transition. *Sci. Rep.* **2019**, *9*, 8295. [CrossRef]

20. Alidadiani, N.; Ghaderi, S.; Dilaver, N.; Bakhshamin, S.; Bayat, M. Epithelial mesenchymal transition Transcription Factor (TF): The structure, function and microRNA feedback loop. *Gene* **2018**, *674*, 115–120. [CrossRef]

21. Yamamura, S.; Imai-Sumida, M.; Tanaka, Y.; Dahiya, R. Interaction and cross-talk between non-coding RNAs. *Cell. Mol. Life Sci.* **2018**, *75*, 467–484. [CrossRef] [PubMed]

22. Bánfai, B.; Jia, H.; Khatun, J.; Wood, E.; Risk, B.; Gundling, W.E.; Kundaje, A.; Gunawardena, H.P.; Yu, Y.; Xie, L.; et al. Long noncoding RNAs are rarely translated in two human cell lines. *Genome Res.* **2012**, *22*, 1646–1657. [CrossRef] [PubMed]

23. Guttman, M.; Amit, I.; Garber, M.; French, C.; Lin, M.F.; Feldser, D.; Huarte, M.; Zuk, O.; Carey, B.W.; Cassady, J.P.; et al. Chromatin signature reveals over a thousand highly conserved large non-coding RNAs in mammals. *Nature* **2009**, *458*, 223–227. [CrossRef] [PubMed]

24. Quinn, J.J.; Chang, H.Y. Unique features of long non-coding RNA biogenesis and function. *Nat. Rev. Genet.* **2016**, *17*, 47–62. [CrossRef] [PubMed]

25. Sigova, A.A.; Mullen, A.C.; Molinie, B.; Gupta, S.; Orlando, D.A.; Guenther, M.G.; Almada, A.E.; Lin, C.; Sharp, P.A.; Giallourakis, C.C.; et al. Divergent transcription of long noncoding RNA/mRNA gene pairs in embryonic stem cells. *Proc. Natl. Acad. Sci. USA* **2013**, *110*, 2876–2881. [CrossRef] [PubMed]

26. Zheng, G.X.Y.; Do, B.T.; Webster, D.E.; Khavari, P.A.; Chang, H.Y. Dicer-microRNA-Myc circuit promotes transcription of hundreds of long noncoding RNAs. *Nat. Struct. Mol. Biol.* **2014**, *21*, 585–590. [CrossRef] [PubMed]

27. Dhir, A.; Dhir, S.; Proudfoot, N.J.; Jopling, C.L. Microprocessor mediates transcriptional termination of long noncoding RNA transcripts hosting microRNAs. *Nat. Struct. Mol. Biol.* **2015**, *22*, 319–327. [CrossRef]

28. Chi, Y.; Wang, D.; Wang, J.; Yu, W.; Yang, J. Long Non-Coding RNA in the Pathogenesis of Cancers. *Cells* **2019**, *8*, 1015. [CrossRef]

29. Conn, S.J.; Pillman, K.A.; Toubia, J.; Conn, V.M.; Salmanidis, M.; Phillips, C.A.; Roslan, S.; Schreiber, A.W.; Gregory, P.A.; Goodall, G.J. The RNA binding protein quaking regulates formation of circRNAs. *Cell* **2015**, *160*, 1125–1134. [CrossRef]

30. Zhang, Y.; Zhang, X.O.; Chen, T.; Xiang, J.F.; Yin, Q.F.; Xing, Y.H.; Zhu, S.; Yang, L.; Chen, L.L. Circular Intronic Long Noncoding RNAs. *Mol. Cell* **2013**, *51*, 792–806. [CrossRef]

31. Li, Z.; Huang, C.; Bao, C.; Chen, L.; Lin, M.; Wang, X.; Zhong, G.; Yu, B.; Hu, W.; Dai, L.; et al. Exon-intron circular RNAs regulate transcription in the nucleus. *Nat. Struct. Mol. Biol.* **2015**, *22*, 256–264. [CrossRef] [PubMed]

32. Su, Y.; Zhong, G.; Jiang, N.; Huang, M.; Lin, T. Circular RNA, a novel marker for cancer determination. *Int. J. Mol. Med.* **2018**, *42*, 1786–1798. [CrossRef] [PubMed]

33. Schneider, T.; Bindereif, A. Circular RNAs: Coding or noncoding? *Cell Res.* **2017**, *27*, 724–725. [CrossRef] [PubMed]

34. Vasudevan, S.; Tong, Y.; Steitz, J.A. Switching from Repression to Activation: MicroRNAs Can Up-Regulate Translation. *Science* **2007**, *318*, 1931–1934. [CrossRef] [PubMed]

35. Stavast, C.J.; Erkeland, S.J. The Non-Canonical Aspects of MicroRNAs: Many Roads to Gene Regulation. *Cells* **2019**, *8*, 1465. [CrossRef]

36. Jing, Q.; Huang, S.; Guth, S.; Zarubin, T.; Motoyama, A.; Chen, J.; Di Padova, F.; Lin, S.-C.; Gram, H.; Han, J. Involvement of microRNA in AU-rich element-mediated mRNA instability. *Cell* **2005**, *120*, 623–634. [CrossRef]

37. Lim, L.P.; Lau, N.C.; Weinstein, E.G.; Abdelhakim, A.; Yekta, S.; Rhoades, M.W.; Burge, C.B.; Bartel, D.P. The microRNAs of Caenorhabditis elegans. *Genes Dev.* **2003**, *17*, 991–1008. [CrossRef]

38. Grimson, A.; Farh, K.K.-H.; Johnston, W.K.; Garrett-Engele, P.; Lim, L.P.; Bartel, D.P. MicroRNA targeting specificity in mammals: Determinants beyond seed pairing. *Mol. Cell* **2007**, *27*, 91–105. [CrossRef]

39. Nielsen, C.B.; Shomron, N.; Sandberg, R.; Hornstein, E.; Kitzman, J.; Burge, C.B. Determinants of targeting by endogenous and exogenous microRNAs and siRNAs. *RNA* **2007**, *13*, 1894–1910. [CrossRef]

40. Baskerville, S.; Bartel, D.P. Microarray profiling of microRNAs reveals frequent coexpression with neighboring miRNAs and host genes. *RNA* **2005**, *11*, 241–247. [CrossRef]

41. Lin, S.-L.; Miller, J.D.; Ying, S.-Y. Intronic MicroRNA (miRNA). *J. Biomed. Biotechnol.* **2006**, *2006*, 26818. [CrossRef] [PubMed]

42. Treiber, T.; Treiber, N.; Meister, G. Regulation of microRNA biogenesis and its crosstalk with other cellular pathways. *Nat. Rev. Mol. Cell Biol.* **2019**, *20*, 5–20. [CrossRef] [PubMed]

43. Lee, Y.; Jeon, K.; Lee, J.-T.; Kim, S.; Kim, V.N. MicroRNA maturation: Stepwise processing and subcellular localization. *EMBO J.* **2002**, *21*, 4663–4670. [CrossRef] [PubMed]

44. Lewis, B.P.; Shih, I.; Jones-Rhoades, M.W.; Bartel, D.P.; Burge, C.B. Prediction of Mammalian MicroRNA Targets. *Cell* **2003**, *115*, 787–798. [CrossRef]

45. Calin, G.A.; Dumitru, C.D.; Shimizu, M.; Bichi, R.; Zupo, S.; Noch, E.; Aldler, H.; Rattan, S.; Keating, M.; Rai, K.; et al. Frequent deletions and down-regulation of micro- RNA genes miR15 and miR16 at 13q14 in chronic lymphocytic leukemia. *Proc. Natl. Acad. Sci. USA* **2002**, *99*, 15524–15529. [CrossRef]

46. Michael, M.Z.; O'Connor, S.M.; van Holst Pellekaan, N.G.; Young, G.P.; James, R.J. Reduced accumulation of specific microRNAs in colorectal neoplasia. *Mol. Cancer Res.* **2003**, *1*, 882–891.

47. Ciesla, M.; Skrzypek, K.; Kozakowska, M.; Loboda, A.; Jozkowicz, A.; Dulak, J. MicroRNAs as biomarkers of disease onset. *Anal. Bioanal. Chem.* **2011**, *401*, 2051–2061. [CrossRef]

48. Lu, J.; Getz, G.; Miska, E.A.; Alvarez-Saavedra, E.; Lamb, J.; Peck, D.; Sweet-Cordero, A.; Ebert, B.L.; Mak, R.H.; Ferrando, A.A.; et al. MicroRNA expression profiles classify human cancers. *Nature* **2005**, *435*, 834–838. [CrossRef]

49. Lin, S.; Gregory, R.I. MicroRNA biogenesis pathways in cancer. *Nat. Rev. Cancer* **2015**, *15*, 321–333. [CrossRef]

50. Sandberg, R.; Neilson, J.R.; Sarma, A.; Sharp, P.A.; Burge, C.B. Proliferating cells express mRNAs with shortened 3′ untranslated regions and fewer microRNA target sites. *Science* **2008**, *320*, 1643–1647. [CrossRef]

51. Calin, G.A.; Sevignani, C.; Dumitru, C.D.; Hyslop, T.; Noch, E.; Yendamuri, S.; Shimizu, M.; Rattan, S.; Bullrich, F.; Negrini, M.; et al. Human microRNA genes are frequently located at fragile sites and genomic regions involved in cancers. *Proc. Natl. Acad. Sci. USA* **2004**, *101*, 2999–3004. [CrossRef] [PubMed]

52. Hayashita, Y.; Osada, H.; Tatematsu, Y.; Yamada, H.; Yanagisawa, K.; Tomida, S.; Yatabe, Y.; Kawahara, K.; Sekido, Y.; Takahashi, T. A polycistronic microRNA cluster, miR-17-92, is overexpressed in human lung cancers and enhances cell proliferation. *Cancer Res.* **2005**, *65*, 9628–9632. [CrossRef] [PubMed]

53. Fang, L.L.; Wang, X.H.; Sun, B.F.; Zhang, X.D.; Zhu, X.H.; Yu, Z.J.; Luo, H. Expression, regulation and mechanism of action of the miR-17-92 cluster in tumor cells. *Int. J. Mol. Med.* **2017**, *40*, 1624–1630. [CrossRef] [PubMed]

54. Shi, L.; Cheng, Z.; Zhang, J.; Li, R.; Zhao, P.; Fu, Z.; You, Y. hsa-mir-181a and hsa-mir-181b function as tumor suppressors in human glioma cells. *Brain Res.* **2008**, *1236*, 185–193. [CrossRef] [PubMed]
55. Gao, W.; Yu, Y.; Cao, H.; Shen, H.; Li, X.; Pan, S.; Shu, Y. Deregulated expression of miR-21, miR-143 and miR-181a in non small cell lung cancer is related to clinicopathologic characteristics or patient prognosis. *Biomed. Pharmacother.* **2010**, *64*, 399–408. [CrossRef]
56. Hagman, Z.; Larne, O.; Edsjö, A.; Bjartell, A.; Ehrnström, R.A.; Ulmert, D.; Lilja, H.; Ceder, Y. miR-34c is downregulated in prostate cancer and exerts tumor suppressive functions. *Int. J. Cancer* **2010**, *127*, 2768–2776. [CrossRef]
57. Dacic, S.; Kelly, L.; Shuai, Y.; Nikiforova, M.N. miRNA expression profiling of lung adenocarcinomas: Correlation with mutational status. *Mod. Pathol.* **2010**, *23*, 1577–1582. [CrossRef]
58. Collet, G.; Skrzypek, K.; Grillon, C.; Matejuk, A.; El Hafni-Rahbi, B.; Lamerant-Fayel, N.; Kieda, C. Hypoxia control to normalize pathologic angiogenesis: Potential role for endothelial precursor cells and miRNAs regulation. *Vascul. Pharmacol.* **2012**, *56*, 1–10. [CrossRef]
59. Musavi Shenas, M.H.; Eghbal-Fard, S.; Mehrisofiani, V.; Abd Yazdani, N.; Rahbar Farzam, O.; Marofi, F.; Yousefi, M. MicroRNAs and signaling networks involved in epithelial-mesenchymal transition. *J. Cell. Physiol.* **2019**, *234*, 5775–5785. [CrossRef]
60. Subhra Das, S.; James, M.; Paul, S.; Chakravorty, N. miRnalyze: An interactive database linking tool to unlock intuitive microRNA regulation of cell signaling pathways. *Database* **2017**, *2017*, bax015. [CrossRef]
61. Li, J.; Yu, H.; Xi, M.; Ma, D.; Lu, X. The SNAI1 3'UTR functions as a sponge for multiple migration-/invasion-related microRNAs. *Tumor Biol.* **2015**, *36*, 1067–1072. [CrossRef] [PubMed]
62. Kumarswamy, R.; Mudduluru, G.; Ceppi, P.; Muppala, S.; Kozlowski, M.; Niklinski, J.; Papotti, M.; Allgayer, H. MicroRNA-30a inhibits epithelial-to-mesenchymal transition by targeting Snail and is downregulated in non-small cell lung cancer. *Int. J. Cancer* **2012**, *130*, 2044–2053. [CrossRef] [PubMed]
63. Xiao, B.; Shi, X.; Bai, J. miR-30a regulates the proliferation and invasion of breast cancer cells by targeting Snail. *Oncol. Lett.* **2018**, *17*, 406–413. [CrossRef] [PubMed]
64. Xiong, Y.; Wang, Y.; Wang, L.; Huang, Y.; Xu, Y.; Xu, L.; Guo, Y.; Lu, J.; Li, X.; Zhu, M.; et al. MicroRNA-30b targets Snail to impede epithelial-mesenchymal transition in pancreatic cancer stem cells. *J. Cancer* **2018**, *9*, 2147–2159. [CrossRef] [PubMed]
65. Noori, J.; Haghjooy Javanmard, S.; Sharifi, M. The role of microRNA-30a and downstream snail1 on the growth and metastasis of melanoma tumor. *Iran. J. Basic Med. Sci.* **2019**, *22*, 534–540. [PubMed]
66. Ma, T.; Zhao, Y.; Lu, Q.; Lu, Y.; Liu, Z.; Xue, T.; Shao, Y. MicroRNA-30c functions as a tumor suppressor via targeting SNAI1 in esophageal squamous cell carcinoma. *Biomed. Pharmacother.* **2018**, *98*, 680–686. [CrossRef] [PubMed]
67. Liu, Z.; Tu, K.; Liu, Q. Effects of microRNA-30a on migration, invasion and prognosis of hepatocellular carcinoma. *FEBS Lett.* **2014**, *588*, 3089–3097. [CrossRef]
68. Zhang, J.; Zhang, H.; Liu, J.; Tu, X.; Zang, Y.; Zhu, J.; Chen, J.; Dong, L.; Zhang, J. miR-30 inhibits TGF-β1-induced epithelial-to-mesenchymal transition in hepatocyte by targeting Snail1. *Biochem. Biophys. Res. Commun.* **2012**, *417*, 1100–1105. [CrossRef]
69. Cheng, Y.; Zhou, M.; Zhou, W. MicroRNA-30e regulates TGF-β-mediated NADPH oxidase 4-dependent oxidative stress by Snail in atherosclerosis. *Int. J. Mol. Med.* **2019**, *43*, 1806–1816. [CrossRef]
70. Chang, C.-W.; Yu, J.-C.; Hsieh, Y.-H.; Yao, C.-C.; Chao, J.-I.; Chen, P.-M.; Hsieh, H.-Y.; Hsiung, C.-N.; Chu, H.-W.; Shen, C.-Y.; et al. MicroRNA-30a increases tight junction protein expression to suppress the epithelial-mesenchymal transition and metastasis by targeting Slug in breast cancer. *Oncotarget* **2016**, *7*, 16462–16478. [CrossRef]
71. Kim, N.H.; Kim, H.S.; Li, X.-Y.; Lee, I.; Choi, H.-S.; Kang, S.E.; Cha, S.Y.; Ryu, J.K.; Yoon, D.; Fearon, E.R.; et al. A p53/miRNA-34 axis regulates Snail1-dependent cancer cell epithelial-mesenchymal transition. *J. Cell Biol.* **2011**, *195*, 417–433. [CrossRef] [PubMed]
72. Dong, P.; Xiong, Y.; Watari, H.; Hanley, S.J.B.; Konno, Y.; Ihira, K.; Yamada, T.; Kudo, M.; Yue, J.; Sakuragi, N. MiR-137 and miR-34a directly target Snail and inhibit EMT, invasion and sphere-forming ability of ovarian cancer cells. *J. Exp. Clin. Cancer Res.* **2016**, *35*, 132. [CrossRef] [PubMed]
73. Siemens, H.; Jackstadt, R.; Hünten, S.; Kaller, M.; Menssen, A.; Götz, U.; Hermeking, H. miR-34 and SNAIL form a double-negative feedback loop to regulate epithelial-mesenchymal transitions. *Cell Cycle* **2011**, *10*, 4256–4271. [CrossRef] [PubMed]

74. Tang, Y.; Tang, Y.; Cheng, Y. miR-34a inhibits pancreatic cancer progression through Snail1-mediated epithelial–mesenchymal transition and the Notch signaling pathway. *Sci. Rep.* **2017**, *7*, 38232. [CrossRef]

75. Zhang, B.; Fu, T.; Zhang, L. MicroRNA-153 suppresses human laryngeal squamous cell carcinoma migration and invasion by targeting the SNAI1 gene. *Oncol. Lett.* **2018**, *16*, 5075–5083. [CrossRef]

76. Zeng, H.F.; Yan, S.; Wu, S.F. MicroRNA-153-3p suppress cell proliferation and invasion by targeting SNAI1 in melanoma. *Biochem. Biophys. Res. Commun.* **2017**, *487*, 140–145. [CrossRef]

77. Zuo, J.; Wang, D.; Shen, H.; Liu, F.; Han, J.; Zhang, X. MicroRNA-153 inhibits tumor progression in esophageal squamous cell carcinoma by targeting SNAI1. *Tumor Biol.* **2016**, *37*, 16135–16140. [CrossRef]

78. Wang, Z.; Liu, C. MiR-153 regulates metastases of gastric cancer through Snail. *Tumor Biol.* **2016**, *37*, 15509–15515. [CrossRef]

79. Xia, W.; Ma, X.; Li, X.; Dong, H.; YI, J.; Zeng, W.; Yang, Z. miR-153 inhibits epithelial-to-mesenchymal transition in hepatocellular carcinoma by targeting Snail. *Oncol. Rep.* **2015**, *34*, 655–662. [CrossRef]

80. Bai, Z.; Sun, J.; Wang, X.; Wang, H.; Pei, H.; Zhang, Z. MicroRNA-153 is a prognostic marker and inhibits cell migration and invasion by targeting SNAI1 in human pancreatic ductal adenocarcinoma. *Oncol. Rep.* **2015**, *34*, 595–602. [CrossRef]

81. Zhang, K.; Li, X.Y.; Wang, Z.M.; Han, Z.F.; Zhao, Y.H. MiR-22 inhibits lung cancer cell EMT and invasion through targeting Snail. *Eur. Rev. Med. Pharmacol. Sci.* **2017**, *21*, 3598–3604. [PubMed]

82. Xu, M.; Li, J.; Wang, X.; Meng, S.; Shen, J.; Wang, S.; Xu, X.; Xie, B.; Liu, B.; Xie, L. MiR-22 suppresses epithelial–mesenchymal transition in bladder cancer by inhibiting Snail and MAPK1/Slug/vimentin feedback loop. *Cell Death Dis.* **2018**, *9*, 209. [CrossRef] [PubMed]

83. Luan, W.; Li, L.; Shi, Y.; Bu, X.; Xia, Y.; Wang, J.; Djangmah, H.S.; Liu, X.; You, Y.; Xu, B. Long non-coding RNA MALAT1 acts as a competing endogenous RNA to promote malignant melanoma growth and metastasis by sponging miR-22. *Oncotarget* **2016**, *7*, 63901–63912. [CrossRef]

84. Zuo, Q.-F.; Cao, L.-Y.; Yu, T.; Gong, L.; Wang, L.-N.; Zhao, Y.-L.; Xiao, B.; Zou, Q.-M. MicroRNA-22 inhibits tumor growth and metastasis in gastric cancer by directly targeting MMP14 and Snail. *Cell Death Dis.* **2015**, *6*, e2000. [CrossRef] [PubMed]

85. Dong, H.; Hu, J.; Zou, K.; Ye, M.; Chen, Y.; Wu, C.; Chen, X.; Han, M. Activation of LncRNA TINCR by H3K27 acetylation promotes Trastuzumab resistance and epithelial-mesenchymal transition by targeting MicroRNA-125b in breast Cancer. *Mol. Cancer* **2019**, *18*, 3. [CrossRef]

86. Moes, M.; Le Béchec, A.; Crespo, I.; Laurini, C.; Halavatyi, A.; Vetter, G.; Del Sol, A.; Friederich, E. A novel network integrating a miRNA-203/SNAI1 feedback loop which regulates epithelial to mesenchymal transition. *PLoS ONE* **2012**, *7*, e35440. [CrossRef]

87. Zhang, Y.-F.; Yu, Y.; Song, W.-Z.; Zhang, R.-M.; Jin, S.; Bai, J.-W.; Kang, H.-B.; Wang, X.; CaoO, X.-C. miR-410-3p suppresses breast cancer progression by targeting Snail. *Oncol. Rep.* **2016**, *36*, 480–486. [CrossRef]

88. Zhan, Y.; Li, X.; Liang, X.; Li, L.; Cao, B.; Wang, B.; Ma, J.; Ding, F.; Wang, X.; Pang, D.; et al. MicroRNA-182 drives colonization and macroscopic metastasis via targeting its suppressor SNAI1 in breast cancer. *Oncotarget* **2017**, *8*, 4629–4641. [CrossRef]

89. Yu, T.; Wang, L.; Li, W.; Zuo, Q.; Li, M.; Zou, Q.; Xiao, B. Downregulation of miR-491-5p promotes gastric cancer metastasis by regulating SNAIL and FGFR4. *Cancer Sci.* **2018**, *109*, 1393–1403. [CrossRef]

90. Liu, Z.; Long, J.; Du, R.; Ge, C.; Guo, K.; Xu, Y. miR-204 regulates the EMT by targeting snail to suppress the invasion and migration of gastric cancer. *Tumor Biol.* **2016**, *37*, 8327–8335. [CrossRef]

91. Suzuki, T.; Mizutani, K.; Minami, A.; Nobutani, K.; Kurita, S.; Nagino, M.; Shimono, Y.; Takai, Y. Suppression of the TGF-β1-induced protein expression of SNAI1 and N-cadherin by miR-199a. *Genes Cells* **2014**, *19*, 667–675. [CrossRef] [PubMed]

92. Jiang, K.; Zhao, T.; Shen, M.; Zhang, F.; Duan, S.; Lei, Z.; Chen, Y. MiR-940 inhibits TGF-β-induced epithelial-mesenchymal transition and cell invasion by targeting Snail in non-small cell lung cancer. *J. Cancer* **2019**, *10*, 2735–2744. [CrossRef] [PubMed]

93. Ma, S.; Jia, W.; Ni, S. miR-199a-5p inhibits the progression of papillary thyroid carcinoma by targeting SNAI1. *Biochem. Biophys. Res. Commun.* **2018**, *497*, 181–186. [CrossRef] [PubMed]

94. Cao, L.; Wan, Q.; Li, F.; Tang, C.E. MiR-363 Inhibits Cisplatin Chemoresistance of Epithelial Ovarian Cancer by Regulating Snail-Induced Epithelial-Mesenchymal Transition. *BMB Rep.* **2018**, *5*, 456–461. [CrossRef]

95. Jin, Y.; Wang, J.; Han, J.; Luo, D.; Sun, Z. MiR-122 inhibits epithelial-mesenchymal transition in hepatocellular carcinoma by targeting Snail1 and Snail2 and suppressing WNT/β-cadherin signaling pathway. *Exp. Cell Res.* **2017**, *360*, 210–217. [CrossRef]

96. Guo, J.; Duan, H.; Li, Y.; Yang, L.; Yuan, L. A novel circular RNA circ-ZNF652 promotes hepatocellular carcinoma metastasis through inducing snail-mediated epithelial-mesenchymal transition by sponging miR-203/miR-502-5p. *Biochem. Biophys. Res. Commun.* **2019**, *513*, 812–819. [CrossRef]

97. Zhang, X.; Zhang, T.; Yang, K.; Zhang, M.; Wang, K. miR-486-5p suppresses prostate cancer metastasis by targeting Snail and regulating epithelial–mesenchymal transition. *Onco. Targets. Ther.* **2016**, *9*, 6909–6914. [CrossRef]

98. Wang, K.; Jin, W.; Jin, P.; Fei, X.; Wang, X.; Chen, X. miR-211-5p Suppresses Metastatic Behavior by Targeting SNAI1 in Renal Cancer. *Mol. Cancer Res.* **2017**, *15*, 448–456. [CrossRef]

99. Muraoka, N.; Yamakawa, H.; Miyamoto, K.; Sadahiro, T.; Umei, T.; Isomi, M.; Nakashima, H.; Akiyama, M.; Wada, R.; Inagawa, K.; et al. MiR-133 promotes cardiac reprogramming by directly repressing Snail and silencing fibroblast signatures. *EMBO J.* **2014**, *33*, 1565–1581. [CrossRef]

100. Bai, X.; Geng, J.; Zhou, Z.; Tian, J.; Li, X. MicroRNA-130b improves renal tubulointerstitial fibrosis via repression of Snail-induced epithelial-mesenchymal transition in diabetic nephropathy. *Sci. Rep.* **2016**, *6*, 20475. [CrossRef]

101. Zhang, J.-P.; Zeng, C.; Xu, L.; Gong, J.; Fang, J.-H.; Zhuang, S.-M. MicroRNA-148a suppresses the epithelial–mesenchymal transition and metastasis of hepatoma cells by targeting Met/Snail signaling. *Oncogene* **2014**, *33*, 4069–4076. [CrossRef] [PubMed]

102. Jia, Z.; Zhang, Y.; Xu, Q.; Guo, W.; Guo, A. miR-126 suppresses epithelial-to-mesenchymal transition and metastasis by targeting PI3K/AKT/Snail signaling of lung cancer cells. *Oncol. Lett.* **2018**, *15*, 7369–7375. [CrossRef] [PubMed]

103. Han, J.; Zhang, M.; Nie, C.; Jia, J.; Wang, F.; Yu, J.; Bi, W.; Liu, B.; Sheng, R.; He, G.; et al. miR-215 suppresses papillary thyroid cancer proliferation, migration, and invasion through the AKT/GSK-3β/Snail signaling by targeting ARFGEF1. *Cell Death Dis.* **2019**, *10*, 195. [CrossRef] [PubMed]

104. Chen, F.; Yang, D.; Ru, Y.; Cao, S.; Gao, A. MicroRNA-101 targets CXCL12-mediated Akt and Snail signaling pathways to inhibit cellular proliferation and invasion in papillary thyroid carcinoma. *Oncol. Res. Featur. Preclin. Clin. Cancer Ther.* **2019**, *27*, 691–701. [CrossRef]

105. He, Z.-J.; Li, W.; Chen, H.; Wen, J.; Gao, Y.-F.; Liu, Y.-J. miR-1306–3p targets FBXL5 to promote metastasis of hepatocellular carcinoma through suppressing snail degradation. *Biochem. Biophys. Res. Commun.* **2018**, *504*, 820–826. [CrossRef]

106. Yoo, J.-O.; Kwak, S.-Y.; An, H.-J.; Bae, I.-H.; Park, M.-J.; Han, Y.-H. miR-181b-3p promotes epithelial–mesenchymal transition in breast cancer cells through Snail stabilization by directly targeting YWHAG. *Biochim. Biophys. Acta Mol. Cell Res.* **2016**, *1863*, 1601–1611. [CrossRef]

107. Kwak, S.-Y.; Yoo, J.-O.; An, H.-J.; Bae, I.-H.; Park, M.-J.; Kim, J.; Han, Y.-H. miR-5003-3p promotes epithelial-mesenchymal transition in breast cancer cells through Snail stabilization and direct targeting of E-cadherin. *J. Mol. Cell Biol.* **2016**, *8*, 372–383. [CrossRef]

108. Liu, S.; Kumar, S.M.; Lu, H.; Liu, A.; Yang, R.; Pushparajan, A.; Guo, W.; Xu, X. MicroRNA-9 up-regulates E-cadherin through inhibition of NF-κB1-Snail1 pathway in melanoma. *J. Pathol.* **2012**, *226*, 61–72. [CrossRef]

109. Li, N.; Yang, L.; Sun, Y.; Wu, X. MicroRNA-16 inhibits migration and invasion via regulation of the Wnt/β-catenin signaling pathway in ovarian cancer. *Oncol. Lett.* **2019**, *17*, 2631–2638. [CrossRef]

110. Che, G.; Gao, H.; Tian, J.; Hu, Q.; Xie, H.; Zhang, Y. MicroRNA-483-3p Promotes Proliferation, Migration, and Invasion and Induces Chemoresistance of Wilms' Tumor Cells. *Pediatr. Dev. Pathol.* **2019**. [CrossRef]

111. Zhang, Z.; Zhang, M.; Chen, Q.; Zhang, Q. Downregulation of microRNA-145 promotes epithelial–mesenchymal transition via regulating Snail in osteosarcoma. *Cancer Gene Ther.* **2017**, *24*, 83–88. [CrossRef] [PubMed]

112. Zhang, L.; Pang, Y.; Cui, X.; Jia, W.; Cui, W.; Liu, Y.; Liu, C.; Li, F. MicroRNA-410-3p upregulation suppresses proliferation, invasion and migration, and promotes apoptosis in rhabdomyosarcoma cells. *Oncol. Lett.* **2019**, *18*, 936–943. [CrossRef] [PubMed]

113. Tang, T.; Yang, Z.; Zhu, Q.; Wu, Y.; Sun, K.; Alahdal, M.; Zhang, Y.; Xing, Y.; Shen, Y.; Xia, T.; et al. Up-regulation of miR-210 induced by a hypoxic microenvironment promotes breast cancer stem cell metastasis, proliferation, and self-renewal by targeting E-cadherin. *FASEB J.* **2018**, *32*, 6965–6981. [CrossRef] [PubMed]

114. Hwang, W.-L.; Jiang, J.-K.; Yang, S.-H.; Huang, T.-S.; Lan, H.-Y.; Teng, H.-W.; Yang, C.-Y.; Tsai, Y.-P.; Lin, C.-H.; Wang, H.-W.; et al. MicroRNA-146a directs the symmetric division of Snail-dominant colorectal cancer stem cells. *Nat. Cell Biol.* **2014**, *16*, 268–280. [CrossRef]

115. Qu, Y.; Li, W.-C.; Hellem, M.R.; Rostad, K.; Popa, M.; McCormack, E.; Oyan, A.M.; Kalland, K.-H.; Ke, X.-S. MiR-182 and miR-203 induce mesenchymal to epithelial transition and self-sufficiency of growth signals via repressing SNAI2 in prostate cells. *Int. J. Cancer* **2013**, *133*, 544–555. [CrossRef]

116. Yu, C.-C.; Chen, P.-N.; Peng, C.-Y.; Yu, C.-H.; Chou, M.-Y. Suppression of miR-204 enables oral squamous cell carcinomas to promote cancer stemness, EMT traits, and lymph node metastasis. *Oncotarget* **2016**, *7*, 20180. [CrossRef]

117. Liao, H.; Bai, Y.; Qiu, S.; Zheng, L.; Huang, L.; Liu, T.; Wang, X.; Liu, Y.; Xu, N.; Yan, X.; et al. MiR-203 downregulation is responsible for chemoresistance in human glioblastoma by promoting epithelial-mesenchymal transition via SNAI2. *Oncotarget* **2015**, *6*, 8914–8928. [CrossRef]

118. Tominaga, E.; Yuasa, K.; Shimazaki, S.; Hijikata, T. MicroRNA-1 targets Slug and endows lung cancer A549 cells with epithelial and anti-tumorigenic properties. *Exp. Cell Res.* **2013**, *319*, 77–88. [CrossRef]

119. Liang, Y.-J.; Wang, Q.-Y.; Zhou, C.-X.; Yin, Q.-Q.; He, M.; Yu, X.-T.; Cao, D.-X.; Chen, G.-Q.; He, J.-R.; Zhao, Q. MiR-124 targets Slug to regulate epithelial–mesenchymal transition and metastasis of breast cancer. *Carcinogenesis* **2013**, *34*, 713–722. [CrossRef]

120. Du, S.; Li, H.; Sun, X.; Li, D.; Yang, Y.; Tao, Z.; Li, Q.; Liu, K. MicroRNA-124 inhibits cell proliferation and migration by regulating SNAI2 in breast cancer. *Oncol. Rep.* **2016**, *36*, 3259–3266. [CrossRef]

121. Wu, Z.; Li, X.; Cai, X.; Huang, C.; Zheng, M. miR-497 inhibits epithelial mesenchymal transition in breast carcinoma by targeting Slug. *Tumor Biol.* **2016**, *37*, 7939–7950. [CrossRef] [PubMed]

122. Liu, B.-W.; Yu, Z.-H.; Chen, A.-X.; Chi, J.-R.; Ge, J.; Yu, Y.; Cao, X.-C. Estrogen receptor-α-miR-1271-SNAI2 feedback loop regulates transforming growth factor-β-induced breast cancer progression. *J. Exp. Clin. Cancer Res.* **2019**, *38*, 109. [CrossRef] [PubMed]

123. Zhang, Z.; Zhang, B.; Li, W.; Fu, L.; Fu, L.; Zhu, Z.; Dong, J.-T. Epigenetic Silencing of miR-203 Upregulates SNAI2 and Contributes to the Invasiveness of Malignant Breast Cancer Cells. *Genes Cancer* **2011**, *2*, 782–791. [CrossRef] [PubMed]

124. Ding, X.; Park, S.I.; McCauley, L.K.; Wang, C.-Y. Signaling between Transforming Growth Factor β (TGF-β) and Transcription Factor SNAI2 Represses Expression of MicroRNA miR-203 to Promote Epithelial-Mesenchymal Transition and Tumor Metastasis. *J. Biol. Chem.* **2013**, *288*, 10241–10253. [CrossRef] [PubMed]

125. Chen, D.-D.; Cheng, J.-T.; Chandoo, A.; Sun, X.-W.; Zhang, L.; Lu, M.-D.; Sun, W.-J.; Huang, Y.-P. microRNA-33a prevents epithelial-mesenchymal transition, invasion, and metastasis of gastric cancer cells through the Snail/Slug pathway. *Am. J. Physiol. Liver Physiol.* **2019**, *317*, G147–G160. [CrossRef] [PubMed]

126. Shi, Z.-M.; Wang, L.; Shen, H.; Jiang, C.-F.; Ge, X.; Li, D.-M.; Wen, Y.-Y.; Sun, H.-R.; Pan, M.-H.; Li, W.; et al. Downregulation of miR-218 contributes to epithelial–mesenchymal transition and tumor metastasis in lung cancer by targeting Slug/ZEB2 signaling. *Oncogene* **2017**, *36*, 2577–2588. [CrossRef] [PubMed]

127. Xiao, H.; Zeng, J.; Li, H.; Chen, K.; Yu, G.; Hu, J.; Tang, K.; Zhou, H.; Huang, Q.; Li, A.; et al. MiR-1 downregulation correlates with poor survival in clear cell renal cell carcinoma where it interferes with cell cycle regulation and metastasis. *Oncotarget* **2015**, *6*, 13201–13215. [CrossRef]

128. Huang, J.; Liang, Y.; Xu, M.; Xiong, J.; Wang, D.; Ding, Q. MicroRNA-124 acts as a tumor-suppressive miRNA by inhibiting the expression of Snail2 in osteosarcoma. *Oncol. Lett.* **2018**, *15*, 4979–4987. [CrossRef]

129. Lin, T.; Yu, C.-C.; Hsieh, P.-L.; Liao, Y.-W.; Yu, C.-H.; Chen, C.-J. Down-regulation of miR-200b-targeting Slug axis by cyclosporine A in human gingival fibroblasts. *J. Formos. Med. Assoc.* **2018**, *117*, 1072–1077. [CrossRef]

130. Xia, H.; Cheung, W.K.C.; Ng, S.S.; Jiang, X.; Jiang, S.; Sze, J.; Leung, G.K.K.; Lu, G.; Chan, D.T.M.; Bian, X.-W.; et al. Loss of Brain-enriched miR-124 MicroRNA Enhances Stem-like Traits and Invasiveness of Glioma Cells. *J. Biol. Chem.* **2012**, *287*, 9962–9971. [CrossRef]

131. Sun, Y.; Cai, J.; Yu, S.; Chen, S.; Li, F.; Fan, C. MiR-630 Inhibits Endothelial-Mesenchymal Transition by Targeting Slug in Traumatic Heterotopic Ossification. *Sci. Rep.* **2016**, *6*, 22729. [CrossRef] [PubMed]

132. Dai, X.; Liang, Z.; Liu, L.; Guo, K.; Xu, S.; Wang, H. Silencing of MALAT1 inhibits migration and invasion by sponging miR-1-3p in prostate cancer cells. *Mol. Med. Rep.* **2019**, *20*, 3499–3508. [CrossRef] [PubMed]

133. Jin, C.; Yan, B.; Lu, Q.; Lin, Y.; Ma, L. The role of MALAT1/miR-1/slug axis on radioresistance in nasopharyngeal carcinoma. *Tumor Biol.* **2016**, *37*, 4025–4033. [CrossRef] [PubMed]

134. Gan, L.; Lv, L.; Liao, S. Long non-coding RNA H19 regulates cell growth and metastasis via the miR-22-3p/Snail1 axis in gastric cancer. *Int. J. Oncol.* **2019**, *54*, 2157–2168. [CrossRef]

135. Zhang, Y.; Yuan, Y.; Zhang, Y.; Cheng, L.; Zhou, X.; Chen, K. SNHG7 accelerates cell migration and invasion through regulating miR-34a-Snail-EMT axis in gastric cancer. *Cell Cycle* **2020**, *19*, 142–152. [CrossRef]

136. Yang, P.; Chen, T.; Xu, Z.; Zhu, H.; Wang, J.; He, Z. Long noncoding RNA GAPLINC promotes invasion in colorectal cancer by targeting SNAI2 through binding with PSF and NONO. *Oncotarget* **2016**, *7*, 42183–42194. [CrossRef]

137. Ge, X.; Li, G.; Jiang, L.; Jia, L.; Zhang, Z.; Li, X.; Wang, R.; Zhou, M.; Zhou, Y.; Zeng, Z.; et al. Long noncoding RNA CAR10 promotes lung adenocarcinoma metastasis via miR-203/30/SNAI axis. *Oncogene* **2019**, *38*, 3061–3076. [CrossRef]

138. Jiang, L.; Wang, R.; Fang, L.; Ge, X.; Chen, L.; Zhou, M.; Zhou, Y.; Xiong, W.; Hu, Y.; Tang, X.; et al. HCP5 is a SMAD3-responsive long non-coding RNA that promotes lung adenocarcinoma metastasis via miR-203/SNAI axis. *Theranostics* **2019**, *9*, 2460–2474. [CrossRef]

139. Xiao, J.-N.; Yan, T.-H.; Yu, R.-M.; Gao, Y.; Zeng, W.-L.; Lu, S.-W.; Que, H.-X.; Liu, Z.-P.; Jiang, J.-H. Long non-coding RNA UCA1 regulates the expression of Snail2 by miR-203 to promote hepatocellular carcinoma progression. *J. Cancer Res. Clin. Oncol.* **2017**, *143*, 981–990. [CrossRef]

140. Xiao, C.; Wan, X.; Yu, H.; Chen, X.; Shan, X.; Miao, Y.; Fan, R.; Cha, W. LncRNA-AB209371 promotes the epithelial-mesenchymal transition of hepatocellular carcinoma cells. *Oncol. Rep.* **2019**, *41*, 2957–2966. [CrossRef]

141. Wang, Y.Q.; Jiang, D.M.; Hu, S.S.; Zhao, L.; Wang, L.; Yang, M.H.; Ai, M.L.; Jiang, H.J.; Han, Y.; Ding, Y.Q.; et al. SATB2-AS1 suppresses colorectal carcinoma aggressiveness by inhibiting SATB2-dependent Snail transcription and epithelial–mesenchymal transition. *Cancer Res.* **2019**, *79*, 3542–3556. [CrossRef] [PubMed]

142. Li, Y.; Cheng, C. Long noncoding RNA NEAT1 promotes the metastasis of osteosarcoma via interaction with the G9a-DNMT1-Snail complex. *Am. J. Cancer Res.* **2018**, *8*, 81–90. [PubMed]

143. Jiang, H.; Li, T.; Qu, Y.; Wang, X.; Li, B.; Song, J.; Sun, X.; Tang, Y.; Wan, J.; Yu, Y.; et al. Long non-coding RNA SNHG15 interacts with and stabilizes transcription factor Slug and promotes colon cancer progression. *Cancer Lett.* **2018**, *425*, 78–87. [CrossRef] [PubMed]

144. Luan, W.; Shi, Y.; Zhou, Z.; Xia, Y.; Wang, J. circRNA_0084043 promote malignant melanoma progression via miR-153-3p/Snail axis. *Biochem. Biophys. Res. Commun.* **2018**, *502*, 22–29. [CrossRef] [PubMed]

145. Chen, X.; Chen, R.X.; Wei, W.S.; Li, Y.H.; Feng, Z.H.; Tan, L.; Chen, J.W.; Yuan, G.J.; Chen, S.L.; Guo, S.J.; et al. PRMT5 circular RNA promotes metastasis of urothelial carcinoma of the bladder through sponging miR-30c to induce epithelial–mesenchymal transition. *Clin. Cancer Res.* **2018**, *24*, 6319–6330. [CrossRef] [PubMed]

146. Ma, H.-B.; Yao, Y.-N.; Yu, J.-J.; Chen, X.-X.; Li, H.-F. Extensive profiling of circular RNAs and the potential regulatory role of circRNA-000284 in cell proliferation and invasion of cervical cancer via sponging miR-506. *Am. J. Transl. Res.* **2018**, *10*, 592–604. [PubMed]

147. Wang, L.; Tong, X.; Zhou, Z.; Wang, S.; Lei, Z.; Zhang, T.; Liu, Z.; Zeng, Y.; Li, C.; Zhao, J.; et al. Circular RNA hsa-circ-0008305 (circPTK2) inhibits TGF-β-induced epithelial-mesenchymal transition and metastasis by controlling TIF1γ in non-small cell lung cancer. *Mol. Cancer* **2018**, *17*, 140. [CrossRef]

148. Chi, Y.; Luo, Q.; Song, Y.; Yang, F.; Wang, Y.; Jin, M.; Zhang, D. Circular RNA circPIP5K1A promotes non-small cell lung cancer proliferation and metastasis through miR-600/HIF-1α regulation. *J. Cell. Biochem.* **2019**, *120*, 19019–19030. [CrossRef]

149. Shen, T.; Cheng, X.; Liu, X.; Xia, C.; Zhang, H.; Pan, D.; Zhang, X.; Li, Y. Circ_0026344 restrains metastasis of human colorectal cancer cells via miR-183. *Artif. Cells Nanomed. Biotechnol.* **2019**, *47*, 4038–4045. [CrossRef]

150. Kumar, A.S.; Jagadeeshan, S.; Pitani, R.S.; Ramshankar, V.; Venkitasamy, K.; Venkatraman, G.; Rayala, S.K. Snail-Modulated MicroRNA 493 Forms a Negative Feedback Loop with the Insulin-Like Growth Factor 1 Receptor Pathway and Blocks Tumorigenesis. *Mol. Cell. Biol.* **2017**, *37*, e00510-16. [CrossRef]

151. Zhu, Y.; Wang, C.; Becker, S.A.; Hurst, K.; Nogueira, L.M.; Findlay, V.J.; Camp, E.R. miR-145 Antagonizes SNAI1-Mediated Stemness and Radiation Resistance in Colorectal Cancer. *Mol. Ther.* **2018**, *26*, 744–754. [CrossRef] [PubMed]

152. Xu, Y.; Jin, J.; Liu, Y.; Huang, Z.; Deng, Y.; You, T.; Zhou, T.; Si, J.; Zhuo, W. Snail-Regulated MiR-375 Inhibits Migration and Invasion of Gastric Cancer Cells by Targeting JAK2. *PLoS ONE* **2014**, *9*, e99516. [CrossRef] [PubMed]

153. Hsieh, C.-H.; Tai, S.-K.; Yang, M.-H. Snail-overexpressing Cancer Cells Promote M2-Like Polarization of Tumor-Associated Macrophages by Delivering MiR-21-Abundant Exosomes. *Neoplasia* **2018**, *20*, 775–788. [CrossRef] [PubMed]

154. Dong, Q.; Cai, N.; Tao, T.; Zhang, R.; Yan, W.; Li, R.; Zhang, J.; Luo, H.; Shi, Y.; Luan, W.; et al. An Axis Involving SNAI1, microRNA-128 and SP1 Modulates Glioma Progression. *PLoS ONE* **2014**, *9*, e98651. [CrossRef]

155. Qian, P.; Banerjee, A.; Wu, Z.-S.; Zhang, X.; Wang, H.; Pandey, V.; Zhang, W.-J.; Lv, X.-F.; Tan, S.; Lobie, P.E.; et al. Loss of SNAIL Regulated miR-128-2 on Chromosome 3p22.3 Targets Multiple Stem Cell Factors to Promote Transformation of Mammary Epithelial Cells. *Cancer Res.* **2012**, *72*, 6036–6050. [CrossRef]

156. Yu, W.-W.; Jiang, H.; Zhang, C.-T.; Peng, Y. The SNAIL/miR-128 axis regulated growth, invasion, metastasis, and epithelial-to-mesenchymal transition of gastric cancer. *Oncotarget* **2017**, *8*, 39280–39295. [CrossRef]

157. Tao, T.; Li, G.; Dong, Q.; Liu, D.; Liu, C.; Han, D.; Huang, Y.; Chen, S.; Xu, B.; Chen, M. Loss of SNAIL inhibits cellular growth and metabolism through the miR-128-mediated RPS6KB1/HIF-1α/PKM2 signaling pathway in prostate cancer cells. *Tumor Biol.* **2014**, *35*, 8543–8550. [CrossRef]

158. Díaz-López, A.; Díaz-Martín, J.; Moreno-Bueno, G.; Cuevas, E.P.; Santos, V.; Olmeda, D.; Portillo, F.; Palacios, J.; Cano, A. Zeb1 and Snail1 engage miR-200f transcriptional and epigenetic regulation during EMT. *Int. J. Cancer* **2015**, *136*, E62–E73. [CrossRef]

159. Gill, J.G.; Langer, E.M.; Lindsley, R.C.; Cai, M.; Murphy, T.L.; Murphy, K.M. Snail promotes the cell-autonomous generation of Flk1(+) endothelial cells through the repression of the microRNA-200 family. *Stem Cells Dev.* **2012**, *21*, 167–176. [CrossRef]

160. Liu, Z.; Liu, H.; Desai, S.; Schmitt, D.C.; Zhou, M.; Khong, H.T.; Klos, K.S.; McClellan, S.; Fodstad, O.; Tan, M. miR-125b Functions as a Key Mediator for Snail-induced Stem Cell Propagation and Chemoresistance. *J. Biol. Chem.* **2013**, *288*, 4334–4345. [CrossRef]

161. Lerner, R.G.; Petritsch, C. A microRNA-operated switch of asymmetric-to-symmetric cancer stem cell divisions. *Nat. Cell Biol.* **2014**, *16*, 212–214. [CrossRef] [PubMed]

162. Yi, M.; Liu, B.; Tang, Y.; Li, F.; Qin, W.; Yuan, X. Irradiated Human Umbilical Vein Endothelial Cells Undergo Endothelial-Mesenchymal Transition via the Snail/miR-199a-5p Axis to Promote the Differentiation of Fibroblasts into Myofibroblasts. *BioMed Res. Int.* **2018**, *2018*, 1–10. [CrossRef] [PubMed]

163. He, J.-H.; Li, B.-X.; Han, Z.-P.; Zou, M.-X.; Wang, L.; Lv, Y.-B.; Zhou, J.-B.; Cao, M.-R.; Li, Y.-G.; Zhang, J. Snail-activated long non-coding RNA PCA3 up-regulates PRKD3 expression by miR-1261 sponging, thereby promotes invasion and migration of prostate cancer cells. *Tumor Biol.* **2016**, *37*, 16163–16176. [CrossRef] [PubMed]

164. Mazzolini, R.; Gonzàlez, N.; Garcia-Garijo, A.; Millanes-Romero, A.; Peiró, S.; Smith, S.; De Herreros, A.G.; Canudas, S. Snail1 transcription factor controls telomere transcription and integrity. *Nucleic Acids Res.* **2018**, *46*, 146–158. [CrossRef] [PubMed]

165. Battistelli, C.; Cicchini, C.; Santangelo, L.; Tramontano, A.; Grassi, L.; Gonzalez, F.J.; De Nonno, V.; Grassi, G.; Amicone, L.; Tripodi, M. The Snail repressor recruits EZH2 to specific genomic sites through the enrollment of the lncRNA HOTAIR in epithelial-to-mesenchymal transition. *Oncogene* **2017**, *36*, 942–955. [CrossRef]

166. Findlay, V.J.; Wang, C.; Nogueira, L.M.; Hurst, K.; Quirk, D.; Ethier, S.P.; Staveley O'Carroll, K.F.; Watson, D.K.; Camp, E.R. SNAI2 Modulates Colorectal Cancer 5-Fluorouracil Sensitivity through miR145 Repression. *Mol. Cancer Ther.* **2014**, *13*, 2713–2726. [CrossRef]

167. Chang, T.-H.; Tsai, M.-F.; Gow, C.-H.; Wu, S.-G.; Liu, Y.-N.; Chang, Y.-L.; Yu, S.-L.; Tsai, H.-C.; Lin, S.-W.; Chen, Y.-W.; et al. Upregulation of microRNA-137 expression by Slug promotes tumor invasion and metastasis of non-small cell lung cancer cells through suppression of TFAP2C. *Cancer Lett.* **2017**, *402*, 190–202. [CrossRef]

168. Liu, Y.-N.; Yin, J.J.; Abou-Kheir, W.; Hynes, P.G.; Casey, O.M.; Fang, L.; Yi, M.; Stephens, R.M.; Seng, V.; Sheppard-Tillman, H.; et al. MiR-1 and miR-200 inhibit EMT via Slug-dependent and tumorigenesis via Slug-independent mechanisms. *Oncogene* **2013**, *32*, 296–306. [CrossRef]
169. Pan, Y.; Li, J.; Zhang, Y.; Wang, N.; Liang, H.; Liu, Y.; Zhang, C.-Y.; Zen, K.; Gu, H. Slug-upregulated miR-221 promotes breast cancer progression through suppressing E-cadherin expression. *Sci. Rep.* **2016**, *6*, 25798. [CrossRef]
170. Zheng, M.; Jiang, Y.; Chen, W.; Li, K.; Liu, X.; Gao, S.; Feng, H.; Wang, S.; Jiang, J.; Ma, X.; et al. Snail and Slug collaborate on EMT and tumor metastasis through miR-101-mediated EZH2 axis in oral tongue squamous cell carcinoma. *Oncotarget* **2015**, *6*, 6797–6810. [CrossRef]
171. Xiong, W.-C.; Han, N.; Wu, N.; Zhao, K.-L.; Han, C.; Wang, H.-X.; Ping, G.-F.; Zheng, P.-F.; Feng, H.; Qin, L.; et al. Interplay between long noncoding RNA ZEB1-AS1 and miR-101/ZEB1 axis regulates proliferation and migration of colorectal cancer cells. *Am. J. Transl. Res.* **2018**, *10*, 605–617. [PubMed]
172. Hahn, S.; Jackstadt, R.; Siemens, H.; Hünten, S.; Hermeking, H. SNAIL and miR-34a feed-forward regulation of ZNF281/ZBP99 promotes epithelial-mesenchymal transition. *EMBO J.* **2013**, *32*, 3079–3095. [CrossRef] [PubMed]
173. Hahn, S.; Hermeking, H. ZNF281/ZBP-99: A new player in epithelial–mesenchymal transition, stemness, and cancer. *J. Mol. Med.* **2014**, *92*, 571–581. [CrossRef] [PubMed]
174. Fumagalli, M.R.; Lionetti, M.C.; Zapperi, S.; La Porta, C.A.M. Cross-Talk Between circRNAs and mRNAs Modulates MiRNA-mediated Circuits and Affects Melanoma Plasticity. *Cancer Microenviron.* **2019**, *12*, 95–104. [CrossRef] [PubMed]
175. Liu, Y.; Xue, M.; Du, S.; Feng, W.; Zhang, K.; Zhang, L.; Liu, H.; Jia, G.; Wu, L.; Hu, X.; et al. Competitive endogenous RNA is an intrinsic component of EMT regulatory circuits and modulates EMT. *Nat. Commun.* **2019**, *10*, 1637. [CrossRef] [PubMed]
176. Zhang, L.; Liao, Y.; Tang, L. MicroRNA-34 family: A potential tumor suppressor and therapeutic candidate in cancer. *J. Exp. Clin. Cancer Res.* **2019**, *38*, 53. [CrossRef]

Review

Viral miRNAs as Active Players and Participants in Tumorigenesis

Alessia Gallo [1,*], Vitale Miceli [1] , Matteo Bulati [1], Gioacchin Iannolo [1], Flavia Contino [1,2] and Pier Giulio Conaldi [1]

[1] Department of Research, IRCCS ISMETT (Istituto Mediterraneo per i Trapianti e Terapie ad alta specializzazione), 90100 Palermo, Italy; vmiceli@ismett.edu (V.M.); mbulati@ismett.edu (M.B.); giannolo@ismett.edu (G.I.); fcontino@ismett.edu (F.C.); pgconaldi@ismett.edu (P.G.C.)

[2] Scienze Mediche Chirurgiche E Sperimentali, Università degli Studi di Sassari, Piazza Universita, 07100 Sassari, Italy

* Correspondence: agallo@ismett.edu; Tel.: +39-0912192649

Received: 12 December 2019; Accepted: 31 January 2020; Published: 4 February 2020

Abstract: The theory that viruses play a role in human cancers is now supported by scientific evidence. In fact, around 12% of human cancers, a leading cause of morbidity and mortality in some regions, are attributed to viral infections. However, the molecular mechanism remains complex to decipher. In recent decades, the uncovering of cellular miRNAs, with their invaluable potential as diagnostic and prognostic biomarkers, has increased the number of studies being conducted regarding human cancer diagnosis. Viruses develop clever mechanisms to succeed in the maintenance of the viral life cycle, and some viruses, especially herpesviruses, encode for miRNA, v-miRNAs. Through this viral miRNA, the viruses are able to manipulate cellular and viral gene expression, driving carcinogenesis and escaping the host innate or adaptive immune system. In this review, we have discussed the main viral miRNAs and virally influenced cellular pathways, and their capability to drive carcinogenesis.

Keywords: viral miRNAs; EBV; HHV-8; HPV; HCV; HBV; MCPyV

---

1. Introduction

According to the IARC (International Agency for Research on Cancer), 13% of new cancer cases worldwide in 2018 were the result of a chronic infection, most of which were caused by viruses [1]. The burden of viral infections in cancer, even if considered high, is still undervalued [2]. Viruses implement multiple strategies to pursue their final goals: viral survival, proliferation, and transmission. Moore and Chang masterfully emphasized that the event of cancer caused by viruses is a biological accident, since it does not increase transmissibility or enhance replication fitness [2]. Moreover, in the particular cases of immunosuppressed populations, cancers generated by tumor virus carcinogens have an increased incidence [3], suggesting the deep relationship between viruses and the immune system. Innate immune signaling shares many key effector proteins with tumor suppressor signaling, such as the p21 cyclin-dependent kinase inhibitor [4] and p53 [5]. This may imply the crucial role of tumor suppressor pathways in inadvertently placing the infected cell at risk for cancerous transformation [6,7]. One of the main roles in this process might be played by virally encoded miRNAs, ideal and non-immunogenic tools for viruses, able to modulate viral as well as host gene expression and lead to immune invisibility of infected cells [8]. V-miRNAs (v-miRNAs) seem to have a leading role in viral persistence and propagation, enacting different immune evasion strategies. V-miRNAs and host miRNAs can both regulate the expression of multiple host- and virus-derived transcripts [9]. An appealing theory suggests the use of v-miRNA orthologues of cellular miRNAs, with which they share a seed sequence and thus regulate the same targets. Still, among oncoviruses, only a few viral orthologues miRNAs have been discovered: kshv-miR-K12-11, which shows significant homology to cellular hsa-miR-155 [10]; kshv-mir-K12-10, which is a viral orthologue of hsa-mir-142-3p [11]; kshv-mir-K3, a homolog of hsa-mir-23 [12]; and ebv-miR-BART-5, which shows significant "seed" sequence homology to hsa-miR-18 [13].

V-miRNAs have slowly evolved and adapted within their specific hosts. In fact, viral miRNA biogenesis involves only cellular factors, as no viral proteins have been described [9]. V-miRNAs are also exported via the exosomal route, rendering them able to enter into cells even at distant sites, thus allowing the virus to manipulate cellular and tissue immunity [14]. Considering that the survival ability of a virus depends on its capacity to escape host immunosurveillance [14], viruses encode multiple miRNAs that show immunomodulatory functions involved in the regulation of critical innate and adaptive immune mechanisms used by the host to defend himself [14]. Moreover, it has been described that v-miRNAs allow viruses to enter the latent phase of their life cycle and become undetected by the host's immune system, with this being a further risk factor for cancer development [15]. Here we have examined the current knowledge of miRNAs encoded by six oncoviruses, Epstein–Barr virus (EBV), Kaposi's Sarcoma Herpesvirus/Human Herpesvirus-8 (KSHV/HHV8), Human Papillomavirus (HPV), Hepatitis C Virus (HCV), Hepatitis B Virus (HBV) and Merkel Cell Polyomavirus (MCPyV) (Figure 1), and the virally influenced cellular pathways (Table 1) and their relationships with the immune system.

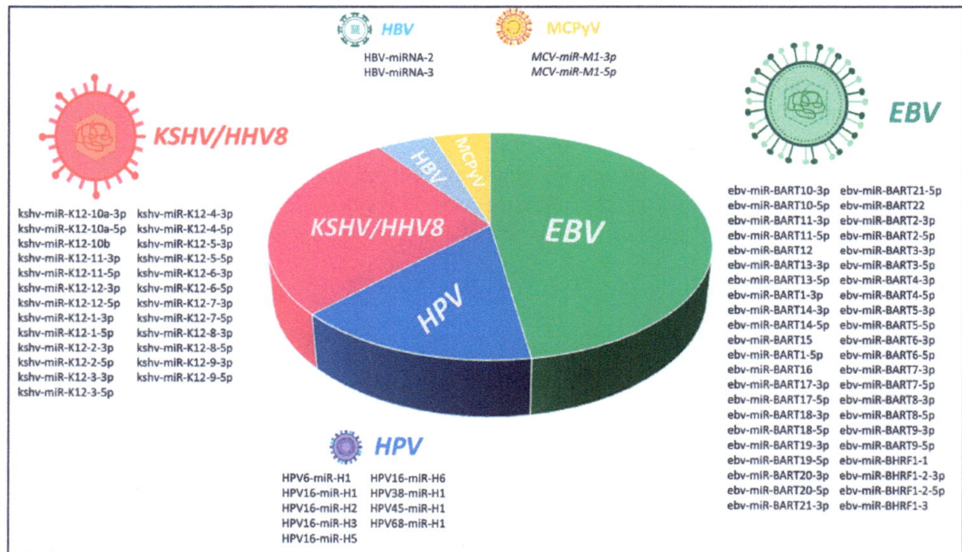

Figure 1. Graphical representation of the relative abundance of viral miRNA production by Epstein–Barr virus (EBV), Kaposi's Sarcoma Herpesvirus/Human Herpesvirus-8 (KSHV/HHV8), Human Papillomavirus (HPV), Hepatitis C Virus (HCV), Hepatitis B Virus (HBV) and Merkel Cell Polyomavirus (MCPyV) viruses.

Table 1. Overview of viral miRNA regulatory functions and targets.

VIRUS	miRNA	Targets	Effects of miRNAs
EBV	ebv-BHRF1-2	IL-12, CATHEPSIN B, AEP, GILT	Immune evasion
	ebv-BHRF1-3	BHRF1-3, TAP	
	ebv-BART1-5p	IL12, CATHEPSIN B, AEP, GILT	
	ebv-BART2-5p	MICB, IL-12, CATHEPSIN B, AEP, GILT	
	ebv-BART3-3p	IPO7	
	ebv-BART5-5p	LMP1	
	ebv-BART6-3p	RIG-1	
	ebv-BART15	NLRP3	
	ebv-BART16	CREB-BP	
	ebv-BART17-5p	TAP	
	ebv-BART22	LMP2A	
	ebv-BHRF1-1	P53	Anti-apoptosis
	ebv-BHRF1-2	PRDM1/Blimp1	
	ebv-BHRF1-3	PTEN	
	ebv-BART1-3p	CASP3	
	ebv-BART4-5p	Bid	
	ebv-BART5-5p	PUMA	
	ebv-BART6-5p	OCT1	
	ebv-BART8	STAT1	
	ebv-BART13-3p	CAPRIN2	
	ebv-BART16	CREB-BP, TOMM22, CASP3, SH2B3	
	ebv-BART22	MAP3K5, CASP3, PAK2, TP53INP1	
	ebv-BART22	NDRG1	Promote metastasis
	ebv-BHRF1-1	RNF4	Promote viral production
	ebv-BHRF1-2	BHRF1	Maintain latency
	ebv-BART2-5p	BALF5	
	ebv-BART6-5p	DICER	
	ebv-BART18-5p	MAP3K2	
	ebv-BART20-5p	BZLF1, BRLF1	
	ebv-BART1-5p	LMP1	Promote cancer development
	ebv-BART16	LMP1	
	ebv-BART17-5p	LMP1	
	ebv-BART1-5p	PTEN	Promote tumor metastasis
	ebv-BART7-3p	PTEN	
	ebv-BART9	E-Cadherin	
	ebv-BART10-3p	BTRC	
	ebv-BART6-3p	PTEN	Promote proliferation
	ebv-BART11	FOXP1	

Table 1. *Cont.*

VIRUS	miRNA	Targets	Effects of miRNAs
KSHV	kshv-miR-K12-1	Casp3	Apoptosis
	kshv-miR-K12-3	Casp3	
	kshv-miR-K12-4	Casp4	
	kshv-miR-K12-5	Tmskα1	
	kshv-miR-K12-10a	TWEAK	
	kshv-miR-K12-12	CASP3, CASP7	
	kshv-miR-K12-1	NF-κB signaling/IκBα	KSHV latency
	kshv-miR-K12-3	nuclear factor I/B, GRK2	
	kshv-miR-K12-4	Rbl2	
	kshv-miR-K12-7	RTA (KSHV ORF50)	
	kshv-miR-K12-9	RTA (KSHV ORF50), BCLAF1	
	kshv-miR-K12-10a	BCLAF1	
	kshv-miR-K12-11	MYB, IKKε	
	kshv-miR-K12-1	THBS1	Cell adhesion, migration, and angiogenesis
	kshv-miR-K12-3	THBS1	
	kshv-miR-K12-6	THBS1, Bcr, SH3BGR	
	kshv-miR-K12-11	THBS1	
	kshv-miR-K12-1	CASTOR1, STAT3,p21	Promote tumorigenesis, Cell survival
	kshv-miR-K12-4	CASTOR1	
	kshv-miR-K12-10a	TGFBR2	
	kshv-miR-K12-10b	TGFBR2	
	kshv-miR-K12-11	SMAD5	
	kshv-miR-K12-1	MICB	Immune evasion
	kshv-miR-K12-3	C/EBPβ p20 (LIP)	
	kshv-miR-K12-5	MYD88	
	kshv-miR-K12-7	C/EBPβ p20 (LIP), MICB	
	kshv-miR-K12-9	IRAK1	
	kshv-miR-K12-11	C/EBPβ	
	kshv-miR-K12-1	MAF	Differentiation of infected cells
	kshv-miR-K12-6	MAF	
	kshv-miR-K12-11	MAF/BACH-1	
HPV	HPV16-miR-H1	BCL11A, CHD7, ITGAM, RAG1, TCEA1	Immune evasion
	HPV16-miR-H2	SP3, XRCC4, JAK2, PKNOX1, FOXP1	
HBV	HBV-mir-2	TRIM35	Promote tumorigenesis
	HBV-mir-3	HBsAg, HBeAg, HBc	Self-replication
MCPyV	MCV-miR-M1-5p	SP100	Immune evasion
	MCV-miR-M1	RUNX1	Viral proliferation

2. Epstein–Barr Virus (EBV)

The Epstein–Barr virus (EBV) is a ubiquitous lymphotropic gamma herpesvirus able to infect >95% of individuals during childhood and early adolescence. It usually causes an asymptomatic infection without significant illness, except some cases in which it may cause infectious mononucleosis [16]. After primary infection, EBV silently inhabits mainly the long-lived memory B cells of infected individuals. In immunocompromised patients, including organ transplant and AIDS patients, in southern Chinese patients, sub-Saharan African children, and other particularly susceptible groups, EBV is linked to a range of cancers and other disorders [16]. In the context of immunosuppression, post-transplant lymphoma, Hodgkin's disease, African Burkitt's lymphoma, and nasopharyngeal carcinoma are the malignancies most consistently and significantly associated with EBV [17]. The study of EBV and its potent growth-transforming action on infected lymphocytes associated with tumors points to specific interactions between environmental, genetic, and viral factors [18]. The EBV life cycle consists of a latent and lytic phase [19]. During the lytic cycle, EBV expresses its full set of viral genes [20]. In the lifelong latent phase, EBV infection establishes different transcription programs expressing a limited quantity of viral genes [21]. The pattern of EBV gene latency expression is essential for its genome maintenance and could be a key component of the puzzle of EBV-induced cancers [16,22]. V-miRNAs were detected for the first time in EBV [23] and, to date, a total of 44 mature miRNAs from 25 miRNA precursors have been encoded [24]. Recent evidence strongly suggests a role of EBV-encoded miRNAs in driving the initiation and progression of EBV-associated malignancies [8]. EBV encodes 44 miRNAs transcribed from two regions: the BamHI-A region rightward transcript (BART), with 22 miRNA precursors (ebv-mir-BART1-22) producing 40 mature miRNAs; and the BamHI fragment H rightward open reading frame 1 (BHRF1) clusters [25], with three miRNA precursors (ebv-mir-BHRF1-1, -2, and -3) producing four mature miRNAs [24]. As for the EBV genes, v-miRNA expression is infection-stage-dependent. BART miRNAs are transcribed in all stages of latency, although more associated with latency types I and II [26]. In contrast, BHRF1 miRNAs are amply expressed in type III latency, but nearly undetectable in latency types I and II [27]. All EBV-infection-associated human tumors display latency programs and related v-miRNA expression, spanning from latency I in BL, NK/T-cell lymphoma, and EBV aGC, to latency II in Hodgkin's disease and nasopharyngeal carcinoma (NPC), and latency III in EBV-associated B lymphoma and Post-transplant lymphoproliferative disorder (PTLD) [28]. Several studies have pointed out that EBV miRNA clusters take part in tumor progression by targeting tumor-suppressing genes and repressing anti-proliferation genes. For example, ebv-mir-BART1 in NPC samples, is able to significantly up- and downregulate a number of genes fundamental for cell metabolism, including PAST1, PHGDH, DHRS3, ASS1, IDH2, PISD, UGT8, and LDHB [29]. Ebv-mir-BART5-5p plays an anti-apoptotic role by directly acting on the pro-apoptotic gene PUMA [30] and suppressing the p53 protein in various stomach cancer and NPC cell lines [31]. Other evidence has shown that ebv-mir-BART5-5p, together with ebv-mir-BART1-3p and ebv-mir-BART7-3p, promotes NPC cell metastasis by targeting the cellular tumor suppressor PTEN [32,33]. In BL, the regulation of PTEN by ebv-mir-BART6-3p promotes cell proliferation and inhibits cell death [34]. In addition, in NPC cells, ebv-mir-BART22 promotes metastasis by targeting N-myc downstream regulated gene 1 (NDRG1), known to be a tumor suppressor [35]. In gastric cancer cell lines, ebv-mir-BART4-5p reduces Bid expression, leading to reduced apoptosis [36]. In nasopharyngeal carcinoma cells, ebv-mir-BART22 has been shown to have proliferative and invasive abilities through the regulation of MAP3K5 [37]. Ebv-mir-BART20-5p has been found to shorten apoptosis, strengthen cell growth, and contribute to carcinogenesis of EBVaGC by directly acting on BAD [38]. In NPC and GC, ebv-mir-BART11-3p and 5-p induce cancer cell proliferation through the suppression of forkhead box P1 (FOXP1) [39]. Interestingly, ebv-mir-BART9, through the regulation of of E-cadherin expression, has been confirmed to induce epithelial–mesenchymal transition (EMT) and thus promote metastasis [39,40]. In addition, ebv-mir-BART10-3p, through the inhibition of BTRC, essential in the ubiquitination and degradation of β-catenin, induces invasion and metastasis and is thus associated with poor prognosis in NPC patients [41]. Among the mechanisms by which EBV may influence host cell equilibrium, it has been postulated that EBV-infected cells can transfer viral miRNAs via exosomes and thus influence host gene expression in uninfected recipient cells [42]. It has been previously demonstrated that ebv-mir-BART13 can be transferred from B cells to salivary epithelial cells where it downregulates STIM1 protein and decreases activation of NFAT and NFAT-dependent transcriptional activity [43]. In addition, we showed the selective packaging of two v-miRNAs, ebv-mir-BART3 and ebv-mir-BHRF1-1, into the exosomes in a lymphoblastoid cell line [44]. The first, ebv-mir-BART3, addressed importin 7 (IPO7), inducing the pro-inflammatory cytokine IL-6 [24]. The second, ebv-mir-BHRF1-1, downregulated host p53 [45].

EBV produces different miRNAs involved in the immunomodulation of both aspecific and specific immunity. Concerning innate immunity, different EBV miRNAs seem to have the ability to influence inflammation and chemotaxis (Figure 2). Ebv-mir-BART6-3p directly binds to RIG-1 mRNA, causing the impaired production of different antiviral cytokines [46]. Moreover, ebv-mir-BART-6-3p, in association with host-derived miR-197, acts on IL-6R mRNA and is involved in the impairment of IL-6 signaling [47]. Host mRNA CREBBP is the target of ebv-mir-BART16 with consequent inhibition of type I interferon signaling [48], which favors the enhancement of viral replication. Ebv-mir-BHRF-1-2-5p acts on interleukin-1 (IL-1) signaling by targeting IL-1 Receptor 1, blocking the activation of host innate immune responses following virus infection [49]. Another EBV-derived miRNA that limits inflammation, an advantage for its own purposes, is ebv-mir-BART15, which regulates both IL-1β production and the NLRP3 inflammasome [50]. Most of the effects of EBV-derived miRNA on the acquired immune system are related to MHC-restricted antigen processing and presentation. Indeed, ebv-mir-BART2 acts on CTSB mRNA and interferes with MHC-I antigen processing, while ebv-mir-BHRF1-3, targeting TAP2, blocks peptide transport to MHC-I [51]. Ebv-mir-BART1-5p also causes an impaired antigen presentation because of its action on LY75 mRNA, which encodes an endocytic receptor involved in antigen capture and processing [52]. Ebv-mir-BART1 and BART2, respectively, act on IFI30 mRNA and LGMN mRNAs, inducing the inhibition of MHC-II-restricted antigen processing [52]. Moreover, EBV-derived miRNA targets genes involved in T cell chemotaxis and polarization; ebv-mir-BART1, -BART2, -BART10, -BART22, and -BHRF1 act on IL12B mRNA preventing the polarization of CD4$^+$ T helper cells toward antiviral Th1 subtype [49], while BHRF1-3 targets CXCL11 mRNA, with the consequence of inhibiting the activated T cells' chemotaxis [53].

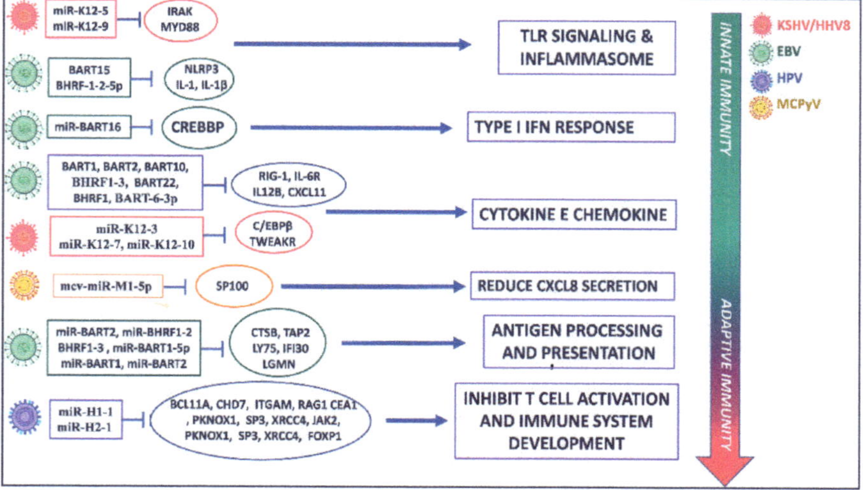

Figure 2. Immunoevasive functions of viral miRNAs. Target cellular components of EBV, KSHV, and HPV miRNAs and the relavant antiviral responses of innate and adaptive immunity.

Collectively, these studies have highlighted the capability of EBV miRNAs to modulate tumor cell proliferation through complicated regulatory networks including tumor suppressor genes, cell apoptosis, and control of the viral oncogenic protein functions.

3. Kaposi's Sarcoma Herpesvirus/Human Herpesvirus-8 (KSHV/HHV8)

HHV-8, a member of the Herpesviridae family, is an oncogenic virus. HHV-8 shares with EBV the ability to establish a chronic infection in lymphocytes, which are its main reservoir [54], but also in macrophages, keratinocytes, and endothelial cells [55], and to induce cellular transformation [56]. HHV-8 infections are notably threatening in immunocompromised patients, such as those with AIDS or patients with transplants or under chemotherapy treatment [8]. Alongside Kaposi's sarcoma, from which its alias KSHV (Kaposi's-sarcoma-associated herpesvirus) was taken [57], HHV-8 is considered the etiological cause of primary effusion lymphoma (PEL) [58] and multicentric Castleman's disease [59]. HHV-8 has a dsDNA genome encoding for more than 90 open reading frames (ORFs). In addition, HHV-8 encodes for 25 mature miRNAs, deriving from 12 viral pre-miRNAs [60]. All the HHV-8 miRNAs are under the control of latent kaposin promoter. Except for kshv-miR-K10 and kshv-miR-K12, which are expressed more during the lytic phase [61], the majority of the pre-miRNA genes are expressed during the latent phase of virus infection [62] and are located between the sequence for kaposin and open reading frame 71 [62]. As for EBV, HHV-8 latency is the phase of KSHV infection where the v-RNAs cooperate in viral replication and thus contribute to oncogenesis. As an example of this mechanism the KSHV miRNAs kshv-miR-K5, kshv-miR-K7-5p, kshv-miR-K9-5p, kshv-miR-K3, and kshv-miR-K4 have been shown to endorse latency by targeting the KSHV lytic switch protein, either directly or indirectly [63]. The final goals of KSHV miRNAs are immune evasion, avoidance apoptosis, and contribution to tumorigenesis. For example, kshv-miR-K12-1, -3, and -4-3p target and inactivate the inducer of apoptosis, Casp3, blocking apoptosis [64], while kshv-K12-1 functions as an oncogene by activating NF-κB/IL-6/STAT3 signaling [65]. Kshv-mir-K12-3 has been demonstrated to be a promoter of cell migration and invasion by targeting GRK2/CXCR2/AKT signaling [66], and kshv-miR-k12-1-5p has been shown to be a promoter of the proliferation, migration, and invasion of KS cells by suppressing cytokine signaling 6 (SOCS6) [67]. It has also been demonstrated that kshv-miR-K1-5p and kshv-miR-K4-5p directly target CASTOR1, inhibiting its expression and activating the mTORC1 pathway with the final result of promoting tumorigenesis [68]. It has also been shown that kshv-mir-K10a targets tumor necrosis factor-like weak inducer of apoptosis receptor protein (TWEAKR), thus preventing TWEAK-induced apoptosis and inflammatory cytokine (IL8) expression [69]. Interestingly, HHV-8 encodes for miRNAs which share perfect seed homology with cellular oncomiRNAs such as kshv-mir-K12-10, which is a viral orthologue of hsa-mir-142-3p [11]; both miRNAs have been shown to inhibit the TGF-β pathway by targeting the TGF-β type II receptor [70]. Another example of this mechanism comes from kshv-mir-K12-11, a homolog of hsa-mir-155 [71], which targets IKKε, BACH-1, and SMAD5 and downregulates the expression of the basic region/leucine zipper motif transcription factor C/EBPb, a regulator of interleukin-6 [10]. The last known, so far, is kshv-mir-K3, a homolog of hsa-mir-23 with which it shares anti-apoptotic functions by targeting caspase 3 and caspase 7 [12].

Some of the 25 miRNAs encoded by KSHV/HHV8 play an important role in viral latency infection in host cells, targeting key genes and their signaling pathways (Figure 2), interfering with immune surveillance and thus contributing to the development of KS [72]. Two KSHV/HHV8 -derived miRNAs, kshv-mir-K12-5 and kshv-mir-K12-9, affect the secretion of inflammatory cytokines, targeting MYD88 and IRAK1, respectively, which are components of TLR/IL-1R-mediated signaling [73]. Kshv-mir-K12-11 acts on the primary response to antiviral immunity by targeting IKKε with the consequent impairment of type I-IFN signaling [74], while kshv-mir-K12-10 reduces the production of IL-6 and IL-10 by targeting TWEAKR [73]. Kshv-mir-K12-3 and kshv-mir-K12-7, targeting C/EBPβ, modulate cytokine secretion by immune cells such as monocytes or NK lymphocytes [75]. RAB3B and RAB3D are two genes targeted by kshv-mir-K12-3, which not only alter cytokine production, but also attenuate bacterial phagocytosis [76].

4. Human Papillomavirus (HPV)

Papillomaviridae is a taxonomic family of non-enveloped DNA tumor viruses that infect both mucosa and cutaneous epithelial cells [77]. To date, 226 genotypes of human papillomavirus (HPV) have been identified [78]. Its circular DNA genome encodes for six nonstructural genes (E1, E2, E4, E5, E6, and E7) and two viral assembly genes (L1 and L2). In particular, E1, E2, and E4 are involved in viral replication, while E5, E6, and E7 are involved in HPV-induced cellular transformation [79,80]. In relation to the ability of HPVs to trigger malignant cellular progression, these viruses are classified as high-risk (HR) and low-risk (LR)-HPVs. LR-HPVs typically cause benign epithelial lesions that may progress to malignant lesions [81], whereas HR-HPVs are associated with cervical carcinomas [82].

Different works have shown that some viruses, including different HPV genotypes, are able to code miRNA-like species, and these miRNAs may be involved in virus-induced carcinogenesis. Unfortunately, very little information is available on the mechanisms by which HPV-encoded miRNAs play a role in the promotion and/or progression of human cancer; this is probably partly due to a lack of proper study models for the various HPV types.

Recently, to characterize new HPV-encoded miRNAs, Chirayil et al. applied a new approach for miRNA discovery based on forced genome expression. They showed that four different HPV genotypes are mainly involved in the synthesis of miRNAs: HPV17, 37, 41, and a Fringilla coelebs HPV (FcPV1). These data were validated by in vitro assays on cell cultures, and two FcPV1 miRNAs were also found in vivo in a natural host. Interestingly, HPV41-miRNAs and FcPV1-miRNAs are involved in the control of HPV life cycle [83]. In other recent studies, Weng et al. showed that HPV-miRNAs differ in their number depending on the HPV species [15], and Virtanem et al. revealed that miRNAs belonging to HPV-16 species such as miR-H1, miR-H3, miR-H5, and miR-H6 were found in tumor samples [84]. Qian et al. recognized nine putative HPV-encoded miRNAs, (HPV6-mir-H1, HPV16-mir-H1, HPV16-mir-H2, HPV16-mir-H3, HPV16-mir-H5, HPV16-mir-H6, HPV38-mir-H1, HPV45-mir-H1, and HPV68-mir-H1) through tissue sequencing of human cervical lesion and cell lines. These miRNAs are upregulated in chronic infection, interfering with different pathways including regulation of the cell cycle, immune responses, and cell adhesion/migration, and therefore they may be involved in the susceptibility of tissue to transformation [85]. In particular, as depicted in Figure 2, HPV16-derived miRNAs such as miR-H1-1 and miR-H2-1 target different genes involved in immune system regulation. Indeed, miR-H1-1 is able to both inhibit T-cell activation and immune system development, targeting BCL11A, CHD7, ITGAM, RAG1, and TCEA1 genes [85]. Similarly, miR-H2-1 targets different protein involved in both T-cell activation (PKNOX1, SP3, XRCC4) and immune system development (JAK2, PKNOX1, SP3, XRCC4, FOXP1) [85]. Due to the central role of the viral immune escape for the development of virus-related cancers, this could represent a crucial mechanism for HPV-associated cancers. Qian et al. also showed that transcriptional enhancer factor 1 (TEF-1) is a target of HPV-16-encoded miRNAs [85]. Interestingly, this gene regulates cell proliferation and migration, and it binds and activates the early HPV 16 promoter of E6 and E7 [86,87]. E6 is able to induce the degradation of p53 protein [88], while E7 induces the degradation of the retinoblastoma protein, causing release of E2F, with consequent activation of transcription of the target genes and overcoming of proliferation arrest [89]. Therefore, downregulation of E6 and E7 by HPV-encoded miRNA through TEF-1 could thus lead to increased cell cycle arrest of HPV-infected cells for cell cycle normalization controls and persistent HPV infection (Figure 3). These mechanisms could represent indirect processes by which HPV-encoded miRNAs affect tumorigenesis through the controls of major pathways involved in the carcinogenesis of host cells.

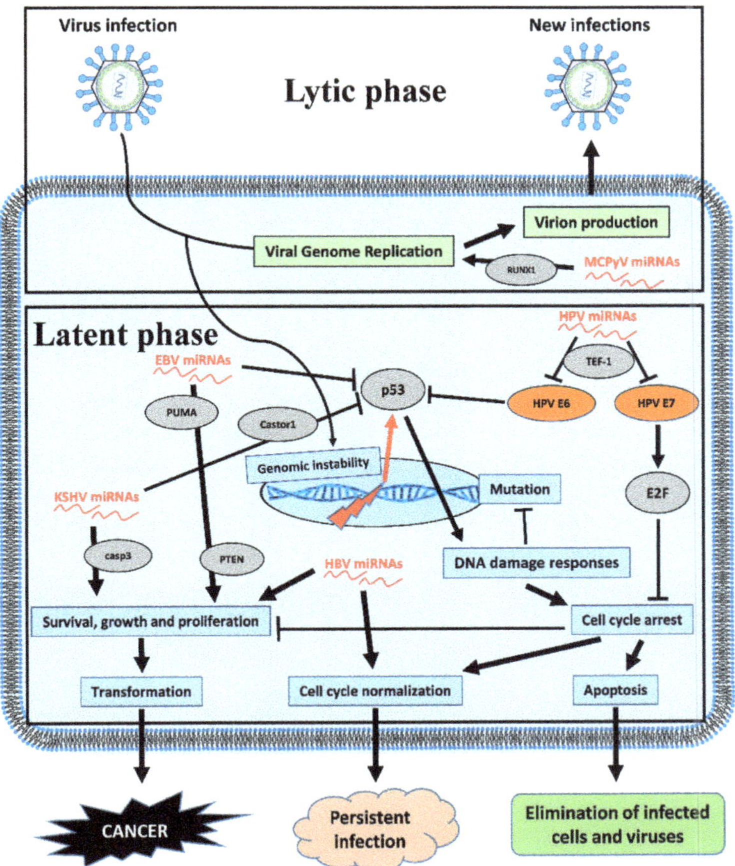

Figure 3. Viral miRNAs affect the pathways involved in the carcinogenesis of host cells. The picture depicts changes to cellular fate because of viral miRNAs. Genomic instability due to viral infection can induce activation of the p53 pathway, which in turn supports both DNA damage responses and cell cycle arrest. In relation to viral infection context, viral miRNAs can affect abortive cell fates such as programmed cell death with consequent cancer transformation. Arrows signify that the factor or process promotes the effect it points to, while blocking arrows signify inhibition. Orange ellipses represent viral proteins, grey ellipses represent host cell proteins, green boxes represent stages of the lytic phase of viral life cycle and blue boxes represent cellular processes affected by viral miRNAs during the viral latent phase.

Conversely, it has been shown that HR-HPVs are also responsible for both upregulation of oncogenic host miRNAs and downregulation of tumor-suppressive host miRNAs, influencing both viral replication and HPV-induced carcinogenesis [90]. This is because HPVs are able to integrate into the genome at the sites where specific miRNAs are frequently located [90]. Furthermore, the expression of many host miRNAs is regulated by HPV proteins (E6 and E7) and the E2F protein of the host cells [91]. As mentioned above, HPV-16-encoded miRNAs are able to regulate E6, E7, and E2F and, therefore, they can potentially regulate the expression of host miRNAs. Different results obtained using the K14-HPV16 transgenic mice model showed that the expression of deregulated host miRNAs in non-neoplastic samples could regulate the vulnerability to oncogenesis induced by HPV-associated mechanisms [92,93].

Based on current knowledge, the action of HPV-encoded miRNAs in the carcinogenesis process seems to be indirect, regulating, for example, HPV oncogenic proteins such as E6 and E7 which, in turn, regulate the tumorigenesis pathways of host cells. Moreover, HPV-encoded miRNAs could regulate key proteins for inhibition of the immune system allowing viral immune escape and the development of HPV-related cancer.

The characterization of the complete genomes of HPV subtypes by bioinformatics methods also allows the prediction of potential v-miRNAs and their target genes, providing a fundamental tool for understanding the role of HPV-encoded miRNAs in the carcinogenesis process of HPV-associated cancers [15]. Further studies are needed to discover new HPV-encoded miRNAs and their implications for both virus infection and carcinogenesis.

5. Hepatitis B Virus (HBV)

In contrast to oncoviruses such as EBV, KSV, and HPV, which dysregulate cellular tumor suppressor activities inducing oncogenicity by affecting p53 and pRB, other viruses do not have a straight correlation with neoplastic transformation. Among them Hepatitis B (HBV) and Hepatitis C viruses (HCV) have been associated, respectively, with 53% and 25% of hepatocellular carcinomas [94]. Hepatocellular carcinoma (HCC) represents the third most common cause of cancer-related death worldwide [95]. Liver cancer is often associated with HBV and/or HCV infection; however, the molecular mechanisms whereby these viruses induce HCC are still not completely elucidated.

HBV belongs to the Hepadnavirus family, and is a hepatotropic virus that possesses a partially double-stranded DNA genome in a relaxed circular DNA (rcDNA) form. This DNA is converted into a covalently closed circular DNA (cccDNA) via the cellular apparatus and organized as a viral minichromosome [96]. Recently, a new viral miRNA involved in HCC was isolated from HBV: HBV-mir-2 [97]. HBV-mir-2 acts in a dual mode by suppressing TRIM35 (tripartite motif containing 35) and stabilizing RAN (ras-related nuclear protein) mRNA by binding to their 3′UTRs [97]. TRIM35 in hematopoietic cells induces apoptosis and inhibits cell proliferation, and for this reason it is described as tumor suppressor [97]. RAN is a protein that is correlated with cell cycle control, nucleocytoplasmic shuttle, and cell transformation. Overexpression in hepatic cell lines (HepG2, Huh7) induces epithelial–mesenchymal transition (EMT), whereas its silencing decreases EMT [97]. RAN is correlated with proliferation, migration, and invasion in ovarian cancer [98] and in HCC [97]. Moreover, these hepatotropic viruses contribute to HCC in an indirect manner by establishing chronic infection; this effect is linked to the long-lasting infection that induces recurrent liver inflammation by the host immune system. In a recent report, it was demonstrated that HBV itself modulates its replicative activity via an endogenous miRNA: HBV-mir-3 [99]. The authors demonstrated that this miRNA represses HBV protein expression and viral production to avoid major damage to the infected cells. In particular, HBV-mir-3 suppresses HBsAg, HBeAg, and HBc protein expression and the intermediate HBV's DNA replication. The same authors pointed out that the infection persistence has a carcinogenic implication mainly because chronic HBV infection contributes to the occurrence of mutations. In this way, HBV-mir-3 indirectly promotes HCC onset [99].

6. Hepatitis C Virus (HCV)

HCV belongs to the Flaviviridae family and it was first discovered in 1989, when a complementary DNA clone encoding for an antigen was associated specifically with NANBH (non-A, non-B hepatitis) infections [100]. It is an RNA virus with a single-stranded positive genome (+) that principally infects hepatocytes [101]. After infection, viral RNA is translated via the cellular ribosomal apparatus and the HCV positive strand is copied, generating a replicative intermediate (RI) with a negative strand (−). One of the mechanisms by which HCV can induce oncogenic transformation is the induction of double strand brakes (DSBs) [102] and oxidative stress [103].

The HCV genome is targeted by cellular miRNA; in particular, has-miR-122, one of the most abundant liver miRNAs, binds two sites (one in the 5′UTR and one in the 3′UTR) in the viral RNA, enhancing its replication [104]. Has-miR-122 inhibition by LNA reduces HCV in vitro and in vivo and has been tested in several clinical trials [105,106], demonstrating the efficacy of this treatment without adverse effects. HCV/has-miR-122 binding does, however, have an effect on the induction of HCC [107]. Luna and coworkers demonstrated that miR-122 co-immunoprecipitates with Ago in association with HCV RNA, which thus acts as a has-miR-122 sponge. It is noteworthy that miR-122 knockout induces liver diseases and HCC [108,109].

Like miR-122, mir-199a* has been described as an HCV genomic RNA binder. Mir-199a* has been demonstrated to inhibit HCV RNA replicative activity by binding a region in the HCV 5′-UTR (domain II of the IRES region), a highly conserved region among the different HCV genotypes [110]. In addition, mir-199a* has been demonstrated to co-immunoprecipitate with Ago2 [110] and to inhibit HCV RNA replicative activity, inducing Ago2 binding to HCV RNA. This mechanism is responsible for a HCV feedback loop that can control the persistence or downmodulation of the virus in the infected cells. In other studies, it has been demonstrated that HCV RNA is not found in cells with HCV-induced mutations, suggesting a "hit and run" mechanism for neoplastic transformation [102]. Surprisingly, and yet to be validated, mir-199a* was recently described as an HCV miRNA by VIRmiRNA, a database for experimentally validated v-miRNAs and their targets [111]. For this reason, miR-199a* could be an interesting objective for targeted HCC therapy.

7. Merkel Cell Polyomavirus (MCPyV)

Merkel cell polyomavirus (MCPyV) is a mammalian, double-stranded DNA polyomavirus [112], which causes a lifelong but inoffensive infection [113]. In immunosuppressed patients, such as solid-organ-transplanted or autoimmune-condition-affected patients, the presence of MCPyV increases the occurrence of Merkel cell carcinoma, a neuroendocrine skin cancer [114,115]. The MCPyV genome is long (5386 bp) and encodes for two early antigens, long LT and small sT-Ag, and two late structural antigens, VP1 and VP2 [116,117]. In addition, it has been reported that two miRNAs, mcv-miR-M1-5p and -3p, edited from a single miRNA precursor [118], are the only miRNAs expressed by actively replicating MCPyV genomes [119]. The miRNA precursor is expressed from the antisense strand of the LT ORF and shows perfect sequence complementarity to a region in exon 2 of the MCPyV LT mRNA transcript [120]. The first evidence for the roles played by mcv-miR-M1 miRNAs indicates the capability to autoregulate early viral gene expression at late stages post infection [118]. The demonstrated regulation of expression of the large T-antigen could potentially lead to the evasion of immune surveillance [118]. Akhbari et al. ran an in silico analysis of the mcv-miR-M1-5p seed sequence and found the direct targeting of SP100, an intrinsic antiviral protein, leading to a reduction in the secretion of CXCL8 with a final effect of the subversion of the host-cell immune response [120]. By analyzing the seed region, Lee at al. built a list of predicted human target genes of the experimentally observed mature mcv-miR-M1 which could be relevant for tumorigenesis processes [121]. Predicted targets include PIK3CD and PSME3, responsible for antigen presentation by the host cell [122,123], which are potentially involved in mediating the host immune response against MCPyV [121]. Another predicted target of mcv-miR-M1 is RUNX1, a transcription factor known to play the roles of an oncogene and an anti-oncogene in epithelial tumors. By downregulating RUNX1, it has been suggested that mcv-miR-M1 could aid the viral life cycle transition from early to late [121] and affect polyomavirus replication [124].

8. Conclusions

Viruses have developed a complex symbiotic system by which to access and regulate host transcriptional machinery. Among these, the discovery of the first miRNAs coded by a virus by Pfeffer et al. [23] paved the way for a very new field of interest: viral mechanism interpretation. Viral miRNAs are ideal tools, because of their non-immunogenicity, to induce immune invisibility of infected cells [8]. The best-known and studied viral miRNAs are the herpetic ones, as is their transcriptional process. Less is known about newly discovered miRNAs, such as for HBV, HVC, and MCPyV, but this fascinating and new mechanism of transcription regulation will surely be the subject of further studies.

In general, viral latency is the phase in which the majority of v-miRNAs are transcribed, serving as an immune evasion strategy and thus regulating host processes in order to promote cell survival. The occurrence of cancers driven by so-called oncoviruses must be considered an unexpected, unfortunate side effect of the infection itself, decreasing transmissibility and replication fitness. In the case of immunosuppressed populations, cancers caused by tumor viruses have an even more increased occurrence [3] suggesting the deep relationship between viruses and the immune system. In this review, we focused our attention on the miRNAs encoded by the viruses EBV, KSHV/HHV-8, HPV, HBV, HCV, and MCPyV These viruses are been related to cancer occurrence in many different populations, but especially in immunosuppressed populations such as HIV or transplant patients. Recently, it has been shown that these viruses produce different oncoviral proteins that cause functional impairment of p53 activity, which is a crucial mechanism of virus-related carcinogenesis [125]. Interestingly, v-miRNAs also seem to have p53 as the main target, which therefore becomes an even more central mechanism by which these viruses can induce oncogenetic processes (Figure 3).

Due to v-miRNA effects, ranging from increasing viral proliferation and increased virulence to tuning the host immune responses, it is not surprising that viral miRNA expression shows great therapeutic potential and represents an appealing antiviral strategy for the miRNA-based treatment of viral infections. Moreover, v-mirRNAs are now the most promising tool for measuring virus infective and reproductive status, with significant value as diagnostic and prognostic biomarkers. Once the mechanism behind v-miRNA actions is more elucidated and clarified, they may really be used for the early diagnosis of virus-related tumors.

Author Contributions: Conceptualization, A.G., and P.G.C. Writing—original draft preparation, A.G., V.M., M.B., F.C. and G.I.; Writing—review and editing, A.G., V.M., M.B., F.C., G.I. and P.G.C. Visualization, V.M. and F.C.; Supervision, A.G. and P.G.C. All authors have read and agreed to the published version of the manuscript.

Funding: This research received no external funding.

Acknowledgments: We would like to thank Warren Blumberg of IRCCS ISMETT's Language Services Department for assistance in editing the manuscript.

Conflicts of Interest: The authors declare no conflict of interest.

References

1. De Martel, C.; Georges, D.; Bray, F.; Ferlay, J.; Clifford, G.M. Global burden of cancer attributable to infections in 2018: A worldwide incidence analysis. *Lancet Glob. Health* **2019**. [CrossRef]
2. Moore, P.S.; Chang, Y. Why do viruses cause cancer? Highlights of the first century of human tumour virology. *Nat. Rev. Cancer* **2010**, *10*, 878–889. [CrossRef] [PubMed]
3. Grulich, A.E.; van Leeuwen, M.T.; Falster, M.O.; Vajdic, C.M. Incidence of cancers in people with HIV/AIDS compared with immunosuppressed transplant recipients: A meta-analysis. *Lancet* **2007**, *370*, 59–67. [CrossRef]
4. Chin, Y.E.; Kitagawa, M.; Su, W.C.; You, Z.H.; Iwamoto, Y.; Fu, X.Y. Cell growth arrest and induction of cyclin-dependent kinase inhibitor p21 WAF1/CIP1 mediated by STAT1. *Science* **1996**, *272*, 719–722. [CrossRef] [PubMed]
5. Takaoka, A.; Hayakawa, S.; Yanai, H.; Stoiber, D.; Negishi, H.; Kikuchi, H.; Sasaki, S.; Imai, K.; Shibue, T.; Honda, K.; et al. Integration of interferon-alpha/beta signalling to p53 responses in tumour suppression and antiviral defence. *Nature* **2003**, *424*, 516–523. [CrossRef] [PubMed]
6. Moore, P.S.; Chang, Y. Kaposi's sarcoma-associated herpesvirus immunoevasion and tumorigenesis: Two sides of the same coin? *Annu. Rev. Microbiol.* **2003**, *57*, 609–639. [CrossRef] [PubMed]
7. Moore, P.S.; Chang, Y. Antiviral activity of tumor-suppressor pathways: Clues from molecular piracy by KSHV. *Trends Genet.* **1998**, *14*, 144–150. [CrossRef]
8. Vojtechova, Z.; Tachezy, R. The Role of miRNAs in Virus-Mediated Oncogenesis. *Int. J. Mol. Sci.* **2018**, *19*, 1217. [CrossRef]

9. Skalsky, R.L.; Cullen, B.R. Viruses, microRNAs, and host interactions. *Annu. Rev. Microbiol.* **2010**, *64*, 123–141. [CrossRef]

10. Gottwein, E.; Mukherjee, N.; Sachse, C.; Frenzel, C.; Majoros, W.H.; Chi, J.T.; Braich, R.; Manoharan, M.; Soutschek, J.; Ohler, U.; et al. A viral microRNA functions as an orthologue of cellular miR-155. *Nature* **2007**, *450*, 1096–1099. [CrossRef]

11. Gottwein, E.; Corcoran, D.L.; Mukherjee, N.; Skalsky, R.L.; Hafner, M.; Nusbaum, J.D.; Shamulailatpam, P.; Love, C.L.; Dave, S.S.; Tuschl, T.; et al. Viral microRNA targetome of KSHV-infected primary effusion lymphoma cell lines. *Cell Host Microbe* **2011**, *10*, 515–526. [CrossRef] [PubMed]

12. Manzano, M.; Shamulailatpam, P.; Raja, A.N.; Gottwein, E. Kaposi's sarcoma-associated herpesvirus encodes a mimic of cellular miR-23. *J. Virol.* **2013**, *87*, 11821–11830. [CrossRef] [PubMed]

13. Babu, S.G.; Ponia, S.S.; Kumar, D.; Saxena, S. Cellular oncomiR orthologue in EBV oncogenesis. *Comput. Biol. Med.* **2011**, *41*, 891–898. [CrossRef]

14. Naqvi, A.R. Immunomodulatory roles of human herpesvirus-encoded microRNA in host-virus interaction. *Rev. Med. Virol.* **2020**, *30*, e2081. [CrossRef]

15. Weng, S.L.; Huang, K.Y.; Weng, J.T.; Hung, F.Y.; Chang, T.H.; Lee, T.Y. Genome-wide discovery of viral microRNAs based on phylogenetic analysis and structural evolution of various human papillomavirus subtypes. *Brief. Bioinform.* **2018**, *19*, 1102–1114. [CrossRef] [PubMed]

16. Kutok, J.L.; Wang, F. Spectrum of Epstein-Barr virus-associated diseases. *Annu. Rev. Pathol.* **2006**, *1*, 375–404. [CrossRef] [PubMed]

17. Fries, K.L.; Miller, W.E.; Raab-Traub, N. Epstein-Barr virus latent membrane protein 1 blocks p53-mediated apoptosis through the induction of the A20 gene. *J. Virol.* **1996**, *70*, 8653–8659. [CrossRef] [PubMed]

18. Raab-Traub, N. EBV-induced oncogenesis. In *Human Herpesviruses: Biology, Therapy, and Immunoprophylaxis*; Arvin, A., Campadelli-Fiume, G., Mocarski, E., Moore, P.S., Roizman, B., Whitley, R., Yamanishi, K., Eds.; Cambridge University Press: Cambridge, UK, 2007; Chapter 55.

19. Murata, T. Regulation of Epstein-Barr virus reactivation from latency. *Microbiol. Immunol.* **2014**, *58*, 307–317. [CrossRef] [PubMed]

20. Sinclair, A.J. Epigenetic control of Epstein-Barr virus transcription—Relevance to viral life cycle? *Front. Genet.* **2013**, *4*, 161. [CrossRef]

21. Andrei, G.; Trompet, E.; Snoeck, R. Novel Therapeutics for Epstein(-)Barr Virus. *Molecules* **2019**, *24*, 997. [CrossRef]

22. Cesarman, E.; Mesri, E.A. Kaposi sarcoma-associated herpesvirus and other viruses in human lymphomagenesis. *Curr. Top. Microbiol. Immunol.* **2007**, *312*, 263–287. [PubMed]

23. Pfeffer, S.; Zavolan, M.; Grasser, F.A.; Chien, M.; Russo, J.J.; Ju, J.; John, B.; Enright, A.J.; Marks, D.; Sander, C.; et al. Identification of virus-encoded microRNAs. *Science* **2004**, *304*, 734–736. [CrossRef] [PubMed]

24. Barth, S.; Meister, G.; Grasser, F.A. EBV-encoded miRNAs. *Biochim. Biophys. Acta* **2011**, *1809*, 631–640. [CrossRef] [PubMed]

25. Klinke, O.; Feederle, R.; Delecluse, H.J. Genetics of Epstein-Barr virus microRNAs. *Semin. Cancer Biol.* **2014**, *26*, 52–59. [CrossRef]

26. Kang, M.S.; Kieff, E. Epstein-Barr virus latent genes. *Exp. Mol. Med.* **2015**, *47*, e131. [CrossRef]

27. Amoroso, R.; Fitzsimmons, L.; Thomas, W.A.; Kelly, G.L.; Rowe, M.; Bell, A.I. Quantitative studies of Epstein-Barr virus-encoded microRNAs provide novel insights into their regulation. *J. Virol.* **2011**, *85*, 996–1010. [CrossRef]

28. Zhang, X.; Ye, Y.; Fu, M.; Zheng, B.; Qiu, Q.; Huang, Z. Implication of viral microRNAs in the genesis and diagnosis of Epstein-Barr virus-associated tumors. *Oncol. Lett.* **2019**, *18*, 3433–3442. [CrossRef]

29. Ye, Y.; Zhou, Y.; Zhang, L.; Chen, Y.; Lyu, X.; Cai, L.; Lu, Y.; Deng, Y.; Wang, J.; Yao, K.; et al. EBV-miR-BART1 is involved in regulating metabolism-associated genes in nasopharyngeal carcinoma. *Biochem. Biophys. Res. Commun.* **2013**, *436*, 19–24. [CrossRef]

30. Choy, E.Y.; Siu, K.L.; Kok, K.H.; Lung, R.W.; Tsang, C.M.; To, K.F.; Kwong, D.L.; Tsao, S.W.; Jin, D.Y. An Epstein-Barr virus-encoded microRNA targets PUMA to promote host cell survival. *J. Exp. Med.* **2008**, *205*, 2551–2560. [CrossRef]

31. Zheng, X.; Wang, J.; Wei, L.; Peng, Q.; Gao, Y.; Fu, Y.; Lu, Y.; Qin, Z.; Zhang, X.; Lu, J.; et al. Epstein-Barr Virus MicroRNA miR-BART5-3p Inhibits p53 Expression. *J. Virol.* **2018**, *92*, e01022-18. [CrossRef]

32. Cai, L.; Ye, Y.; Jiang, Q.; Chen, Y.; Lyu, X.; Li, J.; Wang, S.; Liu, T.; Cai, H.; Yao, K.; et al. Epstein-Barr virus-encoded microRNA BART1 induces tumour metastasis by regulating PTEN-dependent pathways in nasopharyngeal carcinoma. *Nat. Commun.* **2015**, *6*, 7353. [CrossRef] [PubMed]

33. Cai, L.M.; Lyu, X.M.; Luo, W.R.; Cui, X.F.; Ye, Y.F.; Yuan, C.C.; Peng, Q.X.; Wu, D.H.; Liu, T.F.; Wang, E.; et al. EBV-miR-BART7-3p promotes the EMT and metastasis of nasopharyngeal carcinoma cells by suppressing the tumor suppressor PTEN. *Oncogene* **2015**, *34*, 2156–2166. [CrossRef] [PubMed]

34. Ambrosio, M.R.; Navari, M.; Di Lisio, L.; Leon, E.A.; Onnis, A.; Gazaneo, S.; Mundo, L.; Ulivieri, C.; Gomez, G.; Lazzi, S.; et al. The Epstein Barr-encoded BART-6-3p microRNA affects regulation of cell growth and immuno response in Burkitt lymphoma. *Infect. Agents Cancer* **2014**, *9*, 12. [CrossRef] [PubMed]

35. Kanda, T.; Miyata, M.; Kano, M.; Kondo, S.; Yoshizaki, T.; Iizasa, H. Clustered microRNAs of the Epstein-Barr virus cooperatively downregulate an epithelial cell-specific metastasis suppressor. *J. Virol.* **2015**, *89*, 2684–2697. [CrossRef] [PubMed]

36. Shinozaki-Ushiku, A.; Kunita, A.; Isogai, M.; Hibiya, T.; Ushiku, T.; Takada, K.; Fukayama, M. Profiling of Virus-Encoded MicroRNAs in Epstein-Barr Virus-Associated Gastric Carcinoma and Their Roles in Gastric Carcinogenesis. *J. Virol.* **2015**, *89*, 5581–5591. [CrossRef] [PubMed]

37. Chen, R.; Zhang, M.; Li, Q.; Xiong, H.; Liu, S.; Fang, W.; Zhang, Q.; Liu, Z.; Xu, X.; Jiang, Q. The Epstein-Barr Virus-encoded miR-BART22 targets MAP3K5 to promote host cell proliferative and invasive abilities in nasopharyngeal carcinoma. *J. Cancer* **2017**, *8*, 305–313. [CrossRef]

38. Kim, H.; Choi, H.; Lee, S.K. Epstein-Barr virus miR-BART20-5p regulates cell proliferation and apoptosis by targeting BAD. *Cancer Lett.* **2015**, *356*, 733–742. [CrossRef]

39. Song, Y.; Li, X.; Zeng, Z.; Li, Q.; Gong, Z.; Liao, Q.; Li, X.; Chen, P.; Xiang, B.; Zhang, W.; et al. Epstein-Barr virus encoded miR-BART11 promotes inflammation-induced carcinogenesis by targeting FOXP1. *Oncotarget* **2016**, *7*, 36783–36799. [CrossRef]

40. Hsu, C.Y.; Yi, Y.H.; Chang, K.P.; Chang, Y.S.; Chen, S.J.; Chen, H.C. The Epstein-Barr virus-encoded microRNA MiR-BART9 promotes tumor metastasis by targeting E-cadherin in nasopharyngeal carcinoma. *Plos Pathog.* **2014**, *10*, e1003974. [CrossRef]

41. Yan, Q.; Zeng, Z.; Gong, Z.; Zhang, W.; Li, X.; He, B.; Song, Y.; Li, Q.; Zeng, Y.; Liao, Q.; et al. EBV-miR-BART10-3p facilitates epithelial-mesenchymal transition and promotes metastasis of nasopharyngeal carcinoma by targeting BTRC. *Oncotarget* **2015**, *6*, 41766–41782. [CrossRef]

42. Pegtel, D.M.; Cosmopoulos, K.; Thorley-Lawson, D.A.; van Eijndhoven, M.A.; Hopmans, E.S.; Lindenberg, J.L.; de Gruijl, T.D.; Wurdinger, T.; Middeldorp, J.M. Functional delivery of viral miRNAs via exosomes. *Proc. Natl. Acad. Sci. USA* **2010**, *107*, 6328–6333. [CrossRef] [PubMed]

43. Gallo, A.; Jang, S.I.; Ong, H.L.; Perez, P.; Tandon, M.; Ambudkar, I.; Illei, G.; Alevizos, I. Targeting the Ca(2+) Sensor STIM1 by Exosomal Transfer of Ebv-miR-BART13-3p is Associated with Sjogren's Syndrome. *EBioMedicine* **2016**, *10*, 216–226. [CrossRef] [PubMed]

44. Gallo, A.; Vella, S.; Miele, M.; Timoneri, F.; Di Bella, M.; Bosi, S.; Sciveres, M.; Conaldi, P.G. Global profiling of viral and cellular non-coding RNAs in Epstein-Barr virus-induced lymphoblastoid cell lines and released exosome cargos. *Cancer Lett.* **2017**, *388*, 334–343. [CrossRef] [PubMed]

45. Li, Z.; Chen, X.; Li, L.; Liu, S.; Yang, L.; Ma, X.; Tang, M.; Bode, A.M.; Dong, Z.; Sun, L.; et al. EBV encoded miR-BHRF1-1 potentiates viral lytic replication by downregulating host p53 in nasopharyngeal carcinoma. *Int. J. Biochem. Cell Biol.* **2012**, *44*, 275–279. [CrossRef] [PubMed]

46. Lu, Y.; Qin, Z.; Wang, J.; Zheng, X.; Lu, J.; Zhang, X.; Wei, L.; Peng, Q.; Zheng, Y.; Ou, C.; et al. Epstein-Barr Virus miR-BART6-3p Inhibits the RIG-I Pathway. *J. Innate Immun.* **2017**, *9*, 574–586. [CrossRef]

47. Zhang, Y.M.; Yu, Y.; Zhao, H.P. EBVBART63p and cellular microRNA197 compromise the immune defense of host cells in EBVpositive Burkitt lymphoma. *Mol. Med. Rep.* **2017**, *15*, 1877–1883. [CrossRef]

48. Hooykaas, M.J.G.; van Gent, M.; Soppe, J.A.; Kruse, E.; Boer, I.G.J.; van Leenen, D.; Groot Koerkamp, M.J.A.; Holstege, F.C.P.; Ressing, M.E.; Wiertz, E.; et al. EBV MicroRNA BART16 Suppresses Type I IFN Signaling. *J. Immunol.* **2017**, *198*, 4062–4073. [CrossRef]

49. Skinner, C.M.; Ivanov, N.S.; Barr, S.A.; Chen, Y.; Skalsky, R.L. An Epstein-Barr Virus MicroRNA Blocks Interleukin-1 (IL-1) Signaling by Targeting IL-1 Receptor 1. *J. Virol.* **2017**, *91*, e00530-17. [CrossRef]

50. Haneklaus, M.; Gerlic, M.; Kurowska-Stolarska, M.; Rainey, A.A.; Pich, D.; McInnes, I.B.; Hammerschmidt, W.; O'Neill, L.A.; Masters, S.L. Cutting edge: miR-223 and EBV miR-BART15 regulate the NLRP3 inflammasome and IL-1beta production. *J. Immunol.* **2012**, *189*, 3795–3799. [CrossRef]

51. Tagawa, T.; Albanese, M.; Bouvet, M.; Moosmann, A.; Mautner, J.; Heissmeyer, V.; Zielinski, C.; Lutter, D.; Hoser, J.; Hastreiter, M.; et al. Epstein-Barr viral miRNAs inhibit antiviral CD4+ T cell responses targeting IL-12 and peptide processing. *J. Exp. Med.* **2016**, *213*, 2065–2080. [CrossRef] [PubMed]

52. Skalsky, R.L.; Corcoran, D.L.; Gottwein, E.; Frank, C.L.; Kang, D.; Hafner, M.; Nusbaum, J.D.; Feederle, R.; Delecluse, H.J.; Luftig, M.A.; et al. The viral and cellular microRNA targetome in lymphoblastoid cell lines. *PLoS Pathog.* **2012**, *8*, e1002484. [CrossRef] [PubMed]

53. Xia, T.; O'Hara, A.; Araujo, I.; Barreto, J.; Carvalho, E.; Sapucaia, J.B.; Ramos, J.C.; Luz, E.; Pedroso, C.; Manrique, M.; et al. EBV microRNAs in primary lymphomas and targeting of CXCL-11 by ebv-mir-BHRF1-3. *Cancer Res.* **2008**, *68*, 1436–1442. [CrossRef] [PubMed]

54. Ganem, D. Human herpesvirus 8 and its role in the genesis of Kaposi's sarcoma. *Curr. Clin. Top. Infect. Dis.* **1998**, *18*, 237–251. [PubMed]

55. Rappocciolo, G.; Jenkins, F.J.; Hensler, H.R.; Piazza, P.; Jais, M.; Borowski, L.; Watkins, S.C.; Rinaldo, C.R., Jr. DC-SIGN is a receptor for human herpesvirus 8 on dendritic cells and macrophages. *J. Immunol.* **2006**, *176*, 1741–1749. [CrossRef]

56. Ganem, D. KSHV and the pathogenesis of Kaposi sarcoma: Listening to human biology and medicine. *J. Clin. Investig.* **2010**, *120*, 939–949. [CrossRef]

57. Chang, Y.; Cesarman, E.; Pessin, M.S.; Lee, F.; Culpepper, J.; Knowles, D.M.; Moore, P.S. Identification of herpesvirus-like DNA sequences in AIDS-associated Kaposi's sarcoma. *Science* **1994**, *266*, 1865–1869. [CrossRef]

58. Nador, R.G.; Cesarman, E.; Chadburn, A.; Dawson, D.B.; Ansari, M.Q.; Sald, J.; Knowles, D.M. Primary effusion lymphoma: A distinct clinicopathologic entity associated with the Kaposi's sarcoma-associated herpes virus. *Blood* **1996**, *88*, 645–656. [CrossRef]

59. Soulier, J.; Grollet, L.; Oksenhendler, E.; Cacoub, P.; Cazals-Hatem, D.; Babinet, P.; d'Agay, M.F.; Clauvel, J.P.; Raphael, M.; Degos, L.; et al. Kaposi's sarcoma-associated herpesvirus-like DNA sequences in multicentric Castleman's disease. *Blood* **1995**, *86*, 1276–1280. [CrossRef]

60. Arias, C.; Weisburd, B.; Stern-Ginossar, N.; Mercier, A.; Madrid, A.S.; Bellare, P.; Holdorf, M.; Weissman, J.S.; Ganem, D. KSHV 2.0: A comprehensive annotation of the Kaposi's sarcoma-associated herpesvirus genome using next-generation sequencing reveals novel genomic and functional features. *PLoS Pathog.* **2014**, *10*, e1003847. [CrossRef]

61. Lin, Y.T.; Kincaid, R.P.; Arasappan, D.; Dowd, S.E.; Hunicke-Smith, S.P.; Sullivan, C.S. Small RNA profiling reveals antisense transcription throughout the KSHV genome and novel small RNAs. *RNA* **2010**, *16*, 1540–1558. [CrossRef]

62. Cai, X.; Lu, S.; Zhang, Z.; Gonzalez, C.M.; Damania, B.; Cullen, B.R. Kaposi's sarcoma-associated herpesvirus expresses an array of viral microRNAs in latently infected cells. *Proc. Natl. Acad. Sci. USA* **2005**, *102*, 5570–5575. [CrossRef] [PubMed]

63. Piedade, D.; Azevedo-Pereira, J.M. The Role of microRNAs in the Pathogenesis of Herpesvirus Infection. *Viruses* **2016**, *8*, 156. [CrossRef] [PubMed]

64. Suffert, G.; Malterer, G.; Hausser, J.; Viiliainen, J.; Fender, A.; Contrant, M.; Ivacevic, T.; Benes, V.; Gros, F.; Voinnet, O.; et al. Kaposi's sarcoma herpesvirus microRNAs target caspase 3 and regulate apoptosis. *PLoS Pathog.* **2011**, *7*, e1002405. [CrossRef] [PubMed]

65. Chen, M.; Sun, F.; Han, L.; Qu, Z. Kaposi's sarcoma herpesvirus (KSHV) microRNA K12-1 functions as an oncogene by activating NF-kappaB/IL-6/STAT3 signaling. *Oncotarget* **2016**, *7*, 33363–33373. [PubMed]

66. Hu, M.; Wang, C.; Li, W.; Lu, W.; Bai, Z.; Qin, D.; Yan, Q.; Zhu, J.; Krueger, B.J.; Renne, R.; et al. A KSHV microRNA Directly Targets G Protein-Coupled Receptor Kinase 2 to Promote the Migration and Invasion of Endothelial Cells by Inducing CXCR2 and Activating AKT Signaling. *PLoS Pathog.* **2015**, *11*, e1005171. [CrossRef] [PubMed]

67. Zhang, J.; Pu, X.M.; Xiong, Y. kshv-mir-k12-1-5p promotes cell growth and metastasis by targeting SOCS6 in Kaposi's sarcoma cells. *Cancer Manag. Res.* **2019**, *11*, 4985–4995. [CrossRef] [PubMed]

68. Li, T.; Ju, E.; Gao, S.J. Kaposi sarcoma-associated herpesvirus miRNAs suppress CASTOR1-mediated mTORC1 inhibition to promote tumorigenesis. *J. Clin. Investig.* **2019**, *129*, 3310–3323. [CrossRef]

69. Abend, J.R.; Uldrick, T.; Ziegelbauer, J.M. Regulation of tumor necrosis factor-like weak inducer of apoptosis receptor protein (TWEAKR) expression by Kaposi's sarcoma-associated herpesvirus microRNA prevents TWEAK-induced apoptosis and inflammatory cytokine expression. *J. Virol.* **2010**, *84*, 12139–12151. [CrossRef]

70. Lei, X.; Zhu, Y.; Jones, T.; Bai, Z.; Huang, Y.; Gao, S.J. A Kaposi's sarcoma-associated herpesvirus microRNA and its variants target the transforming growth factor beta pathway to promote cell survival. *J. Virol.* **2012**, *86*, 11698–11711. [CrossRef]

71. Skalsky, R.L.; Samols, M.A.; Plaisance, K.B.; Boss, I.W.; Riva, A.; Lopez, M.C.; Baker, H.V.; Renne, R. Kaposi's sarcoma-associated herpesvirus encodes an ortholog of miR-155. *J. Virol.* **2007**, *81*, 12836–12845. [CrossRef]

72. Qin, J.; Li, W.; Gao, S.J.; Lu, C. KSHV microRNAs: Tricks of the Devil. *Trends Microbiol.* **2017**, *25*, 648–661. [CrossRef]

73. Abend, J.R.; Ramalingam, D.; Kieffer-Kwon, P.; Uldrick, T.S.; Yarchoan, R.; Ziegelbauer, J.M. Kaposi's sarcoma-associated herpesvirus microRNAs target IRAK1 and MYD88, two components of the toll-like receptor/interleukin-1R signaling cascade, to reduce inflammatory-cytokine expression. *J. Virol.* **2012**, *86*, 11663–11674. [CrossRef]

74. Liang, D.; Gao, Y.; Lin, X.; He, Z.; Zhao, Q.; Deng, Q.; Lan, K. A human herpesvirus miRNA attenuates interferon signaling and contributes to maintenance of viral latency by targeting IKKepsilon. *Cell Res.* **2011**, *21*, 793–806. [CrossRef] [PubMed]

75. Qin, Z.; Kearney, P.; Plaisance, K.; Parsons, C.H. Pivotal advance: Kaposi's sarcoma-associated herpesvirus (KSHV)-encoded microRNA specifically induce IL-6 and IL-10 secretion by macrophages and monocytes. *J. Leukoc. Biol.* **2010**, *87*, 25–34. [CrossRef] [PubMed]

76. Naqvi, A.R.; Shango, J.; Seal, A.; Shukla, D.; Nares, S. Viral miRNAs Alter Host Cell miRNA Profiles and Modulate Innate Immune Responses. *Front. Immunol.* **2018**, *9*, 433. [CrossRef] [PubMed]

77. Bravo, I.G.; Felez-Sanchez, M. Papillomaviruses: Viral evolution, cancer and evolutionary medicine. *Evol. Med. Public Health* **2015**, *2015*, 32–51. [CrossRef]

78. Bzhalava, D.; Eklund, C.; Dillner, J. International standardization and classification of human papillomavirus types. *Virology* **2015**, *476*, 341–344. [CrossRef]

79. Buck, C.B.; Trus, B.L. The papillomavirus virion: A machine built to hide molecular Achilles' heels. *Adv. Exp. Med. Biol.* **2012**, *726*, 403–422.

80. Graham, S.V.; Faizo, A.A.A. Control of human papillomavirus gene expression by alternative splicing. *Virus Res.* **2017**, *231*, 83–95. [CrossRef]

81. Zur Hausen, H. Papillomaviruses in the causation of human cancers—A brief historical account. *Virology* **2009**, *384*, 260–265. [CrossRef]

82. Mirghani, H.; Amen, F.; Moreau, F.; Lacau St Guily, J. Do high-risk human papillomaviruses cause oral cavity squamous cell carcinoma? *Oral Oncol.* **2015**, *51*, 229–236. [CrossRef] [PubMed]

83. Chirayil, R.; Kincaid, R.P.; Dahlke, C.; Kuny, C.V.; Dalken, N.; Spohn, M.; Lawson, B.; Grundhoff, A.; Sullivan, C.S. Identification of virus-encoded microRNAs in divergent Papillomaviruses. *PLoS Pathog.* **2018**, *14*, e1007156. [CrossRef] [PubMed]

84. Virtanen, E.; Pietila, T.; Nieminen, P.; Qian, K.; Auvinen, E. Low expression levels of putative HPV encoded microRNAs in cervical samples. *SpringerPlus* **2016**, *5*, 1856. [CrossRef] [PubMed]

85. Qian, K.; Pietila, T.; Ronty, M.; Michon, F.; Frilander, M.J.; Ritari, J.; Tarkkanen, J.; Paulin, L.; Auvinen, P.; Auvinen, E. Identification and validation of human papillomavirus encoded microRNAs. *PLoS ONE* **2013**, *8*, e70202. [CrossRef] [PubMed]

86. Bernard, H.U.; Apt, D. Transcriptional control and cell type specificity of HPV gene expression. *Arch. Dermatol.* **1994**, *130*, 210–215. [CrossRef]

87. Ishiji, T.; Lace, M.J.; Parkkinen, S.; Anderson, R.D.; Haugen, T.H.; Cripe, T.P.; Xiao, J.H.; Davidson, I.; Chambon, P.; Turek, L.P. Transcriptional enhancer factor (TEF)-1 and its cell-specific co-activator activate human papillomavirus-16 E6 and E7 oncogene transcription in keratinocytes and cervical carcinoma cells. *EMBO J.* **1992**, *11*, 2271–2281. [CrossRef]

88. Martinez-Zapien, D.; Ruiz, F.X.; Poirson, J.; Mitschler, A.; Ramirez, J.; Forster, A.; Cousido-Siah, A.; Masson, M.; Vande Pol, S.; Podjarny, A.; et al. Structure of the E6/E6AP/p53 complex required for HPV-mediated degradation of p53. *Nature* **2016**, *529*, 541–545. [CrossRef]

89. Ganguly, N.; Parihar, S.P. Human papillomavirus E6 and E7 oncoproteins as risk factors for tumorigenesis. *J. Biosci.* **2009**, *34*, 113–123. [CrossRef]

90. Gomez-Gomez, Y.; Organista-Nava, J.; Gariglio, P. Deregulation of the miRNAs expression in cervical cancer: Human papillomavirus implications. *Biomed. Res. Int.* **2013**, *2013*, 407052. [CrossRef]

91. Zheng, Z.M.; Wang, X. Regulation of cellular miRNA expression by human papillomaviruses. *Biochim. Biophys. Acta* **2011**, *1809*, 668–677. [CrossRef]

92. Paiva, I.; da Costa, R.M.G.; Ribeiro, J.; Sousa, H.; Bastos, M.M.; Faustino-Rocha, A.; Lopes, C.; Oliveira, P.A.; Medeiros, R. MicroRNA-21 expression and susceptibility to HPV-induced carcinogenesis—Role of microenvironment in K14-HPV16 mice model. *Life Sci.* **2015**, *128*, 8–14. [CrossRef] [PubMed]

93. Paiva, I.; da Costa, R.M.G.; Ribeiro, J.; Sousa, H.; Bastos, M.; Faustino-Rocha, A.; Lopes, C.; Oliveira, P.A.; Medeiros, R. A role for microRNA-155 expression in microenvironment associated to HPV-induced carcinogenesis in K14-HPV16 transgenic mice. *PLoS ONE* **2015**, *10*, e0116868. [CrossRef] [PubMed]

94. Krump, N.A.; You, J. Molecular mechanisms of viral oncogenesis in humans. *Nat. Rev. Microbiol.* **2018**, *16*, 684–698. [CrossRef] [PubMed]

95. Jemal, A.; Bray, F.; Center, M.M.; Ferlay, J.; Ward, E.; Forman, D. Global cancer statistics. *CA A Cancer J. Clin.* **2011**, *61*, 69–90. [CrossRef] [PubMed]

96. Pollicino, T.; Saitta, C. Occult hepatitis B virus and hepatocellular carcinoma. *World J. Gastroenterol.* **2014**, *20*, 5951–5961. [CrossRef] [PubMed]

97. Yao, L.; Zhou, Y.; Sui, Z.; Zhang, Y.; Liu, Y.; Xie, H.; Gao, H.; Fan, H.; Zhang, Y.; Liu, M.; et al. HBV-encoded miR-2 functions as an oncogene by downregulating TRIM35 but upregulating RAN in liver cancer cells. *EBioMedicine* **2019**, *48*, 117–129. [CrossRef] [PubMed]

98. Zaoui, K.; Boudhraa, Z.; Khalife, P.; Carmona, E.; Provencher, D.; Mes-Masson, A.M. Ran promotes membrane targeting and stabilization of RhoA to orchestrate ovarian cancer cell invasion. *Nat. Commun.* **2019**, *10*, 2666. [CrossRef]

99. Yang, X.; Li, H.; Sun, H.; Fan, H.; Hu, Y.; Liu, M.; Li, X.; Tang, H. Hepatitis B Virus-Encoded MicroRNA Controls Viral Replication. *J. Virol.* **2017**, *91*, e01919-16. [CrossRef]

100. Choo, Q.L.; Kuo, G.; Weiner, A.J.; Overby, L.R.; Bradley, D.W.; Houghton, M. Isolation of a cDNA clone derived from a blood-borne non-A, non-B viral hepatitis genome. *Science* **1989**, *244*, 359–362. [CrossRef]

101. Russelli, G.; Pizzillo, P.; Iannolo, G.; Barbera, F.; Tuzzolino, F.; Liotta, R.; Traina, M.; Vizzini, G.; Gridelli, B.; Badami, E.; et al. HCV replication in gastrointestinal mucosa: Potential extra-hepatic viral reservoir and possible role in HCV infection recurrence after liver transplantation. *PLoS ONE* **2017**, *12*, e0181683. [CrossRef]

102. Machida, K.; Cheng, K.T.; Sung, V.M.; Shimodaira, S.; Lindsay, K.L.; Levine, A.M.; Lai, M.Y.; Lai, M.M. Hepatitis C virus induces a mutator phenotype: Enhanced mutations of immunoglobulin and protooncogenes. *Proc. Natl. Acad. Sci. USA* **2004**, *101*, 4262–4267. [CrossRef] [PubMed]

103. Dionisio, N.; Garcia-Mediavilla, M.V.; Sanchez-Campos, S.; Majano, P.L.; Benedicto, I.; Rosado, J.A.; Salido, G.M.; Gonzalez-Gallego, J. Hepatitis C virus NS5A and core proteins induce oxidative stress-mediated calcium signalling alterations in hepatocytes. *J. Hepatol.* **2009**, *50*, 872–882. [CrossRef] [PubMed]

104. Jopling, C.L. Targeting microRNA-122 to Treat Hepatitis C Virus Infection. *Viruses* **2010**, *2*, 1382–1393. [CrossRef] [PubMed]

105. Ottosen, S.; Parsley, T.B.; Yang, L.; Zeh, K.; van Doorn, L.J.; van der Veer, E.; Raney, A.K.; Hodges, M.R.; Patick, A.K. In vitro antiviral activity and preclinical and clinical resistance profile of miravirsen, a novel anti-hepatitis C virus therapeutic targeting the human factor miR-122. *Antimicrob. Agents Chemother.* **2015**, *59*, 599–608. [CrossRef] [PubMed]

106. Van der Ree, M.H.; de Vree, J.M.; Stelma, F.; Willemse, S.; van der Valk, M.; Rietdijk, S.; Molenkamp, R.; Schinkel, J.; van Nuenen, A.C.; Beuers, U.; et al. Safety, tolerability, and antiviral effect of RG-101 in patients with chronic hepatitis C: A phase 1B, double-blind, randomised controlled trial. *Lancet* **2017**, *389*, 709–717. [CrossRef]

107. Luna, J.M.; Scheel, T.K.; Danino, T.; Shaw, K.S.; Mele, A.; Fak, J.J.; Nishiuchi, E.; Takacs, C.N.; Catanese, M.T.; de Jong, Y.P.; et al. Hepatitis C virus RNA functionally sequesters miR-122. *Cell* **2015**, *160*, 1099–1110. [CrossRef]

108. Hsu, S.H.; Wang, B.; Kota, J.; Yu, J.; Costinean, S.; Kutay, H.; Yu, L.; Bai, S.; La Perle, K.; Chivukula, R.R.; et al. Essential metabolic, anti-inflammatory, and anti-tumorigenic functions of miR-122 in liver. *J. Clin. Investig.* **2012**, *122*, 2871–2883. [CrossRef]

109. Tsai, W.C.; Hsu, S.D.; Hsu, C.S.; Lai, T.C.; Chen, S.J.; Shen, R.; Huang, Y.; Chen, H.C.; Lee, C.H.; Tsai, T.F.; et al. MicroRNA-122 plays a critical role in liver homeostasis and hepatocarcinogenesis. *J. Clin. Investig.* **2012**, *122*, 2884–2897. [CrossRef]

110. Murakami, Y.; Aly, H.H.; Tajima, A.; Inoue, I.; Shimotohno, K. Regulation of the hepatitis C virus genome replication by miR-199a. *J. Hepatol.* **2009**, *50*, 453–460. [CrossRef]

111. Qureshi, A.; Thakur, N.; Monga, I.; Thakur, A.; Kumar, M. Virmirna: A Comprehensive Resource for Experimentally Validated Viral Mirnas and Their Targets. *Database* **2014**, *2014*, bau103. [CrossRef]

112. Buck, C.B.; Van Doorslaer, K.; Peretti, A.; Geoghegan, E.M.; Tisza, M.J.; An, P.; Katz, J.P.; Pipas, J.M.; McBride, A.A.; Camus, A.C.; et al. The Ancient Evolutionary History of Polyomaviruses. *PLoS Pathog.* **2016**, *12*, e1005574. [CrossRef] [PubMed]

113. DeCaprio, J.A. Merkel cell polyomavirus and Merkel cell carcinoma. *Philos. Trans. R. Soc. Lond. Ser. B Biol. Sci.* **2017**, *372*, 20160276. [CrossRef] [PubMed]

114. Clarke, C.A.; Robbins, H.A.; Tatalovich, Z.; Lynch, C.F.; Pawlish, K.S.; Finch, J.L.; Hernandez, B.Y.; Fraumeni, J.F., Jr.; Madeleine, M.M.; Engels, E.A. Risk of merkel cell carcinoma after solid organ transplantation. *J. Natl. Cancer Inst.* **2015**, *107*, dju382. [CrossRef] [PubMed]

115. Lanoy, E.; Engels, E.A. Skin cancers associated with autoimmune conditions among elderly adults. *Br. J. Cancer* **2010**, *103*, 112–114. [CrossRef] [PubMed]

116. Shuda, M.; Feng, H.; Kwun, H.J.; Rosen, S.T.; Gjoerup, O.; Moore, P.S.; Chang, Y. T antigen mutations are a human tumor-specific signature for Merkel cell polyomavirus. *Proc. Natl. Acad. Sci. USA* **2008**, *105*, 16272–16277. [CrossRef] [PubMed]

117. Feng, H.; Shuda, M.; Chang, Y.; Moore, P.S. Clonal integration of a polyomavirus in human Merkel cell carcinoma. *Science* **2008**, *319*, 1096–1100. [CrossRef]

118. Seo, G.J.; Chen, C.J.; Sullivan, C.S. Merkel cell polyomavirus encodes a microRNA with the ability to autoregulate viral gene expression. *Virology* **2009**, *383*, 183–187. [CrossRef]

119. Theiss, J.M.; Gunther, T.; Alawi, M.; Neumann, F.; Tessmer, U.; Fischer, N.; Grundhoff, A. A Comprehensive Analysis of Replicating Merkel Cell Polyomavirus Genomes Delineates the Viral Transcription Program and Suggests a Role for mcv-miR-M1 in Episomal Persistence. *PLoS Pathog.* **2015**, *11*, e1004974. [CrossRef]

120. Akhbari, P.; Tobin, D.; Poterlowicz, K.; Roberts, W.; Boyne, J.R. MCV-miR-M1 Targets the Host-Cell Immune Response Resulting in the Attenuation of Neutrophil Chemotaxis. *J. Investig. Dermatol.* **2018**, *138*, 2343–2354. [CrossRef]

121. Lee, S.; Paulson, K.G.; Murchison, E.P.; Afanasiev, O.K.; Alkan, C.; Leonard, J.H.; Byrd, D.R.; Hannon, G.J.; Nghiem, P. Identification and validation of a novel mature microRNA encoded by the Merkel cell polyomavirus in human Merkel cell carcinomas. *J. Clin. Virol.* **2011**, *52*, 272–275. [CrossRef]

122. Okkenhaug, K.; Bilancio, A.; Farjot, G.; Priddle, H.; Sancho, S.; Peskett, E.; Pearce, W.; Meek, S.E.; Salpekar, A.; Waterfield, M.D.; et al. Impaired B and T cell antigen receptor signaling in p110delta PI 3-kinase mutant mice. *Science* **2002**, *297*, 1031–1034. [CrossRef] [PubMed]

123. Groettrup, M.; Soza, A.; Eggers, M.; Kuehn, L.; Dick, T.P.; Schild, H.; Rammensee, H.G.; Koszinowski, U.H.; Kloetzel, P.M. A role for the proteasome regulator PA28alpha in antigen presentation. *Nature* **1996**, *381*, 166–168. [CrossRef] [PubMed]

124. Murakami, Y.; Chen, L.F.; Sanechika, N.; Kohzaki, H.; Ito, Y. Transcription factor Runx1 recruits the polyomavirus replication origin to replication factories. *J. Cell. Biochem.* **2007**, *100*, 1313–1323. [CrossRef] [PubMed]

125. Tornesello, M.L.; Annunziata, C.; Tornesello, A.L.; Buonaguro, L.; Buonaguro, F.M. Human Oncoviruses and p53 Tumor Suppressor Pathway Deregulation at the Origin of Human Cancers. *Cancers* **2018**, *10*, 213. [CrossRef] [PubMed]

Review

The Role of miRNA for the Treatment of MGMT Unmethylated Glioblastoma Multiforme

Anna Kirstein [1,2], Thomas E. Schmid [1,2] and Stephanie E. Combs [1,2,3,*]

1   Institute of Radiation Medicine (IRM), Department of Radiation Sciences (DRS), Helmholtz Zentrum München, 85764 Neuherberg, Germany; anna.kirstein@helmholtz-muenchen.de (A.K.); thomas.schmid@helmholtz-muenchen.de (T.E.S.)
2   Department of Radiation Oncology, Technical University of Munich (TUM), Klinikum Rechts der Isar, 81675 Munich, Germany
3   Deutsches Konsortium für Translationale Krebsforschung (DKTK), Partner Site Munich, 81675 Munich, Germany
*   Correspondence: Stephanie.Combs@tum.de; Tel.: +49-89-4140-4501

Received: 7 April 2020; Accepted: 26 April 2020; Published: 28 April 2020

Abstract: Glioblastoma multiforme (GBM) is the most common high-grade intracranial tumor in adults. It is characterized by uncontrolled proliferation, diffuse infiltration due to high invasive and migratory capacities, as well as intense resistance to chemo- and radiotherapy. With a five-year survival of less than 3% and an average survival rate of 12 months after diagnosis, GBM has become a focus of current research to urgently develop new therapeutic approaches in order to prolong survival of GBM patients. The methylation status of the promoter region of the O^6-methylguanine–DNA methyltransferase (MGMT) is nowadays routinely analyzed since a methylated promoter region is beneficial for an effective response to temozolomide-based chemotherapy. Furthermore, several miRNAs were identified regulating MGMT expression, apart from promoter methylation, by degrading MGMT mRNA before protein translation. These miRNAs could be a promising innovative treatment approach to enhance Temozolomide (TMZ) sensitivity in MGMT unmethylated patients and to increase progression-free survival as well as long-term survival. In this review, the relevant miRNAs are systematically reviewed.

Keywords: glioblastoma; miRNA; MGMT; survival; radiotherapy; chemotherapy; temozolomide; translational medicine

---

1. Introduction

Cancer is one of the leading causes of death worldwide with 14 million new cases diagnosed and eight million deaths every year. With 256,000 cases per year, tumors of the central nervous system account for 2% of all diagnosed cancers and are therefore the 17th most common cancer [1]. Each year the American Cancer Society compiles a cancer statistic for the USA, estimating the annual cancer incidences and mortalities based upon mortality data from 1930 to 2017 and incidence data from 1975 to 2016 [2]. For 2020, they estimate 23,890 new brain and other nervous system cases with 18,020 deaths [2]. Although the yearly incidence rate with 4-8 cases per 100,000 worldwide is relatively low, mortality rates are significantly high, making it the 12th most frequent cause of cancer-related deaths. The most common primary malignancy of the central nervous system is Glioblastoma multiforme (GBM) [1].

Glioblastoma multiforme is one of the most common and most aggressive primary brain tumors with a five-year survival of less than 3% [3] and an increasing incidence rate [4]. In recent years, the mechanisms explaining the radio- and chemoresistance of glioblastoma have been extensively studied but are still poorly understood. Radiotherapy with concomitant or adjuvant temozolomide-based chemotherapy following surgery has become the standard treatment for GBM [5]. However, short median survival is still observed in patients with an unmethylated promoter region of the O^6-methylguanine–DNA methyltransferase (MGMT) [6]. Despite the minor benefit of additional Temozolomide (TMZ) to unmethylated patients and regardless of the treatment regimen, MGMT promoter methylation status is routinely investigated in all patients after surgery as an independent prognostic biomarker. Therefore, an unmethylated MGMT promoter is an inherent prognostic indicator for poor overall-survival, which demonstrates the urgent need for the identification of new prognostic factors, especially for these patients. For this specific patient group, tailored study concepts have been performed with intensified TMZ or with concepts omitting TMZ but adding novel potentially effective substances such as Vascular Endothelial Growth Factor (VEGF)-inhibitors, integrin-antagonists or other molecular targeted substances. To date, all of these studies were negative and have not offered additional benefit [7–10].

In recent years, circulating microRNAs have been extensively studied as tumor biomarkers to predict therapy outcome and to follow up therapy response. miRNAs are endogenous, single-stranded, non-coding small RNA molecules with a length of about 22 nucleotides [11]. The interaction of miRNAs with the target mRNA leads to the degradation or translational repression of the target mRNA, which ultimately results in the down-regulation of the designated protein. This regulatory network of miRNAs affects many different biological functions and therefore represents a great potential for clinical applications [12]. The up- or down-regulation of miRNAs in tumor cells is deterministic for either a tumor-suppressive or an oncogenic characteristic of the respective miRNA [12].

In this manuscript, the relevant literature investigating the relationship between different miRNAs and glioblastoma was systematically reviewed and the results were analyzed to evaluate the value of different miRNAs in the treatment of GBM.

2. Glioblastoma Multiforme

The 2007 World Health Organization (WHO) classification of tumors of the central nervous system [13] is mainly based on microscopic analyses of hematoxylin and eosin-stained sections, immunohistochemistry of lineage-associated proteins and characterization of ultrastructures. Important characteristics include nuclear atypia, mitotic activity, vascularization, necrosis, pleomorphism and microvascular proliferation [14]. In the updated version from 2016 [15], molecular markers are taken into account proposing a more detailed classification of glioblastoma. GBM is classified as grade IV diffuse astrocytic tumor and is characterized by uncontrolled cellular proliferation, diffuse infiltration and intense resistance to radiotherapy [3]. Due to increasing evidence towards a different origin of primary and secondary GBM [16–19], GBM is now subdivided into isocitrate dehydrogenase (IDH)-wildtype, IDH-mutant and not otherwise specified (NOS) glioblastoma. NOS glioblastoma are either primary or secondary glioblastoma, but a full evaluation of the IDH status is either inconclusive or not performed due to the patient's age [15]. IDH-wildtype or primary glioblastoma develops rapidly within 3-6 months directly from glial progenitor cells and is characterized by diffuse infiltration, extensive necrosis and a unique mutation pattern [1]. EGFR amplification [17], PTEN mutation [18,20,21] and loss of chromosome 10 [17,18,20] are particular features of primary GBM, as well as the older age of the patients. The median age at diagnosis is 62 years with a male-to-female ratio of 1.46 [14,20], and the median overall-survival is 15 months [15]. A total of 90% of all glioblastoma are primary glioblastoma [14,15]. IDH-mutant or secondary glioblastomas, in contrast, develop over several years from low-grade astrocytomas (WHO grade II) and anaplastic astrocytomas (WHO grade III) and feature a different, unique mutation pattern, which was postulated to be the result of a sequential acquisition [1,19,22,23]. This pattern includes a TP53 mutation [1,17,20], LOH on chromosomes 10q and 19q [1,17,18,20,24] as well as deletion of *p16* [20,25] and inactivation of RB [20,23,25]. Median age at diagnosis is 44 years with a median overall-survival of 31 months and a male-to-female ratio of 1.12 [14,15,19,20]. Although there is no universally accepted glioblastoma stem cell marker and there might be several stem cell markers [26], CD133 expression is significantly higher in primary, compared to secondary glioblastoma [27]. This might explain the intense resistance to chemo- and radiotherapy of primary glioblastoma due to the presence of potential glioblastoma stem cells.

2.1. Current Treatment of GBM

Treatment of patients with GBM is always interdisciplinary. For all treatments, the strongest prognostic factors are patient's age, performance score, tumor volume as well as molecular characterization. Imaging information from magnetic resonance imaging (MRI), computer tomography (CT), positron-emission tomography (PET) as well as other functional imaging, such as 5ALA, provide a basis for solid characterization of tumor extension. After imaging diagnosis, surgical resection of the tumor mass is crucial to relieve symptoms such as headache, vision and memory problems as well as nausea [28] and should be performed following the rules of maximal-safe resection. Resection allows for pathological examinations to confirm the diagnosis and to investigate several molecular markers, such as MGMT and IDH status. The diffuse infiltrative characteristic, as well as extensive vascularization into the surrounding healthy tissue, limits the complete resection of GBM and makes recurrence highly possible [3]. Hence, complete surgical resection is almost impossible and, therefore, surgery is followed by radiotherapy, generally concomitant with chemotherapy to eliminate tumor cells in the microenvironment as well.

In the 1970s, BCNU (bis-chloroethylnitrosourea—carmustine) was discovered and since then administered as an alkylating antineoplastic agent as it was shown to penetrate the blood brain barrier (BBB) and to be effective in treating intracranial neoplasms [29]. However, the combination of BCNU and radiotherapy did not significantly enhance median survival [29].

Since 2005, administration of the oral alkylating agent temozolomide (TMZ) presents the standard agent for GBM patients, as it causes only mild side-effects and efficacy has been proven in clinical trials [5]. It is given as a daily dose of 75 mg per m^2 body-surface area for five consecutive days for six weeks [28]. After four weeks, the dose is increased to 150 mg per m^2. Adjuvant, conventional radiotherapy is given in 30 fractions at 2 Gy to a total dose of 60 Gy over a period of six weeks [28]. Alternatively, hyperfractionated radiotherapy is given for 15 days with a total dose of 34 Gy in 3.4 Gy fractions or in 15 daily fractions to a total dose of 10 Gy in 2.6 Gy fractions [28].

After radiochemotherapy with TMZ was introduced, it has been shown that patients with an unmethylated MGMT promoter as well as older patients benefit less from TMZ [30]. However, it has also been shown that even in elderly patients treated with short course radiotherapy concomitant treatment improves outcome [31]. These inconclusive data argue for more accurate discrimination of patient subgroups. A 4-miRNA signature consisting of let-7b-5p, miR-125a-5p, miR-615-5p and let-7a-5p was proposed to assign patients into high- and low-risk groups [32]. Three of the four miRNAs—let-7b-5p, let-7a-5p and miR-125a-5p—are tumor suppressive in GBM and are higher expressed in the low-risk GBM group [32]. Only miR-615-5p does not show a tendency towards a certain expression level in either risk group [32]. This leads to the promising conclusion that this 4-miRNA signature is associated with overall survival of GBM patients. This 4-miRNA could be used to differentiate GBM patients and predict therapy outcome. Still, all possibilities should be evaluated in newly diagnosed as well as recurrent patients, including surgery, radiotherapy and chemotherapy. Again, the extent of surgical resection is crucial [33] and the benefit of radiotherapy for recurrent GBM is evident for resected as well as unresected lesions [34–37].

Recurrence or progression is almost inevitable and is postulated after a median time of 32 to 36 weeks after treatment completion and a final mortality rate close to 100% [38]. This alone describes the urgent need for treatment improvement and the discovery of alternative treatment regimes.

2.2. TMZ and MGMT

Since 2005, the standard treatment of glioblastoma involves early adjuvant chemotherapy with the administration of TMZ [5,39]. TMZ is a prodrug from an imidazotetrazine derivative, which is stable in acidic pH and rapidly hydrolyzes by passing through neutral to basic pH [40–42]. Therefore, it survives the gastric acid enabling an oral administration. Due to the lipophilic character of the prodrug, it is able to penetrate the BBB [41]. Only in the brain, where the pH is around 7, spontaneous ring-opening hydrolysis of the imidazotetrazine leads to the formation of the active alkylating metabolite 3-methyl-(triazen-1-yl) imidazole-4-carboximide (MTIC) intermediate [40]. MTIC, in turn, is unstable at pH values below 7 but stable in an alkaline environment [42]. Further hydrolysis of MTIC forms 5-amino-imidazole-4-carboxamide (AIC) and methyl diazonium ions, which react with nucleophilic sites on the DNA producing methyl adducts [41]. There are several sites for DNA methylation, such as N^7 (70%) and O^6 (5%) of the base guanine as well as the N^3 (9%) site of adenine [41,43]. However, only the relatively rare site of the O^6 position at the base guanine is of importance for the anti-cancer activity of TMZ [41,42] and this site is, therefore, speculated to be mutagenic and cytotoxic [44,45].

During DNA replication, O^6-methylguanine pairs with thymine creating a wobble base pair. This mismatch is repaired by the DNA mismatch repair (MMR) pathway, which involves the recognition of the mismatch via several mismatch recognition complexes [46]. Single-stranded DNA nicks are created in close proximity to the wobble base pair allowing accessibility to the mismatched base thymine, which is digested by the 5′-3′ exonuclease I [46]. Eventually, DNA polymerase δ fills the gap with a new thymine [46,47]. Continuous rounds of thymine deletion and insertion eventually lead to a depletion of deoxythymidine triphosphates (dTTP). Lack of dTTP will result in a lack of DNA synthesis and ultimately causes cell death via DNA double-strand breaks [47].

MGMT or sometimes also called the O^6-alkylguanine-DNA-alkyltransferase is a nuclear protein involved in this mismatch repair pathway [48]. MGMT, therefore, protects not only normal cells from apoptosis but also tumor cells. It removes alkyl groups, preferably methyl groups, from the O^6-methylguanine to counteract the futile circles of thymine deletion and insertion [49]. The removed methyl groups are covalently transferred to a cysteine acceptor residue contained within the active site of MGMT [50]. This results in a conformational change, which leads to degradation of the MGMT protein. As the cysteine site is not regenerated, the reaction is a suicide reaction [50], which makes MGMT a protein and not an enzyme [49]. Hence, the amount of methyl groups that can be removed is limited to the amount of MGMT present in the cell, which is dependent on the MGMT promoter methylation status. So, the absence or presence of MGMT mainly contributes to the chemoresistant character of GBM [48,49].

MGMT, therefore, counteracts the therapeutic efficacy of TMZ and promotes treatment failure. Stupp et al. discovered in their studies from 2000 to 2002 that administration of TMZ starting early in the treatment course and adjuvant to radiotherapy increases median survival to 2.5 months and a resulting survival rate of 27% [5]. This constant treatment regime makes dose escalation possible as well as depletion of MGMT.

In 2005, Hegi et al. published that the promoter methylation status of MGMT is an important prognostic biomarker to predict the TMZ chemotherapy outcome [39]. Overall survival of patients with a methylated MGMT promoter who received radiotherapy plus temozolomide was significantly increased compared to patients with an unmethylated MGMT promoter [39]. MGMT promoter unmethylated patients have no or only little benefit from TMZ adjuvant to radiotherapy, which suggests that other mechanisms play a role to overcome TMZ resistance. Since then, the MGMT promoter methylation status in GBM patients is routinely investigated after surgery to predict which patients would benefit most from TMZ.

Recent studies have shown that MGMT expression does not always correlate with MGMT promoter status and that some individual patients with an unmethylated MGMT promoter show comparable long-term survival [51]. This leads to the assumption that other mechanisms are active in regulating MGMT expression, which includes miRNAs [52]. Therefore, new innovative and personalized treatment options need to be developed, especially for patients with an unmethylated MGMT promoter. Some compounds were already tested or are currently tested in clinical trials for the treatment of unmethylated patients.

2.3. Current Diagnostic and Prognostic Biomarkers for GBM

The most commonly analyzed biomarkers in GBM are currently IDH status, MGMT status, 1p/19q co-deletion and ATRX loss [53]. There are, however, several classes of molecules, proposed to aim as biomarkers for GBM detection, which are found in the blood, cerebrospinal fluid (CSF) and urine.

Proteins are detectable in all kinds of body fluids and can be easily withdrawn from the patient. GBM-specific protein markers include VEGF, angiogenesis-associated proteins, extracellular matrix proteins, matrix metalloproteinases, cell line associated proteins, macrophage migration inhibitory factor (MIF) as well as functionally-related proteins, such as CD44 [53,54]. CD44 was shown as a potential marker for survival outcome and treatment resistance [54]. All these have shown deviating amounts and compositions in patients where tumor progression was observed [53].

Another class used for biomarkers are small molecules, such as lipids and metabolites. Due to their low specificity and small size, they can only be used to verify a diagnosis after other markers were tested positive [53].

Circulating tumor cells (CTCs), which are primary tumor cells circulating in the body via the blood stream, for example, might be important in other cancers apart from GBM [53]. As GBM rarely metastasizes and is described as a cranial-restricted tumor, CTCs might not be found in GBM patients in blood samples [53].

Extracellular vesicles, secreted by the tumor and containing material characteristic of the parental cells, can be found in the serum as well as the CSF. It is known, that GBM secrete exosomes, microvesicles, apoptotic bodies and oncosomes containing the glioma-specific receptor of epidermal growth factor (EGFRvIII), miR-21 as well as mutant IDH1 mRNA [53].

Circulating miRNAs have recently gained attention in research and present promising new biomarkers [55]. They can usually be found in peripheral blood of GBM patients and plasma levels of some miRNAs were already shown to be altered [56]. Some of these circulating miRNAs seem predictive in early diagnosis and helpful during treatment monitoring [55].

2.4. Innovative Treatment Options for MGMT Unmethylated Patients

Apart from TMZ, other compounds and therapeutic candidates have also been discovered and are currently tested for the treatment of unmethylated patients. Most of these compounds aim for radiosensitization [10,57,58] affecting the DNA repair pathway or other related pathways. However, in the following, two therapy alternatives will be presented, which target MGMT for radiosensitization.

2.4.1. O^6-Benzylguanine

O^6-benzylguanine is a guanine analog with antineoplastic activity and has been proposed to serve as a therapeutic agent to improve efficiency of alkylating agents [59]. Since benzyl groups get displaced faster compared to methyl groups, O^6-benzylguanine would serve as an effective agent to inactivate MGMT [60]. O^6-benzylguanine binds to the active site of MGMT, thereby transferring the benzyl moiety to the cysteine residue blocking the active site for methyl groups [47]. Dolan et al. have shown that O^6-benzylguanine enhances the cytotoxicity of alkylating agents, which specifically produce O^6-methylguanine [61]. They observed a direct correlation in vitro between increased effectiveness of methylating agents upon O^6-benzylguanine addition and depletion of MGMT [61]. Furthermore, Dolan et al. have shown in vivo that already low doses of O^6-benzylguanine completely deplete MGMT activity [60]. However, to achieve long-lasting efficiencies, higher doses were required, which exhibited increased acute cytotoxicities, especially to the hematopoietic system. The assumption that due to the already low levels of MGMT in the bone marrow, the toxicity in the bone marrow would not significantly increase should later be proven wrong [61].

Quinn et al. reported in a phase I trial [62] and in a phase II trial [63], where TMZ plus O^6-benzylguanine was administered to patients with recurrent, TMZ-resistant glioblastoma, that myelosuppression was most commonly identified. Patients experienced grade 4 neutropenia, grade

4 thrombocytopenia, grade 4 lymphopenia and grade 3 and 4 anemia, which required a TMZ dose reduction in several patients. Although they observed MGMT depletion after O^6-benzylguanine administration in blood samples [62], they did not observe a TMZ sensitization in MGMT unmethylated patients [63]. Therefore, O^6-benzylguanine was not included in the standard therapy of GBM patients.

2.4.2. PARP Inhibitors

The poly(ADP-ribose) polymerase (PARP) family consists of 18 PARP enzymes mainly involved in DNA damage repair and programmed cell death. PARP-1 and PARP-2 are activated upon DNA damages caused, for example, by ionizing radiation or alkylating agents to repair the DNA damage via the base-excision repair (BER) pathway [64]. Both, PARP-1 and PARP-2, were found to increase the antitumor effects of cytotoxic agents and offer treatment options for chemo- and radiosensitization.

PARP-1 binds to the damage on the DNA and generates poly ADP-ribose (PAR) polymers using NAD^+. Further polymers are transferred to histones and chromatin-associated proteins on the DNA [65]. Once the repair enzymes are recruited, PARP-1 is released from the DNA break to give way for XRCC1. XRCC1 assembles the repair enzymes and factors onto the DNA to repair the break. While the DNA is repaired, PARP-1 gets reactivated by the glycohydrolase PARG removing the PARylations [65]. Therefore, PARP-1 enhances cell survival and mediates resistance to radio- and chemotherapy.

PARP inhibitors either inhibit NAD^+ binding and following PARylation or trap PARP, thereby, blocking the damaged site for repair enzyme assembly [65]. Both lead to replicative stress and DNA double-strand breaks [66].

Several PARP inhibitors are currently tested in phase I, II and III clinical trials, including olaparib, iniparib, pamiparib, niraparib, veliparib, and talazoparib. Dungey et al. showed that olaparib increased radiosensitivity of GBM cells in vitro due to collapsed replication forks after radiation treatment [67]. They propose that the radiosensitizing effect occurs due to the replicating cells necessitating a fractionated treatment regimen [67]. Here, the PARP inhibitor does not directly have an effect on MGMT but rather on DNA replication, making it a good example for the radiosensitizing effects of PARP inhibitors.

Veliparib, in contrast, was found an alternative treatment option for MGMT unmethylated GBM patients as a combination of veliparib with irradiation inhibited cell proliferation in MGMT unmethylated primary cell lines as well as increased survival and apoptosis and decreased cell proliferation in vivo [68]. However, a randomized phase I/II study from 2016 combining TMZ and veliparib in recurrent GBM patients did not significantly increase overall survival and progression-free survival [69]. However, the results of a more recent published phase II trial (2019) comparing standard of care to veliparib concomitant to radio- as well as chemotherapy indicate an advantage of veliparib compared to standard of care treatment with an extended six months progression-free survival [70]

PARP inhibitors present a novel, innovative and personalized treatment option for MGMT unmethylated GBM patients; however, clinical trials are currently ongoing and analyses need to be completed before adding PARP inhibitors to the standard treatment of GBM.

3. miRNA

microRNAs (miRNA) are small non-coding RNA molecules consisting of 19-22 nucleotides first described in Caenorhabditis elegans in 1993 [71]. Lee et al. discovered that the *lin-4* gene produces short RNAs that are complementary to the 3'UTR of *lin-14* mRNA and further observed a down-regulation of LIN-14 protein. This led them to the assumption that the direct RNA-RNA interaction between the *lin-4* transcript and the *lin-14* 3'UTR leads to LIN-14 protein down-regulation [71]. Further, they proposed the existence of a class of regulatory genes producing small antisense RNAs influencing gene expression later to be known as microRNAs [71].

In 2001, the word microRNA was first introduced by Lagos-Quintana et al. [72] who could show that many miRNAs are expressed in several species and are highly conserved. The main role of miRNAs is posttranscriptional regulation by sequence-specific repression of mRNAs [72]. To date, more than 2000 miRNAs have been discovered in the human genome [73], which each regulates hundreds of targets including genetic pathways, indicating their role in gene regulation, disease development and also tumorigenesis [74].

3.1. miRNA Biogenesis

miRNAs are initially produced in the nucleus from large hairpin looped RNA precursors by the RNA polymerase II [75,76]. These precursors are termed pri-miRNAs and are processed to pre-miRNAs of varying length [75] by the RNase III enzyme Drosha [77,78] and the double-stranded RNA-binding protein Pasha [79]. Via exportin 5 [80], the pre-miRNAs get exported into the cytoplasm [75], where the RNAse III enzyme Dicer processes it to 22 nucleotides long double-stranded RNAs that form the miRNA: miRNA*duplex. The mature miRNA is unwound and released from Dicer [75] and Argonaut protein 2 (Ago2) [81] mediates the assembly to the multiprotein RNA-induced-silencing complex (miRISC) [81]. Which strand eventually enters the miRISC depends on the internal strand stability [82]. The end of the strand with the lowest stability is likely to be the target of a helicase-like enzyme, which unwinds the duplex [82]. Here, the 5′ end exhibits the lowest internal stability. Perfect complementarity between the mature miRNA and the mRNA target leads to the cleavage of the target mRNA, whereas imperfect complementarity only leads to translational repression [83].

Dysregulation of miRNAs due to gene deletions, amplifications and translocations or defects in the miRNA biogenesis machinery seem to be the mechanisms contributing to the malignant cell types eventually leading to cancer.

3.2. miRNA in Cancer

In 2002, Calin et al. were the first to discover an association between miRNA dysregulation and cancer: a deletion on chromosome 13q14 coding for the *miR15* and *miR16* genes was observed in more than half of the B-cell chronic lymphocytic leukemia (CLL) and deletions or down-regulations of miR-15 and miR-16 were observed in 68% of the B-cell chronic lymphocytic leukemia [84].

Further, in 2004, they published that miRNAs are either tumor suppressive or oncogenic depending on their location; located at regions of loss of heterozygosity suggests tumor suppressors, while located at regions of amplifications suggests oncogenes [85]. In their genome-wide examination, they discovered an association between miRNA location and cancer. miRNAs are commonly found at cancer-associated regions, in which loss of heterozygosity regions may contain tumor suppressor genes and amplifications harbor oncogenes or the other way around [85]. An example for tumor suppressive miRNAs are miR-15 and miR-16, as their absence due to deletions on chromosome 13 leads to CLL. Further, it was shown by Cimmino et al. that the deletion of miR-15 and miR-16 leads to increased expression of Bcl-2 resulting in the formation of leukemias and lymphomas [86]. From this discovery, they proposed tumor suppressive miRNAs as inhibitors of their oncogenic targets in cancer therapy. Another mechanism for dysregulation of miRNAs in cancer apart from deletions and amplifications is the control of the transcription factors. The dysregulation of transcription factors regulating, for example, cell cycle progression, apoptosis, autophagy, invasion, and neoangiogenesis is tightly linked to cancer development. A key regulator of cell cycle progression and a commonly known tumor suppressor gene is p53. Mutation of p53 is frequently found in many cancers and its interaction with miRNAs suggests tumor suppressive features. Yamakuchi and Lowenstein discovered that miR-34a expression is induced by p53, which in turn suppresses p53, negatively regulating SIRT1 to induce apoptosis [87].

Therefore, miRNAs play an important role in tumor development, progression and recurrence. However, miRNAs also represent an innovative treatment option as prognostic and diagnostic biomarkers as well as therapeutic targets in cancer therapy [11,88,89].

The most common upregulated miRNA in many cancers is miR-21. miR-21 is an oncogenic miRNA inhibiting key regulator of apoptotic genes [90]. It was first found to be significantly upregulated in human glioblastoma and its inhibition leads to increased caspase activation followed by apoptotic cell death [90]. Therefore, miR-21 is an example of an oncogenic miRNA, in which upregulation is associated with cancerogenesis. Various bioinformatics and experimental studies have tried to identify a set of de-regulated miRNAs in glioblastoma that are responsible for this tumor.

3.3. miRNA in GBM

The dysregulation of miRNAs in several cancers was shown to contribute to cancer development and progression. These miRNAs, their targets, prognostic and diagnostic value, as well as their potential in the treatment of GBM, need to be identified. Table 1 gives an overview of some miRNAs already discovered in GBM, their targets (if known) and their prognostic value (if available). This table gives a small insight into some of the most important miRNAs in GBM and, by far, does not include all up to date identified miRNA dysregulated in GBM.

Table 1. miRNAs in Glioblastoma multiforme, their targets, function, and prognostic value (↓ = decreased, ↑ = increased)

microRNA	Regulation	Type	Target	Function	Prognosis	Ref.
miR-10b	up	oncogenic	uPAR, RhoC	↑invasion		[91]
miR-7	down	tumor suppressor	EGFR	↑apoptosis, ↓cell proliferation, ↓migration, ↓invasion		[92,93]
miR-17	up	oncogenic	DFFA, PI2KCA, E2F3m VEGFA, ATG7	↑autophagy		[94,95]
miR-21	up	oncogenic	HNRPK, TAp63, PTEN, EGFR, E2F3, PDCD4, WNT5A	↓apoptosis, ↓autophagy, ↑invasion		[88,93–96]
miR-26a	up	oncogenic	PTEN	↑tumor growth ↑angiogenesis	high level = longer OS and PFS with carmustine ↑TMZ resistance	[97–99]
miR-34a	down	tumor suppressor	E2F3, PI2KCA, EGFR, DFFA, CSL2, BAX, c-Met, Notch	↑cell cycle arrest, ↓invasion ↓migration ↓cell proliferation		[95,100]
miR-92b-3p	up	oncogenic	PTEN	↑migration, ↑invasion ↓apoptosis	low level = shorter OS	[101,102]
miR-124	down	tumor suppressor	CDK6	↓cell cycle progression		[103]
miR-125b	up	oncogenic	p53, p38MAPK, Bmf	↑proliferation, ↑cell cycle progression,	high level = higher grade	[104,105]
miR-128	down	tumor suppressor	RTKs, EGFR, PDGF-R, E2F3a	↓proliferation, ↑differentiation, ↓migration		[93,106,107]
miR-130a	up	tumor suppressor	E2F8	ROS production	high level = extended survival without progression predictor for TMZ response	[108,109]
miR-137	down	tumor suppressor	CDK6	↓cell cycle progression		[103]

Table 1. *Cont.*

microRNA	Regulation	Type	Target	Function	Prognosis	Ref.
miR-142-3p	down	tumor suppressor	IL-6, HMGA2	↓cell viability	high levels = low MGMT low levels = high MGMT	[110,111]
miR-155	up	oncogenic	FOXO3a	↑proliferation, ↑migration, ↑invasion	low level = long OS	[108,112]
miR-181a	down	tumor suppressor	Bcl-2	↑apoptosis		[113]
miR-181b	down	tumor suppressor	SALL4	↓proliferation, ↓migration, ↓invasion		[114]
miR-181d	down	tumor suppressor	MGMT, Bcl-2, KRAS	↓proliferation, ↓cell cycle progression, ↑apoptosis	high level = improved OS	[115,116]
miR-210	up	oncogenic	SIN3A	↑proliferation, ↓apoptosis	low level = long OS	[108,117]
miR-218	down	tumor suppressor	IKK-ß, LEF1, Bmi1	↓invasion, ↓migration, ↓proliferation, ↑apoptosis		[93,118–120]
miR-221/222	up	oncogenic	p27, AKT, TIMP-3, PTEN, E2F3	↑proliferation, ↑invasion	up in short-term, down in long-term survivors, ↑TMZ sensitivity ↓↓long-term survival	[121–123]
miR-326	down	tumor suppressor	WNT5A, TOM34		high level = extended survival without progression	[95,108]
miR-335	up	oncogenic	DAAM1, PAX6	↑proliferation, ↑invasion		[124,125]
miR-339	up	oncogenic		↑migration, ↑invasion ↓apoptosis		[101]
miR-370-3p	down and up	tumor suppressor	ß-catenin, FOXM1	↓cell proliferation ↓cell cycle progression	upregulation = inhibition of GBM growth long upregulation = longer survival	[126,127]
miR-409	down	tumor suppressor	HMGN5, cyclin D1, MMP2	↑invasion, ↑proliferation		[128]
miR-451	down	tumor suppressor	Cyclin D1, p27, Bcl-2, MMP-2, MMP-9	↓cell cycle ↓cell growth ↑apoptosis		[129]
miR-603	up	oncogenic	WIF1, CTNNBIP1	↓cell growth ↑proliferation, ↑cell cycle progression		[130]

3.4. miRNAs Targeting MGMT

Since the discovery of the importance of the MGMT promoter methylation status in GBM therapy outcome [39], it is now known that the promoter methylation is not the only deterministic factor for MGMT protein expression. In 2013, Kreth et al. discovered the presence of two MGMT transcripts, which are both expressed in GBM [52]. In normal brain tissue, only the shorter transcript with a size of 440 bp is found, which contains a canonical poly(A) signal as well as a 3'UTR of 105 nt. The longer transcript of about 850 bp contains an alternative poly(A) signal of 522 nt and is found only in GBM. In patient samples, they discovered that the length of the transcript is associated with high or low MGMT expression; high MGMT expression correlated with the normal 3'UTR length, whereas reduced MGMT expression levels were associated with the MGMT transcript containing the elongated 3'UTR [52]. Analysis of potential miRNA binding sites revealed 29 miRNAs specific for the long 3'UTR and only two for the short 3' UTR; miR-181d was found in both. This led to the conclusion that longer UTRs render transcripts more accessibility to miRNA targets.

First, an in silico analysis of miRNAs targeting MGMT using the TarBase v.8 online tool (DIANA-LAB, Biomedical Sciences Research Center Alexander Fleming, Vari, Greece) [131] was done. The 43 found miRNAs are present in Table 2 below.

Table 2. miRNAs targeting O^6-methylguanine–DNA methyltransferase (MGMT) derived from an in silico analysis.

microRNAs	
let-7a-2-3p	miR-342-3p
let-7f-2-3p	miR-361-5p
let-7i-3p	miR-3619-5p
miR-1-3p	miR-370-3p
miR-16-5p	miR-371a-3p
miR-17-5p	miR-374a-5p
miR-20a-3p	miR-379-5p
miR-21-3p	miR-423-3p
miR-27a-3p	miR-429
miR-27a-5p	miR-497-3p
miR-30d-5p	miR-548a-3p
miR-30e-3p	miR-561-3p
miR-155-5p	miR-589-3p
miR-181b-5p	miR-603
miR-181d-5p	miR-612
miR-183-5p	miR-616-3p
miR-184	miR-648
miR-191-5p	miR-651-5p
miR-194-5p	miR-661
miR-324-5p	miR-767-3p
miR-325	miR-2115-5p
miR-338-5p	

Table 3 gives an overview of those miRNAs regulating MGMT expression, which were identified experimentally and which exhibit significant effects in cell lines. Detailed descriptions follow in the sections below.

Table 3. miRNAs involved in MGMT regulation in Glioblastoma multiforme.

microRNA	Regulation	Type	Prognosis	Ref.
miR-142-3p	down	tumor suppressor	suppression of MGMT protein ↑TMZ sensitivity	[110,111]
miR-181d	down	tumor suppressor	degradation of MGMT mRNA; high level = improved OS	[116]
miR-221/222	up	oncogenic	suppression of MGMT; ↑TMZ sensitivity	[123]
miR-370-3p	down and up	tumor suppressor	regulatory effects on MGMT; ↑TMZ sensitivity	[127,132]
miR-409-3p	up	oncogenic	repression of MGMT	[133]
miR-603	up	oncogenic	suppression of MGMT ↑TMZ sensitivity	[134]
miR-648	up	tumor suppressor	inhibition of MGMT protein translation	[135]
miR-767-3p	up	tumor suppressor	degradation of MGMT mRNA	[135]

3.4.1. miR-142-3p

Lee et al. determined an inverse correlation between MGMT and miR-142-3p expression levels in GBM cell lines: high MGMT expressing cell lines show low levels of miR-142-3p and low MGMT expressing cell lines show high levels of miR-142-3p [110]. In miR-142-3p overexpression experiments, no change in MGMT mRNA expression was observed, but a reduction in MGMT protein expression. This leads to the assumption that miR-142-3p directly interacts with the 3′UTR of MGMT, which was further proven by luciferase reporter assay experiments [110]. Additionally, an increased sensitivity towards alkylating agents was determined using TMZ and BCNU in miR-142-3p overexpressing cell lines with a stronger effect when BCNU was added [110].

Previously, the same workgroup reported that miR-142-3p is suppressed by the oncogenic cytokine IL6 promoting GBM propagation, suggesting that miR-142-3p is a tumor suppressive miRNA [111].

Taken together, miR-142-3p regulates MGMT protein expression and sensitizes cells in vitro to alkylating agents, which might indicate a potential biomarker for individual GBM treatment [111].

3.4.2. miR-181d

Zhang et al. were the first to identify a miRNA regulating MGMT. miR-181d post-transcriptionally regulates MGMT by direct interaction with the long 3′UTR MGMT transcript [116,135]. In vitro experiments could show that transfection with miR-181d significantly downregulated MGMT mRNA as well as MGMT protein expression and sensitized cells to TMZ [116]. Further analysis of glioblastoma patient samples indicated that miR-181d is usually down-regulated and that transfection with a mimic in vitro inhibits cell proliferation by targeting K-ras, promotes G1 cell cycle arrest and induces apoptosis by targeting Bcl-2 [115]. Evaluation of clinical data also revealed that a higher miR-181d expression was associated with improved overall survival [116] and that miR-181d expression levels increased after either TMZ or irradiation alone and significantly increased after irradiation and TMZ treatment combined [136]. This suggests that miR-181d could act as a predictive biomarker for chemo- and radiotherapy outcomes.

Several studies have investigated the effect of miR-181d and MGMT expression and discovered similar results to Zhang et al. Interaction between miR-181d and other miRNAs, such as miR-409-3p [133], miR-648 and miR-661 [135], have been found to enhance the effect of MGMT down-regulation, suggesting that miR-181d is the key miRNA regulating MGMT expression.

Taken these factors together miR-181d as a tumor suppressive miRNA could be of great use in treating glioblastoma patients to increase sensitivity to TMZ by directly targeting MGMT mRNA [135]. To the best of our knowledge, miR-181d is the only miRNA that regulates MGMT and is associated with overall survival. Up to date, there are no clinical trials ongoing investigating miR-181d as an innovative treatment option.

3.4.3. miR-221/222

miR-221/222 have been extensively studied in various cancers and were shown to be overexpressed in glioblastoma, prostate carcinoma, papillary thyroid carcinoma, hepatocellular cancer and pancreatic cancer [122]. Gillies and Lorimer demonstrated that miR-221/222 are upregulated in human glioblastoma and target p27, a cell cycle regulator [121]. Further targets include the Akt signaling pathway, PTEN, TIMP-3, as well as MMP-2 and MMP-9 [122,137]. In vitro overexpression of miR-221/222 resulted in the induction of p-Akt, MMP-2, and MMP-9 protein expression and hence increased cell proliferation and invasion. These results were confirmed in in vivo overexpression experiments, which also led to increased tumor growth as well as morphological changes towards a malignant phenotype [122].

A binding site of miR-221/222 was found at the 3′UTR of MGMT and further confirmed in in vitro experiments [123]. Overexpression of miR-221/222 reduced MGMT levels in transfected human glioblastoma cell lines and increased the cells' sensitivity to TMZ [123]. It can be concluded that miR-221/222 are oncogenic miRNAs negatively influencing patients' survival, however, increasing sensitivity to TMZ in vitro by directly targeting MGMT.

3.4.4. miR-370-3p

Peng et al. were the first to discover a suppressive potential of miR-370-3p in human glioblastoma [114]. miR-370-3p is significantly down-regulated in low- and high-grade gliomas (Grade II and IV) and also in glioblastoma cell lines. Upon transfection with a miR-370-3p mimic cell viability decreased, long-term proliferation was suppressed as well as the percentage of cells arrested in S and G2/M phase of the cell cycle decreased [114]. A direct post-transcriptional target was found in the 3'UTR of ß-catenin, which is involved in the Wnt signaling pathway promoting cell proliferation and migration [114].

Gao et al. found similar results: in recurrent GBM miR-370-3p expression was significantly decreased compared to normal brain tissue, GBM cell lines showed low levels of miR-370-3p, as well as miR-370-3p-transfected cells showed decreased proliferation [132]. Cell lines expressing the lowest miR-370-3p were more resistant to TMZ compared to cell lines expressing higher levels of miR-370-3p [132]. Additionally, they could demonstrate a negative correlation between MGMT mRNA and miR-370-3p expression.

In 2018, Nadaradjane et al. postulated that miR-370-3p is a biomarker for the prediction of GBM treatment planning and therapy outcome. However, they found out that the expression level of miR-370-3p in the blood of GBM patients varies during standard treatment and is not associated with overall survival [127]. Still, they observed a longer patient survival when miR-370-3p overexpression lasted longer before relapse occurred. In vitro they could show that miR-370-3p overexpression leads to decreased MGMT mRNA and decreased MGMT protein levels. Further, miR-370-3p increased the cells' sensitivity to TMZ indicated by increased cell death after treatment. Subcutaneous tumors grown in mice and treated with a combination of TMZ and miR-370-3p significantly decreased in volume. In the resected tumors, a significant reduction of MGMT expression was observed [127].

Another target of miR-370-3p is FOXM1, which is involved in cell cycle progression. Upon miR-370-3p overexpression, FOXM1 expression reduced as well [127]. Hence, cell cycle progression was inhibited and cell death induced.

It can be concluded that miR-370-3p is a tumor suppressive miRNA in GBM by downregulating the mRNA and protein expression of MGMT as well as FOXM1 expression. miR-370-3p is not deterministic for patients' survival but can be used to sensitize to TMZ especially in MGMT unmethylated patients. However, no clinical trials are currently ongoing investigating miR-370-3p as an innovative treatment option.

3.4.5. miR-409-3p

miR-409-3p was found 5-fold upregulated in human GBM samples compared to healthy brain tissue with an inverse correlation between MGMT and miR-409-3p expression [133]. Patient samples with low MGMT show high miR-409-3p levels, while high MGMT expressing samples show low miR-409-3p levels. In vitro transfection with a miR-409-3p mimic of the high MGMT expressing cell line T98G demonstrated a significant down-regulation of MGMT mRNA as well as MGMT protein. This suggests that miR-409-3p is a strong inhibitor of MGMT by the degradation of MGMT mRNA as well as by translational repression [133]. An even more enhanced effect of MGMT suppression was observed when miR-409-3p mimics were cotransfected with miR-181d mimics [133].

As miR-409-3p was found significantly downregulated in human GBM samples repressing MGMT expression, it can be concluded that miR-409-3p might be a potential therapeutic approach to sensitize MGMT unmethylated patients to alkylating chemotherapeutics. However, other targets of miR-409-3p are still unknown, but Khalil et al. suggested a possible protective role in pro-angiogenic and pro-metastatic processes [133].

3.4.6. miR-603

miR-603 is found upregulated in glioblastoma samples and promotes cell proliferation as well as cell cycle progression [130]. Targets of miR-603 include WIF1 and CTNNBIP1 activating the Wnt/ß-catenin signaling pathway and promoting cell proliferation and migration [130]. Therefore, miR-603 can be considered an oncogenic miRNA.

MGMT is directly suppressed by the interaction of miR-603 with the 3′UTR of MGMT [134]. Transfection with a miR-603 mimic significantly reduced MGMT mRNA levels and protein expression and further enhanced sensitivity to TMZ in vitro as well as in vivo [134]. Kushwaha et al. also showed that the combination of miR-181d and miR-603 most effectively regulated MGMT expression compared to either alone [134].

For innovative treatment options, miR-603 might be a promising candidate to inhibit MGMT and Wnt/ß-catenin signaling pathway activation. No clinical trials have been proposed yet.

3.4.7. miR-648, miR-661 and miR-767-3p

When Kreth et al. discovered the presence of two MGMT isoforms either containing a long or a short UTR, they used target prediction software to determine miRNA with a binding site within the UTRs of MGMT [52]. They assumed that these miRNAs are expressed in human GBM and negatively correlate with MGMT expression. In human GBM samples, six miRNAs (miR-184, miR-183, miR-661, miR-370, miR-767-3p, and miR-648) were found binding exclusively in the long UTR, two (miR-1197 and miR-655) within the short UTR and one (miR-181-d) in both UTRs [135]. Upon cloning both UTRs into a reporter vector containing two luciferases and co-transfection with the miRNAs, they observed that the short UTR-binding miRNAs (miR-181d, miR-665, and miR-1197) did not show regulatory activity [135]. This indicates that the short UTR of MGMT is not regulated by these miRNAs. Five (miR-661, miR-370, miR-181d, miR-767-39, and miR-648) of the seven miRNAs possibly regulating the long UTR showed significant luciferase repression but only three miRNAs (miR-181d, miR-767-3p and miR-648) showed decreased MGMT protein expression. Here, miR-648 exerted the strongest MGMT protein reduction. In qPCR experiments, only two miRNAs (miR-181d and miR-767-3p) significantly reduced MGMT mRNA expression, indicating that those two regulate MGMT expression via direct degradation of the mRNA transcript and miR-648 might act via translational repression [135].

Further, they could show that miR-767-3p and miR-648 are significantly upregulated in human GBM samples and that cotransfection with all three miRNAs (miR-181d, miR-767-3p and miR-648) significantly increased the sensitivity to TMZ treatment [135]. These data correlate with data from Jesionek-Kupnicka et al., who also found an association between MGMT and miR-181d and miR-648 expression [138].

3.5. miRNAs as Innovative Treatment Option for GBM

To the best of our knowledge, no clinical trials are currently ongoing investigating the above-mentioned miRNAs as innovative treatment options for GBM patients nor have any miRNA-based therapies been approved by the FDA. Target specificity and tissue toxicities are major problems in the delivery of miRNA or miRNA inhibitors to their mRNA target.

Several invasive strategies have been postulated to enhance drug delivery across the BBB including intracerebral implants, disruption of the BBB, intra-arterial and intrathecal drug delivery, direct injections into the brain, catheters, pumps or microdialysis [139]. As all of these strategies require invasion into the brain tissue or tumor tissue, there is an increased risk for brain damage and other side effects, including toxicities, indicating the urgent need for non-invasive strategies. Therefore, biological strategies have been developed as innovative tools for drug delivery. These strategies include RNA interference, viral vectors, exosomes, antisense therapy, gene therapy, antibody conjugates, peptide carriers and other carriers [139,140]. Also, chemical systems have been developed, such as lipophilic analogues, prodrugs, efflux transporter inhibition, liposomes, nanoparticles, polymeric micelles and dendrimers [139,140]. Both biological and chemical strategies allow for target specific delivery, as it is most important and challenging at the same time to deliver and internalize the drug or miRNA specifically to the tumor. The challenges of designing nanoparticles are reviewed elsewhere [141].

However, the most limiting factor in delivering these compounds into the brain and promoting restricted bioavailability is the BBB. Major issues are the enzymatic degradation of the miRNA or miRNA inhibitors themselves before the target can be reached as well as the inability of packaging molecules due to high molecular weight and polar functional groups [139]. The BBB is a natural barrier against toxins, harmful substances, and fluctuations in chemical concentrations [139]. It consists of endothelial cells, forming the walls of the capillaries and epithelial cells, creating the blood-cerebrospinal fluid barrier (BCSFB) [142]. The cerebrospinal fluid is secreted into the brain, while the interstitial fluid is secreted by the capillary endothelium [142]. These two fluids can communicate in order to regulate fluctuations and maintain a stable environment [142]. The avascular arachnoid epithelium is the enclosing layer sealing the extracellular fluids from the rest of the body [142]. Physical barriers such as tight junctions, transport barriers such as transporters, and metabolic barriers including enzymes, are found at all interfaces representing the protecting characteristics of the BBB. The most important factor thereby are tight junctions, significantly reducing the trespassing of polar solutes by blocking their penetration [142]. The only routes molecules and solutes can penetrate the BBB are via passive diffusion and ABC transporter efflux (lipid, soluble, non-polar molecules), via solute carriers (e.g., glucose, amino acids, small peptides), via transcytosis or receptor- and adsorptive-mediated (e.g., lipoproteins, insulin, glycosylated proteins, histones) or leukocytes via diapedesis [142].

However, all these strategies need further characterization, experimentation, and clinical trials to safely deliver molecules, miRNAs and other compounds to specific target sites. Up to date, only some miRNAs, including miR-122, -21, -155, -92 and -29 are currently tested in clinical trials as targeted therapy for Hepatitis C (HCV), nephritis, CLL, wounds and fibrosis [143,144]. Only two miRNAs are currently tested for the use in cancer therapy: a miR-16 mimic is involved in a Phase I trial for non-small-cell lung cancer [144] and another clinical trial testing a miR-34a mimic for hepatocellular carcinoma (HCC) has recently been terminated [145]. MRX34, a synthetic, 23 nt long double-stranded RNA encapsulated in a liposomal nanoparticle was administered to patients mainly suffering from HCC. Although pre-clinical studies in non-human primates showed promising results, severe adverse effects and also death of four patients due to the drug forced the phase I trial to be terminated [145]. Severe adverse effects were unlikely due to the liposomal carrier, but rather due to severe immune-related toxicities, which have yet to be resolved [145].

4. Conclusions

In the last decade, miRNAs have become promising tools as prognostic and diagnostic biomarkers as well as therapeutic targets for innovative and personalized cancer treatment [11,89,90]. Several miRNAs have been found differentially expressed and predictive for overall survival, progression-free survival or treatment outcome in several cancer entities. Some miRNAs such as miR-21, the miR-17 cluster and miR-221/222 are dysregulated in several cancer types, but most importantly, cancer type-specific miRNA signatures were also discovered [89,146–149].

With a survival rate of less than 3% [3], Glioblastoma multiforme presents an urgent need for new innovative and personalized treatment options. Patients with a wildtype IDH and an unmethylated MGMT promoter region have the poorest prognosis and the shortest survival [150], identifying these patients with the most urgent need for new treatment options. In this review, we focused on MGMT unmethylated patients and tried to identify possible miRNAs regulating MGMT expression, which could be used for personalized treatment in the future.

We identified eight promising miRNAs—miR-142-3p, -181d, -221/222, -370-3p, -409-3p, -603, -648, and -767-3p—negatively regulating MGMT expression either via mRNA degradation or translational repression. Five of these miRNAs (miR-142-3p, -181d, -221/222, -370-3p and -603) were positively tested to increase sensitivity to alkylating agents such as BCNU and TMZ in vitro as well as in vivo [110]. miR-181d was the only miRNA found predictive for overall survival [116,136].

We present here miRNAs that could help reduce and repress MGMT expression by targeted treatment to sensitize the tumors against alkylating agents. However, target-specific delivery, especially into the brain, represents a challenging task, which has yet to be overcome.

Funding: This research received no external funding.

Acknowledgments: This project was funded in part by the German Consortium for Translational Cancer Research, Munich/TUM site, as well as by the Medical Faculty of TUM.

Conflicts of Interest: The authors declare no conflict of interest.

References

1. Stewart, B.; Wild, C.P. World cancer report 2014. *Public Health* **2019**, *16*, 511–519.
2. Siegel, R.L.; Miller, K.D.; Jemal, A. Cancer statistics, 2020. *CA* **2020**, *70*, 7–30. [CrossRef] [PubMed]
3. Furnari, F.B.; Fenton, T.; Bachoo, R.M.; Mukasa, A.; Stommel, J.M.; Stegh, A.; Hahn, W.C.; Ligon, K.L.; Louis, D.N.; Brennan, C. Malignant astrocytic glioma: Genetics, biology, and paths to treatment. *Genes Dev.* **2007**, *21*, 2683–2710. [CrossRef] [PubMed]
4. Crocetti, E.; Trama, A.; Stiller, C.; Caldarella, A.; Soffietti, R.; Jaal, J.; Weber, D.C.; Ricardi, U.; Slowinski, J.; Brandes, A. Epidemiology of glial and non-glial brain tumours in Europe. *Eur. J. Cancer* **2012**, *48*, 1532–1542. [CrossRef] [PubMed]
5. Stupp, R.; Mason, W.P.; Van Den Bent, M.J.; Weller, M.; Fisher, B.; Taphoorn, M.J.; Belanger, K.; Brandes, A.A.; Marosi, C.; Bogdahn, U. Radiotherapy plus concomitant and adjuvant temozolomide for glioblastoma. *N. Engl. J. Med.* **2005**, *352*, 987–996. [CrossRef] [PubMed]
6. Stupp, R.; Hegi, M.E.; Mason, W.P.; Van Den Bent, M.J.; Taphoorn, M.J.; Janzer, R.C.; Ludwin, S.K.; Allgeier, A.; Fisher, B.; Belanger, K. Effects of radiotherapy with concomitant and adjuvant temozolomide versus radiotherapy alone on survival in glioblastoma in a randomised phase III study: 5-year analysis of the EORTC-NCIC trial. *Lancet Oncol.* **2009**, *10*, 459–466. [CrossRef]
7. Wick, W.; Gorlia, T.; Bady, P.; Platten, M.; Van Den Bent, M.J.; Taphoorn, M.J.; Steuve, J.; Brandes, A.A.; Hamou, M.-F.; Wick, A. Phase II study of radiotherapy and temsirolimus versus radiochemotherapy with temozolomide in patients with newly diagnosed glioblastoma without MGMT promoter hypermethylation (EORTC 26082). *Clin. Cancer Res.* **2016**, *22*, 4797–4806. [CrossRef]
8. Weiler, M.; Hartmann, C.; Wiewrodt, D.; Herrlinger, U.; Gorlia, T.; Bähr, O.; Meyermann, R.; Bamberg, M.; Tatagiba, M.; von Deimling, A. Chemoradiotherapy of newly diagnosed glioblastoma with intensified temozolomide. *Int. J. Radiat. Oncol. Biol. Phys.* **2010**, *77*, 670–676. [CrossRef]
9. Khasraw, M.; Lee, A.; McCowatt, S.; Kerestes, Z.; Buyse, M.E.; Back, M.; Kichenadasse, G.; Ackland, S.; Wheeler, H. Cilengitide with metronomic temozolomide, procarbazine, and standard radiotherapy in patients with glioblastoma and unmethylated MGMT gene promoter in ExCentric, an open-label phase II trial. *J. Neuro-Oncol.* **2016**, *128*, 163–171. [CrossRef]
10. Nabors, L.B.; Fink, K.L.; Mikkelsen, T.; Grujicic, D.; Tarnawski, R.; Nam, D.H.; Mazurkiewicz, M.; Salacz, M.; Ashby, L.; Zagonel, V. Two cilengitide regimens in combination with standard treatment for patients with newly diagnosed glioblastoma and unmethylated MGMT gene promoter: Results of the open-label, controlled, randomized phase II CORE study. *Neuro-Oncol.* **2015**, *17*, 708–717. [CrossRef]
11. Esquela-Kerscher, A.; Slack, F.J. Oncomirs—microRNAs with a role in cancer. *Nat. Rev. Cancer* **2006**, *6*, 259–269. [CrossRef] [PubMed]
12. Si, W.; Shen, J.; Zheng, H.; Fan, W. The role and mechanisms of action of microRNAs in cancer drug resistance. *Clin. Epigenetics* **2019**, *11*, 25. [CrossRef] [PubMed]
13. Louis, D.N.; Ohgaki, H.; Wiestler, O.D.; Cavenee, W.K.; Burger, P.C.; Jouvet, A.; Scheithauer, B.W.; Kleihues, P. The 2007 WHO classification of tumours of the central nervous system. *Acta Neuropathol.* **2007**, *114*, 97–109. [CrossRef] [PubMed]
14. Ohgaki, H.; Kleihues, P. The definition of primary and secondary glioblastoma. *Clin. Cancer Res.* **2013**, *19*, 764–772. [CrossRef]

15. Louis, D.N.; Perry, A.; Reifenberger, G.; Von Deimling, A.; Figarella-Branger, D.; Cavenee, W.K.; Ohgaki, H.; Wiestler, O.D.; Kleihues, P.; Ellison, D.W. The 2016 World Health Organization classification of tumors of the central nervous system: A summary. *Acta Neuropathol.* **2016**, *131*, 803–820. [CrossRef]

16. Scherer, H. Cerebral astrocytomas and their derivatives. *Am. J. Cancer* **1940**, *40*, 159–198.

17. Watanabe, K.; Tachibana, O.; Sato, K.; Yonekawa, Y.; Kleihues, P.; Ohgaki, H. Overexpression of the EGF receptor and p53 mutations are mutually exclusive in the evolution of primary and secondary glioblastomas. *Brain Pathol.* **1996**, *6*, 217–223. [CrossRef]

18. Fujisawa, H.; Reis, R.M.; Nakamura, M.; Colella, S.; Yonekawa, Y.; Kleihues, P.; Ohgaki, H. Loss of heterozygosity on chromosome 10 is more extensive in primary (de novo) than in secondary glioblastomas. *Lab. Investig.* **2000**, *80*, 65–72. [CrossRef]

19. Parsons, D.W.; Jones, S.; Zhang, X.; Lin, J.C.-H.; Leary, R.J.; Angenendt, P.; Mankoo, P.; Carter, H.; Siu, I.-M.; Gallia, G.L. An integrated genomic analysis of human glioblastoma multiforme. *Science* **2008**, *321*, 1807–1812. [CrossRef]

20. Ohgaki, H.; Dessen, P.; Jourde, B.; Horstmann, S.; Nishikawa, T.; Di Patre, P.L.; Burkhard, C.; Schüler, D.; Probst-Hensch, N.M.; Maiorka, P.C. Genetic pathways to glioblastoma: A population-based study. *Cancer Res.* **2004**, *64*, 6892–6899. [CrossRef]

21. Ohgaki, H.; Kleihues, P. Genetic pathways to primary and secondary glioblastoma. *Am. J. Pathol.* **2007**, *170*, 1445–1453. [CrossRef] [PubMed]

22. Bögler, O.; Su Huang, H.J.; Kleihues, P.; Cavenee, W.K. The p53 gene and its role in human brain tumors. *Glia* **1995**, *15*, 308–327. [CrossRef] [PubMed]

23. Henson, J.W.; Schnitker, B.L.; Correa, K.M.; von Deimling, A.; Fassbender, F.; Xu, H.J.; Benedict, W.F.; Yandell, D.W.; Louis, D.N. The retinoblastoma gene is involved in malignant progression of astrocytomas. *Ann. Neurol.* **1994**, *36*, 714–721. [CrossRef] [PubMed]

24. Nakamura, M.; Yang, F.; Fujisawa, H.; Yonekawa, Y.; Kleihues, P.; Ohgaki, H. Loss of heterozygosity on chromosome 19 in secondary glioblastomas. *J. Neuropathol. Exp. Neurol.* **2000**, *59*, 539–543. [CrossRef] [PubMed]

25. Ueki, K.; Ono, Y.; Henson, J.W.; Efird, J.T.; Von Deimling, A.; Louis, D.N. CDKN2/p16 or RB alterations occur in the majority of glioblastomas and are inversely correlated. *Cancer Res.* **1996**, *56*, 150–153.

26. Lathia, J.D.; Mack, S.C.; Mulkearns-Hubert, E.E.; Valentim, C.L.; Rich, J.N. Cancer stem cells in glioblastoma. *Genes Dev.* **2015**, *29*, 1203–1217. [CrossRef]

27. Beier, D.; Hau, P.; Proescholdt, M.; Lohmeier, A.; Wischhusen, J.; Oefner, P.J.; Aigner, L.; Brawanski, A.; Bogdahn, U.; Beier, C.P. CD133+ and CD133− glioblastoma-derived cancer stem cells show differential growth characteristics and molecular profiles. *Cancer Res.* **2007**, *67*, 4010–4015. [CrossRef]

28. Hottinger, A.F.; Stupp, R.; Homicsko, K. Standards of care and novel approaches in the management of glioblastoma multiforme. *Chin. J. Cancer* **2014**, *33*, 32. [CrossRef]

29. Walker, M.D.; Alexander, E.; Hunt, W.E.; MacCarty, C.S.; Mahaley, M.S.; Mealey, J.; Norrell, H.A.; Owens, G.; Ransohoff, J.; Wilson, C.B. Evaluation of BCNU and/or radiotherapy in the treatment of anaplastic gliomas: A cooperative clinical trial. *J. Neurosurg.* **1978**, *49*, 333–343. [CrossRef]

30. Wick, W.; Platten, M.; Meisner, C.; Felsberg, J.; Tabatabai, G.; Simon, M.; Nikkhah, G.; Papsdorf, K.; Steinbach, J.P.; Sabel, M. Temozolomide chemotherapy alone versus radiotherapy alone for malignant astrocytoma in the elderly: The NOA-08 randomised, phase 3 trial. *Lancet Oncol.* **2012**, *13*, 707–715. [CrossRef]

31. Perry, J.R.; Laperriere, N.; O'Callaghan, C.J.; Brandes, A.A.; Menten, J.; Phillips, C.; Fay, M.; Nishikawa, R.; Cairncross, J.G.; Roa, W. Short-course radiation plus temozolomide in elderly patients with glioblastoma. *N. Engl. J. Med.* **2017**, *376*, 1027–1037. [CrossRef] [PubMed]

32. Niyazi, M.; Pitea, A.; Mittelbronn, M.; Steinbach, J.; Sticht, C.; Zehentmayr, F.; Piehlmaier, D.; Zitzelsberger, H.; Ganswindt, U.; Rödel, C. A 4-miRNA signature predicts the therapeutic outcome of glioblastoma. *Oncotarget* **2016**, *7*, 45764. [PubMed]

33. Ringel, F.; Pape, H.; Sabel, M.; Krex, D.; Bock, H.C. Clinical benefit from resection of recurrent glioblastomas: Results of a multicenter study including 503 patients with recurrent glioblastomas undergoing surgical resection. *Neuro-Oncol.* **2015**, *18*, 96–104. [CrossRef] [PubMed]

34. Straube, C.; Kessel, K.A.; Zimmer, C.; Schmidt-Graf, F.; Schlegel, J.; Gempt, J.; Meyer, B.; Combs, S.E. A Second Course of Radiotherapy in Patients with Recurrent Malignant Gliomas: Clinical Data on Re-irradiation, Prognostic Factors, and Usefulness of Digital Biomarkers. *Curr. Treat. Options Oncol.* **2019**, *20*, 71. [CrossRef]

35. Combs, S.E.; Thilmann, C.; Edler, L.; Debus, J.R.; Schulz-Ertner, D. Efficacy of fractionated stereotactic reirradiation in recurrent gliomas: Long-term results in 172 patients treated in a single institution. *J. Clin. Oncol.* **2005**, *23*, 8863–8869. [CrossRef]

36. Kessel, K.A.; Hesse, J.; Straube, C.; Zimmer, C.; Schmidt-Graf, F.; Schlegel, J.; Meyer, B.; Combs, S.E. Validation of an established prognostic score after re-irradiation of recurrent glioma. *Acta Oncol.* **2017**, *56*, 422–426. [CrossRef]

37. Combs, S.E.; Kessel, K.A.; Hesse, J.; Straube, C.; Zimmer, C.; Schmidt-Graf, F.; Schlegel, J.; Gempt, J.; Meyer, B. Moving second courses of radiotherapy forward: Early re-irradiation after surgical resection for recurrent gliomas improves efficacy with excellent tolerability. *Neurosurgery* **2018**, *83*, 1241–1248. [CrossRef]

38. Roy, S.; Lahiri, D.; Maji, T.; Biswas, J. Recurrent glioblastoma: Where we stand. *South Asian J. Cancer* **2015**, *4*, 163. [CrossRef]

39. Hegi, M.E.; Diserens, A.C.; Gorlia, T.; Hamou, M.F.; De Tribolet, N.; Weller, M.; Kros, J.M.; Hainfellner, J.A.; Mason, W.; Mariani, L. MGMT gene silencing and benefit from temozolomide in glioblastoma. *N. Engl. J. Med.* **2005**, *352*, 997–1003. [CrossRef]

40. Reid, J.M.; Stevens, D.C.; Rubin, J.; Ames, M.M. Pharmacokinetics of 3-methyl-(triazen-1-yl) imidazole-4-carboximide following administration of temozolomide to patients with advanced cancer. *Clin. Cancer Res.* **1997**, *3*, 2393–2398.

41. Newlands, E.; Stevens, M.; Wedge, S.; Wheelhouse, R.T.; Brock, C. Temozolomide: A review of its discovery, chemical properties, pre-clinical development and clinical trials. *Cancer Treat. Rev.* **1997**, *23*, 35–61. [CrossRef]

42. Denny, B.J.; Wheelhouse, R.T.; Stevens, M.F.; Tsang, L.L.; Slack, J.A. NMR and molecular modeling investigation of the mechanism of activation of the antitumor drug temozolomide and its interaction with DNA. *Biochemistry* **1994**, *33*, 9045–9051. [CrossRef] [PubMed]

43. Barciszewska, A.M.; Gurda, D.; Głodowicz, P.; Nowak, S.; Naskręt-Barciszewska, M.Z. A new epigenetic mechanism of temozolomide action in glioma cells. *PLoS ONE* **2015**, *10*, e0136669. [CrossRef] [PubMed]

44. Mitra, G.; Pauly, G.T.; Kumar, R.; Pei, G.K.; Hughes, S.H.; Moschel, R.C.; Barbacid, M. Molecular analysis of O6-substituted guanine-induced mutagenesis of ras oncogenes. *Proc. Natl. Acad. Sci. USA* **1989**, *86*, 8650–8654. [CrossRef] [PubMed]

45. Brennand, J.; Margison, G.P. Reduction of the toxicity and mutagenicity of alkylating agents in mammalian cells harboring the Escherichia coli alkyltransferase gene. *Proc. Natl. Acad. Sci. USA* **1986**, *83*, 6292–6296. [CrossRef]

46. Kolodner, R.D.; Marsischky, G.T. Eukaryotic DNA mismatch repair. *Curr. Opin. Genet. Dev.* **1999**, *9*, 89–96. [CrossRef]

47. Zhang, J.; Stevens, M.F.G.; Bradshaw, T.D. Temozolomide: Mechanisms of action, repair and resistance. *Curr. Mol. Pharmacol.* **2012**, *5*, 102–114. [CrossRef]

48. Belanich, M.; Randall, T.; Pastor, M.A.; Kibitel, J.T.; Alas, L.G.; Dolan, M.E.; Schold, S.C., Jr.; Gander, M.; Lejeune, F.J.; Li, B.F. Intracellular localization and intercellular heterogeneity of the human DNA repair protein O6-methylguanine-DNA methyltransferase. *Cancer Chemother. Pharmacol.* **1996**, *37*, 547–555. [CrossRef]

49. Pegg, A.; Byers, T. Repair of DNA containing O6-alkylguanine. *FASEB J.* **1992**, *6*, 2302–2310. [CrossRef]

50. Tano, K.; Shiota, S.; Collier, J.; Foote, R.S.; Mitra, S. Isolation and structural characterization of a cDNA clone encoding the human DNA repair protein for O6-alkylguanine. *Proc. Natl. Acad. Sci. USA* **1990**, *87*, 686–690. [CrossRef]

51. Krex, D.; Klink, B.; Hartmann, C.; von Deimling, A.; Pietsch, T.; Simon, M.; Sabel, M.; Steinbach, J.P.; Heese, O.; Reifenberger, G. Long-term survival with glioblastoma multiforme. *Brain* **2007**, *130*, 2596–2606. [CrossRef]

52. Kreth, S.; Thon, N.; Eigenbrod, S.; Lutz, J.; Ledderose, C.; Egensperger, R.; Tonn, J.C.; Kretzschmar, H.A.; Hinske, L.C.; Kreth, F.W. O6-methylguanine-DNA methyltransferase (MGMT) mRNA expression predicts outcome in malignant glioma independent of MGMT promoter methylation. *PLoS ONE* **2011**, *6*, e17156. [CrossRef] [PubMed]

53. Silantyev, A.S.; Falzone, L.; Libra, M.; Gurina, O.I.; Kardashova, K.S.; Nikolouzakis, T.K.; Nosyrev, A.E.; Sutton, C.W.; Mitsias, P.D.; Tsatsakis, A. Current and future trends on diagnosis and prognosis of glioblastoma: From molecular biology to proteomics. *Cells* **2019**, *8*, 863. [CrossRef] [PubMed]

54. Wang, F.; Zheng, Z.; Guan, J.; Qi, D.; Zhou, S.; Shen, X.; Wang, F.; Wenkert, D.; Kirmani, B.; Solouki, T. Identification of a panel of genes as a prognostic biomarker for glioblastoma. *EBioMedicine* **2018**, *37*, 68–77. [CrossRef] [PubMed]

55. Tabibkhooei, A.; Izadpanahi, M.; Arab, A.; Zare-Mirzaei, A.; Minaeian, S.; Rostami, A.; Mohsenian, A. Profiling of novel circulating microRNAs as a non-invasive biomarker in diagnosis and follow-up of high and low-grade gliomas. *Clin. Neurol. Neurosurg.* **2020**, *190*, 105652. [CrossRef] [PubMed]

56. Wang, Q.; Li, P.; Li, A.; Jiang, W.; Wang, H.; Wang, J.; Xie, K. Plasma specific miRNAs as predictive biomarkers for diagnosis and prognosis of glioma. *J. Exp. Clin. Cancer Res.* **2012**, *31*, 97. [CrossRef] [PubMed]

57. Wick, W.; Weller, M.; Van Den Bent, M.; Sanson, M.; Weiler, M.; Von Deimling, A.; Plass, C.; Hegi, M.; Platten, M.; Reifenberger, G. MGMT testing—The challenges for biomarker-based glioma treatment. *Nat. Rev. Neurol.* **2014**, *10*, 372. [CrossRef]

58. Metro, G.; Fabi, A.; Mirri, M.A.; Vidiri, A.; Pace, A.; Carosi, M.; Russillo, M.; Maschio, M.; Giannarelli, D.; Pellegrini, D. Phase II study of fixed dose rate gemcitabine as radiosensitizer for newly diagnosed glioblastoma multiforme. *Cancer Chemother. Pharmacol.* **2010**, *65*, 391. [CrossRef]

59. Pegg, A.E.; Boosalis, M.; Samson, L.; Moschel, R.C.; Byers, T.L.; Swenn, K.; Dolan, M.E. Mechanism of inactivation of human O6-alkylguanine-DNA alkyltransferase by O6-benzylguanine. *Biochemistry* **1993**, *32*, 11998–12006. [CrossRef]

60. Dolan, M.E.; Moschel, R.C.; Pegg, A.E. Depletion of mammalian O6-alkylguanine-DNA alkyltransferase activity by O6-benzylguanine provides a means to evaluate the role of this protein in protection against carcinogenic and therapeutic alkylating agents. *Proc. Natl. Acad. Sci. USA* **1990**, *87*, 5368–5372. [CrossRef]

61. Dolan, M.E.; Mitchell, R.B.; Mummert, C.; Moschel, R.C.; Pegg, A.E. Effect of O6-benzylguanine analogues on sensitivity of human tumor cells to the cytotoxic effects of alkylating agents. *Cancer Res.* **1991**, *51*, 3367–3372. [PubMed]

62. Quinn, J.A.; Desjardins, A.; Weingart, J.; Brem, H.; Dolan, M.E.; Delaney, S.M.; Vredenburgh, J.; Rich, J.; Friedman, A.H.; Reardon, D.A. Phase I trial of temozolomide plus O6-benzylguanine for patients with recurrent or progressive malignant glioma. *J. Clin. Oncol.* **2005**, *23*, 7178–7187. [CrossRef] [PubMed]

63. Quinn, J.A.; Jiang, S.X.; Reardon, D.A.; Desjardins, A.; Vredenburgh, J.J.; Rich, J.N.; Gururangan, S.; Friedman, A.H.; Bigner, D.D.; Sampson, J.H. Phase II trial of temozolomide plus o6-benzylguanine in adults with recurrent, temozolomide-resistant malignant glioma. *J. Clin. Oncol.* **2009**, *27*, 1262. [CrossRef] [PubMed]

64. Curtin, N.J. PARP inhibitors for cancer therapy. *Expert Rev. Mol. Med.* **2005**, *7*, 1–20. [CrossRef]

65. Pommier, Y.; O'Connor, M.J.; de Bono, J. Laying a trap to kill cancer cells: PARP inhibitors and their mechanisms of action. *Sci. Transl. Med.* **2016**, *8*, 317–362. [CrossRef]

66. Gupta, S.K.; Smith, E.J.; Mladek, A.C.; Tian, S.; Decker, P.A.; Kizilbash, S.H.; Kitange, G.J.; Sarkaria, J.N. PARP inhibitors for sensitization of alkylation chemotherapy in glioblastoma: Impact of blood-brain barrier and molecular heterogeneity. *Front. Oncol.* **2019**, *8*, 670. [CrossRef]

67. Dungey, F.A.; Löser, D.A.; Chalmers, A.J. Replication-dependent radiosensitization of human glioma cells by inhibition of poly (ADP-Ribose) polymerase: Mechanisms and therapeutic potential. *Int. J. Radiat. Oncol. Biol. Phys.* **2008**, *72*, 1188–1197. [CrossRef]

68. Jue, T.R.; Nozue, K.; Lester, A.J.; Joshi, S.; Schroder, L.B.; Whittaker, S.P.; Nixdorf, S.; Rapkins, R.W.; Khasraw, M.; McDonald, K.L. Veliparib in combination with radiotherapy for the treatment of MGMT unmethylated glioblastoma. *J. Transl. Med.* **2017**, *15*, 61. [CrossRef]

69. Robins, H.I.; Zhang, P.; Gilbert, M.R.; Chakravarti, A.; De Groot, J.F.; Grimm, S.A.; Wang, F.; Lieberman, F.S.; Krauze, A.; Trotti, A.M. A randomized phase I/II study of ABT-888 in combination with temozolomide in recurrent temozolomide resistant glioblastoma: An NRG oncology RTOG group study. *J. Neuro-Oncol.* **2016**, *126*, 309–316. [CrossRef]

70. Khasraw, M.; McDonald, K.L.; Rosenthal, M.; Lwin, Z.; Ashley, D.M.; Wheeler, H.; Barnes, E.; Foote, M.C.; Koh, E.S.; Sulman, E.P.; et al. A randomized phase II trial of veliparib (V), radiotherapy (RT) and temozolomide (TMZ) in patients (pts) with unmethylated MGMT (uMGMT) glioblastoma (GBM). *J. Clin. Oncol.* **2019**, *37*, 2011. [CrossRef]

71. Lee, R.C.; Feinbaum, R.L.; Ambros, V. The C. elegans heterochronic gene lin-4 encodes small RNAs with antisense complementarity to lin-14. *Cell* **1993**, *75*, 843–854. [CrossRef]

72. Lagos-Quintana, M.; Rauhut, R.; Lendeckel, W.; Tuschl, T. Identification of novel genes coding for small expressed RNAs. *Science* **2001**, *294*, 853–858. [CrossRef] [PubMed]

73. Griffiths-Jones, S.; Grocock, R.J.; Van Dongen, S.; Bateman, A.; Enright, A.J. miRBase: microRNA sequences, targets and gene nomenclature. *Nucleic Acids Res.* **2006**, *34*, D140–D144. [CrossRef] [PubMed]

74. Bartel, D.P. MicroRNAs: Genomics, biogenesis, mechanism, and function. *Cell* **2004**, *116*, 281–297. [CrossRef]

75. Lee, Y.; Jeon, K.; Lee, J.T.; Kim, S.; Kim, V.N. MicroRNA maturation: Stepwise processing and subcellular localization. *EMBO J.* **2002**, *21*, 4663–4670. [CrossRef]

76. Lee, Y.; Kim, M.; Han, J.; Yeom, K.H.; Lee, S.; Baek, S.H.; Kim, V.N. MicroRNA genes are transcribed by RNA polymerase II. *EMBO J.* **2004**, *23*, 4051–4060. [CrossRef]

77. Lee, Y.; Ahn, C.; Han, J.; Choi, H.; Kim, J.; Yim, J.; Lee, J.; Provost, P.; Rådmark, O.; Kim, S. The nuclear RNase III Drosha initiates microRNA processing. *Nature* **2003**, *425*, 415–419. [CrossRef]

78. Basyuk, E.; Suavet, F.; Doglio, A.; Bordonné, R.; Bertrand, E. Human let-7 stem–loop precursors harbor features of RNase III cleavage products. *Nucleic Acids Res.* **2003**, *31*, 6593–6597. [CrossRef]

79. Denli, A.M.; Tops, B.B.; Plasterk, R.H.; Ketting, R.F.; Hannon, G.J. Processing of primary microRNAs by the Microprocessor complex. *Nature* **2004**, *432*, 231–235. [CrossRef]

80. Yi, R.; Qin, Y.; Macara, I.G.; Cullen, B.R. Exportin-5 mediates the nuclear export of pre-microRNAs and short hairpin RNAs. *Genes Dev.* **2003**, *17*, 3011–3016. [CrossRef]

81. Hammond, S.M.; Bernstein, E.; Beach, D.; Hannon, G.J. An RNA-directed nuclease mediates post-transcriptional gene silencing in Drosophila cells. *Nature* **2000**, *404*, 293–296. [CrossRef] [PubMed]

82. Khvorova, A.; Reynolds, A.; Jayasena, S.D. Functional siRNAs and miRNAs exhibit strand bias. *Cell* **2003**, *115*, 209–216. [CrossRef]

83. Hutvágner, G.; Zamore, P.D. A microRNA in a multiple-turnover RNAi enzyme complex. *Science* **2002**, *297*, 2056–2060. [CrossRef] [PubMed]

84. Calin, G.A.; Dumitru, C.D.; Shimizu, M.; Bichi, R.; Zupo, S.; Noch, E.; Aldler, H.; Rattan, S.; Keating, M.; Rai, K. Frequent deletions and down-regulation of micro-RNA genes miR15 and miR16 at 13q14 in chronic lymphocytic leukemia. *Proc. Natl. Acad. Sci. USA* **2002**, *99*, 15524–15529. [CrossRef]

85. Calin, G.A.; Sevignani, C.; Dumitru, C.D.; Hyslop, T.; Noch, E.; Yendamuri, S.; Shimizu, M.; Rattan, S.; Bullrich, F.; Negrini, M. Human microRNA genes are frequently located at fragile sites and genomic regions involved in cancers. *Proc. Natl. Acad. Sci. USA* **2004**, *101*, 2999–3004. [CrossRef]

86. Cimmino, A.; Calin, G.A.; Fabbri, M.; Iorio, M.V.; Ferracin, M.; Shimizu, M.; Wojcik, S.E.; Aqeilan, R.I.; Zupo, S.; Dono, M. miR-15 and miR-16 induce apoptosis by targeting BCL2. *Proc. Natl. Acad. Sci. USA* **2005**, *102*, 13944–13949. [CrossRef] [PubMed]

87. Yamakuchi, M.; Lowenstein, C.J. MiR-34, SIRT1, and p53: The feedback loop. *Cell Cycle* **2009**, *8*, 712–715. [CrossRef]

88. Bautista-Sánchez, D.; Arriaga-Canon, C.; Pedroza-Torres, A.; De La Rosa-Velázquez, I.A.; González-Barrios, R.; Contreras-Espinosa, L.; Montiel-Manríquez, R.; Castro-Hernández, C.; Fragoso-Ontiveros, V.; Álvarez-Gómez, R.M. The promising role of miR-21 as a cancer biomarker and its importance in RNA-based therapeutics. *Mol. Ther. Nucleic Acids* **2020**. [CrossRef]

89. Calin, G.A.; Croce, C.M. MicroRNA-cancer connection: The beginning of a new tale. *Cancer Res.* **2006**, *66*, 7390–7394. [CrossRef]

90. Chan, J.A.; Krichevsky, A.M.; Kosik, K.S. MicroRNA-21 is an antiapoptotic factor in human glioblastoma cells. *Cancer Res.* **2005**, *65*, 6029–6033. [CrossRef]

91. Sasayama, T.; Nishihara, M.; Kondoh, T.; Hosoda, K.; Kohmura, E. MicroRNA-10b is overexpressed in malignant glioma and associated with tumor invasive factors, uPAR and RhoC. *Int. J. Cancer* **2009**, *125*, 1407–1413. [CrossRef] [PubMed]

92. Kefas, B.; Godlewski, J.; Comeau, L.; Li, Y.; Abounader, R.; Hawkinson, M.; Lee, J.; Fine, H.; Chiocca, E.A.; Lawler, S. microRNA-7 inhibits the epidermal growth factor receptor and the Akt pathway and is down-regulated in glioblastoma. *Cancer Res.* **2008**, *68*, 3566–3572. [CrossRef] [PubMed]

93. Candido, S.; Lupo, G.; Pennisi, M.; Basile, M.S.; Anfuso, C.D.; Petralia, M.C.; Gattuso, G.; Vivarelli, S.; Spandidos, D.A.; Libra, M. The analysis of miRNA expression profiling datasets reveals inverse microRNA patterns in glioblastoma and Alzheimer's disease. *Oncol. Rep.* **2019**, *42*, 911–922. [CrossRef] [PubMed]

94. Comincini, S.; Allavena, G.; Palumbo, S.; Morini, M.; Durando, F.; Angeletti, F.; Pirtoli, L.; Miracco, C. microRNA-17 regulates the expression of ATG7 and modulates the autophagy process, improving the sensitivity to temozolomide and low-dose ionizing radiation treatments in human glioblastoma cells. *Cancer Biol. Ther.* **2013**, *14*, 574–586. [CrossRef] [PubMed]

95. Toraih, E.A.; Aly, N.M.; Abdallah, H.Y.; Al-Qahtani, S.A.; Shaalan, A.A.; Hussein, M.H.; Fawzy, M.S. MicroRNA–target cross-talks: Key players in glioblastoma multiforme. *Tumor Biol.* **2017**, *39*, 1010428317726842. [CrossRef] [PubMed]

96. Papagiannakopoulos, T.; Shapiro, A.; Kosik, K.S. MicroRNA-21 targets a network of key tumor-suppressive pathways in glioblastoma cells. *Cancer Res.* **2008**, *68*, 8164–8172. [CrossRef]

97. Ge, X.; Pan, M.H.; Wang, L.; Li, W.; Jiang, C.; He, J.; Abouzid, K.; Liu, L.Z.; Shi, Z.; Jiang, B.H. Hypoxia-mediated mitochondria apoptosis inhibition induces temozolomide treatment resistance through miR-26a/Bad/Bax axis. *Cell Death Dis.* **2018**, *9*, 1–16. [CrossRef]

98. Sippl, C.; Ketter, R.; Braun, L.; Teping, F.; Schoeneberger, L.; Kim, Y.J.; List, M.; Nakhoda, A.; Wemmert, S.; Oertel, J. miRNA-26a expression influences the therapy response to carmustine wafer implantation in patients with glioblastoma multiforme. *Acta Neurochir.* **2019**, *161*, 2299–2309. [CrossRef]

99. Huse, J.T.; Brennan, C.; Hambardzumyan, D.; Wee, B.; Pena, J.; Rouhanifard, S.H.; Sohn-Lee, C.; Le Sage, C.; Agami, R.; Tuschl, T. The PTEN-regulating microRNA miR-26a is amplified in high-grade glioma and facilitates gliomagenesis in vivo. *Genes Dev.* **2009**, *23*, 1327–1337. [CrossRef]

100. Li, Y.; Guessous, F.; Zhang, Y.; DiPierro, C.; Kefas, B.; Johnson, E.; Marcinkiewicz, L.; Jiang, J.; Yang, Y.; Schmittgen, T.D. MicroRNA-34a inhibits glioblastoma growth by targeting multiple oncogenes. *Cancer Res.* **2009**, *69*, 7569–7576. [CrossRef]

101. Gulluoglu, S.; Tuysuz, E.C.; Sahin, M.; Kuskucu, A.; Yaltirik, C.K.; Ture, U.; Kucukkaraduman, B.; Akbar, M.W.; Gure, A.O.; Bayrak, O.F. Simultaneous miRNA and mRNA transcriptome profiling of glioblastoma samples reveals a novel set of OncomiR candidates and their target genes. *Brain Res.* **2018**, *1700*, 199–210. [CrossRef] [PubMed]

102. Xu, T.; Wang, H.; Jiang, M.; Yan, Y.; Li, W.; Xu, H.; Huang, Q.; Lu, Y.; Chen, J. The E3 ubiquitin ligase CHIP/miR-92b/PTEN regulatory network contributes to tumorigenesis of glioblastoma. *Am. J. Cancer Res.* **2017**, *7*, 289. [PubMed]

103. Silber, J.; Lim, D.A.; Petritsch, C.; Persson, A.I.; Maunakea, A.K.; Yu, M.; Vandenberg, S.R.; Ginzinger, D.G.; James, C.D.; Costello, J.F. miR-124 and miR-137 inhibit proliferation of glioblastoma multiforme cells and induce differentiation of brain tumor stem cells. *BMC Med.* **2008**, *6*, 14. [CrossRef] [PubMed]

104. Wu, N.; Lin, X.; Zhao, X.; Zheng, L.; Xiao, L.; Liu, J.; Ge, L.; Cao, S. MiR-125b acts as an oncogene in glioblastoma cells and inhibits cell apoptosis through p53 and p38MAPK-independent pathways. *Br. J. Cancer* **2013**, *109*, 2853–2863. [CrossRef]

105. Xia, H.F.; He, T.Z.; Liu, C.M.; Cui, Y.; Song, P.P.; Jin, X.H.; Ma, X. MiR-125b expression affects the proliferation and apoptosis of human glioma cells by targeting Bmf. *Cell. Physiol. Biochem.* **2009**, *23*, 347–358. [CrossRef]

106. Zhang, Y.; Chao, T.; Li, R.; Liu, W.; Chen, Y.; Yan, X.; Gong, Y.; Yin, B.; Qiang, B.; Zhao, J. MicroRNA-128 inhibits glioma cells proliferation by targeting transcription factor E2F3a. *J. Mol. Med.* **2009**, *87*, 43–51. [CrossRef]

107. Papagiannakopoulos, T.; Friedmann-Morvinski, D.; Neveu, P.; Dugas, J.; Gill, R.; Huillard, E.; Liu, C.; Zong, H.; Rowitch, D.; Barres, B. Pro-neural miR-128 is a glioma tumor suppressor that targets mitogenic kinases. *Oncogene* **2012**, *31*, 1884–1895. [CrossRef]

108. Qiu, S.; Lin, S.; Hu, D.; Feng, Y.; Tan, Y.; Peng, Y. Interactions of miR-323/miR-326/miR-329 and miR-130a/miR-155/miR-210 as prognostic indicators for clinical outcome of glioblastoma patients. *J. Transl. Med.* **2013**, *11*, 10. [CrossRef]

109. Chen, H.; Li, X.; Li, W.; Zheng, H. miR-130a can predict response to temozolomide in patients with glioblastoma multiforme, independently of O6-methylguanine-DNA methyltransferase. *J. Transl. Med.* **2015**, *13*, 69. [CrossRef]

110. Lee, Y.Y.; Yarmishyn, A.A.; Wang, M.L.; Chen, H.Y.; Chiou, S.H.; Yang, Y.P.; Lin, C.F.; Huang, P.I.; Chen, Y.W.; Ma, H.I. MicroRNA-142-3p is involved in regulation of MGMT expression in glioblastoma cells. *Cancer Manag. Res.* **2018**, *10*, 775. [CrossRef]

111. Chiou, G.Y.; Chien, C.S.; Wang, M.L.; Chen, M.T.; Yang, Y.P.; Yu, Y.L.; Chien, Y.; Chang, Y.C.; Shen, C.C.; Chio, C.C. Epigenetic regulation of the miR142-3p/interleukin-6 circuit in glioblastoma. *Mol. Cell* **2013**, *52*, 693–706. [CrossRef] [PubMed]

112. Ling, N.; Gu, J.; Lei, Z.; Li, M.; Zhao, J.; Zhang, H.T.; Li, X. microRNA-155 regulates cell proliferation and invasion by targeting FOXO3a in glioma. *Oncol. Rep.* **2013**, *30*, 2111–2118. [CrossRef] [PubMed]

113. Chen, G.; Zhu, W.; Shi, D.; Lv, L.; Zhang, C.; Liu, P.; Hu, W. MicroRNA-181a sensitizes human malignant glioma U87MG cells to radiation by targeting Bcl-2. *Oncol. Rep.* **2010**, *23*, 997–1003. [PubMed]

114. Zhou, Y.; Peng, Y.; Liu, M.; Jiang, Y. MicroRNA-181b inhibits cellular proliferation and invasion of glioma cells via targeting Sal-like protein 4. *Oncol. Res. Featur. Preclin. Clin. Cancer Ther.* **2017**, *25*, 947–957. [CrossRef]

115. Wang, X.F.; Shi, Z.M.; Wang, X.R.; Cao, L.; Wang, Y.Y.; Zhang, J.X.; Yin, Y.; Luo, H.; Kang, C.S.; Liu, N. MiR-181d acts as a tumor suppressor in glioma by targeting K-ras and Bcl-2. *J. Cancer Res. Clin. Oncol.* **2012**, *138*, 573–584. [CrossRef]

116. Zhang, W.; Zhang, J.; Hoadley, K.; Kushwaha, D.; Ramakrishnan, V.; Li, S.; Kang, C.; You, Y.; Jiang, C.; Song, S.W. miR-181d: A predictive glioblastoma biomarker that downregulates MGMT expression. *Neuro-Oncol.* **2012**, *14*, 712–719. [CrossRef]

117. Shang, C.; Hong, Y.; Guo, Y.; Liu, Y.H.; Xue, Y.X. MiR-210 up-regulation inhibits proliferation and induces apoptosis in glioma cells by targeting SIN3A. *Med Sci. Monit.* **2014**, *20*, 2571.

118. Song, L.; Huang, Q.; Chen, K.; Liu, L.; Lin, C.; Dai, T.; Yu, C.; Wu, Z.; Li, J. miR-218 inhibits the invasive ability of glioma cells by direct downregulation of IKK-β. *Biochem. Biophys. Res. Commun.* **2010**, *402*, 135–140. [CrossRef]

119. Liu, Y.; Yan, W.; Zhang, W.; Chen, L.; You, G.; Bao, Z.; Wang, Y.; Wang, H.; Kang, C.; Jiang, T. MiR-218 reverses high invasiveness of glioblastoma cells by targeting the oncogenic transcription factor LEF1. *Oncol. Rep.* **2012**, *28*, 1013–1021. [CrossRef]

120. Tu, Y.; Gao, X.; Li, G.; Fu, H.; Cui, D.; Liu, H.; Jin, W.; Zhang, Y. MicroRNA-218 inhibits glioma invasion, migration, proliferation, and cancer stem-like cell self-renewal by targeting the polycomb group gene Bmi1. *Cancer Res.* **2013**, *73*, 6046–6055. [CrossRef]

121. Gillies, J.K.; Lorimer, I.A. Regulation of p27Kip1 by miRNA 221/222 in glioblastoma. *Cell Cycle* **2007**, *6*, 2005–2009. [CrossRef] [PubMed]

122. Zhang, J.; Han, L.; Ge, Y.; Zhou, X.; Zhang, A.; Zhang, C.; Zhong, Y.; You, Y.; Pu, P.; Kang, C. miR-221/222 promote malignant progression of glioma through activation of the Akt pathway. *Int. J. Oncol.* **2010**, *36*, 913–920. [PubMed]

123. Quintavalle, C.; Mangani, D.; Roscigno, G.; Romano, G.; Diaz-Lagares, A.; Iaboni, M.; Donnarumma, E.; Fiore, D.; De Marinis, P.; Soini, Y. MiR-221/222 target the DNA methyltransferase MGMT in glioma cells. *PLoS ONE* **2013**, *8*, e74466. [CrossRef]

124. Shu, M.; Zheng, X.; Wu, S.; Lu, H.; Leng, T.; Zhu, W.; Zhou, Y.; Ou, Y.; Lin, X.; Lin, Y. Targeting oncogenic miR-335 inhibits growth and invasion of malignant astrocytoma cells. *Mol. Cancer* **2011**, *10*, 59. [CrossRef] [PubMed]

125. Cheng, Q.; Cao, H.; Chen, Z.; Ma, Z.; Wan, X.; Peng, R.; Jiang, B. PAX6, a novel target of miR-335, inhibits cell proliferation and invasion in glioma cells. *Mol. Med. Rep.* **2014**, *10*, 399–404. [CrossRef] [PubMed]

126. Peng, Z.; Wu, T.; Li, Y.; Xu, Z.; Zhang, S.; Liu, B.; Chen, Q.; Tian, D. MicroRNA-370-3p inhibits human glioma cell proliferation and induces cell cycle arrest by directly targeting β-catenin. *Brain Res.* **2016**, *1644*, 53–61. [CrossRef]

127. Nadaradjane, A.; Briand, J.; Bougras-Cartron, G.; Disdero, V.; Vallette, F.M.; Frenel, J.S.; Cartron, P.F. miR-370-3p is a therapeutic tool in anti-glioblastoma therapy but is not an Intratumoral or cell-free circulating biomarker. *Mol. Ther. Nucleic Acids* **2018**, *13*, 642–650. [CrossRef]

128. Cao, Y.; Zhang, L.; Wei, M.; Jiang, X.; Jia, D. MicroRNA-409-3p represses glioma cell invasion and proliferation by targeting high-mobility group nucleosome-binding domain 5. *Oncol. Res. Featur. Preclin. Clin. Cancer Ther.* **2017**, *25*, 1097–1107. [CrossRef]

129. Nan, Y.; Han, L.; Zhang, A.; Wang, G.; Jia, Z.; Yang, Y.; Yue, X.; Pu, P.; Zhong, Y.; Kang, C. MiRNA-451 plays a role as tumor suppressor in human glioma cells. *Brain Res.* **2010**, *1359*, 14–21. [CrossRef]

130. Guo, M.; Zhang, X.; Wang, G.; Sun, J.; Jiang, Z.; Khadarian, K.; Yu, S.; Zhao, Y.; Xie, C.; Zhang, K. miR-603 promotes glioma cell growth via Wnt/β-catenin pathway by inhibiting WIF1 and CTNNBIP1. *Cancer Lett.* **2015**, *360*, 76–86. [CrossRef]

131. Karagkouni, D.; Paraskevopoulou, M.D.; Chatzopoulos, S.; Vlachos, I.S.; Tastsoglou, S.; Kanellos, I.; Papadimitriou, D.; Kavakiotis, I.; Maniou, S.; Skoufos, G. DIANA-TarBase v8: A decade-long collection of experimentally supported miRNA–gene interactions. *Nucleic Acids Res.* **2018**, *46*, D239–D245. [CrossRef] [PubMed]

132. Gao, Y.T.; Chen, X.B.; Liu, H.L. Up-regulation of miR-370-3p restores glioblastoma multiforme sensitivity to temozolomide by influencing MGMT expression. *Sci. Rep.* **2016**, *6*, 1–9. [CrossRef] [PubMed]

133. Khalil, S.; Fabbri, E.; Santangelo, A.; Bezzerri, V.; Cantù, C.; Di Gennaro, G.; Finotti, A.; Ghimenton, C.; Eccher, A.; Dechecchi, M. miRNA array screening reveals cooperative MGMT-regulation between miR-181d-5p and miR-409-3p in glioblastoma. *Oncotarget* **2016**, *7*, 28195. [CrossRef] [PubMed]

134. Kushwaha, D.; Ramakrishnan, V.; Ng, K.; Steed, T.; Nguyen, T.; Futalan, D.; Akers, J.C.; Sarkaria, J.; Jiang, T.; Chowdhury, D. A genome-wide miRNA screen revealed miR-603 as a MGMT-regulating miRNA in glioblastomas. *Oncotarget* **2014**, *5*, 4026. [CrossRef]

135. Kreth, S.; Limbeck, E.; Hinske, L.C.; Schütz, S.V.; Thon, N.; Hoefig, K.; Egensperger, R.; Kreth, F.W. In human glioblastomas transcript elongation by alternative polyadenylation and miRNA targeting is a potent mechanism of MGMT silencing. *Acta Neuropathol.* **2013**, *125*, 671–681. [CrossRef]

136. Neto, F.S.L.; Rodrigues, A.R.; Trevisan, F.A.; de Assis Cirino, M.L.; Matias, C.C.M.S.; Pereira-da-Silva, G.; Peria, F.M.; da Cunha Tirapelli, D.P.; Carlotti, C.G., Jr. microRNA-181d associated with the methylation status of the MGMT gene in Glioblastoma multiforme cancer stem cells submitted to treatments with ionizing radiation and temozolomide. *Brain Res.* **2019**, *1720*, 146302. [CrossRef]

137. Garofalo, M.; Di Leva, G.; Romano, G.; Nuovo, G.; Suh, S.-S.; Ngankeu, A.; Taccioli, C.; Pichiorri, F.; Alder, H.; Secchiero, P. miR-221&222 regulate TRAIL resistance and enhance tumorigenicity through PTEN and TIMP3 downregulation. *Cancer Cell* **2009**, *16*, 498–509.

138. Jesionek-Kupnicka, D.; Braun, M.; Trąbska-Kluch, B.; Czech, J.; Szybka, M.; Szymańska, B.; Kulczycka-Wojdala, D.; Bieńkowski, M.; Kordek, R.; Zawlik, I. MiR-21, miR-34a, miR-125b, miR-181d and miR-648 levels inversely correlate with MGMT and TP53 expression in primary glioblastoma patients. *Arch. Med. Sci.* **2019**, *15*, 504. [CrossRef]

139. Garg, T.; Bhandari, S.; Rath, G.; Goyal, A.K. Current strategies for targeted delivery of bio-active drug molecules in the treatment of brain tumor. *J. Drug Target.* **2015**, *23*, 865–887. [CrossRef]

140. Dong, X. Current strategies for brain drug delivery. *Theranostics* **2018**, *8*, 1481. [CrossRef]

141. Lee, S.W.L.; Paoletti, C.; Campisi, M.; Osaki, T.; Adriani, G.; Kamm, R.D.; Mattu, C.; Chiono, V. MicroRNA delivery through nanoparticles. *J. Control. Release* **2019**. [CrossRef] [PubMed]

142. Abbott, N.J.; Patabendige, A.A.; Dolman, D.E.; Yusof, S.R.; Begley, D.J. Structure and function of the blood–brain barrier. *Neurobiol. Dis.* **2010**, *37*, 13–25. [CrossRef] [PubMed]

143. Nana-Sinkam, S.; Croce, C. Clinical applications for microRNAs in cancer. *Clin. Pharmacol. Ther.* **2013**, *93*, 98–104. [CrossRef] [PubMed]

144. Bajan, S.; Hutvagner, G. RNA-Based Therapeutics: From Antisense Oligonucleotides to miRNAs. *Cells* **2020**, *9*, 137. [CrossRef] [PubMed]

145. Hong, D.S.; Kang, Y.K.; Borad, M.; Sachdev, J.; Ejadi, S.; Lim, H.Y.; Brenner, A.J.; Park, K.; Lee, J.L.; Kim, T.Y. Phase 1 study of MRX34, a liposomal miR-34a mimic, in patients with advanced solid tumours. *Br. J. Cancer* **2020**, 1–8.

146. Gits, C.M.; Van Kuijk, P.F.; Jonkers, M.B.; Boersma, A.W.; Van Ijcken, W.; Wozniak, A.; Sciot, R.; Rutkowski, P.; Schöffski, P.; Taguchi, T. MiR-17-92 and miR-221/222 cluster members target KIT and ETV1 in human gastrointestinal stromal tumours. *Br. J. Cancer* **2013**, *109*, 1625–1635. [CrossRef]

147. Falzone, L.; Romano, G.L.; Salemi, R.; Bucolo, C.; Tomasello, B.; Lupo, G.; Anfuso, C.D.; Spandidos, D.A.; Libra, M.; Candido, S. Prognostic significance of deregulated microRNAs in uveal melanomas. *Mol. Med. Rep.* **2019**, *19*, 2599–2610. [CrossRef]

148. Pedroza-Torres, A.; Romero-Córdoba, S.L.; Justo-Garrido, M.; Salido-Guadarrama, I.; Rodríguez-Bautista, R.; Montaño, S.; Muñiz-Mendoza, R.; Arriaga-Canon, C.; Fragoso-Ontiveros, V.; Álvarez-Gómez, R.M. microRNAs in tumor cell metabolism: Roles and therapeutic opportunities. *Front. Oncol.* **2019**, *9*, 1404. [CrossRef]

149. Hermansen, S.K.; Sørensen, M.D.; Hansen, A.; Knudsen, S.; Alvarado, A.G.; Lathia, J.D.; Kristensen, B.W. A 4-miRNA signature to predict survival in glioblastomas. *PLoS ONE* **2017**, *12*, e0188090. [CrossRef]
150. Molenaar, R.J.; Verbaan, D.; Lamba, S.; Zanon, C.; Jeuken, J.W.; Boots-Sprenger, S.H.; Wesseling, P.; Hulsebos, T.J.; Troost, D.; Van Tilborg, A.A. The combination of IDH1 mutations and MGMT methylation status predicts survival in glioblastoma better than either IDH1 or MGMT alone. *Neuro-Oncol.* **2014**, *16*, 1263–1273. [CrossRef]

Review

Key MicroRNA's and Their Targetome in Adrenocortical Cancer

Marthe Chehade [1,2], **Martyn Bullock** [1,2], **Anthony Glover** [1,2,3], **Gyorgy Hutvagner** [4,*] and **Stan Sidhu** [1,2,3,*]

1   Cancer Genetics Laboratory, Kolling Institute, Northern Sydney Local Health District, St. Leonards, NSW 2065, Australia; mche2952@uni.sydney.edu.au (M.C.); martyn.bullock@sydney.edu.au (M.B.); anthony.glover@sydney.edu.au (A.G.)
2   Sydney Medical School Northern, Royal North Shore Hospital, University of Sydney, Sydney, NSW 2065, Australia
3   Endocrine Surgery Unit, Royal North Shore Hospital, Northern Clinical School, Faculty of Medicine and Health, The University of Sydney, St. Leonards, Sydney, NSW 2007, Australia
4   School of Biomedical Engineering, Faculty of Engineering and Information Technology, University of Technology Sydney, Sydney, NSW 2007, Australia
*   Correspondence: gyorgy.hutvagner@uts.edu.au (G.H.); stansidhu@nebsc.com.au (S.S.)

Received: 14 June 2020; Accepted: 28 July 2020; Published: 6 August 2020

Abstract: Adrenocortical Carcinoma (ACC) is a rare but aggressive malignancy with poor prognosis and limited response to available systemic therapies. Although complete surgical resection gives the best chance for long-term survival, ACC has a two-year recurrence rate of 50%, which poses a therapeutic challenge. High throughput analyses focused on characterizing the molecular signature of ACC have revealed specific micro-RNAs (miRNAs) that are associated with aggressive tumor phenotypes. MiRNAs are small non-coding RNA molecules that regulate gene expression by inhibiting mRNA translation or degrading mRNA transcripts and have been generally implicated in carcinogenesis. This review summarizes the current insights into dysregulated miRNAs in ACC tumorigenesis, their known functions, and specific targetomes. In addition, we explore the possibility of particular miRNAs to be exploited as clinical biomarkers in ACC and as potential therapeutics.

Keywords: adrenocortical carcinoma; micro RNA; non-coding RNA

---

1. Introduction

Adrenal tumors are very common, affecting up to 10% of the general population, of which the large majority are benign non-functional adenomas [1]. Adrenocortical cancer (ACC), in contrast, is a rare endocrine malignancy with an incidence of 0.76 per million in the general population [2]. Approximately 60% of patients with ACC present with signs and symptoms of hormone excess [3], and approximately 20% present with mass associated symptoms, such as abdominal pain, early satiety, or abdominal fullness [4]. The remaining patients are incidentally diagnosed on abdominal imaging for other medical indications. As the clinical manifestations of hormone excess may be subtle and mass effect symptoms are vague, ACC is often diagnosed late. The median size of the primary tumor is 12 cm at diagnosis [5], and the rate of unresectable metastatic disease at diagnosis ranges between 30% [6] and 70% [7]. Complete surgical resection with oncologically clear margins affords the best chance of cure in ACC but even despite this, the rate of disease recurrence is high, and the prognosis is generally poor with five-year survival of less than 40% [8].

Management options for metastatic ACC are limited as cytotoxic chemotherapy affords only a marginal survival benefit, and mitotane, an adrenolytic agent, which is the only other approved systemic therapy for metastatic ACC, is poorly tolerated. The first and only randomized controlled chemotherapy-based phase III clinical trial for advanced ACC (First International Randomized Trial in Locally Advanced and Metastatic Adrenocortical Carcinoma Treatment (FIRM-ACT)) was completed in 2010. This study compared mitotane administered in combination with either streptozocin or etoposide, doxorubicin, and cisplatin (EDPM) and demonstrated a modest improvement in progression-free survival in the EDPM arm, but no benefit in the overall survival [9]. ACC is, therefore, an orphan disease that presents challenges on both diagnostic and management fronts. ACC research is currently focused on developing methods for early detection and effective management of a metastatic disease. In particular, the discovery of novel approaches to the management of metastatic ACC is crucial to improving patient outlook.

MicroRNA (miRNA) are small non-protein-coding RNA molecules whose deregulation has been implicated in the pathogenesis of many human diseases, particularly cancer. Over the past two decades, miRNA research in cancer has focused on determining the miRNA expression signatures of different tumors in order to identify potential biomarkers for early diagnosis, as well as functional studies of specific miRNAs to determine their targets and function. The set of mRNAs targeted by a defined miRNA is known as its targetome. While multiple studies have profiled the miRNA signature of childhood and adult ACCs using various techniques and shown consistent deregulation in a set of candidate miRNAs, relatively fewer have demonstrated miRNA-target interactions. MiRNAs have also been identified to have both diagnostic and therapeutic potential in the cancer literature, broadening our understanding of their roles in tumor biology. In this review, we present a current summary of the mounting body of work describing miRNA dysfunction in ACC with the aims of highlighting their potential function and roles in modulating key oncogenic pathways.

2. ACC Genetic Landscape and Associated Genetic Disorders

ACC has a bimodal distribution with a worldwide childhood incidence of 0.2 per million [10], and an adult peak in the fifth decade of life. Childhood ACC differs from adult ACC, as 50–80% of childhood cases are associated with germline *TP53* mutations [11,12]. In contrast, most cases of adult ACC are sporadic, with germline *TP53* mutations being present in around three percent of patients [13].

In rare cases, ACC can be associated with specific germline mutations that cause hereditary cancer syndromes. ACC is a core malignancy in Li Fraumeni Syndrome (LFS) caused by the germline *TP53* mutation and affects ten percent of cases [14]. Notably, in a study by Soon and colleagues, sporadic ACC was associated with loss of heterozygosity (LOH) at the *TP53* gene locus 17p13.1 in 74% of cases compared with only 14% of adrenal adenomas [15]. ACC affects approximately seven percent of children with Beckwith–Weidemann Syndrome (BWS) [16], which is caused by mutations or epigenetic modifications at the genetic locus 11p15 containing the *Insulin-Like Growth Factor 2* (*IGF2*) gene. 11p15 LOH or *IGF2* overexpression were demonstrated in 93.1% of sporadic ACCs compared with only 8.6% of benign adrenal tumors [17] in a study by Gicquel and colleagues, highlighting the importance of this imprinted locus in the pathogenesis of ACC. Approximately three percent of patients with Familial Adenomatous Polyposis (FAP), caused by mutations in the Adenomatous Polyposis Coli (APC) gene, develop adrenocortical cancer as adults [18]. The APC protein is a negative regulator of β-Catenin, whose accumulation in the nuclei of primary ACCs has been associated with advanced tumor stage and poor prognosis [19]. ACC is also rarely associated with Lynch Syndrome, Neurofibromatosis Type 1, and Carney Complex, as well as Multiple Endocrine Neoplasia Type 1 (MEN1) in adults [20] (Table 1). Although mutations in *TP53*, *IGF-2*, and *β-catenin* genes have been established as drivers of sporadic ACC, the low penetrance of ACC in these genetic cancer syndromes indicates that mutations or the epigenetic regulation of the expression of nearby genes may play an important role in its etiology.

Table 1. Genetic syndromes associated with Adrenocortical Carcinoma.

Genetic Syndrome	Inheritance	Mutated Gene/s	Cellular Pathway/s Affected	Gene Locus	ACC Penetrance	Reference/s
Li Fraumeni Syndrome	Autosomal Dominant	*TP53*	Cell cycle	17p13.1	10%	[14]
Beckwith-Weidemann Syndrome	Sporadic	*IGF2/H19* * *CDKN1C/KCNQ1OT1* *	PI3K/TGF-β	11p15	7%	[16]
Familial Adenomatous Polyposis	Autosomal Dominant	*APC*	Wnt/β-Catenin	5q22.2	3%	[18]
Multiple Endocrine Neoplasia Type 1	Autosomal Dominant	*MEN1*	Cell Cycle	11q13	1–5%	[21,22]
Lynch Syndrome	Autosomal Dominant	*MLH1* *MSH2* *MSH6* *PMS2*	DNA mismatch repair	3p22.2 2p21 2p16.3 7p22	14 case reports	[23]
Neurofibromatosis Type 1	Autosomal Dominant	*NF1*	MAPK/ERK	17q11.2	9 case reports	[24]
Carney Complex	Autosomal Dominant	*CNC1 (PRKAR1A)*	cAMP	17q22-24	2 case reports	[25]

* Epigenetic modifications to methylation of imprinting control regions or paternal uniparental disomy are more common than gene mutations. IGF2, Insulin-like growth factor 2; CDKN1C, Cyclin-dependent kinase inhibitor 1C; KCNQ1OT, Potassium voltage-gated channel subfamily Q member 1 antisense gene; APC, Adenomatous polyposis coli; MEN1, Multiple endocrine neoplasia Type 1; MLH1, MutL homolog 1; MutS homolog 2; MSH6, MutS homolog 6; PMS2, PMS1 homolog 2; NF1, Neurofibromatosis 1; CNC1, Carney complex type 1; PRKAR1A, Protein kinase A regulatory subunit 1-alpha; PI3K, Phosphatidylinositol-3-kinase; TGF-β, Transforming growth factor-beta; Wnt, Wingless-related integration site; MAPK, Mitogen-activated protein kinase; ERK, Extracellular signal-related kinase; cAMP, cyclic adenosine monophosphate.

3. Key Genetic Drivers of ACC and Their Cellular Pathways

It is now known that mutations in gene drivers alone do not completely explain the pathogenesis of ACC, and therefore non-coding gene mutations that lead to aberrant regulation of driver genes through their pathways can also contribute to tumor biogenesis. The following summary of key genetic drivers in ACC, therefore, serves to explore the extent to which ACC pathogenesis could be explained by them, and contextualize the importance of known miRNA targets within these pathways.

3.1. Tumor Suppressor Protein 53 (TP53)

The *TP53* gene encodes a homo-tetrameric transcription factor that mediates the cellular response to genotoxic stress and the activation of oncogenes by transcriptionally targeting many genes to ultimately activate cellular pathways involved in cell-cycle arrest and DNA damage repair. Where the cell fails to repair this damage, p53 induces cellular apoptosis via a p53-upregulated modulator of apoptosis (PUMA) to avoid propagating genetic mistakes. P53 is regulated by Human Double Minute 2 homolog (HDM2), an E3 ubiquitin-protein ligase, which targets p53 for cytosolic translocation, or proteosomal degradation when it is polyubiquitinated. HDM2 is, in turn, regulated by p53, forming a negative feedback loop [26].

TP53 mutations in cancer are common and are present in more than half of human tumors [27]. ACC, despite its rarity, accounts for 11.9% of all human tumors harboring germline *TP53* mutations, after breast, soft-tissue, and brain tumors [28]. ACC's harboring somatic TP53 mutations are on average larger, more advanced in stage, and associated with shorter disease-free survival [29]. The majority of *TP53* mutations associated with ACC are loss-of-function mutations; however, many of these are predicted to result only in partial loss of p53 function [30]. Curiously, transgenic *TP53* knockout and mutant mouse models do not develop ACC despite developing multiple tumors [31,32]. Else and colleagues showed that transgenic mice carrying an inactivation mutation in the *tripeptidyl peptidase*

1/ACD sheltering complex subunit and telomerase recruitment factor (Tpp/Acd) in addition to a single wild type *TP53* allele do develop ACCs at low frequencies [33].

3.2. Insulin-Like Growth Factor 2 (IGF2)

IGF2 is a paternally imprinted critical growth factor in the development of many organ systems, including the adrenal cortex, where it is highly expressed in early fetal development [34]. Multiple studies have confirmed *IGF2* overexpression in between 83.3% and 90.9% of ACC's when compared with ACA and NAC [35–38]. The maternally imprinted long non-coding RNA *H19* gene located on the antisense strand of the *IGF2* gene is shown to be underexpressed in ACC compared with ACA and NAC in multiple studies [39–42]. Both LOH [17] and paternal uniparental disomy at the 11p15 locus result in *IGF2* overexpression and reduced expression of *H19* and *Cyclin-dependent kinase inhibitor 1C (CDKN1C)* in ACC [36], which are associated with poor prognosis and increased rates of recurrence [43].

IGF2 binds to the membrane tyrosine kinase receptor IGF Receptor Type 1 (IGF-1R), leading to receptor autophosphorylation and binding of the insulin receptor substrate 1 (IRS-1). Tyrosine phosphorylation of IRS-1 activates the phosphatidylinositol-3-kinase(PI3K)/serine/threonine protein kinase B (Akt) and mammalian target of rapamycin(mTOR) pathway as well as the Ras/Raf/mitogen-activated protein kinase (MEK)/extracellular signal-related kinase (ERK) pathways, potentiating cellular proliferation and viability in ACC cell models [44].

Transgenic mouse models of *IGF2* overexpression [45] and adrenal cortex-specific loss of imprinting at the *IGF2/H19* region [46] have demonstrated that these factors alone are not sufficient to initiate tumorigenesis. From a therapeutic approach, clinical trials of the IGF-1R small molecule inhibitors, Linsitinib and Figitumumab to treat advanced ACC failed to show benefit in progression-free survival or overall survival [47,48]. Another trial involving 26 ACC patients treated with the IGF-1R antibody Cixutumumab in combination with the mTOR inhibitor Temsirolimus achieved stable disease for at least six months in 42% of patients but did not lead to any partial or complete responses to therapy [49].

3.3. Wnt/β-Catenin Signalling Pathway

The Wnt signaling pathways are activated by Wnt-protein ligand binding extracellularly to a membrane Frizzled receptor. Canonical Wnt pathway activation leads to the accumulation of β-catenin in the cytoplasm, which ultimately translocates to the nucleus where it activates transcription. In the absence of Wnt, β-catenin is degraded by a protein complex formed by Axin, APC, protein phosphatase 2A (PP2A), glycogen synthase kinase 3 (GSK3) and casein kinase 1α (CK1α), by targeting it for ubiquitination and ultimate proteosomal degradation [50].

Assie and colleagues performed exome sequencing and single nucleotide polymorphism (SNP) analysis of 77 ACC tissues and showed alterations in the β-catenin pathway associated genes *zinc and ring finger 3 (ZNRF3)* (21%), *cadherin-associated protein β1 (CTNNB1)* (16%) and *APC* (2%) [51]. More strikingly, in a series of 50 ACC tissues, Maharjan and colleagues reported that the Wnt/β-catenin pathway was aberrantly activated in 62% [52].

Transgenic mouse models of constitutive *β-catenin* activation in the adrenal cortex produced aggressive adrenal tumors only in a subset of 17-month-old mice [53], while *APC* knockout mice displayed hyperplasia progressing to adrenal adenoma but not carcinoma [46].

Together these findings indicate that multiple genetic aberrations are required for the development of ACC, and combinatorial therapeutic strategies targeting multiple pathways may be effective.

4. Overview of microRNA Structure, Biogenesis, and Function

MiRNAs are short, single-stranded non-coding RNA molecules spanning between 17–25 nucleotides. Approximately 2,300 miRNAs have been identified in the human genome [54]. Their expression is tissue-specific, and they broadly act to negatively regulate the gene expression of at least 60% of human RNA transcripts through either translational inhibition or transcript decay [55].

While the majority of miRNA targets are mRNAs, other classes of RNA, including rRNA, tRNA, lncRNA, and other miRNAs make up 30% of all miRNA targets [56].

The RNAse enzymes DICER and DROSHA involved in miRNA maturation have been implicated in ACC. In a study which compared the expression of key miRNA processing factors between 29 ACC and 43 adrenocortical adenoma (ACA) tissues, Caramuta and colleagues showed that DICER, DROSHA, and TAR RNA-binding protein 2 (TARBP2) (a DICER cofactor required for miRNA processing [57]) were overexpressed at the mRNA and protein levels in ACC compared with ACA [58]. In addition, in vitro inhibition of *TARBP2* expression in the ACC cell line H295R resulted in decreased cellular proliferation and increased apoptosis [58]. This evidence suggests that dysregulation of the miRNA biogenesis pathway may potentiate tumorigenesis in ACC.

In the cytoplasm, the mature miRNA duplex unwinds from the thermodynamically less stable end, and the RNA strand that orients its 5′ end in this direction known as the guide strand is loaded onto the RNA-induced silencing complex (RISC) [59]. RISC is a multiprotein complex containing one member of the Argonaute protein family [60]. miRNAs contain a seed region spanning 2–8 nucleotides at their 5′ ends, which allows them to guide RISC mainly to the 3′ untranslated region (UTR) of their target mRNA through complementary base pairing. The degree of complementarity between miRNA and mRNA and the enzymatic properties of the Ago-2 protein determine whether mRNA silencing will be achieved through target cleavage or translational inhibition [61].

In cancer, the dysregulation of miRNA expression results from various mechanisms, including amplification or deletion of miRNA genes, dysregulation of transcriptional machinery, changes in methylation and histone modifications, as well as mutations and changes in expression of miRNA biogenesis-related proteins [62,63]. As miRNAs play an important role in the regulation of gene expression, their aberrant expression can lead to significant alterations in cellular phenotype. In their dysregulated state miRNAs can act as either oncogenes or tumor suppressors, affecting the cellular processes required for tumor initiation and progression.

IsomiRs and Their Emerging Significance in Cancer

Inaccurate cleavage by either DICER or DROSHA, nucleotide additions at the 3′ end, and nucleotide modifications could result in the production of miRNA isoforms (isomiRs) [64]. Increasing data shows that isomiRs have significant impacts on miRNA-mediated gene regulation [65]. Variations in the 5′ seed sequence could impact the specificity of miRNAs to their targets [66], and variations at the 3′ ends determine the stability of miRNA-mRNA binding [67]. The impact of a particular isomiR on miRNA function depends on the relative abundance and stability of the isomiR relative to its canonical miRNA, as well as its binding efficiency to its target. Chan and colleagues demonstrated this by showing that isomiRs of *miR-31* regulated the expression of known targets to varying degrees *in vitro*, and their binding capacity to the RISC complex determined using Ago-2 immunoprecipitation (IP) was also varied [68].

Recent advances in high-throughput RNA sequencing technologies have allowed tissue transcriptome profiling at the isomiR level. Telonis and colleagues [69] used The Cancer Gene Atlas (TCGA) miRNA sequencing data across 32 cancer types, including ACC, to determine whether the presence or absence of isomiRs could discriminate between the cancer types. By binarizing the isomiR expression data, they were able to successfully classify tumor datasets with an average sensitivity of 90% and false discovery rate (FDR) of 3%, which was superior to wild type miRNA expression (average sensitivity 83%, FDR 5%) [69]. More recently, Wang and colleagues used the same data to demonstrate that isomiRs that share their seed region (5′ isoforms) could similarly discriminate between tumors [70]. In 2018, Lan and colleagues published the first study to use breast cancer TCGA small RNA sequencing expression data to show that the isomiR expression-based classification was superior to gene expression profiling at distinguishing between breast cancer subtypes [71]. Together these findings suggest that miRNAs may carry out significant functional roles in tumorigenesis at the isomiR level.

A study that investigated isomiR expression in adrenal tissues using RNA sequencing technology published in 2017 found that 411 miRNAs existed as 1763 various isoforms in a cohort of 14 ACC, 18 ACA, and 18 NAC samples [72]. These isomiRs contained 520 various seed sequences, of which 38% were non-canonical. Over a quarter of all expressed miRNAs in the ACC, ACA, and NAC groups produced isomiRs with two or more seed regions that were predicted to target a different set of mRNAs, but these were not investigated further. Therefore, the diagnostic and clinical significance of differentially expressed isomiRs in ACC remains unknown, and further research is needed to elucidate this.

While technical challenges related to the identification and quantification of isomiRs exist, recent developments in data sharing and technology are helping to overcome these. Often isomiR sequencing data is drawn from small sample numbers making it difficult to make general conclusions, but this is being overcome by the release of small RNA sequencing data from large sample databases such as TCGA. This has allowed researchers to study isomiR expression in more detail than previously possible. Also, recently developed specialized techniques such as photoactivatable ribonucleoside-enhanced crosslinking and Ago immunoprecipitation (Ago PAR-CLIP) ensures that identified isomiRs are biologically active and not degradation products, improving the reliability of the experimental data in this field. The utilization of such advances will facilitate research that will unlock a deeper understanding of the role of isomiRs in ACC.

5. The Unique microRNA Expression Signature of ACC and Its Clinical Significance

5.1. The microRNA Expression Signature of ACC Tissues

Sporadic ACC is a genetically heterogeneous malignancy that can be classified into distinct groups based on transcriptomics and clinical behavior. Several published studies have profiled differential miRNA expression in ACC tissue samples compared with either ACA and/or normal adrenal cortex (NAC) tissue using microarray data [73–77], TaqMan Low-Density Arrays (TLDA) [78,79], RT-q-PCR [80], or RNA sequencing [51,72,81] (Table 2). Across these studies, *miR-483-5p*, *miR-503-5p*, *miR-210*, and *miR-483-3p* were overexpressed in ACC compared with ACA or NAC in multiple datasets, and *miR-195*, *miR-497*, and *miR-335* were underexpressed. Of note, *miR-483-5p* was overexpressed in eight of eleven studies, and has been associated with poor prognosis in ACC [73]. Earlier studies using microarray and RT-q-PCR techniques could only investigate known miRNAs, whereas later studies which utilized RNA sequencing could identify differentially expressed miRNAs which had not previously been characterized. Hence, more numerous microRNA candidates have been identified with RNA sequencing, and novel candidates like *miR-508-3p* were only identified and validated in these later studies [51,81].

Table 2. Studies investigating differential expression of microRNAs in ACC Tissues compared with ACA and NAC.

Year of Publication	Methodology	Tissue Samples	Upregulated miRNA		Downregulated miRNA		Reference
2009	TLDA	7 ACC, 19 ACA, 10 NAC	*miR-184* *miR-210* *miR-503*		*miR-214* *miR-511* *miR-375*		[78]
2009	Microarray VC: RT-q-PCR	22 ACC, 27 ACA, 6 NAC VC (10 ACC, 9 ACA)	*miR-483-5p* *miR-503*		*miR-7* *miR-195*	*miR-335*	[73]
2011	Microarray VC: RT-q-PCR	25 ACC, 43 ACA, 10 NAC	*miR-483-3p* *miR-483-5p*	*miR-210* *miR-21*	*miR-195* *miR-497*		[74]
2011	Microarray VC: RT-q-PCR	10 ACC, 26 ACA VC (31 ACC, 35 ACA, 21 NAC)	*miR-483-5p*		*miR-195* *miR-125b*	*miR-100*	[75]
2011	TLDA VC: RT-q-PCR	7 ACC, 9 ACA, 4 NAC VC (16 ACC)	*miR-139-5p*		*miR-139-3p* *miR-675*	*miR-335*	[79]
2013	Microarray	12 ACC, 6 NAC VC (18 ACC, 10 ACA, 3 NAC)	*miR-483-5p* *miR-503* *miR-210* *miR-542-5p*	*miR-320a* *miR-93* *miR-148b*	*miR-195* *miR-335* *miR-497*	*miR-199a-5p* *miR-199a-3p*	[76]
2014	RT-q-PCR	51 ACC, 47 ACA	*miR-483-3p* *miR-483-5p* *miR-210*		*miR-195*		[80]
2014	RNA sequencing	45 ACC, 3 NAC	*miR-34b-5p* *miR-410* *miR-483-3p* *miR-483-5p* *miR-503*	*miR-506-3p* *miR-506-5p* *miR-508-3p* *miR-508-5p* *miR-510*	*miR-511* *miR-214-3p* *miR-485-3p* *miR-497* *miR-195*		[51]
2015	Microarray VC: RT-q-PCR	8 ACC, 25 ACA VC (11 ACC, 4 ACA)	*miR-503*		*miR-34a*	*miR-497*	[77]
2016	RNA sequencing	79 ACC, 120 NAC	*miR-10-5p* *miR-483-5p* *miR-22-3p* *miR-508-3p* *miR-509-3p*	*miR-509-5p* *miR-340* *miR-146a* *miR-21-3p* *miR-21-5p*			[81]
2017	RNA sequencing	7 ACC, 8 ACA, 8 NAC VC (8 ACC, 10 ACA, 10 NAC)	*miR-503-5p* *miR-450a-5p* *miR-542-5p* *miR-483-3p* *miR-542-3p* *miR-450b-5p* *miR-210* *miR-483-5p*	*miR-421* *miR-424-3p* *miR-424-5p* *miR-598* *miR-148b-3p* *miR-184* *miR-128*			[72]

TLDA, TaqMan Low Density Array; VC, Validation Cohort; ACC, adrenocortical carcinoma; ACA, adrenal adenoma; NAC, normal adrenal cortex.

5.2. Circulating microRNAs as Diagnostic Biomarkers in ACC

Diagnosing patients with ACC continues to present challenges as no preoperative blood-borne tumor marker for the disease exists, and suspicion is often raised on imaging. MiRNAs are stable in bodily fluids within extracellular vesicles shed from tumor cells or in protein complexes and are relatively protected from enzymatic degradation in the circulation [82], making them attractive as potential noninvasive diagnostic markers for cancer. Despite a significant body of research confirming differentially expressed circulating miRNAs in various diseases, none have translated into clinical use.

A number of studies have reported on the expression and diagnostic utility of circulating miRNAs in ACC, whose findings have been summarized in a recent review by Decmann and colleagues [83]. Several of these studies have attempted to define candidate diagnostic circulating miRNA biomarkers for ACC in serum [76,84,85], but both relatively low sensitivity and specificity values have limited their clinical application. Although circulating *miR-483-5p* relative expression is a reliable differentiator between aggressive and non-aggressive ACC in serum, as demonstrated by Chabre and colleagues (AUC 0.929) [76], it is unable to differentiate between ACC and adrenocortical adenoma (AUC 0.74) [85]. Circulating *miR-483-5p* is also overexpressed in hepatocellular [86] as well as head and neck cancers [87], and has been proposed as a diagnostic marker in these diseases as in ACC. The lack of specificity of this miRNA as a biomarker further limits its clinical utility in ACC, and is characteristic of oncological miRNAs across different tumor types [88].

At present, various methods are in use for the quantification of circulating miRNAs in preclinical research. The lack of standardization in sample collection, storage, and processing introduces significant variation in the data, which limits the generalizability of differential expression results [89]. In addition, common reference RNA genes used as calibrators in comparing miRNA expressions are differentially expressed in the serum of patients with various diseases, leading to challenges in data normalization for analysis [90]. To overcome this, the addition of synthetic RNA during RNA extraction as a spike-in control is a method widely used for technical normalization. Such an exogenous control is helpful as it undergoes the same processing as endogenous RNA in the sample, but this does not correct for variables such as the serum miRNA fraction [91]. Standardization of sample collection techniques, developments in vesicle-associated miRNA quantification, and the use of absolute quantification methods that do not rely on housekeeper genes could help overcome the obstacles to the clinical adoption of circulating miRNAs as diagnostic biomarkers in the future.

5.3. Tissue microRNA Expression as a Prognostic Tool in ACC

In patients with ACC, the prescribed adjuvant clinical management and follow-up regimen is informed by the estimated risk of disease recurrence. Even in patients with unresectable disease, systemic therapy depends on tumor biology. While clinical and pathological prognostic indicators such as tumor stage, pathological grade, Ki67 proliferation index, and resection status are helpful in estimating survival, more recently, genomic and transcriptome based studies have identified molecular markers that can help predict recurrence-free survival as well [92]. Various RT-q-PCR studies have shown that tissue relative expression levels of *miR-210*, *miR-483-5p*, *miR-195*, *miR-503*, *miR-1202*, and *miR-1275* are associated with overall survival in ACC [73,74,80]. In addition, ACC tissue *miR-9* relative expression has been shown to correlate with recurrence-free survival as well as overall survival [93], as has the serum relative expression of *miR-483-5p* and *miR-195* [76].

In an RNA sequencing study of 45 ACC and three NAC tissues, Assie and colleagues used consensus clustering to classify tumors into three groups based on their microRNA profiles. The cluster most distinct from NAC, Mi1, was also characterized by consistent 14q32 LOH and *maternally expressed 3* (*MEG3*) long non-coding RNA promotor methylation. The 14q32 cytogenetic band contains 54 miRNAs, one of the largest miRNA clusters in the human genome, and 38 of these were underexpressed in the good prognosis of Mi1 tumors. The Mi2 group was characterized by weak overexpression of the *miR-506-514* cluster, known to have an oncogenic role in melanoma [94], while the Mi3 group was strongly correlated with the poor prognosis transcriptome cluster C1A. Interestingly, while miRNA expression was maximally deregulated in Mi1 and Mi2 cluster tumors, ACC driver pathway alterations were more consistently associated with Mi3 cluster tumors [51]. This study suggests that the integrated analysis of miRNA expression is likely to be a superior approach to single miRNA prognostic biomarkers for ACC.

6. Computational and Experimental Methods of miRNA Target Identification

Based on the existing understanding of the interactions between miRNAs and target mRNAs, various software tools have been developed to predict endogenous miRNA targets for experimental validation [95]. Recently, Ab Mutalib and colleagues reviewed the thirty-nine computational tools currently available for miRNA target prediction [96], of which only one, DeAnnIso, allows for target prediction of isomiRs [97]. These bioinformatic prediction tools are limited as they cannot predict miRNA binding to non-coding RNAs, nor do they account for non-canonical mRNA binding sites [98]. In addition, bioinformatics methods may give false-positive results, and miRNA target predictions do not always account for the tissue specificity of miRNA expression [99]; therefore, computational target predictions should always be validated experimentally.

The various experimental methods available for miRNA target validation in biological systems have been comprehensively reviewed elsewhere [100–102]. They include indirect methods such as expression profiling or stable isotope labeling by amino acids in cell culture (SILAC) following miRNA overexpression or inhibition, as well as direct methods such as reporter assays, biotinylated miRNA pulldown assays, and RISC component pulldown assays. In reporter assays, direct evidence of miRNA regulation is established when a mutated mRNA target site results in loss of miRNA regulation. In these experimental approaches, miRNAs are often overexpressed to supraphysiological levels, resulting in the saturation of RISC complexes at the expense of other endogenous miRNAs, and false-positive results that allow low-affinity targets to appear functionally relevant [103]. Nevertheless, these low-affinity targets, while irrelevant to the endogenous functioning of the miRNA in question, continue to be important when considering the cellular effects of miRNAs as potential therapeutics. Caution must also be exercised when extrapolating results from miRNA target validation experiments in particular cellular environments across tissue types, as the failure to detect cell-specific natural targets may ensue [100].

7. Functional miRNA Target Relationships in ACC

In ACC, the evidence for miRNA functional targets comes largely from reporter assays in combination with the cellular effects of modulation of miRNA expression in cell culture.

7.1. Overexpressed miRNAs and Their ACCs

Several miRNAs that are overexpressed in ACC relative to NAC have proven oncogenic roles *in vitro*, as well as defined molecular targets, that they regulate (Table 3). *miR-9* [104], *miR-21* [105], *miR-483-3p* [106], and *miR-483-5p* [106] have been well described in the literature as 'oncomiRs' across multiple mammalian cell types, which is consistent with their role in ACC. In contrast, *miR-139-5p* is an established tumor suppressor in head and neck/oral, breast, and gastric cancers [107], but is overexpressed in aggressive ACCs compared with non-aggressive ACCs [108].

Table 3. Overexpressed microRNAs and their regulated targets in ACC.

microRNA	Functional Role in ACC	Molecular Target	Evidence for Regulatory Interaction	Reference/s
miR-9	Associated with reduced DFS and increased recurrence in clinical data	*LIN28*	Weak protein expression pattern in aggressive ACC Established reporter assays in non-ACC models	[109–111]
miR-21	↑ cellular proliferation	*PCDC4*	Inverse correlation in expression Reporter assays in non-ACC cell models	[112–115]
miR-139-5p	Associated with anchorage-independent colony formation	*NDRG4*	Inverse correlation in expression Reporter assays in non-ACC cell models	[116]
miR-483-3p	↑ cellular proliferation ↓ apoptosis	*PUMA*	Inverse correlation in expression Reporter assays in non-ACC models	[74,117]
miR-483-5p	Associated with anchorage-independent colony formation	*NDRG2*	Inverse correlation in expression Reporter assays in non-ACC models	[116]

DFS, Disease-Free Survival; PDCD4, Programmed cell death protein 4; NDRG, N-myc downstream-regulated gene; PUMA, p53 upregulated modulator of apoptosis.

7.1.1. miR-9 Regulates LIN28

miR-9 has diverse actions in cancer, and whether it acts as a tumor suppressor or oncomiR is tissue dependent [118]. In ACC, aggressive phenotypes overexpress *miR-9* in comparison with non-aggressive phenotypes [51], and *miR-9* overexpression is associated with poor prognosis in clinical datasets [93]. Luciferase-based assays have demonstrated direct binding between *miR-9* and LIN28, an RNA binding protein that regulates miRNA biogenesis via the miRNA *let-7* in HeLa and A2780 ovarian carcinoma cell lines [111]. Aggressive ACCs have been demonstrated to have weak LIN28 protein expression on immunostaining [93], lending evidence to the hypothesis that LIN28 expression is regulated by *miR-9* in this cellular environment.

7.1.2. MiR-21 Regulates PDCD4

MiR-21 is the most commonly overexpressed miRNA in cancer and is generally associated with an aggressive phenotype and poor prognosis. *miR-21* expression is negatively correlated with *Programmed Cell Death Protein 4 (PCDC4)* expression across many solid tumors [105]. *PCDC4* is upregulated during apoptosis and inhibits translation of particular genes, including *p53*, by competitively binding translation initiation factors [119]. Luciferase reporter assays in HeLa cells as well as colorectal and thyroid cell lines, have established *miR-21* as a direct regulator of *PCDC4* expression [114,115]. In ACC, in vitro gene-specific silencing of *miR-21* resulted in increased *PCDC4* expression and reduced cellular proliferation [113], which suggests that this regulatory relationship between *miR-21* and *PCDC4* is also present in ACC.

7.1.3. miR-483-3p Regulates PUMA

miR-483-3p is an oncomiR in ACC, promoting cellular proliferation and inhibiting apoptosis in in vitro cell models [74]. Reporter assays in three different cell lines, including human embryonic kidney (HEK293), liver cancer (HepG2), and colon cancer (HCT116), demonstrated that *miR-483-3p* directly inhibits *PUMA* expression [117]. *PUMA* is a downstream target of p53, which antagonizes the anti-apoptotic B-cell lymphoma 2 (Bcl-2) family proteins and consequently induces apoptosis [120]. *PUMA* expression was found to be inversely correlated with *miR-483-3p* expression in ACC, but not in ACA or NAC tissue [74]. Given that *miR-483-3p* is a proven regulator of *PUMA* expression in various cell models, this relationship can be extrapolated to ACC.

7.2. Underexpressed miRNAs and Their ACC Targets

The molecular targets of a number of underexpressed tumor suppressor miRNAs in ACC have also been characterized (Table 4). These miRNAs include *miR-7* and *miR-205* which have demonstrated tumor suppressor activity in in vivo xenograft ACC models [121,122]. They also include *miR-195* and *miR-497*, which are members of the tumor suppressor *miR-15* family and share the same seed sequence [123], as well as *miR-99* family members *miR-99a* and *miR-100*, which are known to target the mTOR signaling pathway [124].

Table 4. Underexpressed microRNAs and their regulated targets in ACC.

MicroRNA	Functional Role in ACC	Molecular Target	Evidence for Regulatory Interaction	Reference/s
miR-7	↓ cellular proliferation ↑ G1 cell cycle arrest ↓ H295R xenograft growth *in vivo*	*Raf-1* * *EGFR* * *CDK1* *PAK1* *CKS2*	Reporter assays * Inverse correlation in expression	[121]
miR-99a		*IGFR1* *mTOR*	Inverse correlation in expression Reporter assays in non-ACC models	[125]
miR-100		*IGFR1* *mTOR*	Inverse correlation in expression Reporter assays in non-ACC models	[125]
miR-195	↓ cellular proliferation ↑ cellular invasion ↑ apoptosis	*TARBP2* *DICER1*	Inverse correlation in expression Ago-2 IP	[58,74]
		ZNF367	Reporter assays in SW13 cells	[126]
miR-205	↓ cellular proliferation ↑ apoptosis ↓ SW13 xenograft tumor growth *in vivo*	*Bcl-2*	Reporter assays	[122]
miR-375	↓ cellular proliferation	*MTDH*	Reporter assays Inverse correlation in expression	[127]
miR-431	↑ cellular sensitivity to doxorubicin and mitotane	*ZEB1*	Inverse correlation in expression in miRNA overexpressing doxorubicin treated H295R cells. Reporter assays in non-ACC models	[128]
miR-497	↓ cellular proliferation ↑ apoptosis ↑ G1 cell cycle arrest	*TARBP2* *DICER1*	Inverse correlation in expressionAgo-2 IP	[58,74]
		MALAT1 *eIF4E* *SFPQ*	Inverse correlation in expression Reporter assays	[129]

* Regulatory interaction demonstrated by reporter assays for Raf-1 and EGFR only. Raf-1, Rapidly Accelerated Fibrosarcoma-1; EGFR, Epidermal Growth Factor Receptor; PAK1, p21 activated kinase 1; CKS2, CDC28 Protein Kinase Regulatory Subunit 2; CDK1, Cyclin-Dependent Kinase 1; IGFR1, Insulin-like Growth Factor 1 Receptor; mTOR, Mechanistic Target of Rapamycin Kinase; TARBP2, TAR RNA-binding protein 2; DICER1, Dicer 1 Ribonuclease III; Ago-2 IP, Argonaute-2 immunoprecipitation; ZNF367, Zinc Finger Protein 367; Bcl-2, B-cell lymphoma 2; MTDH, metadherin; ZEB1, Zinc finger E-box binding homeobox 1; MALAT1, metastasis-associated lung adenocarcinoma transcript 1; eIF4E, eukaryotic translation initiation factor 4E; SFPQ, splicing factor proline and glutamine-rich.

7.2.1. *miR-7* Regulates *Raf-1, EGFR, CDK1, PAK1, CKS2*

In the human genome, *miR-7* is encoded on three separate loci whose different DNA sequences can all be processed into the same mature *miR-7* sequence [130]. *miR-7* is an almost ubiquitous tumor suppressor, being underexpressed in malignancies that range from those derived from brain tissue to a myriad of solid tumors as well as leukemias [131]. In ACC, in vitro overexpression of *miR-7* decreases proliferation and induces G1 cell cycle arrest and decreases the expression of *p21 activated kinase 1 (PAK1), CDC28 protein kinase regulatory subunit 2 (CKS2)*, and *cyclin-dependent kinase 1 (CDK1)* mRNA [121]. PAK1 activation induces apoptosis, while CKS2 is an essential co-factor for CDK proteins that regulate the cell cycle. Luciferase reporter assays in H295R cells demonstrated the regulatory relationship between *miR-7* and *rapidly accelerated fibrosarcoma-1 (Raf-1)* as well as the *epidermal growth factor receptor (EGFR)* [121]. *miR-7* targeting of *EGFR* has also been demonstrated in breast, lung, gastric, and ovarian cancers, as well as glioma and schwannoma tumors. In schwannomas and breast cancer, *miR-7* has also been shown to target *PAK1* [131].

7.2.2. miR-99a/100 Regulates IGFR1, mTOR

Both *miR-99a* and *miR-100* were discovered as underexpressed relative to NAC in childhood ACC tissue samples, where their expression was inversely correlated with both *mTOR* and *insulin-like growth factor 1 receptor (IGFR1)* mRNA. *miR-100* specific knockdown in in vitro ACC cell models was associated with increased mTOR and IGFR1 protein expression, and furthermore, luciferase reporter assays in HEK293 cells showed that both miRNAs could regulate mTOR and IGFR1 [125]. The regulatory relationship between the *miR-99* family and mTOR has been well studied in cardiovascular disease [132] as well as in wound healing [133]. In cancer, *miR-99* regulation of mTOR has been demonstrated to enhance radiation sensitivity in urothelial carcinoma [134] as well as non-small cell lung cancer [135].

7.2.3. miR-205 Regulates Bcl-2

In ACC, *miR-205* was shown to be underexpressed in a clinical cohort with RT-q-PCR. The subsequent gain of function studies carried out using SW13 cells (a cell line derived from adrenocortical metastasis of unknown origin) showed that *miR-205* promoted apoptosis and impaired cellular proliferation *in vitro*, and in vivo mouse SW13 xenograft studies showed that it could inhibit tumor growth [122]. Luciferase reporter assays also carried out in SW13 demonstrated a direct regulatory relationship between *miR-205* and B-cell lymphoma 2 (Bcl-2) protein, which is known to regulate the intrinsic apoptotic pathway in cancer [136].

7.2.4. *miR-375* Regulates MTDH

miR-375, a known tumor suppressor in multiple cancers, is underexpressed in ACC [78], and in aldosterone-producing adrenal adenomas, its expression is correlated with the tumor size [127]. In vitro overexpression of *miR-375* reduces cellular proliferation and suppresses metadherin (MTDH), which functions to promote tumor invasion, metastasis, and chemoresistance. It also acts via the PI3K/Akt and Wnt/β-catenin pathways to promote cellular proliferation, invasion, and survival [137]. Luciferase reporter assays in H295R cells show that *miR-375* directly binds *MTDH* mRNA and regulates its expression in vitro in ACC.

7.2.5. miR-431 Regulates ZEB1

In ACC clinical samples, *miR-431* is differentially expressed in chemosensitive tumors compared with chemoresistant tumors. The gain of function studies in H295R and primary ACC cells showed that *miR-431* overexpression decreased the IC50 of both doxorubicin and mitotane to inhibit cellular proliferation. In cells treated with doxorubicin, *miR-431* reversed epithelial-to-mesenchymal transition (EMT) [128]. Zinc finger E-box binding homeobox 1 (ZEB1), a protein that induces EMT in cancer cells, had already been established as a direct target of *miR-431* in hepatocellular carcinoma [138].

Both *ZEB1* mRNA and protein expression decreased in doxorubicin treated H295R cells overexpressing *miR-431* [128], indicating that this regulatory relationship between *miR-431* and ZEB1 is active in ACC.

7.2.6. miR-497 Regulates TARBP2, DICER1, MALAT1, eIF4E, SFPQ

miR-497 expression is dysregulated in many solid organ tumors, which suggests that it may play an important tumor suppressor role. Multiple studies have confirmed *miR-497* underexpression in ACC and its genomic location in a region of frequent LOH (17p13.1-13.3), in close proximity to the *p53* locus, indicates that it may play a role in ACC tumorigenesis [15]. In vitro H295R gain of function studies have shown that *miR-497* decreases cellular proliferation, increases apoptosis, and also induces G1 cell cycle arrest [74,129]. *miR-497* has been shown to directly target the miRNA biogenesis related proteins DICER1 and TARBP2 in Ago-2 IP assays, along with *miR-195*. This was confirmed with gain of function studies that showed an inverse correlation between *miR-497* expression and *DICER1*, as well as *TARBP2* mRNA and protein expression in H295R cells [58]. In a separate study, which was the first to demonstrate miRNA targeting of long non-coding RNAs in ACC, luciferase reporter assays demonstrated that *miR-497* regulates the expression of *metastasis-associated lung adenocarcinoma transcript 1 (MALAT1)* [129]. *MALAT1* is overexpressed in numerous types of tumors, including ACC, and is known to promote cellular proliferation, apoptosis, migration, and invasion [138]. In H295R, *miR-497* overexpression and *MALAT1* knockdown inhibit the expression of *eukaryotic translation initiation factor 4E (eIF4E)*, which directs ribosomes to the cap structure of mRNAs and is, therefore, essential for protein synthesis [129]. *miR-497* gain of function and *MALAT1* knockdown studies further demonstrated the reciprocal inhibitory relationship between them in in vitro ACC models.

Within the limits of the caveats previously outlined, we can infer from the above studies that miRNAs modulate many protein targets that are involved in key driver pathways in ACC.

8. miRNA Modulation of ACC Driver Pathways

A significant proportion of the identified ACC miRNA molecular targets play various roles in the established ACC driver pathways. This supports the notion that miRNA modulation of protein expression, which in healthy cells helps to finetune and maintain the homeostatic balance, can potentiate oncogenesis when dysregulated.

8.1. miRNA Modulators of the p53 Pathway in ACC

The overexpressed oncomiR *miR-483-3p* and the underexpressed tumor suppressors *miR-7* and *miR-205* all regulate downstream targets of p53 (Figure 1). *miR-483-3p* suppression of PUMA expression and the alleviation of *miR-205* modulated Bcl-2 inhibition of Bax, act synergistically to inhibit p53-mediated apoptosis. P53 is known to transcriptionally downregulate CDK1, and thus, initiates G2 cell cycle arrest. Constitutive activation of CDK1 resulting from the loss of *miR-7* targeted suppression overrides p53 mediated G2 arrest, leading to uncontrolled proliferation.

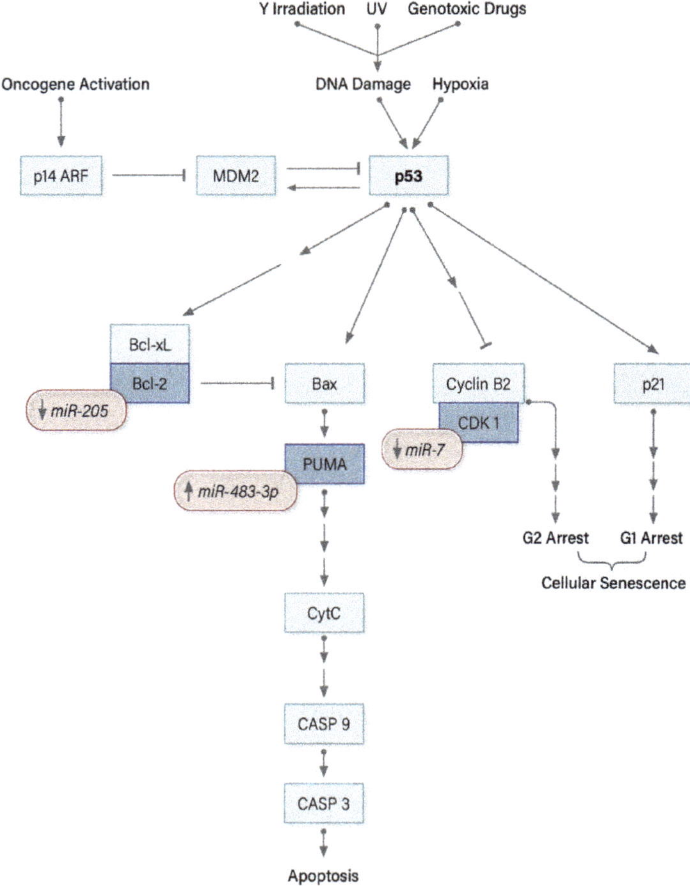

Figure 1. miRNA modulators of the p53 signaling pathway in Adrenocortical Carcinoma.

8.2. miRNA Modulators of the mTOR Pathway in ACC

The loss of *miR-7* regulation of Raf-1 and EGFR expression leads to downstream mTOR activation in ACC. The underexpression of *miR-99a/100* also leads to mTOR activation, ultimately potentiating protein synthesis, which is further enhanced by the loss of *miR-497* mediated eIF4E regulation. The loss of *miR-99a/100* mediated IGFR1 expression also promotes cell survival via the PI3K/AKT signaling pathway (Figure 2).

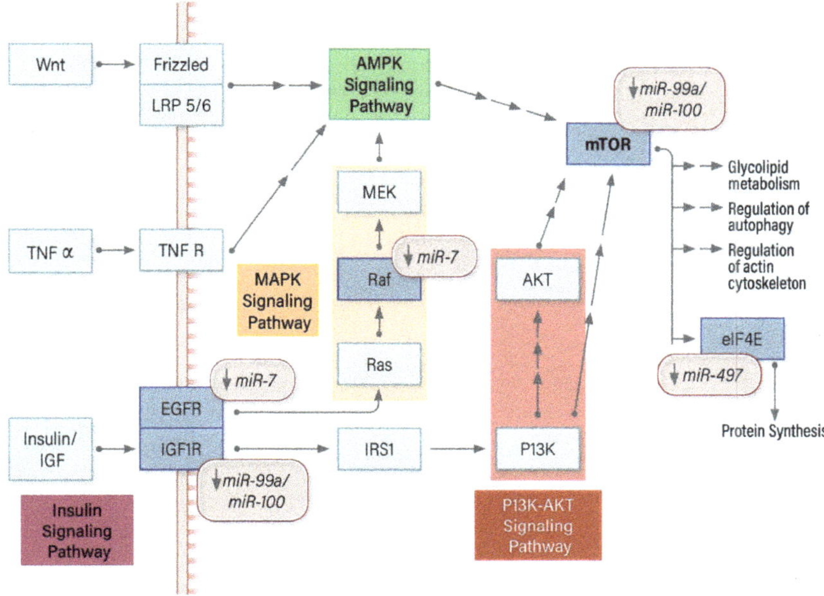

Figure 2. miRNA modulators of the mTOR signaling pathway in Adrenocortical Carcinoma.

8.3. miRNA Modulators of the Wnt/B-Catenin Pathway in ACC

In ACC, the loss of *miR-431* regulation allows ZEB1 to activate Wnt, consequently activating β-catenin, which potentiates cell cycling. The loss of *miR-375* mediated MTDH suppression upstream of MAPK modulates the Wnt/β-catenin pathway to promote cell cycling (Figure 3).

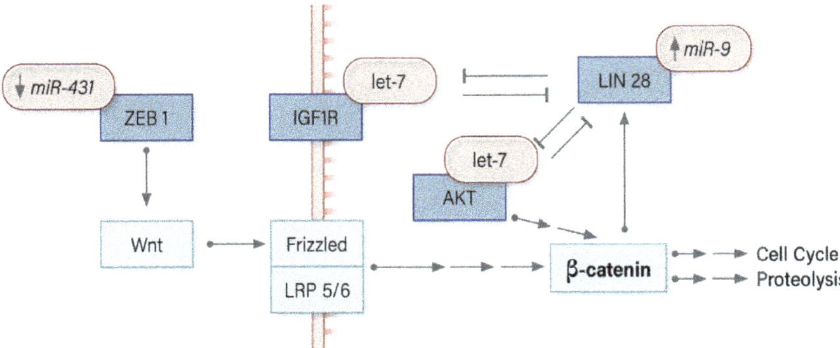

Figure 3. miRNA modulators of the Wnt/β-catenin signaling pathway in Adrenocortical Carcinoma.

9. Future Directions and Conclusions

miRNAs play an important role in the modulation of ACC related target protein expression. The dysregulation of miRNA expression disturbs the homeostatic balance of proteins that participate in the pathways controlling cell cycle progression, cellular proliferation, apoptosis, and chemoresistance. The overexpression of oncogenic miRNAs and underexpression of tumor suppressor miRNAs thus potentiate tumorigenesis. The role of miRNA regulation in ACC remains an area of active research with the potential to further our understanding of its tumor biology and the molecular pathways involved. Small RNA sequencing of isomiRs and further refining our understanding of the miRNA signature of

ACC provides the opportunity to improve diagnostic accuracy with techniques such as miRNA liquid biopsy. With continuing advances in functional techniques that allow molecular interactions to be clearly established, it will be possible to explore novel miRNA-based therapeutic approaches with the aim of improving the current poor prognosis of these patients.

Author Contributions: Conceptualization, M.C., S.S. and G.H.; writing—original draft preparation, M.C.; writing—review and editing, M.B., A.G., S.S., G.H.; visualization, M.C.; supervision, G.H. and S.S. All authors have read and agreed to the published version of the manuscript.

Funding: This research received no external funding.

Acknowledgments: Charbel Azzi for graphics support.

References

1. Hedeland, H.; Östberg, G.; Hökfelt, B. On the Prevalence of Adrenocortical Adenomas In An Autopsy Material In Relation To Hypertension And Diabetes. *Acta Medica Scand.* **2009**, *184*, 211–214. [CrossRef] [PubMed]

2. Kebebew, E.; Reiff, E.; Duh, Q.-Y.; Clark, O.H.; McMillan, A. Extent of Disease at Presentation and Outcome for Adrenocortical Carcinoma: Have We Made Progress? *World J. Surg.* **2006**, *30*, 872–878. [CrossRef] [PubMed]

3. Koschker, A.-C.; Fassnacht, M.; Hahner, S.; Weismann, D.; Allolio, B. Adrenocortical Carcinoma—Improving Patient Care by Establishing New Structures. *Exp. Clin. Endocrinol. Diabetes* **2006**, *114*, 45–51. [CrossRef] [PubMed]

4. Luton, J.-P.; Cerdas, S.; Billaud, L.; Thomas, G.; Guilhaume, B.; Bertagna, X.; Laudat, M.-H.; Louvel, A.; Chapuis, Y.; Blondeau, P.; et al. Clinical Features of Adrenocortical Carcinoma, Prognostic Factors, and the Effect of Mitotane Therapy. *N. Engl. J. Med.* **1990**, *322*, 1195–1201. [CrossRef] [PubMed]

5. Sturgeon, C.; Shen, W.T.; Clark, O.H.; Duh, Q.-Y.; Kebebew, E. Risk Assessment in 457 Adrenal Cortical Carcinomas: How Much Does Tumor Size Predict the Likelihood of Malignancy? *J. Am. Coll. Surg.* **2006**, *202*, 423–430. [CrossRef] [PubMed]

6. Pommier, R.F.; Brennan, M.F. An eleven-year experience with adrenocortical carcinoma. *Surgery* **1992**, *112*, 963–971.

7. Baur, J.; Büntemeyer, T.-O.; Megerle, F.; Deutschbein, T.; Spitzweg, C.; Quinkler, M.; Nawroth, P.P.; Kroiss, M.; Germer, C.-T.; Fassnacht, M.; et al. Outcome after resection of Adrenocortical Carcinoma liver metastases: A retrospective study. *BMC Cancer* **2017**, *17*, 1–10. [CrossRef]

8. Bilimoria, K.Y.; Shen, W.T.; Elaraj, D.; Bentrem, D.J.; Winchester, D.J.; Kebebew, E.; Sturgeon, C. Adrenocortical carcinoma in the United States. *Cancer* **2008**, *113*, 3130–3136. [CrossRef]

9. Fassnacht, P.D.M.M.; Terzolo, M.; Allolio, B.; Baudin, E.; Haak, H.; Berruti, A.; Welin, S.; Schade-Brittinger, C.; Lacroix, A.; Jarząb, M.; et al. Combination Chemotherapy in Advanced Adrenocortical Carcinoma. *N. Engl. J. Med.* **2012**, *366*, 2189–2197. [CrossRef]

10. McAteer, J.P.; Huaco, J.A.; Gow, K.W. Predictors of survival in pediatric adrenocortical carcinoma: A Surveillance, Epidemiology, and End Results (SEER) program study. *J. Pediatr. Surg.* **2013**, *48*, 1025–1031. [CrossRef]

11. Wasserman, J.; Novokmet, A.; Eichler-Jonsson, C.; Ribeiro, R.C.; Rodriguez-Galindo, C.; Zambetti, G.P.; Malkin, D. Prevalence and Functional Consequence of TP53 Mutations in Pediatric Adrenocortical Carcinoma: A Children's Oncology Group Study. *J. Clin. Oncol.* **2015**, *33*, 602–609. [CrossRef] [PubMed]

12. Varley, J.M.; McGown, G.; Thorncroft, M.; James, L.A.; Margison, G.P.; Forster, G.; Evans, D.G.R.; Harris, M.; Kelsey, A.M.; Birch, J.M. Are There Low-Penetrance TP53 Alleles? Evidence from Childhood Adrenocortical Tumors. *Am. J. Hum. Genet.* **1999**, *65*, 995–1006. [CrossRef] [PubMed]

13. Herrmann, L.J.M.; Heinze, B.; Fassnacht, M.; Willenberg, H.S.; Quinkler, M.; Reisch, N.; Zink, M.; Allolio, B.; Hahner, S. TP53Germline Mutations in Adult Patients with Adrenocortical Carcinoma. *J. Clin. Endocrinol. Metab.* **2012**, *97*, E476–E485. [CrossRef]

14. Figueiredo, B.C.; Sandrini, R.; Zambetti, G.P.; Pereira, R.M.; Cheng, C.; Liu, W.; Lacerda, L.; A Pianovski, M.; Michalkiewicz, E.; Jenkins, J.; et al. Penetrance of adrenocortical tumours associated with the germline TP53 R337H mutation. *J. Med Genet.* **2005**, *43*, 91–96. [CrossRef] [PubMed]

15. Soon, P.S.H.; Libé, R.; Benn, D.E.; Gill, A.; Shaw, J.; Sywak, M.S.; Groussin, L.; Bertagna, X.; Gicquel, C.; Bertherat, J.; et al. Loss of Heterozygosity of 17p13, With Possible Involvement of ACADVL and ALOX15B, in the Pathogenesis of Adrenocortical Tumors. *Ann. Surg.* **2008**, *247*, 157–164. [CrossRef] [PubMed]

16. Lapunzina, P. Risk of Tumorigenesis in Overgrowth Syndromes: A Comprehensive Review. *Am. J. Med. Genet. Part C Semin. Med. Genet.* **2005**, *137C*, 53–71. [CrossRef] [PubMed]

17. Gicquel, C.; Raffin-Sanson, M.-L.; Gaston, V.; Bertagna, X.; Plouin, P.F.; Schlumberger, M.; Louvel, A.; Luton, J.P.; Le Bouc, Y. Structural and Functional Abnormalities at 11p15 are associated with the Malignant Phenotype in Sporadic Adrenocortical Tumors: Study on a Series of 82 Tumors. *J. Clin. Endocrinol. Metab.* **1997**, *82*, 2559–2565. [CrossRef]

18. Shiroky, J.S.; Lerner-Ellis, J.P.; Govindarajan, A.; Urbach, D.R.; Devon, K.M. Characteristics of Adrenal Masses in Familial Adenomatous Polyposis. *Dis. Colon Rectum* **2018**, *61*, 679–685. [CrossRef]

19. Gaujoux, S.; Grabar, S.; Fassnacht, M.; Ragazzon, B.; Launay, P.; Libé, R.; Chokri, I.; Audebourg, A.; Royer, B.; Sbiera, S.; et al. β-Catenin Activation is Associated with Specific Clinical and Pathologic Characteristics and a Poor Outcome in Adrenocortical Carcinoma. *Clin. Cancer Res.* **2011**, *17*, 206–211. [CrossRef]

20. Else, T. Association of adrenocortical carcinoma with familial cancer susceptibility syndromes. *Mol. Cell. Endocrinol.* **2012**, *351*, 66–70. [CrossRef]

21. Gatta-Cherifi, B.; Chabre, O.; Murat, A.; Niccoli, P.; Cardot-Bauters, C.; Rohmer, V.; Young, J.; Delemer, B.; Du Boullay, H.; Verger, M.F.; et al. Adrenal involvement in MEN1. Analysis of 715 cases from the Groupe d'etude des Tumeurs Endocrines database. *Eur. J. Endocrinol.* **2011**, *166*, 269–279. [CrossRef] [PubMed]

22. Langer, P.D.P.; Cupisti, K.; Bartsch, D.K.; Nies, C.; Goretzki, P.E.; Rothmund, M.; Röher, H.D. Adrenal Involvement in Multiple Endocrine Neoplasia Type 1. *World J. Surg.* **2002**, *26*, 891–896. [CrossRef] [PubMed]

23. Kaur, R.J.; Pichurin, P.N.; Hines, J.M.; Singh, R.J.; Grebe, S.K.; Bancos, I. Adrenal Cortical Carcinoma Associated With Lynch Syndrome: A Case Report and Review of Literature. *J. Endocr. Soc.* **2019**, *3*, 784–790. [CrossRef] [PubMed]

24. Menon, R.K.; Ferraù, F.; Kurzawinski, T.R.; Rumsby, G.; Freeman, A.; Amin, Z.; Korbonits, M.; Chung, T.-T.L.L. Adrenal cancer in neurofibromatosis type 1: Case report and DNA analysis. *Endocrinol. Diabetes Metab. Case Rep.* **2014**, *2014*, 140074. [CrossRef] [PubMed]

25. Bertherat, J. Adrenocortical cancer in Carney complex: A paradigm of endocrine tumor progression or an association of genetic predisposing factors? *J. Clin. Endocrinol. Metab.* **2012**, *97*, 387–390. [CrossRef] [PubMed]

26. Hafner, A.; Bulyk, M.L.; Jambhekar, A.; Lahav, G. The multiple mechanisms that regulate p53 activity and cell fate. *Nat. Rev. Mol. Cell Boil.* **2019**, *20*, 199–210. [CrossRef]

27. Baugh, E.H.; Ke, H.; Levine, A.J.; A Bonneau, R.; Chan, C.S. Why are there hotspot mutations in the TP53 gene in human cancers? *Cell Death Differ.* **2017**, *25*, 154–160. [CrossRef]

28. Petitjean, A.; Mathe, E.; Kato, S.; Ishioka, C.; Tavtigian, S.V.; Hainaut, P.; Olivier, M. Impact of mutant p53 functional properties onTP53mutation patterns and tumor phenotype: Lessons from recent developments in the IARC TP53 database. *Hum. Mutat.* **2007**, *28*, 622–629. [CrossRef]

29. Libé, R.; Groussin, L.; Tissier, F.; Elie, C.; René-Corail, F.; Fratticci, A.; Jullian, E.; Beck-Peccoz, P.; Bertagna, X.; Gicquel, C.; et al. Somatic TP53 Mutations Are Relatively Rare among Adrenocortical Cancers with the Frequent 17p13 Loss of Heterozygosity. *Clin. Cancer Res.* **2007**, *13*, 844–850. [CrossRef]

30. Wasserman, J.D.; Zambetti, G.P.; Malkin, D. Towards an understanding of the role of p53 in adrenocortical carcinogenesis. *Mol. Cell. Endocrinol.* **2012**, *351*, 101–110. [CrossRef]

31. Jacks, T.; Remington, L.; Williams, B.O.; Schmitt, E.M.; Halachmi, S.; Bronson, R.T.; Weinberg, R.A. Tumor spectrum analysis in p53-mutant mice. *Curr. Boil.* **1994**, *4*, 1–7. [CrossRef]

32. Lang, G.A.; Iwakuma, T.; Suh, Y.-A.; Liu, G.; Rao, V.; Parant, J.M.; Valentin-Vega, Y.A.; Terzian, T.; Caldwell, L.C.; Strong, L.C.; et al. Gain of Function of a p53 Hot Spot Mutation in a Mouse Model of Li-Fraumeni Syndrome. *Cell* **2004**, *119*, 861–872. [CrossRef]

33. Else, T.; Trovato, A.; Kim, A.C.; Wu, Y.; Ferguson, D.O.; Kuick, R.D.; Lucas, P.C.; Hammer, G.D. Genetic p53 Deficiency Partially Rescues the Adrenocortical Dysplasia Phenotype at the Expense of Increased Tumorigenesis. *Cancer Cell* **2009**, *15*, 465–476. [CrossRef]

34. Brice, A.L.; E Cheetham, J.; Bolton, V.N.; Hill, N.C.; Schofield, P.N. Temporal changes in the expression of the insulin-like growth factor II gene associated with tissue maturation in the human fetus. *Development* **1989**, *106*, 543–554.

35. Giordano, T.J.; Thomas, D.G.; Kuick, R.; Lizyness, M.; Misek, D.E.; Smith, A.L.; Sanders, D.; Aljundi, R.T.; Gauger, P.G.; Thompson, N.W.; et al. Distinct Transcriptional Profiles of Adrenocortical Tumors Uncovered by DNA Microarray Analysis. *Am. J. Pathol.* **2003**, *162*, 521–531. [CrossRef]

36. Gicquel, C.; Bertagna, X.; Gaston, V.; Coste, J.; Louvel, A.; Baudin, E.; Bertherat, J.; Chapuis, Y.; Duclos, J.M.; Schlumberger, M.; et al. Molecular markers and long-term recurrences in a large cohort of patients with sporadic adrenocortical tumors. *Cancer Res.* **2001**, *61*, 6762–6767. [PubMed]

37. Giordano, T.J.; Kuick, R.; Else, T.; Gauger, P.G.; Vinco, M.; Bauersfeld, J.; Sanders, D.; Thomas, D.G.; Doherty, G.M.; Hammer, G. Molecular classification and prognostication of adrenocortical tumors by transcriptome profiling. *Clin. Cancer Res.* **2009**, *15*, 668–676. [CrossRef] [PubMed]

38. De Fraipont, F.; El Atifi, M.; Cherradi, N.; Le Moigne, G.; Defaye, G.; Houlgatte, R.; Bertherat, J.; Bertagna, X.; Plouin, P.-F.; Baudin, E.; et al. Gene Expression Profiling of Human Adrenocortical Tumors Using Complementary Deoxyribonucleic Acid Microarrays Identifies Several Candidate Genes as Markers of Malignancy. *J. Clin. Endocrinol. Metab.* **2005**, *90*, 1819–1829. [CrossRef] [PubMed]

39. Gao, Z.-H.; Suppola, S.; Liu, J.; Heikkilä, P.; Jänne, J.; Voutilainen, R. Association of H19 Promoter Methylation with the Expression of H19 and IGF-II Genes in Adrenocortical Tumors. *J. Clin. Endocrinol. Metab.* **2002**, *87*, 1170–1176. [CrossRef] [PubMed]

40. Liu, J.; Kahri, A.I.; Heikkila, P.; Ilvesmäki, V.; Voutilainen, R. H19 and Insulin-Like Growth Factor-II Gene Expression in Adrenal Tumors and Cultured Adrenal Cells. *J. Clin. Endocrinol. Metab.* **1995**, *80*, 492–496. [PubMed]

41. Glover, A.; Zhao, J.T.; Ip, J.C.; Lee, J.C.; Robinson, B.G.; Gill, A.J.; Soon, P.S.H.; Sidhu, S.B. Long noncoding RNA profiles of adrenocortical cancer can be used to predict recurrence. *Endocrine-Related Cancer* **2015**, *22*, 99–109. [CrossRef]

42. Zhou, Y.; Wang, X.; Zhu, X.; Liu, F.-C.; Ye, F.; Wu, D.-H.; Zhong, P. Bioinformatic analysis of long non-coding RNA-associated competing endogenous RNA network in adrenocortical carcinoma. *Transl. Cancer Res.* **2019**, *8*, 2175–2186. [CrossRef]

43. Soon, P.S.; McDonald, K.L.; Robinson, B.G.; Sidhu, S.B. Molecular Markers and the Pathogenesis of Adrenocortical Cancer. *Oncology* **2008**, *13*, 548–561. [CrossRef] [PubMed]

44. Pereira, S.S.; Monteiro, M.P.; Costa, M.M.; Moreira, Â.; Alves, M.G.; Oliveira, P.F.; Jarak, I.; Pignatelli, D. IGF2 role in adrenocortical carcinoma biology. *Endocrine* **2019**, *66*, 326–337. [CrossRef] [PubMed]

45. Drelon, C.; Berthon, A.S.; Ragazzon, B.; Tissier, F.; Bandiera, R.; Sahut-Barnola, I.; De Joussineau, C.; Batisse-Lignier, M.; Lefrançois-Martinez, A.-M.; Bertherat, J.; et al. Analysis of the Role of Igf2 in Adrenal Tumour Development in Transgenic Mouse Models. *PLoS ONE* **2012**, *7*, e44171. [CrossRef] [PubMed]

46. Heaton, J.H.; Wood, M.A.; Kim, A.C.; Lima, L.O.; Barlaskar, F.M.; Almeida, M.Q.; Fragoso, M.; Kuick, R.; Lerario, A.; Simon, D.P.; et al. Progression to adrenocortical tumorigenesis in mice and humans through insulin-like growth factor 2 and β-catenin. *Am. J. Pathol.* **2012**, *181*, 1017–1033. [CrossRef]

47. Fassnacht, M.; Berruti, A.; Baudin, E.; Demeure, M.J.; Gilbert, J.; Haak, H.; Kroiss, M.; Quinn, D.I.; Hesseltine, E.; Ronchi, C.L.; et al. Linsitinib (OSI-906) versus placebo for patients with locally advanced or metastatic adrenocortical carcinoma: A double-blind, randomised, phase 3 study. *Lancet Oncol.* **2015**, *16*, 426–435. [CrossRef]

48. Haluska, P.; Worden, F.; Olmos, D.; Yin, D.; Schteingart, D.; Batzel, G.N.; Paccagnella, M.L.; De Bono, J.S.; Gualberto, A.; Hammer, G.D. Safety, tolerability, and pharmacokinetics of the anti-IGF-1R monoclonal antibody figitumumab in patients with refractory adrenocortical carcinoma. *Cancer Chemother. Pharmacol.* **2009**, *65*, 765–773. [CrossRef]

49. Naing, A.; Lorusso, P.; Fu, S.; Hong, D.; Chen, H.X.; A Doyle, L.; Phan, A.T.; Habra, M.A.; Kurzrock, R. Insulin growth factor receptor (IGF-1R) antibody cixutumumab combined with the mTOR inhibitor temsirolimus in patients with metastatic adrenocortical carcinoma. *Br. J. Cancer* **2013**, *108*, 826–830. [CrossRef]

50. Nusse, R.; Clevers, H. Wnt/β-Catenin Signaling, Disease, and Emerging Therapeutic Modalities. *Cell* **2017**, *169*, 985–999. [CrossRef]

51. Assié, G.; Letouzé, E.; Fassnacht, M.; Jouinot, A.; Luscap, W.; Barreau, O.; Omeiri, H.; Rodriguez, S.; Perlemoine, K.; Rene-Corail, F.; et al. Integrated genomic characterization of adrenocortical carcinoma. *Nat. Genet.* **2014**, *46*, 607–612. [CrossRef] [PubMed]

52. Maharjan, R.; Backman, S.; Åkerström, T.; Hellman, P.; Björklund, P. Comprehensive analysis of CTNNB1 in adrenocortical carcinomas: Identification of novel mutations and correlation to survival. *Sci. Rep.* **2018**, *8*, 8610. [CrossRef] [PubMed]

53. Berthon, A.S.; Sahut-Barnola, I.; Lambert-Langlais, S.; De Joussineau, C.; Damon-Soubeyrand, C.; Louiset, E.; Taketo, M.M.; Tissier, F.; Bertherat, J.; Lefrançois-Martinez, A.-M.; et al. Constitutive β-catenin activation induces adrenal hyperplasia and promotes adrenal cancer development. *Hum. Mol. Genet.* **2010**, *19*, 1561–1576. [CrossRef] [PubMed]

54. Alles, J.; Fehlmann, T.; Fischer, U.; Backes, C.; Galata, V.; Minet, M.; Hart, M.; Abu-Halima, M.; A Grässer, F.; Lenhof, H.-P.; et al. An estimate of the total number of true human miRNAs. *Nucleic Acids Res.* **2019**, *47*, 3353–3364. [CrossRef] [PubMed]

55. Friedman, R.C.; Farh, K.K.-H.; Burge, C.B.; Bartel, B. Most mammalian mRNAs are conserved targets of microRNAs. *Genome Res.* **2008**, *19*, 92–105. [CrossRef] [PubMed]

56. Helwak, A.; Kudla, G.; Dudnakova, T.; Tollervey, D. Mapping the Human miRNA Interactome by CLASH Reveals Frequent Noncanonical Binding. *Cell* **2013**, *153*, 654–665. [CrossRef] [PubMed]

57. Chendrimada, T.P.; Gregory, R.I.; Kumaraswamy, E.; Norman, J.; Cooch, N.; Nishikura, K.; Shiekhattar, R. TRBP recruits the Dicer complex to Ago2 for microRNA processing and gene silencing. *Nature* **2005**, *436*, 740–744. [CrossRef]

58. Caramuta, S.; Lee, L.; Özata, D.M.; Akçakaya, P.; Xie, H.; Höög, A.; Zedenius, J.; Bäckdahl, M.; Larsson, C.; Lui, W.-O. Clinical and functional impact of TARBP2 over-expression in adrenocortical carcinoma. *Endocrine-Related Cancer* **2013**, *20*, 551–564. [CrossRef]

59. Hibio, N.; Hino, K.; Shimizu, E.; Nagata, Y.; Ui-Tei, K. Stability of miRNA 5′terminal and seed regions is correlated with experimentally observed miRNA-mediated silencing efficacy. *Sci. Rep.* **2012**, *2*, srep00996. [CrossRef]

60. Hammond, S.M.; Kuner, R.; Köhr, G.; Grünewald, S.; Eisenhardt, G.; Bach, A.; Kornau, H.-C. Argonaute2, a Link Between Genetic and Biochemical Analyses of RNAi. *Science* **2001**, *293*, 1146–1150. [CrossRef]

61. Zeng, Y.; Yi, R.; Cullen, B.R. MicroRNAs and small interfering RNAs can inhibit mRNA expression by similar mechanisms. *Proc. Natl. Acad. Sci. USA* **2003**, *100*, 9779–9784. [CrossRef] [PubMed]

62. Peng, Y.; Croce, C.M. The role of MicroRNAs in human cancer. *Signal Transduct. Target. Ther.* **2016**, *1*, 15004. [CrossRef] [PubMed]

63. Macfarlane, L.-A.; Murphy, P.R. MicroRNA: Biogenesis, Function and Role in Cancer. *Curr. Genom.* **2010**, *11*, 537–561. [CrossRef] [PubMed]

64. Marti, E.; Pantano, L.; Banez-Coronel, M.; Llorens, F.; Miñones-Moyano, E.; Porta, S.; Sumoy, L.; Ferrer, I.; Estivill, X. A myriad of miRNA variants in control and Huntington's disease brin regions detected by massively parallel sequencing. *Nucleic Acids Res.* **2010**, *38*, 7219–7235. [CrossRef]

65. Neilsen, C.T.; Goodall, G.J.; Bracken, C.P. IsomiRs – the overlooked repertoire in the dynamic microRNAome. *Trends Genet.* **2012**, *28*, 544–549. [CrossRef]

66. Manzano, M.; Forte, E.; Raja, A.N.; Schipma, M.J.; Gottwein, E. Divergent target recognition by coexpressed 5′-isomiRs of miR-142-3p and selective viral mimicry. *RNA* **2015**, *21*, 1606–1620. [CrossRef]

67. Moore, M.J.; Scheel, T.K.H.; Luna, J.M.; Park, C.Y.; Fak, J.J.; Nishiuchi, E.; Rice, C.M.; Darnell, R.B. miRNA-target chimeras reveal miRNA 3′-end pairing as a major determinant of Argonaute target specificity. *Nat. Commun.* **2015**, *6*, 8864. [CrossRef]

68. Chan, Y.-T.; Lin, Y.-C.; Lin, R.-J.; Kuo, H.-H.; Thang, W.C.; Chiu, K.-P.; Yu, A.L. Concordant and Discordant Regulation of Target Genes by miR-31 and Its Isoforms. *PLoS ONE* **2013**, *8*, e58169. [CrossRef]

69. Telonis, A.G.; Magee, R.; Loher, P.; Chervoneva, I.; Londin, E.R.; Rigoutsos, I. Knowledge about the presence or absence of miRNA isoforms (isomiRs) can successfully discriminate amongst 32 TCGA cancer types. *Nucleic Acids Res.* **2017**, *45*, 2973–2985. [CrossRef]

70. Wang, S.; Zheng, Z.; Chen, P.; Wu, M. Tumor classification and biomarker discovery based on the 5′isomiR expression level. *BMC Cancer* **2019**, *19*, 1–10. [CrossRef]

71. Lan, C.; Peng, H.; McGowan, E.M.; Hutvagner, G.; Li, J. An isomiR expression panel based novel breast cancer classification approach using improved mutual information. *BMC Med. Genom.* **2018**, *11*, 118. [CrossRef]

72. Koperski, L.; Kotlarek, M.; Swierniak, M.; Kolanowska, M.; Kubiak, A.; Górnicka, B.; Jazdzewski, K.; Wójcicka, A. Next-generation sequencing reveals microRNA markers of adrenocortical tumors malignancy. *Oncotarget* **2017**, *8*, 49191–49200. [CrossRef]

73. Soon, P.; Tacon, L.J.; Gill, A.J.; Bambach, C.P.; Sywak, M.S.; Campbell, P.R.; Yeh, M.W.; Wong, S.G.; Clifton-Bligh, R.J.; Robinson, B.G.; et al. miR-195 and miR-483-5p Identified as Predictors of Poor Prognosis in Adrenocortical Cancer. *Clin. Cancer Res.* **2009**, *15*, 7684–7692. [CrossRef]

74. Özata, D.M.; Caramuta, S.; Velàzquez-Fernàndez, D.; Akçakaya, P.; Xie, H.; Höög, A.; Zedenius, J.; Bäckdahl, M.; Larsson, C.; Lui, W.-O. The role of microRNA deregulation in the pathogenesis of adrenocortical carcinoma. *Endocrine-Related Cancer* **2011**, *18*, 643–655. [CrossRef]

75. Patterson, E.E.; Holloway, A.K.; Weng, J.; Fojo, T.; Kebebew, E. MicroRNA profiling of adrenocortical tumors reveals miR-483 as a marker of malignancy. *Cancer* **2010**, *117*, 1630–1639. [CrossRef]

76. Chabre, O.; Libé, R.; Assié, G.; Barreau, O.; Bertherat, J.; Bertagna, X.; Feige, J.-J.; Cherradi, N. Serum miR-483-5p and miR-195 are predictive of recurrence risk in adrenocortical cancer patients. *Endocrine-Related Cancer* **2013**, *20*, 579–594. [CrossRef]

77. Feinmesser, M.; Benbassat, C.; Meiri, E.; Benjamin, H.; Lebanony, D.; Lebenthal, Y.; De Vries, L.; Drozd, T.; Spector, Y. Specific MicroRNAs Differentiate Adrenocortical Adenomas from Carcinomas and Correlate With Weiss Histopathologic System. *Appl. Immunohistochem. Mol. Morphol.* **2015**, *23*, 522–531. [CrossRef]

78. Tömböl, Z.; Szabó, P.M.; Molnár, V.; Wiener, Z.; Tölgyesi, G.; Horányi, J.; Riesz, P.; Reismann, P.; Patocs, A.; Liko, I.; et al. Integrative molecular bioinformatics study of human adrenocortical tumors: microRNA, tissue-specific target prediction, and pathway analysis. *Endocrine-Related Cancer* **2009**, *16*, 895–906. [CrossRef]

79. Schmitz, K.J.; Helwig, J.; Bertram, S.; Sheu, S.Y.; Suttorp, A.C.; Seggewiß, J.; Willscher, E.; Walz, M.K.; Worm, K.; Schmid, K.W. Differential expression of microRNA-675, microRNA-139-3p and microRNA-335 in benign and malignant adrenocortical tumours. *J. Clin. Pathol.* **2011**, *64*, 529–535. [CrossRef]

80. Duregon, E.; Rapa, I.; Votta, A.; Giorcelli, J.; Daffara, F.; Terzolo, M.; Scagliotti, G.V.; Volante, M.; Papotti, M. MicroRNA expression patterns in adrenocortical carcinoma variants and clinical pathologic correlations. *Hum. Pathol.* **2014**, *45*, 1555–1562. [CrossRef]

81. Zheng, S.; Cherniack, A.D.; Dewal, N.; Moffitt, R.A.; Danilova, L.; Murray, B.A.; Lerario, A.M.; Else, T.; Knijnenbury, T.A.; Girielli, G.; et al. Comprehensive Pan-Genomic Characterization of Adrenocortical Carcinoma. *Cancer Cell* **2016**, *29*, 723–736. [CrossRef]

82. Mitchell, P.S.; Parkin, R.K.; Kroh, E.M.; Fritz, B.R.; Wyman, S.K.; Pogosova-Agadjanyan, E.L.; Peterson, A.; Noteboom, J.; O'Briant, K.C.; Allen, A.; et al. Circulating microRNAs as stable blood-based markers for cancer detection. *Proc. Natl. Acad. Sci. USA* **2008**, *105*, 10513–10518. [CrossRef]

83. Decmann, A.; Perge, P.; Turai, P.I.; Patocs, A.; Igaz, P. Non-Coding RNAs in Adrenocortical Cancer: From Pathogenesis to Diagnosis. *Cancers* **2020**, *12*, 461. [CrossRef] [PubMed]

84. Szabó, D.R.; Luconi, M.; Szabó, P.M.; Tóth, M.; Szücs, N.; Horányi, J.; Nagy, Z.; Mannelli, M.; Patocs, A.; Rácz, K.; et al. Analysis of circulating microRNAs in adrenocortical tumors. *Lab. Investig.* **2013**, *94*, 331–339. [CrossRef]

85. Patel, D.; Boufraqech, M.; Jain, M.; Zhang, L.; He, M.; Gesuwan, K.; Gulati, N.; Nilubol, N.; Fojo, T.; Kebebew, E. MiR-34a and miR-483-5p are candidate serum biomarkers for adrenocortical tumors. *Surgery* **2013**, *154*, 1224–1229. [CrossRef]

86. Shen, J.; Wang, A.; Wang, Q.; Gurvich, I.; Siegel, A.B.; Remotti, H.; Santella, R.M. Exploration of genome-wide circulating microRNA in hepatocellular carcinoma: MiR-483-5p as a potential biomarker. *Cancer Epidemiol. Biomark. Prev.* **2013**, *22*, 2364–2373. [CrossRef]

87. Lamichhane, S.R.; Thachil, T.; Gee, H.; Milic, N. Circulating MicroRNAs as Prognostic Molecular Biomarkers in Human Head and Neck Cancer: A Systematic Review and Meta-Analysis. *Dis. Markers* **2019**, *2019*, 1–12. [CrossRef]

88. Saliminejad, K.; Khorshid, H.R.K.; Ghaffari, S.H. Why have microRNA biomarkers not been translated from bench to clinic? *Future Oncol.* **2019**, *15*, 801–803. [CrossRef]

89. Farina, N.H.; Wood, M.E.; Perrapato, S.D.; Francklyn, C.S.; Stein, G.S.; Stein, J.L.; Lian, J. Standardizing analysis of circulating microRNA: Clinical and biological relevance. *J. Cell. Biochem.* **2014**, *115*, 805–811. [CrossRef]

90. Benz, F.; Roderburg, C.; Cardenas, D.V.; Vucur, M.; Gautheron, J.; Koch, A.; Zimmermann, H.; Janssen, J.; Nieuwenhuijsen, L.; Luedde, M.; et al. U6 is unsuitable for normalization of serum miRNA levels in patients with sepsis or liver fibrosis. *Exp. Mol. Med.* **2013**, *45*, e42. [CrossRef]

91. Marabita, F.; De Candia, P.; Torri, A.; Tegnér, J.; Abrignani, S.; Rossi, R.L. Normalization of circulating microRNA expression data obtained by quantitative real-time RT-PCR. *Briefings Bioinform.* **2015**, *17*, 204–212. [CrossRef] [PubMed]

92. Jouinot, A.; Bertherat, J. MANAGEMENT OF ENDOCRINE DISEASE: Adrenocortical carcinoma: Differentiating the good from the poor prognosis tumors. *Eur. J. Endocrinol.* **2018**, *178*, R215–R230. [CrossRef] [PubMed]

93. Faria, A.M.; Sbiera, S.; Ribeiro, T.C.; Soares, I.C.; Mariani, B.M.; Freire, D.S.; De Sousa, G.R.; Lerario, A.M.; Ronchi, C.L.; Deutschbein, T.; et al. Expression of LIN28 and its regulatory microRNAs in adult adrenocortical cancer. *Clin. Endocrinol.* **2014**, *82*, 481–488. [CrossRef]

94. Streicher, K.L.; Zhu, W.; Lehmann, K.P.; Georgantas, R.W.; A Morehouse, C.; Brohawn, P.; A Carrasco, R.; Xiao, Z.; A Tice, D.; Higgs, B.W.; et al. A novel oncogenic role for the miRNA-506-514 cluster in initiating melanocyte transformation and promoting melanoma growth. *Oncogene* **2011**, *31*, 1558–1570. [CrossRef] [PubMed]

95. Riffo-Campos, Á.L.; Riquelme, I.; Brebi, P. Tools for Sequence-Based miRNA Target Prediction: What to Choose? *Int. J. Mol. Sci.* **2016**, *17*, 1987. [CrossRef] [PubMed]

96. Ab Mutalib, N.-S.; Sulaiman, S.A.; Jamal, R. Computational Tools for microRNA Target Prediction. In *Computational Epigenetics and Diseases*; Elsevier: Amsterdam, The Netherlands, 2019; pp. 79–105.

97. Zhang, Y.; Zang, Q.; Zhang, H.; Ban, R.; Yang, Y.; Iqbal, F.; Li, A.; Shi, Q. DeAnnIso: A tool for online detection and annotation of isomiRs from small RNA sequencing data. *Nucleic Acids Res.* **2016**, *44*, W166–W175. [CrossRef]

98. Hammell, M. Computational methods to identify miRNA targets. *Semin. Cell Dev. Boil.* **2010**, *21*, 738–744. [CrossRef]

99. Liu, B.; Li, J.; Cairns, M.J. Identifying miRNAs, targets and functions. *Briefings Bioinform.* **2012**, *15*, 1–19. [CrossRef]

100. Thomson, D.W.; Bracken, C.P.; Goodall, G. Experimental strategies for microRNA target identification. *Nucleic Acids Res.* **2011**, *39*, 6845–6853. [CrossRef]

101. Martinez-Sanchez, A.; Murphy, C.L. MicroRNA Target Identification—Experimental Approaches. *Boil.* **2013**, *2*, 189–205. [CrossRef]

102. Li, J.; Zhang, Y. Current experimental strategies for intracellular target identification of microRNA. *ExRNA* **2019**, *1*, 6. [CrossRef]

103. Mockly, S.; Seitz, H. Inconsistencies and Limitations of Current MicroRNA Target Identification Methods. In *Breast Cancer*; Springer Science and Business Media LLC: Berlin/Heidelberg, Germany, 2019; Volume 1970, pp. 291–314.

104. Khafaei, M.; Rezaie, E.; Mohammadi, A.; Gerdehsang, P.S.; Ghavidel, S.; Kadkhoda, S.; Zahra, A.Z.; Forouzanfar, N.; Arabameri, H.; Tavallaei, M. miR-9: From function to therapeutic potential in cancer. *J. Cell. Physiol.* **2019**, *234*, 14651–14665. [CrossRef] [PubMed]

105. Feng, Y.-H.; Tsao, C.-J. Emerging role of microRNA-21 in cancer. *Biomed. Rep.* **2016**, *5*, 395–402. [CrossRef] [PubMed]

106. Zhou, W.; Yang, W.; Ma, J.; Zhang, H.; Li, Z.; Zhang, L.; Liu, J.; Han, Z.; Wang, H.; Hong, L. Role of miR-483 in digestive tract cancers: From basic research to clinical value. *J. Cancer* **2018**, *9*, 407–414. [CrossRef] [PubMed]

107. Huang, L.-L.; Huang, L.; Wang, L.; Tong, B.-D.; Wei, Q.; Ding, X.-S. Potential role of miR-139-5p in cancer diagnosis, prognosis and therapy. *Oncol. Lett.* **2017**, *14*, 1215–1222. [CrossRef]

108. Agosta, C.; Laugier, J.; Guyon, L.; Denis, J.; Bertherat, J.; Libe, R.; Boisson, B.; Sturm, N.; Feige, J.-J.; Chabre, O.; et al. MiR-483-5p and miR-139-5p promote aggressiveness by targeting N-myc downstream-regulated gene family members in adrenocortical cancer. *Int. J. Cancer* **2018**, *143*, 944–957. [CrossRef]

109. De Sousa, G.R.V.; Ribeiro, T.C.; Faria, A.M.; Mariani, B.M.P.; Lerario, A.M.; Zerbini, M.C.N.; Soares, I.C.; Wakamatsu, A.; Alves, V.A.F.; Mendonca, B.B.; et al. Low DICER1 expression is associated with poor clinical outcome in adrenocortical carcinoma. *Oncotarget* **2015**, *6*, 22724–22733. [CrossRef]

110. Zhou, J.; Ng, S.-B.; Chng, W.J. LIN28/LIN28B: An emerging oncogenic driver in cancer stem cells. *Int. J. Biochem. Cell Boil.* **2013**, *45*, 973–978. [CrossRef]

111. Zhong, X.; Li, N.; Liang, S.; Huang, Q.; Coukos, G.; Zhang, L. Identification of MicroRNAs Regulating Reprogramming FactorLIN28in Embryonic Stem Cells and Cancer Cells. *J. Boil. Chem.* **2010**, *285*, 41961–41971. [CrossRef]

112. Lima, C.R.; Gomes, C.C.; Santos, M. Role of microRNAs in endocrine cancer metastasis. *Mol. Cell. Endocrinol.* **2017**, *456*, 62–75. [CrossRef]

113. Romero, D.G.; Plonczynski, M.W.; Carvajal, C.A.; Gomez-Sanchez, E.P.; Gomez-Sanchez, C.E. Microribonucleic Acid-21 Increases Aldosterone Secretion and Proliferation in H295R Human Adrenocortical Cells. *Endocrinology* **2008**, *149*, 2477–2483. [CrossRef] [PubMed]

114. Asangani, I.A.; Rasheed, S.A.K.; A Nikolova, D.; Leupold, J.H.; Colburn, N.H.; Post, S.; Allgayer, H. MicroRNA-21 (miR-21) post-transcriptionally downregulates tumor suppressor Pdcd4 and stimulates invasion, intravasation and metastasis in colorectal cancer. *Oncogene* **2007**, *27*, 2128–2136. [CrossRef] [PubMed]

115. Talotta, F.; Cimmino, A.; Matarazzo, M.R.; Casalino, L.; De Vita, G.; D'Esposito, M.; Di Lauro, R.; Verde, P. An autoregulatory loop mediated by miR-21 and PDCD4 controls the AP-1 activity in RAS transformation. *Oncogene* **2008**, *28*, 73–84. [CrossRef]

116. Veronese, A.; Lupini, L.; Consiglio, J.; Visone, R.; Ferracin, M.; Fornari, F.; Zanesi, N.; Alder, H.; D'Elia, G.; Gramantieri, L.; et al. Oncogenic role of miR-483-3p at the IGF2/483 Locus. *Cancer Res.* **2010**, *70*, 3140–3149. [CrossRef] [PubMed]

117. Nowek, K.; Wiemer, E.A.C.; Jongen-Lavrencic, M. The versatile nature of miR-9/9* in human cancer. *Oncotarget* **2018**, *9*, 20838–20854. [CrossRef]

118. Jiang, Y.; Jia, Y.; Zhang, L. Role of programmed cell death 4 in diseases: A double-edged sword. *Cell. Mol. Immunol.* **2017**, *14*, 884–886. [CrossRef]

119. Hikisz, P.; Kiliańska, Z.M. Puma, a critical mediator of cell death—One decade on from its discovery. *Cell. Mol. Boil. Lett.* **2012**, *17*, 646–669. [CrossRef]

120. Glover, A.; Zhao, J.T.; Gill, A.J.; Weiss, J.; Mugridge, N.; Kim, E.; Feeney, A.L.; Ip, J.C.; Reid, G.; Clarke, S.; et al. microRNA-7 as a tumor suppressor and novel therapeutic for adrenocortical carcinoma. *Oncotarget* **2015**, *6*, 36675–36688. [CrossRef]

121. Wu, Y.; Wang, W.; Hu, W.; Xu, W.; Xiao, G.; Nie, Q.; Ouyang, K.; Chen, S. MicroRNA-205 suppresses the growth of adrenocortical carcinoma SW-13 cells via targeting Bcl-2. *Oncol. Rep.* **2015**, *34*, 3104–3110. [CrossRef]

122. Finnerty, J.R.; Wang, W.-X.; Hebert, S.S.; Wilfred, B.R.; Mao, G.; Nelson, P.T. The miR-15/107 Group of MicroRNA Genes: Evolutionary Biology, Cellular Functions, and Roles in Human Diseases. *J. Mol. Boil.* **2010**, *402*, 491–509. [CrossRef]

123. Zhang, Y.; Huang, B.; Wang, H.-Y.; Chang, A.; Zheng, X.S.; Bo, H. Emerging Role of MicroRNAs in mTOR Signaling. *Cell. Mol. Life Sci.* **2017**, *74*, 2613–2625. [CrossRef] [PubMed]

124. Doghman, M.; El Wakil, A.; Cardinaud, B.; Thomas, E.; Wang, J.; Zhao, W.; Valle, M.P.-D.; Figueiredo, B.; Zambetti, G.; Lalli, E. Regulation of IGF—mTOR Signalling by miRNA in Childhood Adrenocortical Tumors. In Proceedings of the Endocrine Society's 92nd Annual Meeting, San Diego, CA, USA, 19–22 June 2010; Volume 70, p. OR20-4. [CrossRef]

125. Jain, M.; Zhang, L.; Boufraqech, M.; Liu, J.O.; Bussey, K.J.; Demeure, M.J.; Wu, X.; Su, L.; Pacak, K.; Stratakis, C.A.; et al. ZNF367 Inhibits Cancer Progression and Is Targeted by miR-195. *PLoS ONE* **2014**, *9*, e101423. [CrossRef] [PubMed]

126. He, J.; Cao, Y.; Su, T.; Jiang, Y.; Jiang, L.; Zhou, W.; Zhang, C.; Wang, W.; Ning, G. Downregulation of miR-375 in aldosterone-producing adenomas promotes tumour cell growth via MTDH. *Clin. Endocrinol.* **2015**, *83*, 581–589. [CrossRef] [PubMed]

127. Kwok, G.T.; Zhao, J.T.; Glover, A.; Gill, A.J.; Clifton-Bligh, R.; Robinson, B.G.; Ip, J.C.; Sidhu, S.B. microRNA-431 as a Chemosensitizer and Potentiator of Drug Activity in Adrenocortical Carcinoma. *Oncology* **2019**, *24*, e241–e250. [CrossRef]

128. Hassan, N.; Zhao, J.; Glover, A.R.; Robinson, B.G.; Sidhu, S.B. Reciprocal interplay of miR-497 and MALAT1 promotes tumourigenesis of adrenocortical cancer. *Endocrine-Related Cancer* **2019**, *27*, 677–688. [CrossRef]

129. Kalinowski, F.; Brown, R.A.; Ganda, C.; Giles, K.M.; Epis, M.R.; Horsham, J.L.; Leedman, P.J. microRNA-7: A tumor suppressor miRNA with therapeutic potential. *Int. J. Biochem. Cell Boil.* **2014**, *54*, 312–317. [CrossRef]

130. Zhao, J.; Tao, Y.; Zhou, Y.; Qin, N.; Chen, C.; Tian, D.; Xu, L. MicroRNA-7: A promising new target in cancer therapy. *Cancer Cell Int.* **2015**, *15*, 103. [CrossRef]

131. Das, A.; Reis, F.; Mishra, P.K. mTOR Signaling in Cardiometabolic Disease, Cancer, and Aging 2018. *Oxidative Med. Cell. Longev.* **2019**, *2019*, 1–3. [CrossRef]

132. Jin, Y.; Tymen, S.D.; Chen, D.; Fang, Z.J.; Zhao, Y.; Dragas, D.; Dai, Y.; Marucha, P.T.; Zhou, X. MicroRNA-99 Family Targets AKT/mTOR Signaling Pathway in Dermal Wound Healing. *PLoS ONE* **2013**, *8*, e64434. [CrossRef]

133. Tsai, T.-F.; Lin, J.-F.; Chou, K.-Y.; Lin, Y.-C.; Chen, H.-E.; Hwang, T.I.-S. miR-99a-5p acts as tumor suppressor via targeting to mTOR and enhances RAD001-induced apoptosis in human urinary bladder urothelial carcinoma cells. *OncoTargets Ther.* **2018**, *11*, 239–252. [CrossRef]

134. Yin, H.; Ma, J.; Chen, L.; Piao, S.; Zhang, Y.; Zhang, S.; Ma, H.; Li, Y.; Qu, Y.; Wang, X.; et al. MiR-99a Enhances the Radiation Sensitivity of Non-Small Cell Lung Cancer by Targeting mTOR. *Cell. Physiol. Biochem.* **2018**, *46*, 471–481. [CrossRef] [PubMed]

135. Campbell, K.J.; Tait, S.W. Targeting BCL-2 regulated apoptosis in cancer. *Open Boil.* **2018**, *8*, 180002. [CrossRef] [PubMed]

136. Hu, G.; Wei, Y.; Kang, Y. The multifaceted role of MTDH/AEG-1 in cancer progression. *Clin. Cancer Res.* **2009**, *15*, 5615–5620. [CrossRef] [PubMed]

137. Sun, K.; Zeng, T.; Huang, D.; Liu, Z.; Huang, S.; Liu, J.; Qu, Z. MicroRNA-431 inhibits migration and invasion of hepatocellular carcinoma cells by targeting the ZEB1-mediated epithelial-mensenchymal transition. *FEBS Open Bio* **2015**, *5*, 900–907. [CrossRef] [PubMed]

138. Zhao, M.; Wang, S.; Li, Q.; Ji, Q.; Guo, P.; Liu, X. MALAT1: A long non-coding RNA highly associated with human cancers. *Oncol. Lett.* **2018**, *16*, 19–26. [CrossRef]

Review

Non-Coding RNAs: Uncharted Mediators of Thyroid Cancer Pathogenesis

Hossein Tabatabaeian [1], Samantha Peiling Yang [2,3,*] and Yvonne Tay [1,4,*]

1   Cancer Science Institute of Singapore, National University of Singapore, Singapore 117599, Singapore; csiht@nus.edu.sg
2   Endocrinology Division, Department of Medicine, National University Hospital, Singapore 119228, Singapore
3   Department of Medicine, Yong Loo Lin School of Medicine, National University of Singapore, Singapore 117597, Singapore
4   Department of Biochemistry, Yong Loo Lin School of Medicine, National University of Singapore, Singapore 117597, Singapore
*   Correspondence: mdcyp@nus.edu.sg (S.P.Y.); yvonnetay@nus.edu.sg (Y.T.)

Received: 5 October 2020; Accepted: 2 November 2020; Published: 4 November 2020

Simple Summary: Thyroid cancer is the most common type of endocrine system malignancy. The effective diagnosis, precise treatment, and better short and long-term prognosis of thyroid cancer patients have remained challenging. Non-coding RNAs (ncRNAs) are emerging molecules with diverse capabilities in initiating and promoting thyroid cancer upon dysregulation. The expression profile of these molecules could be used to detect thyroid cancer, determine the therapeutic approaches, and predict the patients' survival. Thus, ncRNAs could have clinical significance in precision medicine.

Abstract: Thyroid cancer is the most prevalent malignancy of the endocrine system and the ninth most common cancer globally. Despite the advances in the management of thyroid cancer, there are critical issues with the diagnosis and treatment of thyroid cancer that result in the poor overall survival of undifferentiated and metastatic thyroid cancer patients. Recent studies have revealed the role of different non-coding RNAs (ncRNAs), such as microRNAs (miRNAs), long non-coding RNAs (lncRNAs) and circular RNAs (circRNAs) that are dysregulated during thyroid cancer development or the acquisition of resistance to therapeutics, and may play key roles in treatment failure and poor prognosis of the thyroid cancer patients. Here, we systematically review the emerging roles and molecular mechanisms of ncRNAs that regulate thyroid tumorigenesis and drug response. We then propose the potential clinical implications of ncRNAs as novel diagnostic and prognostic biomarkers for thyroid cancer.

Keywords: thyroid carcinoma; non-coding RNA; radioactive iodine; drug resistance; prognosis

---

1. Introduction

Thyroid carcinoma is the most prevalent malignancy of the endocrine system with a significantly higher incidence in women [1–3] and is the 9th most common cancer globally [3]. The incidence of thyroid cancer has been escalating worldwide in recent years [2,4]. According to the Surveillance, Epidemiology and End Results (SEER) Program [5], ~53,000 newly diagnosed cases of thyroid cancer are anticipated in the United States in 2020 and these will represent 2.9% of all new cancer cases. Globally, there were newly diagnosed 567,233 thyroid cancer cases in 2018 that comprise 3.1% of all cancer incidences [3]. The cause of this rise in incidence is multi-factorial. It is partially contributed by the detection of incidental thyroid cancers with the increasing use of cross-sectional imaging. However, other factors likely play a role in the pathogenesis of these thyroid cancer cases. The well-described risk factors of thyroid cancer are exposure to ionizing radiation, particularly in childhood, and family history of thyroid cancer. The potential role of other factors such as smoking, obesity, hormonal exposures, and environmental toxins/ endocrine-disrupting chemicals have been implicated [6]. Their mechanistic role in the pathogenesis of thyroid cancer has not been well-elucidated. Further environmental and biomolecular studies will be essential to better understand the complex etiology of thyroid cancer [7].

The thyroid gland comprises two specific cell types. Thyroid follicular (epithelial) cells produce and secrete the thyroid hormones, thereby regulating the body's temperature, metabolism, and heart rate. The second cell type is parafollicular cells (C cells) that reside in the thyroid connective tissue and are responsible for calcitonin secretion to regulate the levels of calcium and phosphate in the body [8,9]. Thyroid cancer is classified into four types based on histopathological analysis and site of origin: (1) papillary thyroid carcinoma (PTC) that accounts for 80% of all cases and commonly metastasize to cervical lymph nodes. (2) Follicular thyroid carcinoma (FTC) represents 10% of all cases and the widely invasive sub-type has a tendency for metastasis to distant sites, whereas its minimally invasive sub-type has an overall low risk of recurrence and metastasis. (3) Medullary thyroid carcinoma (MTC) that is more aggressive than PTC and FTC, has higher metastatic rates to cervical lymph nodes and distant sites, and occurs in 4% of thyroid cancer cases. (4) Poorly differentiated thyroid carcinoma (PDTC) and anaplastic thyroid carcinoma (ATC) are the most aggressive and least differentiated types of thyroid carcinoma with the highest rate of spread to other organs. PDTC and ATC account for 5–10% of follicular thyroid cancers [1,10,11]. Among these types, PTC, FTC, PDTC and ATC originate from follicular cells, whereas MTC develops from C cells.

The main measure to detect thyroid cancer is ultrasound, which can be complemented with radioiodine scanning and fine-needle aspiration (FNA) biopsy [12]; and the treatment for thyroid cancer is surgical excision [13]. Post-surgery, other therapeutic strategies including radioactive iodine (RAI) therapy, and systemic therapies might be incorporated into the treatment regime for patients with high risk of recurrent or persistent disease [14]. Improvements in detection and treatment strategies have improved the survival of thyroid cancer. According to the American Society of Clinical Oncology, the 5-year survival for most of the non-metastatic thyroid cancer types is above 95%, whereas this rate falls drastically to 31% for ATC. In metastatic thyroid cancer, however, the 5-year survival rate for PTC, FTC and MTC drops to 78, 63 and 39%, respectively. The 5-year survival rate for metastatic ATC cases is only 4%. These data indicate that there is room for improvement in the current measures for thyroid cancer management, especially for more aggressive subtypes, and highlight the need to identify specific molecular diagnostic/prognostic biomarkers for personalized medicine.

Current diagnostic biomarkers in thyroid cancer include point mutations in the BRAF, NRAS, KRAS, HRAS genes, and rearrangements in paired box 8 (PAX8)/peroxisome proliferator-activated receptor gamma (PPARG) and RET [15–18]. In addition, mutations within the telomerase reverse transcriptase (TERT) promoter region have been detected more frequently in aggressive thyroid cancer cases [19]. Notwithstanding, the present molecular testing of thyroid cancer is chiefly useful to stratify indeterminate nodules, thereby to avoid surgery or prevent unnecessary repeated FNA [20]. The identification of novel molecular biomarkers to facilitate early diagnosis and predict drug responsiveness would be invaluable to improve the survival rate and quality of life for thyroid cancer patients. Non-coding RNAs (ncRNAs) are fast emerging as novel functional molecules and also biomarkers in human cancers [21–24]. Around 98% of the transcriptome in human cells corresponds to ncRNAs that are transcribed from previously considered "junk DNA" sequences such as introns and intergenic nucleotides [25]. These non-protein-coding transcripts are classified as small ncRNAs (sncRNAs) and long non-coding RNAs (lncRNAs) that are less than or more than 200 nucleotides in length, respectively. sncRNAs comprise small nuclear RNAs (snRNAs), microRNAs (miRNAs), piwi interacting RNAs (piRNAs) and small nucleolar RNAs (snoRNAs) [26–30]. However, circular RNAs (circRNAs) belong to both sncRNA and lncRNA classifications due to their variable length, ranging from 100–10,000 nucleotides [31]. The majority of human miRNAs and circRNAs are transcribed from introns, reflecting the importance of non-coding DNA sequences in determining cell fate [32]. Compelling evidence highlights the involvement of ncRNAs in almost all physiological and biological cell processes, such as cell growth, proliferation, senescence, apoptosis, invasion, migration, angiogenesis and inflammation [22,33–38]. ncRNAs impose their functions through different mechanisms—summarized in the following sections and Figure 1.

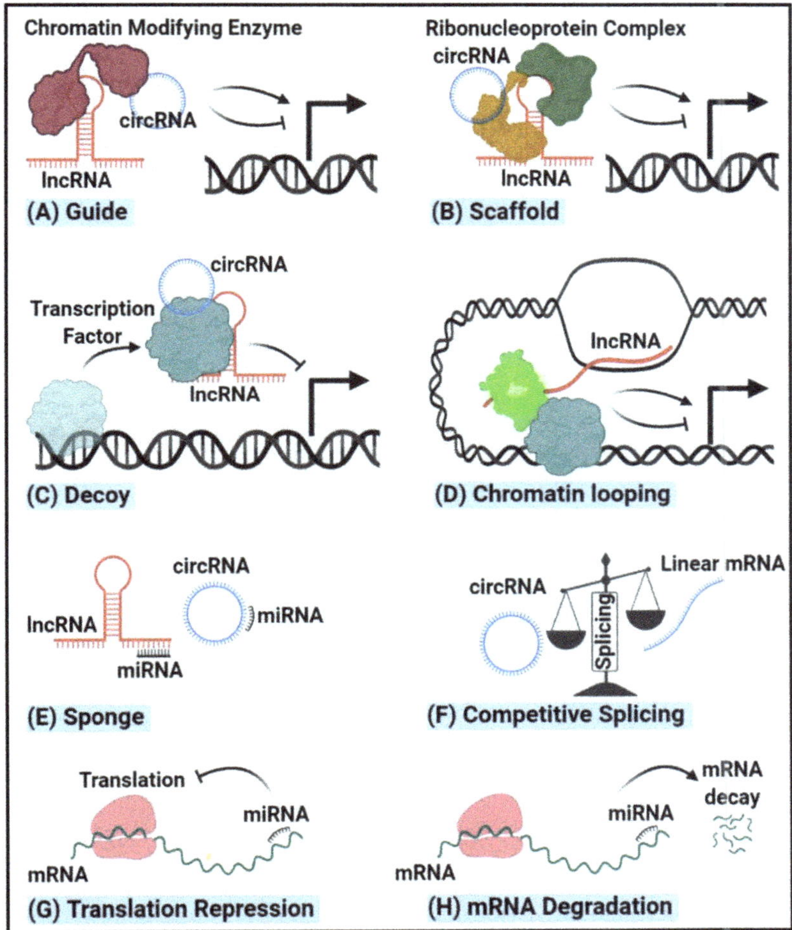

Figure 1. The schematic of mechanisms of action of ncRNAs. (**A**) lncRNAs and circRNAs can guide chromatin remodeling factors to either activate or repress the transcription of target genes. (**B**) lncRNAs and circRNAs, as scaffolds, can facilitate the assembly of ribonucleoprotein complexes to either activate or repress the transcription of target genes. (**C**) lncRNAs and circRNAs can sponge the transcription factors to repress the transcription of the target genes. (**D**) Upon transcription, lncRNAs can facilitate the formation of regulatory complexes and loop the DNA, thereby priming long-range gene transcription. (**E**) lncRNAs and circRNAs can sponge the miRNAs, thereby rescuing the miRNA target transcripts. (**F**) circRNAs can compete with the linear mRNA(s) transcribed from their host gene and repress the canonical splicing over the back splicing. (**G**) miRNAs bind to their target mRNAs and repress the translation efficiency upon non-perfect complementation between the seed region and targeted binding site. (**H**) miRNAs bind to their target mRNAs and result in transcript degradation upon perfect complementation between the seed region and targeted binding site.

Upon dysregulation, oncogenic and tumor suppressor ncRNAs could drive cancer initiation and progression, and/or alter drug response [39–41]. This review systematically focuses on the recent advances in the molecular roles of different ncRNA classes in thyroid cancer, as well as their implications as diagnostic and prognostic biomarkers and therapeutic targets.

2. miRNAs Regulating Thyroid Carcinogenesis

miRNAs are a subclass of sncRNAs with about 19–24 nucleotides in length that mainly regulate gene expression at the post-transcriptional level. miRNAs bind to miRNA response elements on target protein-coding and non-coding transcripts to regulate their expression. miRNA binding may lead to mRNA degradation or translational repression, depending on the extent of complementarity with each target region [42–44] (Figure 1). miRNAs have also been shown to regulate genes transcriptionally via interacting with lncRNAs, leading to the up or down-regulation of target genes [45–47]. The impact of miRNA dysregulation on cancer was first described by Calin et al. in 2002 [48]. Shortly after, He et al. showed the up-regulation of miR-221, miR-222, and miR-146 in PTC patients, which are the first dysregulated miRNAs reported in thyroid cancer [49]. To date, many additional studies have identified miRNAs involved in thyroid cancer initiation and progression [50,51]. Here, we summarize the recent list of dysregulated miRNAs in thyroid cancer, published from 2015 onward, with implications in cancer initiation and progression, briefly listed in Table 1.

A group of dysregulated miRNAs has been shown by in vitro and in vivo assays to inhibit apoptosis and promote proliferation, invasion and migration in thyroid cancer. For instance, the down-regulation of miR-524-5p [68], miR-141-3p [61], miR-9 [67], miR-199b-5p [66], miR-1266 [60], miR-144 [62], miR-150 [64] and miR-7 [69] were demonstrated to play pivotal roles in thyroid tumorigenesis. However, there is lack of inclusive data to know how these alterations could mechanistically impose tumorigenic properties in the context of thyroid cancer. Despite this ambiguity, the dysregulation of some other miRNAs has been reported to promote thyroid carcinomas via different signaling pathways, such as Wnt and phosphatidylinositol-4,5-Bisphosphate 3-Kinase (PI$_3$K)/Akt (Figure 2).

Table 1. Dysregulated miRNAs in thyroid cancer.

miRNA	Alteration	Mechanism	Thyroid Cancer Type	Sample	Ref.
miR-146a/b-5p		↓ RAR-β	PTC		[52]
miR-155		↓ SOCS1	ATC		[53]
miR-21	↑	↓ VHL → ↑ N-cadherin and vimentin		Cell line and tissue	[54]
miR-574-5p		↓ SCAI → ↑ β-catenin	PTC		[55]
miR-625-3p		↑ Bcl-2, ↓ Bax, and cleaved caspase 3/9			[56]
miR-96		↓ FOXO1 → ↓ Bim		Cell line	[57]
Let-7a		Not known	Not known	Tissue	[58]
miR-125a-5p		↑ CD147 → ↓ glucose metabolism	PTC, FTC, MTC and ATC		[59]
miR-1266		↑ FGFR2	PTC	Cell line and tissue	[60]
miR-141-3p		↑ YY1			[61]
miR-144		↑ TGF-α	ATC		[62]
miR-148a		↑ INO80		ATC cancer stem cells	[63]
miR-150	↓	↑ROCK1			[64]
miR-195		↑ CCND1 and FGF2 → ↑ β-catenin, c-Myc, cyclin D1 and MMP-13	PTC	Cell line and tissue	[65]
miR-199b-5p		↑ STON2 → ↑ N-cadherin and fibronectin			[66]
miR-375		↑ ERBB2			[57]
miR-497		↑ AKT3			[67]
miR-524-5p		↑ FOXE1 and ITGA3			[68]
miR-7		↑ PAK1			[69]
miR-873-5p		↑ CXCL16 → ↑ p-p65 and p-Rel-B, MMP1, MMP9 and MMP13			[70]
miR-9-5p		↑ BRAF			[67]

PTC: Papillary Thyroid Cancer; FTC: Follicular Thyroid Cancer; ATC: Anaplastic Thyroid Cancer; MTC: Medullar Thyroid Cancer; FOXN3: Forkhead Box N3; QKI5-7: Quaking protein 5-7; SCAI: Suppressor of Cancer Cell Invasion; CXCL16: C-X-C Motif Chemokine Ligand 16; SOCS1: Suppressor Protein of Cytokine Signaling 1; MMP1: Matrix Metallopeptidase 1; VHL: Von Hippel-Lindau Tumor Suppressor; RAR-β: Retinoic Acid Receptor Beta; FOXE1: Forkhead Box E1; ITGA3: Integrin Subunit Alpha 3; HK2: Hexokinase 2; YY1: Yin Yang 1; STON2: Stonin 2; FGFR2: Fibroblast Growth Factor Receptor 2; TGF-α: Transforming Growth Factor Alpha; AKT3: RAC-γ serine/threonine-protein kinase; CCND1: Cyclin D1; FGF2: Fibroblast Growth Factor; ROCK1: Rho-associated Protein Kinase 1; PAK1: p21 Activated Kinase-1; ERBB2: Erb-B2 Receptor Tyrosine Kinase 2; FOXO1: Forkhead Box O1.

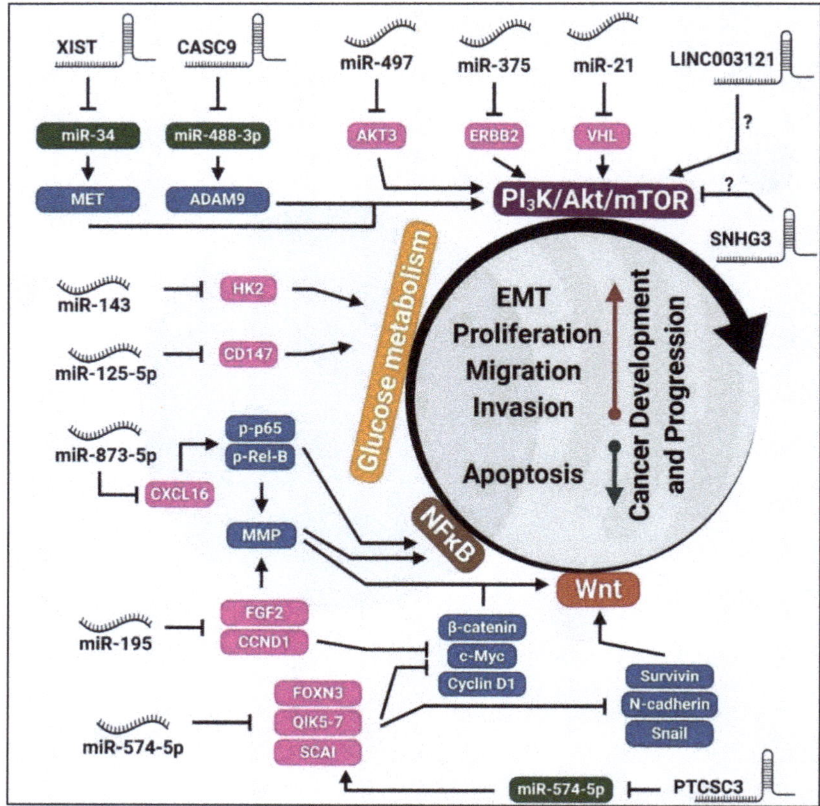

Figure 2. Schematic overview of ncRNA involvement in thyroid cancer-related signaling pathways. Key components of the PI₃K/Akt/mTOR pathway, such as AKT3, ERBB2 and VHL, are regulated by different miRNAs and lncRNAs in thyroid cancer. The metabolism of glucose is regulated by miR-143 and miR-125-5p. Well-known cancer-related pathways, NFκB and Wnt, are tightly regulated by miRNAs and lncRNAs in thyroid cancer. Dysregulation of these miRNAs and lncRNAs in various types of thyroid cancer eventually results in the induction of proliferation, migration and invasion, while apoptosis is suppressed. Note: Pink represents direct targets of miRNAs, green represents direct miRNA targets of lncRNAs and blue represents the indirect target of miRNA or lncRNAs.

2.1. Wnt-Mediated Tumorigenic Effects of Dysregulated miRNAs

Wnt signaling pathway is essential to embryonic development and tissue homeostasis. Hyperactivation of Wnt signaling is abundant in human cancer [71,72] and thyroid carcinoma is not an exception [73]. miR-574-5p is one of the most well-studied miRNAs that is linked to thyroid carcinomas, FTC and PTC in particular, via Wnt pathway. Up-regulation of miR-574-5p was reported by different groups to promote proliferation, Epithelial-Mesenchymal Transition (EMT), invasion, migration and inhibit apoptosis [55,74,75]. miR-574-5p exerts its oncogenic effects via targeting Forkhead Box N3 (FOXN3), Quaking protein 5–7 (QKI5–7), and Suppressor of Cancer Cell Invasion (SCAI) transcripts. Collectively, this results in the activation of the Wnt signaling pathway. This, in turn, up-regulates β-catenin, N-cadherin (a mesenchymal biomarker), Snail, c-Myc, cyclin D1 and survivin proteins [55,74,75], leading to downstream oncogenic phenotypes and thyroid cancer development.

In addition, the down-regulation of miR-195, which is shown in different cancers [76–80], was reported to cause thyroid tumorigenesis in vitro and in vivo via activating Wnt pathway. Through direct targeting cyclin D1 and fibroblast growth factor 2 (FGF2) genes, miR-195 was shown to inhibit cell proliferation, migration, and invasion via down-regulating β-catenin, c-Myc, cyclin D1 and matrix metallopeptidase-13 (MMP-13) [65]. A number of Wnt signaling pathway inhibitors are being examined in clinical trials, such as frizzled receptor antagonist, Vantictumab [81]. It is conceivable that Wnt pathway blockade may neutralize the oncogenic effects of miR-574-5p or miR-195 up-regulation in thyroid cancer, leading to the suppression of cancer development or progression. This hypothesis is yet to be tested experimentally.

2.2. PI₃K/Akt-Mediated Tumorigenic Effects of Dysregulated miRNAs

PI_3K/Akt signaling is a well-established oncogenic signaling pathway in multiple human cancers [82], including thyroid carcinoma [83]. Thus far, dysregulation of several miRNAs has been reported to regulate thyroid cancer by targeting the PI_3K/Akt signaling pathway. For instance, miR-21 overexpression was shown to target Von Hippel-Lindau (VHL) tumor suppressor and activate PI_3K/Akt pathway. This, in turn, resulted in an increase in the EMT markers N-cadherin and vimentin, and promote cell proliferation and invasion [54]. PTEN is another validated target of miR-21, reported in various human cancers, that plays pivotal role in the inhibition of PI_3K/Akt pathway [84–86]. Thus, miR-21 up-regulation could synergistically suppress PI_3K/Akt pathway via targeting VHL and PTEN transcripts. This needs to be examined experimentally. In addition to miR-21, heightened expression of miR-625-3p was reported to activate PI_3K/Akt pathway via an unknown mechanism [56]. miR-497 [67] and miR-375 [57] were demonstrated to be down-regulated in thyroid cancer and target AKT3 and ErbB2 receptor tyrosine kinase 2 (ERBB2), respectively. The latter is reported to be overexpressed in PTC clinical samples [87] and mediates the resistance to mitogen-activated protein kinase (MAPK) inhibitors in BRAF-mutant thyroid cancer cell lines [88], implying that ERBB2 could drastically affect the thyroid tumorigenesis. The AKT3 up-regulation, caused by miR-497 down-regulation, resulted in the activation of PI_3K/Akt pathway that eventually promotes proliferation and invasion in thyroid cancer [67]. Although miR-375 down-regulation similarly led to augmentation of oncogenic properties in vitro and in vivo [57], it is unclear whether these effects were mediated by ERBB2, a key oncogenic protein involved in the PI_3K pathway [89]. Besides, miR-96 is another example of overexpressed ncRNA in thyroid cancer with possible involvement in PI_3K/Akt pathway. miR-96 imposes its proliferative and anti-apoptotic effects via targeting Forkhead Box O1 (FOXO1), a known tumor suppressor in thyroid cancer [90] that is phosphorylated and subsequently degraded by PI_3K/Akt activation [91]. In parallel with PI_3K/Akt, miR-96 could synergistically amplify the FOXO1-mediated oncogenic properties in thyroid cancer. However, this proposed model has not been experimentally tested.

Despite the advances in harnessing the oncogenic properties of thyroid cancer via dual administration of PI3K/Akt pathway inhibitors—palbociclib and omipalisib [92]—it is not known how dysregulated miRNAs in this context could determine the drug response.

2.3. Glucose Metabolism-Mediated Tumorigenic Effects of Dysregulated miRNAs

The higher activity of glycolytic pathway has been shown in thyroid cancer to impose tumorigenesis [93,94]. Of note, miR-143 and miR-125a-5p have been reported to affect thyroid tumorigenesis via regulating glucose metabolism [59,95]. Studies in vitro and in vivo demonstrated that miR-143 directly targets Hexokinase 2 (HK2) and thereby down-regulates its expression. miR-143-mediated HK2 down-regulation resulted in suppressed glycolysis and therefore, decreased proliferation and migration [95]. These support the tumor-suppressive role of miR-143; however, the expression profile of this miRNA remained to be investigated in clinical thyroid cancer samples. A more comprehensive study revealed that miR-125a-5p was down-regulated in both thyroid cancer cell lines and clinical samples. miR-125a-5p was shown to block glucose metabolism via direct targeting CD147 that eventually suppressed cell viability and migration [59]. Given the clinical significance of targeting glucose metabolism in cancer therapy [96,97], a more in-depth study to examine both miR-143 and miR-125a-5p may open a new avenue to control thyroid cancer via regulating glucose metabolism.

2.4. Dysregulated miRNAs in Other Signaling Pathways

The down-regulation of miR-873-5p was shown to increase the expression of its direct target, C-X-C Motif Chemokine Ligand 16 (CXCL16) in PTC cell lines. This resulted in an increase in phosphorylation of p65 and Rel-B, thereby leading to activation of the NFκB pathway., and consequently, up-regulation of MMP1, MMP9 and MMP13 proteins. The overall phenotypic effects of miR-873-5p down-regulation are enhanced thyroid cancer cell proliferation, migration and invasion [70], which is theoretically consistent with the lowered expression of this miRNA in PTC clinical samples.

Overexpression of miR-155 was shown by Zhang et al. in ATC aggressive cells to promote cell proliferation, invasion and migration via directly targeting suppressor of cytokine signaling 1 (SOCS1) transcripts [53]. SOCS1 is a ubiquitin ligase with the tumor-suppressive activity that is silenced in human tumors [98–100]. This protein was demonstrated to be a direct inhibitor of the catalytic activity of Janus kinase 1 (JAK1), JAK2 and tyrosine kinase 2 (TYK2), thereby suppressing the JAK/signal transducer and activator of transcription (STAT) signaling pathway [101]. Given the oncogenic nature of JAK/STAT pathway in different human cancers [102] including thyroid carcinomas [103], targeting miR-155 could inhibit the thyroid cancer development/progression via activating SOCS1 and suppressing JAK/STAT. Notwithstanding, this hypothesis has not been tested in thyroid cancer. The reduced expression of miR-30 and miR-200 in ATC tumors is another remarkable finding that can help distinguish the ATC from PTCs or FTCs. The down-regulation of miR-30 and miR-200 mediates the suppression of mesenchymal-epithelial transition (MET), while inducing EMT [104].

Although the dysregulation of dozens of miRNAs has been reported in thyroid cancer, the majority of these studies focus on specific miRNAs and do not include transcriptome-wide screening approaches. Hence, RNA-seq- and/or microarray-based methods could be utilized to show a clearer map of dysregulated miRNAs. Moreover, despite the comprehensive studies with beneficial attempts to unravel the underlying molecular pathways involved in thyroid cancer, more studies are needed to fully map the mechanisms by which miRNA dysregulation mediates thyroid cancer development and progression. These could reveal the master regulator miRNAs that control key thyroid cancer signaling pathways, which may be used in personalized medicine.

3. lncRNAs Regulating Thyroid Carcinogenesis

LncRNAs are a sub-class of ncRNAs >200 nucleotides in length. LncRNAs regulate target gene expression through different mechanisms. At the transcriptional level, lncRNAs can interact with the polycomb repressive complex 2 (PRC2) and confine its access to particular genomic regions, leading to the suppression of gene expression [67]. For example, HOX transcript antisense RNA (HOTAIR) suppresses homeobox D (HOXD) expression via interacting with the catalytic subunit of PRC2, called enhancer of zeste homolog 2 (EZH2) [105,106]. Alternatively, lncRNAs have been reported to recruit DNA methyltransferases to modify chromatin conformation [107–109]. As an example, PTEN pseudogene (PTENpg1) utilizes DNA methyltransferase 3A (DNMT3a) to regulate the transcription of PTEN [110]. LncRNAs can also regulate gene expression at the post-transcriptional level by complementary sequence-specific mechanisms that affect the mRNA splicing, turnover and translation [44]. For example, metastasis- associated lung adenocarcinoma transcript 1 (MALAT1) was shown to compete with splicing regulatory proteins for binding on target mRNAs [111]; BETA-SECRETASE 1-ANTISENSE (Bace1-AS) hybridizes with Bace1 mRNA to increase its half-life [112], and long intergenic non-coding RNA p21 (lincRNA-p21) recruits translation repressors to catenin beta 1 (CTNNB1) gene to silence it [113]. The schematic of lncRNAs regulatory mechanisms is shown in Figure 1.

A series of lncRNAs have been identified to be abnormally regulated and expressed in thyroid cancer (Table 2).

Table 2. Dysregulated lncRNAs in thyroid cancer.

LncRNA	Alteration	Mechanism	Thyroid Cancer Type	Sample	Ref.
CASC9		↓ miR-488-3p → ↑ ADAM9 → ↑ EGFR/PI$_3$K/Akt pathway activation.	PTC	Cell line and tissue	[114]
DLX6-AS1		Negatively correlated with UPF1	Not known		[115]
ENST00000539653.1 (ENS-653)		Not known	PTC		[116]
H19		Not known	ATC	Tissue	[117]
		Not known	Not known		[58]
		↓ miR-3126-5p → ↑ ER-β	PTCSCs and PTC tissue		[118]
HCP5	↑	↓ miR-22-3p, miR-186-5p and miR-216a-5p → ↑ ST6GAL2	FTC		[119]
LINC00152		↓ miR-497 → ↑ BDNF		Cell line and tissue	[120]
LINC00514		↓ miR-204-3p → ↑ CDC23	PTC		[121]
LINC00941		↓ CDH6			[122]
MALAT1		No mechanism			[123]
n340790		↓ miR-1254	Not known		[119]
NEAT1		↓ miR-9-5p ↑ SPAG9	ATC		[67]
UNC5B-AS1		Not known	PTC	Tissue	[124]
XIST		↓ miR-34a → ↑ MET → PI$_3$K/Akt activation	Not known	Cell line and tissue	[125]
H19		Not known	FTC	Tissue	[126]
H19		Not known	PTC		[127]
LINC003121		↑ PI$_3$K and p-Akt	Not known		[128]
PAR5	↓	↑ EZH2 → ↓ E-cadherin	ATC		[129]
PTCSC3		↑ STAT3 → ↑ INO80		Cell line and tissue	[130]
		↑ miR-574-5p → ↓ SCAI → ↑ β-catenin → ↑ Wnt pathway activation	PTC		[55]
SNHG3		↑ PI$_3$K/Akt/mTOR pathway			[131]

PTC: Papillary Thyroid Cancer; FTC: Follicular Thyroid Cancer; ATC: Anaplastic Thyroid Cancer; PTCSCs: Papillary Thyroid Cancer Stem Cells; SNHG3: Small Nucleolar RNA Host Gene 3; PAR5: Prader Willi/Angelman Region RNA5; CASC9: Cancer Susceptibility 9; DLX6-AS1: Distal-Less Homeobox 6-Antisense 1; NEAT1: Nuclear Paraspeckle Assembly Transcript 1; UNC5B-AS1: Unc-5 Netrin Receptor B-Antisense 1; XIST: X-inactive specific transcript; MALAT1: Metastasis Associated Lung Adenocarcinoma Transcript 1; HCP5: HLA complex P5; PTCSC3: Thyroid Carcinoma Susceptibility Candidate 3; PI$_3$K: Phosphatidylinositol-4,5-Bisphosphate 3-Kinase; mTOR: Mechanistic Target of Rapamycin Kinase; EZH2: Enhancer of Zeste Homolog 2; ADAM9: ADAM Metallopeptidase Domain 9; EGFR: Epidermal Growth Factor Receptor; UPF1: UPF1 RNA Helicase and ATPase; ER-β: Estrogen Receptor Beta; SPAG9: Perm-Associated Antigen 9; BDNF: Brain-Derived Neurotrophic Factor; CDH6: Cadherin 6; STAT3: Signal Transducer And Activator of Transcription 3; ST6GAL2: alpha-2, 6-sialyltransferase 2; SCAI: Suppressor of Cancer Cell Invasion.

LncRNAs play roles in regulating diverse cellular processes [132], thereby causing cancer upon dysregulation [133].

The dysregulation of lncRNA H19 has been studied widely in different cancers such as bladder [134], breast [135,136] and liver [137]. The dysregulation of H19 in thyroid cancer is subject of controversy since the down-regulation was reported in FTC [126] and PTC [127], while the overexpression is reported in PTC, ATC [117] and PTC/PTC stem cells (PTCSCs) [118]. Given the functional studies performed in these reports, the oncogenic nature of H19 in thyroid cancer is more conceivable since H19 knockdown resulted in the suppression of proliferation, migration, and invasion in ATC cells in vitro and inhibited tumorigenesis and metastasis in vivo [117]. Moreover, the depletion of H19 was shown to suppress the sphere formation ability [118], collectively suggest that H19 acts as an oncogene in thyroid cancer.

In contrast to the many studies focusing on specific lncRNAs, Pellecchia et al. performed a transcriptome-wide screening using lncRNA microarray method. Comparing the ATC tumor to the normal samples revealed that Prader Willi/Angelman Region RNA5 (PAR5) was a significant down-regulated lncRNA in tumors. Mechanistically, overexpression of PAR5 expression resulted in the down-regulation and dissociation of EZH2, leading to the relieving of E-cadherin transcription efficiency, which subsequently reduced proliferation and migration of ATC-derived cells [129]. In another unbiased study, utilizing the chromatin immunoprecipitation (ChIP)-seq method, Linc00941 was found as a highly expressed enhancer-associated lncRNA in PTC tumor samples as compared to the paired healthy tissues. Of note, Linc00941 expression was significantly higher in BRAFV600E PTC patients and correlated with extrathyroidal extension in PTC patients, suggesting that the up-regulation might be mediated by BRAFV600E. However, this hypothesis was not tested. Functionally, Linc00941 promoted the proliferation and invasion of PTC cell lines, hypothetically mediated by targeting Cadherin 6 CDH6 transcripts [122].

The up-regulation and corresponding oncogenic properties of a diverse range of lncRNAs are reported in different studies, including distal-less homeobox 6-antisense 1 (DLX6-AS1) [115], nuclear paraspeckle assembly transcript 1 (NEAT1) [67], ENST00000539653.1 (ENS-653) [116], Unc-5 netrin receptor B-antisense 1 (UNC5B-AS1) [124], LINC00514 [121], LINC00152 [120], MALAT1 [123], HLA complex P5 (HCP5) [119] and n340790 [119]. Although the exact modes of action by which these lncRNAs promote/suppress thyroid carcinomas have not been elucidated, we summarize a number of altered lncRNAs with implications in Wnt and PI$_3$K/Akt signaling pathways (Figure 2).

3.1. Wnt-Mediated Tumorigenic Effects of Dysregulated lncRNAs

The putative role of lncRNAs in thyroid tumorigenesis mediated by activating the Wnt signaling pathway has been also tested. Thyroid carcinoma susceptibility candidate 3 (PTCSC3) down-regulation was observed in the clinical samples as well the cell lines resulting in elevation of miR-574-5p and subsequently down-regulation of SCAI and activation of β-catenin (Figure 2). Functionally, overexpression of PTCSC3 inhibited the cell proliferation and migration via suppressing the Wnt pathway. Besides, PTCSC3 suppressed the growth in vivo, indicating that PTCSC3 acts as a tumor suppressor in thyroid cancer [55]. It is unclear whether inhibiting/overexpressing the dysregulated lncRNAs could affect the downstream targetome of thyroid cancer-related signaling pathways. For example, could PTCSC3 overexpression suppress the Wnt activity using TopFlash assay?

3.2. PI$_3$K/Akt-Mediated Tumorigenic Effects of Dysregulated lncRNAs

In the context of the PI$_3$K/Akt pathway, the down-regulation of small nucleolar RNA host gene 3 (SNHG3) was reported to promote growth and invasiveness in vitro and in vivo via activating PI$_3$K/Akt/mechanistic target of rapamycin kinase (mTOR) pathway [131]. However, the mechanism by which such phenotype was observed has not been elucidated. Cancer susceptibility 9 (CASC9) is another example of a dysregulated ncRNA in thyroid cancer. The up-regulation of this lncRNA was reported in PTC patient tissues and cell lines and mechanistically was shown to sponge miR-488-3p to rescue ADAM metallopeptidase domain 9 (ADAM9) oncogene. This consequently resulted in promoting the proliferative, migrative, and invasive abilities of thyroid cancer cells in vitro, and augmenting the tumorigenesis in vivo via activating epidermal growth factor receptor (EGFR)/PI$_3$K/Akt pathway [114].

Similarly, the up-regulation of X-inactive specific transcript (XIST) in thyroid cancer was demonstrated to activate the PI$_3$K/Akt pathway via sponging miR-34a and the subsequent rescue of MET, a well-known oncogene in thyroid cancer. The functional experiments properly proved the oncogenic role of XIST, where its silencing suppressed the proliferation and tumor growth in vitro and in vivo [125].

LINC003121 is an example of down-regulated lncRNA in thyroid cancer. Although the mechanistic evaluations were not performed, the lower expression of LINC003121 was shown to increase PI$_3$K and p-Akt expression, leading to increased cell proliferation and invasion in vitro, and promoting tumorigenicity in thyroid cancer xenograft models in nude mice [128]. Together with reviewed miRNAs, PI$_3$K/Akt pathway plays a pivotal role in thyroid development and progression upstream of a myriad of ncRNAs (Figure 2).

Taken together, recent evidence highlights the molecular and functional relevance of different lncRNAs in thyroid cancer, especially in the context of PI$_3$K/Akt and Wnt signaling pathways. However, it remained to know how and to what extend the alterations in lncRNAs could influence the Wnt and/or PI$_3$K-mediated therapy. Besides, high throughput methods, such as RNA-seq or ChIP-seq and microarray, have not been utilized to uncover the dysregulated lncRNAs in thyroid cancer patients in a transcriptome-wide manner.

4. circRNAs Regulating Thyroid Carcinogenesis

circRNAs, the stably expressed ncRNAs in different cell types with special annular structures, play fundamental regulatory roles in the physiological processes of the cell and have implications in human diseases such as cancer. Mechanistically, circRNAs impose their downstream effects via sponging miRNAs or interacting with proteins [138–140] (Figure 1). A few of oncogenic circRNAs have been reported to drive thyroid cancer (Table 3).

Table 3. Dysregulated circRNAs in thyroid cancer.

circRNA	Alteration	Mechanism	Sample	Thyroid Cancer Type	Ref.
circ_0008274	↑	↓ AMPK/mTOR signaling pathway	Cell line and tissue	PTC	[141]
circEIF6	↑	↓ miR-144-3p → ↑ TGF-α	Cell line and tissue	PTC & ATC	[142]
circFOXM1	↑	↓ miR-1179 → ↑ HMGB1	Cell line and tissue	PTC	[143]

PTC: Papillary Thyroid Cancer; ATC: Anaplastic Thyroid Cancer; circFOXM1: Circular Forkhead Box Protein M1; circEIF6: circular Eukaryotic Translation Initiation Factor 6; HMGB1: High Mobility Group Box Protein 1; AMPK: 5′ AMP-activated Protein Kinase; mTOR: Mammalian Target of Rapamycin (mTOR); TGF-α: Transforming Growth Factor Alpha.

For instance, the up-regulation of circular forkhead box protein M1 (circFOXM1) was reported in PTC tissues, as well as ATC and PTC cell lines [143]. With no effect on the linear FOXM1 transcript, circFOXM1 was demonstrated to modulate cancer progression through sponging miR-1179 and rescuing high mobility group box protein 1 (HMGB1) expression, which eventually promotes tumor growth of PTC in vitro and in vivo [143]. Likewise, the up-regulation of circ_0008274 has been reported in PTC tissues and cell lines. High circ_0008274 expression has been associated with more advanced thyroid cancer TNM staging and lymph node metastases. The in vitro studies revealed that this circRNA promoted cell proliferation and invasion. circ_0008274 imposes its effect via the activation of the mammalian target of rapamycin (mTOR) signaling pathway (increasing p-mTOR) and the inhibition of the 5′ AMP-activated protein kinase (AMPK) (reducing p-AMPK) [141]. Given the known negative regulatory effect of AMPK on mTOR protein [144], it is conceivable that circ_0008274 activates mTOR pathway via inhibiting AMPK protein, leading to thyroid cancer development and progression. However, this notion remained to be tested in depth in thyroid cancer cell lines.

Collectively, circRNAs have been shown to be deregulated and promote thyroid cancer mainly through sponging miRNAs. More studies are needed, particularly in ATC and metastatic thyroid cancers, to develop deeper insights on circRNAs roles in thyroid cancer initiation and promotion.

5. ncRNAs May Regulate the Biology of Thyroid Tumor Microenvironment

The tumor microenvironment is composed of a variety of tumor-associated immune cells as well as growth factors, cytokines, and chemokines. Changes in the tumor microenvironment that occurs during cancer progression could induce various biological processes like angiogenesis, proliferation, invasion and metastasis, immune tolerance and alter the response to therapeutic agents [145,146]. B cells, T cells, mast cells, dendritic cells and macrophages are the main immune cells accumulated in the tumor microenvironment, where machrophages are characterized by plasticity and diversity and play an important role in the immune response [147]. In response to hypoxia, tumor-associated macrophages produce WNT7b, which in turn attributes to up-regulation of vascular endothelial growth factor (VEGF) by adjacent vascular endothelial cells in the tumor microenvironment. This eventually results in the elevated angiogenesis and promotion of tumor growth [148,149]. Of note, the secretion of diverse chemokines and cytokines recruits the regulatory T lymphocytes that result in the inhibition of effector T cells [150]. Although there is lack of conclusive evidence about the roles of ncRNAs in regulating the microenvironment-associated cells/components in thyroid cancer, here we review the current data and propose the potential ncRNA-related mechanisms.

NEAT1, an up-regulated lncRNA in thyroid cancer, has been reported to induce tumor-associated macrophages via sponging miR-214 and inducing the β-catenin/Wnt signaling pathway [151]. This further highlights the therapeutic potential of NEAT1 as a potential target for thyroid cancer therapy [152]. Another highly-expressed lncRNA in thyroid cancer, MALAT1, can activate angiogenesis via increasing fibroblast growth factor 2 (FGF2) expression in tumor-associated macrophages [153]. Another up-regulated lncRNA, PTCSC3, may also regulate the thyroid cancer tumor microenvironment via activating the Wnt signaling pathway [55].The up-regulation of miR-574-5p and miR-195 has been discussed in Section 2.1 to induce proliferation, EMT, invasion and migration via activating the Wnt pathway in thyroid cancer. miR-574-5p and/or miR-195 overexpression could induce the chemokines/cytokines and activate the tumor-associated immune cells. This hypothesis remained to be tested to assess the potential of these miRNAs in modulating the thyroid tumor microenvironment.

6. ncRNAs are Novel Candidates for Early Detection of Thyroid Cancer

In poorly differentiated, medullary, and anaplastic thyroid carcinomas [154], early detection of cancer is key to maximizing the chance of successful treatment and prolonging patient survival. This is achieved by systematic screening and recognizing the warning signs [155]. For thyroid cancer, the current standard of care is to perform neck ultrasonography together with FNA cytology, which are determinants to discriminate benign and malignant thyroid nodules [156]. However, 10–40% of thyroid nodule FNA cytology results are indeterminate, resulting in a repeat FNA several months later [157,158]. In indeterminate thyroid FNA cytology, the estimated incidence of malignancy ranges between 10 and 75% [159]. Furthermore, it is not possible to distinguish the benign follicular adenoma (FA) from the malignant FTC via FNA cytology unlike other forms of thyroid cancers. Hence, the diagnosis of malignancy in this group can only be made post-surgery via histology that demonstrates the presence of capsular or vascular invasion. This difficulty in making a definite diagnosis for cytologically-indeterminate FNA usually results in delayed definitive treatment and management. Treatment for repeatedly indeterminate cases usually takes the form of a diagnostic hemi-thyroidectomy, or a two-stage thyroidectomy for patients with thyroid cancer on diagnostic histology. Furthermore, the majority of patients (76–81%) with indeterminate cytology have benign thyroid nodules and thus were subjected to unnecessary diagnostic thyroid surgeries [160]. As such, there is a critical need for clinicians to improve the ability to predict the risk of thyroid cancer in thyroid nodules.

As reviewed in the introduction section, molecular tests for oncogene mutations such as BRAF, NRAS, KRAS, HRAS, Pax8-PPARG, re-arrangement of RET and TERT have been used to improve benign/malignant differentiation [6,161]. However, they have not increased the specificity of such diagnostic methods [162,163]. Moreover, the performance of these kits may also differ in different populations due to the different prevalence of the genetic alterations of interests. For example, while BRAF^{V600E} is present in 38–58% of thyroid cancers in patients of European ancestry, the prevalence across East Asia can range from 80% in Korea, 62% in China, and 53% in India [164–169]. These suggest that more molecular tools are needed, especially for the timely diagnosis of advanced thyroid cancers. Here, we focus on the significance of diagnostic ncRNAs in thyroid cancer (Table 4).

Table 4. Diagnostic ncRNAs in thyroid cancer.

ncRNA	ncRNA Type	Source	Finding	Thyroid Samples	Ref.
miR-138-1-3p miR-139-5p miR-146b-5p miR-155miR-204-5p miR-222-3pmiR-29b-1-5p miR-31-5p miR-375 miR-551b-3p		FNA	miRNA testing, recently commercialized as ThyraMIR, identified 64% of malignant cases and 98% of benign cases correctly.	Not known	[170]
miR-146b			Higher expression in PTC FNA samples	PTC	[171]
miR-132-3p miR-146a-5pmiR-17-5p miR-183-3p miR-222-3p miR-451a	miRNA		miR-222-3p and miR-17-5p can accurately discriminate MTC from the benign nodule and healthy control groups	PTC, MTC, benign nodules and controls	[172]
miR-146a-5p miR-221-3p miR-222-3p		Serum	High pre-surgical expressionLow post-surgical expression		[173]
miR-146b miR-21 miR-221 miR-222			Higher expression in PTM serum samples	PTC	[174]
miR-146a-5p miR-199b-3p			Lower expression in PTC serum as compared to benign serum samples		[175]
let-7b-5pmiR-10a-5p			Higher expression in PTC serum as compared to benign serum samples		
miR-423-5p			Higher expression in PTC serum samples		[176]
let-7a		Tissue	Lower expression in thyroid tumor samples	Not known	[58]
H19			Higher expression in thyroid tumor samples		
H19	lncRNA		Lower expression in thyroid tumor samples as compared to benign samples	PTC	[127]
MALAT1			Higher expression in thyroid tumor samples		[123]
n340790				Not known	[119]
UNC5B-AS1				PTC	[124]

PTC: Papillary Thyroid Cancer; MTC: Medullar Thyroid Cancer; UNC5B-AS1: Unc-5 Netrin Receptor B-Antisense RNA 1; MALAT1: Metastasis Associated Lung Adenocarcinoma Transcript 1; FNA: Fine Needle Aspiration.

ncRNAs have tremendous potential as diagnostic biomarkers as they are found to be stable and detectable in body fluids [177–179]. For instance, miR-10b-5p, miR-195-5p, miR-132-3p, miR-20a3p, miR-185-5p, and miR-296-5p are reported to be significantly overexpressed in gastric cancer patients' sera [180]. Commercially, a lncRNA named prostate cancer-associated 3 (PCA3) has been approved by the US Food and Drug Administration as a urine biomarker for prostate cancer [181].

The diagnostic significance of miR-222, miR-146a-5p and miR-146b has been shown by different studies. Zhang et al. studied the serum level of miR-222-3p, miR-17-5p, miR-451a, miR-146a-5p, miR-132-3p and miR-183-3p in PTC, MTC, benign nodule and control groups. The results revealed that the serum levels of miR-222-3p, miR-17-5p, and miR-451a were markedly increased, while miR-146a-5p, miR-132-3p, and miR-183-3p were significantly decreased in the PTC and benign nodule groups compared with the control group. There was no difference in the miRNA expression profile between the PTC and the benign nodule group. Nevertheless, the serum levels of miR-222-3p and miR-17-5p were significantly increased in the MTC group than the benign nodule and control groups. Therefore, they concluded that miR-222-3p and miR-17-5p can accurately discriminate MTC from the benign nodule group and healthy controls [172]. Rosignolo et al. studied the serum level of miR-146a-5p, miR-221-3p and miR-222-3p before and 30 days after surgery in PTC patients. The pre-surgical high expression of these miRNAs with a significant post-surgical down-regulation supported the diagnostic significance of miR-146a-5p, miR-221-3p and miR-222-3p [173]. Moreover, the heightened serum level of miR-222, miR-221, miR-146b and miR-21 was detected in the PTC tumors with <10 mm diameter as compared to benign nodules [174]. The importance of miR-146b was supported by another study that showed the escalated expression of this miRNA in FNA biopsy samples of PTC samples [171]. Graham et al. measured the expression level of miRNAs in the serum of PTC samples versus the benign nodule group. The results revealed that miR-146a-5p and miR-199b-3p were highly expressed in the PTC group, whereas let7b-5p and miR-10a-5p were down-regulated [175]. These various studies on the clinical significance of miRNAs have resulted in the advent of a commercial diagnostic product in thyroid cancer. ThyraMIR® miRNA classifier product is developed to be used in combination with the conventional molecular diagnostic platform as a new thyroid cancer diagnostic tool, with promising outcomes showing 89% sensitivity and 85% specificity [170]. This kit is expected to reduce 85% of unnecessary surgeries of benign thyroid nodules [161,182–184]. The miRNAs utilized in this classifier are miR-29b-1-5p, miR-31-5p, miR-138-1-3p, miR-139-5p, miR-146b-5p, miR-155, miR-204-5p, miR-222-3p, miR-375 and miR-551b-3p.

Although the diagnostic values of miRNAs have been widely studied in the serum, the diagnostic significance of lncRNAs is dominantly studied in tissues. This may be because the stability of lncRNA levels is the lowest among several different RNA species [185]. A nonrandomized, retrospective study examining PTC patients and benign thyroid nodes revealed that lower H19 expression levels could distinguish PTC from benign with area under the curve (AUC) of the receiver operating characteristic (ROC) curve of 0.813 [127]. Liu et al. also studied the diagnostic significance of H19 together with let-7a in thyroid cancer patients showing that these ncRNAs could discriminate the thyroid cancer tumors against the healthy samples with AUC of 0.801 and 0.116 for H19 and let-7a, respectively [58]. The higher expression of UNC5B-AS1, MATAL1 and n340790 was shown to distinguish the tumor and normal samples with AUC values of 0.932, 0.632 and 0.845, respectively [119,123,124]. With the increasing identification of diagnostic circulating lncRNAs in different cancers [186–190], we expect to see the utilization of circulating lncRNAs to optimize diagnostic accuracy of thyroid nodules in the future. Collectively, ncRNAs could be clinically relevant biomarkers for thyroid cancer diagnosis (Figure 3).

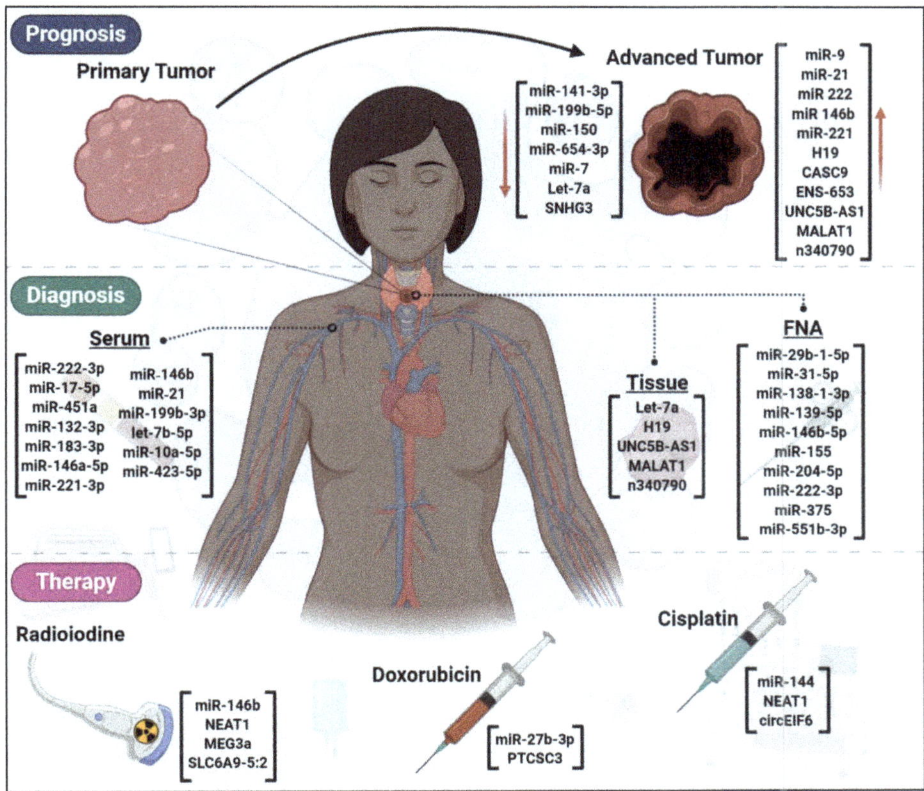

Figure 3. The diagnostic, prognostic and therapeutic significance of dysregulated ncRNAs in thyroid cancer. A variety of ncRNAs are dysregulated during primary to advanced tumor progression. These ncRNAs could have implications for the prognosis of thyroid cancer patients. The expression levels of ncRNAs could distinguish normal from tumor thyroid cells, therefore, acting as diagnostic biomarkers. Several ncRNAs may determine the response/resistance to the routine thyroid cancer treatment options of radioiodine and/or chemotherapy. FNA: fine needle aspiration.

Other than the ncRNAs utilized in the current commercial kit, other potential ncRNAs such as miR-146a-5p and H19 could be evaluated in clinical validation studies as diagnostic biomarkers. Although no circRNA has been reported to convey diagnostic significance in thyroid cancer, these ncRNAs have tremendous potential as diagnostic biomarkers due to the high stability of their circular structure and accumulation in exosomes. These characteristics result in the stable secretion of circRNAs in peripheral body fluids such as plasma and saliva, where they can be detected for early diagnosis of cancer [170,191–193].

7. ncRNAs as Prognostic Factors for Thyroid Cancer

ncRNAs have recently shown a massive capability as prognostic factors in human cancers [194,195]. For instance, the reduced expression of let-7 was shown to be associated with shortened postoperative survival of lung cancer patients [196]; and tumor suppressor candidate 7 (TUSC7) is a prognostic lncRNA that is inversely associated with aggressive stages and shorter survival of gastric cancer patients has been reported [197].

Thus far, BRAF^{V600E} and TERT [13]/p53 [198] mutations are the main molecular prognostic biomarkers [199] used along with clinicopathological factors such as age, extra-thyroid tumor spread, lymph node and distant metastases and increasing tumor size in thyroid cancer [200]. Nonetheless, nearly 30% of thyroid cancer patients may face over- or undertreatment in a condition based on BRAF status alone [200]. Moreover, the impact of BRAF status on the risk of recurrence in the very low-risk patients appears to be small [13]. This suggests that more molecular biomarkers are needed to determine the prognosis of thyroid cancer patients. In this section, we review the recent advances in prognostic ncRNAs in thyroid cancer, that are listed in Table 5.

Table 5. Prognostic ncRNAs in thyroid cancer.

ncRNA	ncRNA Type	Prognostic Significance	Thyroid Cancer Type	Ref.
Let-7a		Negative correlation with higher TNM stages lymph node metastasis and lower 5-year survival	Not known	[58]
miR-141-3p		Negative association with TNM stage and lymph node metastasis		[61]
miR-146b miR-21 miR-222		Poor prognosis		[174]
miR-150	miRNA	Negative association with TNM stage and lymph node metastasis		[64]
miR-199b-5p		Negative association with stage		[66]
miR-21		Poor prognosis		[54]
miR-21 miR-9		Independent prognostic factors of PTC recurrence	PTC	[201]
miR-221		Independent prognostic factors of PTC recurrence		[202]
miR-654-3p		Down-regulation upon a long-term PTC progression in BRAF^{V600E}-transgenic mice		[203]
miR-7		Negative association with stage		[69]
CASC9		Positive association with large tumor size, advanced stage, or lymph node metastasis.		[114]
ENST00000539653.1 (ENS-653)		Positive association with larger tumor size, more advanced clinical stage and poorer disease-free survival		[116]
H19	lncRNA	Positive correlation with higher TNM stages lymph node metastasis and lower 5-year survival	Not known	[58]
		Positive correlation with poor overall survival	PTCSCs and PTC tissue	[118]
		Negative correlation with extrathyroid extension, tumor size, histological aggressive type, pathological lateral node metastasis and poorer disease-free survivalIndependent risk factor for extrathyroidal extension and lymph node metastasis.	PTC	[127]
		Negative association with tumor size, distant metastasis and vascular invasion	FTC	[126]
MALAT1		Positive correlation with tumor size and lymph node metastases	PTC	[123]
n340790		Positive correlation with primary clinicopathological characteristics (good prognostic factor)	Not known	[119]
SNHG3		Negative association with stage and poor prognosis	PTC	[131]
UNC5B-AS1		Positive correlation with lymph node metastasis, tumor size and histological type		[124]

PTC: Papillary Thyroid Cancer; FTC: Follicular Thyroid Cancer; ATC: Anaplastic Thyroid Cancer; PTCSCs: Papillary Thyroid Cancer Stem Cells; SNHG3: Small Nucleolar RNA Host Gene 3; CASC9: Cancer Susceptibility 9; PTCSC3: Thyroid Carcinoma Susceptibility Candidate 3; UNC5B-AS1: Unc-5 Netrin Receptor B-Antisense RNA 1; MALAT1: Metastasis Associated Lung Adenocarcinoma Transcript 1.

The overexpression and oncogenic actions of miR-21 have been widely reported in thyroid cancer [54,204–206], suggesting that this miRNA could be a potential diagnostic factor for early detection of thyroid cancer. Beyond that, the association between miR-21 with the clinicopathological characteristics of thyroid cancer uncovered its capability to be used as a prognostic factor. The survival analysis by Zang et al. revealed that the higher expression of miR-21 can predict poor prognosis of PTC patients [54]. In another study, the multivariate survival analysis of patients for at least 120 months after surgery showed that miR-9 and miR-21 were significant independent prognostic factors for recurrence of PTC patients [201]. Zhang et al. studied miR-21 together with miR-221, miR-222 and miR-146b and depicted that all these miRNAs, except for miR-221, were highly expressed in poor-prognosis PTC patients group [174]. Comparing the expression level of miR-21, miR-222, miR-9, miR-10b, miR-146b, miR-31, miR-220 and miR-221 in recurrent vs. non-recurrent groups in another study showed that although all the miRNAs were dysregulated, only miR-221 overexpression was the independent prognostic factor of PTC recurrence [202]. The different outcomes among these studies could be due to the type of statistical tests used, e.g., multivariate vs. univariate Cox survival analysis. A deeper epidemiological analysis encompassing more dysregulated miRNAs, all the clinicopathological features and controlling for potential confounder parameters in a larger cohort size could better assess the utility of these miRNAs as prognostic markers.

miR-141-3p and miR-150 were separately demonstrated by different studies to be inversely associated with TNM stage and lymph node metastasis in PTC patients [61,64]. Similarly, the inverse association between miR-199b-5p or miR-7 and TNM stage was shown in PTC patients [66,69]. In another study, Geraldo et al. performed a prognosis study using the BRAF^{V600E}-mutant PTC progression model in mice. The results interestingly showed that miR-654-3p levels underwent a significant decrease with long-term PTC progression in mice and negatively correlated with EMT. They further reported the down-regulation of miR-654-3p in PTC cell lines with the subsequent effect on increasing proliferation and migration. This suggests that not only this miRNA undergoes down-regulation in PTC development, but it also continues to be suppressed to progress cancer. Nonetheless, the epidemiological study is required to prove it as a prognostic factor in thyroid cancer [69].

With regard to the prognostic values of lncRNAs in thyroid cancer, several lncRNAs have been investigated. The decreased expression of small nucleolar RNA host gene 3 (SNHG3) was shown to correlate with the higher TNM stages and poorer prognosis of PTC patients [131]. On the contrary, CASC9, ENS-653, MALAT1 and UNC5B-AS1 associated positively with the advanced clinicopathological characteristics including large tumor size, advanced stage, or lymph node metastasis in PTC patients [114,116,123,124]. lncRNA H19 has been widely studied in different thyroid cancer cohorts that showed controversial prognostic results. In PTC, H19 was shown to be inversely associated with tumor size, pathological lateral node metastasis, extrathyroid extension, histological aggressive type and poorer disease-free survival [127]. The multivariate analysis of this retrospective, non-randomised study including 89 patients with benign thyroid nodes and 410 patients with PTC confirmed that H19 could be an independent risk factor for the extrathyroidal extension and lymph node metastasis [127]. Li et al. reported that higher H19 expression correlated with the poorer overall survival of PTC patients [118]. Similarly, Liu et al. studied thyroid cancer patients and showed that H19 positively correlated with higher TNM stages, lymph node metastasis and lower 5-year survival rate. Controlling for another ncRNA surveyed in this study, the higher H19 and lower let-7a along with tumor size, stage and lymph node metastasis were confirmed as the independent prognostic factors of thyroid cancer [58]. In FTC, H19 was revealed as the prognostic factor negatively associated with tumor size, distant metastasis and vascular invasion. However, the multivariate regression demonstrated that only age, primary tumor size ≥4 cm and vascular invasion were the significant prognostic factors of survival [126].

Taken together, the prognostic significance of ncRNAs has been shown in thyroid cancer (Figure 3). Notwithstanding, very little is known about the prognostic ncRNAs in more advanced thyroid cancers of MTC and ATC that show worse prognosis as compared to PTC and FTC. This indicates that the dysregulated ncRNAs could be studied epidemiologically in MTC/ATC cohorts.

8. ncRNAs Could Affect Thyroid Cancer Therapy

Upon thyroid cancer diagnosis, surgical excision of tumors is performed [13]. Patients with more advanced, differentiated thyroid cancers and a higher risk of recurrent or persistent disease undergo adjuvant RAI therapy. The molecular basis of this adjuvant therapy is the uptake of RAI via the plasma membrane sodium iodide symporter (NIS). NIS mediates the influx of RAI via transporting two Na^+ ions and one I^- ion into the cytosol. RAI is then concentrated into the thyroid cells by iodine-metabolizing machinery. This eventually increases the efficiency of RAI therapy and improves the prognosis of thyroid cancer patients.

Notwithstanding, about 25–50% of locally advanced or metastatic thyroid cancers become refractory to RAI therapy. This leads to a poorer outcome with 5-year survival of <50% and 10-year survival of <10% [207,208]. RAI refractory response occurs through a complex de-differentiation process that leads to a diminished or a loss of NIS expression and/or correct localization. These prevent the cytoplasmic influx of RAI in thyroid cells, thereby causing adjuvant therapy resistance [209]. There are multiple mechanisms by which RAI resistance happens. Dysregulation of the MAPK signaling pathway is a well-studied mechanism that represses NIS protein expression. Thyroid cancer cells harboring $BRAF^{V600E}$ mutations exhibit robust activation of MAPK signaling, associated with a de-differentiated state [209]. $BRAF^{V600E}$-induced MAPK-independent repression of NIS has also been reported, where $BRAF^{V600E}$ induces Transforming Growth Factor β (TGF-β) secretion. This resulted in the repression of NIS and elevated oncogenic properties in PTC cells [210]. The mechanistic role of PI_3K and notch signaling pathways have also been demonstrated in RAI resistance [211]. Although the diversity of underlying mechanisms delineates the complexity and difficulty of the restoration of RAI sensitivity in thyroid cancer patients, different approaches have been implemented to overcome this clinical obstacle. Treatment of RAI-refractory thyroid cancer patients with retinoic acid [212], epigenetic transcriptional restoration of NIS expression via histone deacetylase inhibitors (HDACi) [213], or peroxisome proliferator-activated receptor (PPAR)-γ agonist [214] have shown sub-optimal re-differentiation outcomes for the patients.

Among various tested measures, ncRNAs have emerged as potential modulators of NIS restoration. Inhibition of miR-21, an upregulated miRNA in thyroid cancer, resulted in up-regulation of NIS expression, although the detailed mechanisms remain unknown [204]. In addition, miR-146b was reported to be highly up-regulated in dedifferentiated thyroid cancer cells, resulting in the repression of NIS via direct targeting of the NIS mRNA [215,216]. NIS expression can also be regulated by let7f-5p in PTC and FTC cells [217]. These findings highlight the potential clinical significance of inhibiting NIS-targeting miRNAs with respect to the re-differentiation and restoration of RAI sensitivity in RAI-refractory thyroid cancer patients. Additionally, transcriptome-wide approaches will be critical to identify the dysregulated lncRNAs and circRNAs in de-differentiated thyroid cancer cases, providing previously uncharacterized targets for potential ncRNA-based strategies for the restoration of NIS expression. Given the known regulatory effects of miR-375, miR-497, CASC9 and XIST on the PI_3K/Akt signaling pathway (Figure 2)—that is involved in NIS repression—in-depth studies could unravel their potential roles in the redifferentiation process.

Currently, patients with RAI-refractory differentiated thyroid cancer undergo systemic therapy, including targeted therapy and chemotherapy [12]. The chemotherapy drugs commonly used to treat thyroid cancer, in particular, the aggressive medullary and anaplastic thyroid malignancies are dacarbazine, vincristine, cyclophosphamide, doxorubicin, streptozocin, fluorouracil, paclitaxel, docetaxel and carboplatin [218–222]. However, chemotherapy is rarely used for thyroid cancer treatment, except for ATC patients. Doxorubicin was the only chemotherapy approved for the treatment of thyroid cancer patients. Nevertheless, it yielded a complete or partial response rate of <40% with limited durability [223]. Thus far, 2 multi-targeted tyrosine kinase inhibitors, sorafenib and lenvatinib, have been FDA-approved for the treatment of locally advanced or metastatic progressive RAI-refractory differentiated thyroid cancer. Both had been shown to improve progression-free survival but not overall survival [224–226]. For progressive metastatic MTC, 2 multi-targeted tyrosine kinase inhibitors, vandetanib and cabozantinib showed to improve progression-free survival and have been FDA-approved [227,228]. In metastatic BRAF^{V600E}-mutant ATC, combined targeted therapy with dabrafenib and trametinib has been shown to improve progression-free survival and is FDA-approved for this indication [229]. Mutation-selective kinase inhibitor such as RET-inhibitor selpercatinib is FDA-approved for the treatment of metastatic RET-mutant MTC or RET-fusion mutant differentiated thyroid cancers with phase II clinical trial showing an overall response rate of 70%. Another mutation-selective kinase inhibitor that has been FDA-approved is the TRK inhibitor (larotrectinib or entrectinib) that can be used in metastatic thyroid cancers with NTRK-fusion mutation [230,231].

Serum thyroglobulin (Tg) is used as a tumor marker to monitor disease burden with treatment [232]. It detects recurrence in thyroid cancer with a sensitivity of 19–40% and specificity of 92–97% [233]. In addition, the presence of anti-thyroglobulin antibody in 25% of thyroid cancer patients affects the reliability of Tg assay [234,235]. Patients with poorly-differentiated thyroid cancers lose the ability to produce Tg, making the measurement of Tg an unreliable reflection of tumor burden [232]. Given the low survival rate of aggressive thyroid cancer patients and the rather low sensitivity of Tg for detecting thyroid cancer recurrence leave room for the development of molecular tools that are more sensitive, and hopefully equally specific, than Tg.

The other underlying factor playing role in cancer treatment failure could be drug resistance, leading to elevated cancer relapse and mortality in patients [236,237]. ncRNAs are increasingly studied to unravel the complex mechanism of drug resistance development [238,239]. In this section, we discuss the recent advances contributing to the understanding of how ncRNAs contribute to drug sensitivity and/or resistance in thyroid cancer (Table 6).

Table 6. Therapeutic ncRNAs in thyroid cancer.

ncRNA	ncRNA Type	Therapeutic Significance	Thyroid Cancer Type	Ref.
miR-144		↑ Sensitivity to cisplatin	ATC	[62]
miR-146b		↓ Radioiodine uptake	FTC	[215]
miR-27b-3p	miRNA	↑ Resistance to doxorubicin	ATC	[240]
miR-625-3p		Target of icariin anti-tumor substance	PTC	[56]
MEG3a		↑ Resistance to radioactive iodine	FTC and PTC	[241]
NEAT1		↑ Resistance to cisplatin	ATC	[67]
NEAT1	lncRNA	↑ Resistance to radioactive iodine	PTC	[242]
PTCSC3		↑ Resistance to doxorubicin	ATC	[130]
SLC6A9-5:2		↑ Resistance to radioactive iodine	PTC	[243]
circEIF6	circRNA	↑ Resistance to cisplatin	PTC and ATC	[142]

FTC: Follicular Thyroid Cancer; PTC: Papillary Thyroid Cancer; ATC: Anaplastic Thyroid Cancer; NEAT1: Nuclear Paraspeckle Assembly Transcript 1; MEG3a: Maternally Expressed Gene 3; PTCSC3: Papillary Thyroid Carcinoma Susceptibility Candidate 3; SLC6A9-5:2: Solute Carrier Family 6 Member 9-5:2; circEIF6: Circular Eukaryotic Translation Initiation Factor 6.

Icariin, a chemical flavonoid compound isolated from different species of the genus *Epimedium* plant, has recently emerged as an anti-cancer substance [244], e.g., in ovarian cancer by targeting miR-21 [245] or in colorectal cancer through enhancing the NFκB suppression-mediated radiosensitivity [246]. In thyroid cancer, Fang et al. demonstrated that icariin inhibited cell growth, invasion and migration, while promoting apoptosis [56]. Mechanistically, icariin was shown to target miR-625-3p leading to inactivation of the PI$_3$K/Akt and mitogen-activated protein kinase kinase (MEK)/mitogen-activated protein kinase 1 (ERK) signaling pathways [56]. Therefore, targeting miR-625-3p expression that is elevated in thyroid cancer could enhance the therapeutic sensitivity of tumor cells to icariin. In another study, Xu et al. reported that miR-27b-3p expression level was increased in doxorubicin-resistant ATC cells through targeting and suppressing peroxisome proliferator-activated receptor gamma (PPARγ) gene [240]. Their findings indicated that targeted inhibition of miR-27b-3p might be a potential therapeutic approach in doxorubicin-resistant ATC cells. Another study on progressive ATC cells depicted that down-regulation of miR-144 led to the cisplatin resistance. Transfection with miR-144 mimics improved the sensitivity of ATC cells to cisplatin and inhibited tumor growth by suppressing (transforming growth factor alpha) TGF-α both in vitro and in vivo [62]. Collectively, the dysregulation of miRNAs could determine the sensitivity or resistance to the anti-tumor compounds in thyroid cancer (Figure 3). Together with drug screening, the miRNA microarray array clinical studies could provide more insights into the functional miRNAs involved in drug response.

To date, many lncRNAs have been shown in human cancer to play role as therapeutic determinants [247–249]. In thyroid cancer, lncRNAs have recently emerged as important factors involved in the sensitivity of patients to different treatments. LncRNA NEAT1 has been shown to have therapeutic implications in different thyroid cancer types. In ATC, in vitro and in vivo overexpression of NEAT1 was demonstrated to elevate cisplatin resistance via sponging miR-9-5p, which resulted in rescuing perm associated antigen 9 (SPAG9) [67]. In addition, higher expression of NEAT1 was detected in radioactivity iodine-resistant PTC tissues and cell lines, and this was associated with miR-101-3p inhibition, Fibronectin 1 overexpression, and PI$_3$K/Akt pathway activation [242]. These highlight the importance of targeted inhibition of NEAT1 to overcome the chemotherapy and radioactivity iodine-resistance of thyroid cancer patients.

In addition to NEAT1, solute carrier family 6 member 9–5:2 (SLC6A9–5:2) and maternally expressed 3 (MEG3a) were discovered as therapeutic lncRNAs in determining the response to ^{131}I therapy. MEG3a, an important lncRNA in different human diseases [250–252], was shown to be down-regulated in radioactivity iodine-resistant PTC and FTC tissues and cell lines. Mechanistically, overexpression of MEG3a resulted in sponging miR-182 that subsequently suppressed ^{131}I-resistant cell viability and induced DNA damage [241]. Similarly, SLC6A9-5:2 down-regulation was reported in radioactivity iodine-resistant PTC tissues and cell lines. However, overexpression of SLC6A9-5:2 revoked the ^{131}I-resistant sensitivity via up-regulating PARP-1 protein with an unknown mechanism [243]. lncRNA PTCSC3 was observed to be down-regulated in ATC tissues and cell lines leading to increased STAT3 and INO80 expression. This axis consequently conferred resistance to doxorubicin [130]. This indicates that overexpressing PTCSC3 could overcome the resistance to doxorubicin in ATC patients.

The physiological roles of circRNAs are not restricted to their contribution to cancer development. circRNAs, could also determine the drug response and thereby promote drug resistance in human cancers [253–257]. In thyroid cancer, circular eukaryotic translation initiation factor 6 (circEIF6) has been shown to inhibit the response to cisplatin. In the presence of this chemotherapeutic drug, circEIF6 was shown to inhibit the apoptosis via sponging miR-144-3p and the corresponding up-regulation of TGF-α, thereby enhancing the resistance to cisplatin in both PTC and ATC thyroid cancer cells [142].

Collectively, ncRNAs play role as therapeutic factors in thyroid cancer (Figure 3). Altering miR-146b, NEAT1, MEG3a, SLC6A9-5:2 may synergistically improve the resistance to radioactive iodine, while targeted inhibition of miR-27b-3p and PTCSC3 may overcome the resistance to doxorubicin. To heighten the sensitivity to cisplatin, miR-144, NEAT1 and circEIF6 could be studied together in thyroid cancer. Future studies could evaluate the functional role of ncRNAs in determining the response to targeted kinase therapy.

9. Conclusions and Future Perspectives

Over the years, remarkable progress has been achieved in mapping the genetic basis of thyroid cancer and developing more efficient molecular tests for its early detection. Notwithstanding, the overall survival of MTC, ATC, PDTC and metastatic differentiated thyroid cancer patients have not improved satisfactorily, reflecting the need for deciphering pristine molecular determinants that could guide early diagnosis and personalized treatment- utilization of ncRNAs is a promising strategy.

In this review, we discussed the current knowledge of three main subtypes of ncRNAs, including miRNAs, lncRNAs and circRNAs in different histopathological subtypes of thyroid cancer. Dysregulation of ncRNAs play an important role in thyroid cancer pathogenesis. This information could be utilized in the diagnostic, prognostic and therapeutic aspects of thyroid cancer clinical care. A number of clinical trials are ongoing to investigate the potential diagnostic and therapeutic impact of ncRNA molecules in thyroid cancer (Table 7).

Table 7. List of ongoing clinical trials indexed in ClinicalTrials.gov (https://clinicaltrials.gov/ct2/home) portal with diagnostic or prognostic relevance in thyroid cancer.

Identifier	ncRNA Type	Type of Sample	Study Type	Observational Model	Clinical Significance	Status
NCT03469544	HOTAIR	Peripheral blood samples	Observational	Case-Control	Diagnostic biomarker	Not yet recruiting
NCT01964508	miRNAs	FNA samples	Observational	Cohort	Diagnostic biomarker	Not yet recruiting
NCT04594720	lncRNAs	Peripheral blood samples	Observational	Case-Control	Diagnostic biomarker	Recruiting completed
NCT01240590	miRNAs	ATC tumor samples	Interventional	Parallel Assignment	Therapeutic biomarker for Crolibulin and cisplatin	Recruiting completed
NCT04285476	miRNAs	Thyroid carcinoma	Interventional	Single Group Assignment	Diagnostic biomarker	Not yet recruiting
NCT00689065	siRNA	Variety of solid tumors including Thyroid carcinoma	Interventional	Single Group Assignment	RNA-based therapy (CALAA-01)	Recruitment terminated

FNA: fine needle aspiration.

Owing to the tumor-suppressive or oncogenic function, dysregulated ncRNAs could promote tumorigenesis via regulating various physiological and cellular activities leading to proliferation/cell growth or inhibition of cell death. However, little is known about the detailed regulatory mechanisms by which the ncRNAs, especially circRNAs, are dysregulated in thyroid cancer with the corresponding downstream tumorigenic and drug-resistance cascades. Understanding the mode of action by which different ncRNAs, individually or in a network, impose their oncogenic effects could aid in the development of new therapeutic approaches to harness the progression of malignant cells. This could be primarily achieved by rational in vitro RNA-based drug design to target the up-regulated ncRNAs using antagomirs and antisense oligos (ASOs), or by expressing the key down-regulated ncRNAs using agomirs and expression vectors. However, selecting the key target ncRNAs from a large number of candidate ncRNAs remains a big challenge.

The other challenge would be the instability and high immunogenicity of the RNA therapeutics, necessitating chemical modifications of the RNA molecules. One example is using the inverted thymidine residues at the 3′ end of the RNA to protect it against exonucleases, thereby improving stability [258]. In addition, conjugating the RNA with an active targeting moiety, such as an antibody, has been shown to reduce immunogenicity [259]. Although significant progress has been made, the delivery of RNA therapeutics remains a major challenge. Negatively charged phosphate backbones and large molecular weight of RNA molecules hamper RNA uptake through difficulties in passing through the cell membrane, micropinocytosis, endosomal escape and kidney clearance [258,260,261]. In addition to reducing immunogenicity, the conjugation of RNAs with targeting moieties including antibodies, aptamers, lipic nanoparticles and polymers has led to magnificent advances in the delivery efficiency of RNA therapeutics [262]. Attachment of a monoclonal antibody (TCM-9), a specific antibody for human thyroid cancer [263], could be a strategy to improve the targeted delivery of RNA therapeutics to thyroid carcinoma cells.

Combination of ncRNAs-based therapeutic interventions with conventional systemic therapy could emerge as an impactful way to conquer drug resistance in advanced thyroid cancer. Such in-depth studies may prove the way toward pre-clinical and clinical investigations that eventually could provide more impactful therapies. Beyond the understanding of cancer pathogenesis and drug resistance, the alteration in circulating or tissue ncRNAs expression could facilitate the diagnosis of different thyroid malignancies with improved sensitivity and specificity, and a minimized need for diagnostic thyroid surgeries. Taken together, we expect the application of ncRNAs as diagnostic/prognostic biomarkers and therapeutic targets to emerge within the few next years in thyroid cancer.

Author Contributions: Y.T. and S.P.Y. participated in the conception and revision of the manuscript, H.T. participated in writing the manuscript and preparing the figures. All authors read and approved the final version of this manuscript.

Funding: This work is supported by funding from the NUHS clinician scientist program (NCSP), National Medical Research Council (NMRC) Research Fellowship, Singapore National Research Foundation Fellowship, National University of Singapore President's Assistant Professorship, and the RNA Biology Center at CSI Singapore, NUS, from funding by the Singapore Ministry of Education's Tier 3 grant number [MOE2014-T3-1-006].

Acknowledgments: The figures were created using BioRender.

Conflicts of Interest: The authors declare no conflict of interest.

Abbreviations

ADAM9	ADAM Metallopeptidase Domain 9
AKT3	RAC-γ serine/threonine-protein kinase
AMPK	5′ AMP-activated Protein Kinase
ATC	Anaplastic Thyroid Cancer
BDNF	Brain-Derived Neurotrophic Factor
CASC9	Cancer Susceptibility 9
CCND1	Cyclin D1
CDH6	Cadherin 6
circEIF6	Circular Eukaryotic Translation Initiation Factor 6
circFOXM1	Circular Forkhead Box Protein M1
circRNA	Circular RNA
CXCL16	C-X-C Motif Chemokine Ligand 16
DLX6-AS1	Distal-Less Homeobox 6-Antisense 1
EGFR	Epidermal Growth Factor Receptor
ERBB2	Erb-B2 Receptor Tyrosine Kinase 2
ER-β	Estrogen Receptor Beta
EZH2	Enhancer of Zeste Homolog 2
FGF2	Fibroblast Growth Factor

FGFR2	Fibroblast Growth Factor Receptor 2
FNA	Fine Needle Aspiration
FOXE1	Forkhead Box E1
FOXN3	Forkhead Box N3
FOXO1	Forkhead Box O1
FTC	Follicular Thyroid Carcinoma
HCP5	HLA complex P5
HK2	Hexokinase 2
HMGB1	High Mobility Group Box Protein 1
ITGA3	Integrin Subunit Alpha 3
lncRNA	Long Non-coding RNA
MALAT1	Metastasis Associated Lung Adenocarcinoma Transcript 1
MEG3a	Maternally Expressed Gene 3
miRNA	microRNA
MMP1	Matrix Metallopeptidase 1
MTC	Medullary Thyroid Carcinoma
mTOR	Mechanistic Target of Rapamycin Kinase
ncRNA	Non-coding RNA
NEAT1	Nuclear Paraspeckle Assembly Transcript 1
PAK1	p21 Activated Kinase-1
PAR5	Prader Willi/Angelman Region RNA5
PDTC	Poorly differentiated thyroid carcinoma
PI$_3$K	Phosphatidylinositol-4,5-Bisphosphate 3-Kinase
piRNA	PIWI-interacting RNA:
PTC	Papillary Thyroid Cancer
PTCSC3	Thyroid Carcinoma Susceptibility Candidate 3
PTCSCs	Papillary Thyroid Cancer Stem Cells
QKI5-7	Quaking protein 5-7
RAR-β	Retinoic Acid Receptor Beta
ROCK1	Rho-associated Protein Kinase 1
SCAI	Suppressor of Cancer Cell Invasion
SEER	Surveillance, Epidemiology and End Results
SLC6A9-5:2	Solute Carrier Family 6 Member 9-5:2
SNHG3	Small Nucleolar RNA Host Gene 3
snoRNA	Small nuclear RNA
SOCS1	Suppressor Protein of Cytokine Signaling 1
SPAG9	Perm Associated Antigen 9
ST6GAL2	alpha-2, 6-sialyltransferase 2
STAT3	Signal Transducer And Activator of Transcription 3
STON2	Stonin 2
TGF-α	Transforming Growth Factor Alpha
UNC5B-AS1	Unc-5 Netrin Receptor B-Antisense RNA 1
UPF1	UPF1 RNA Helicase and ATPase
VHL	Von Hippel-Lindau Tumor Suppressor
XIST	X-inactive specific transcript
YY1	Yin Yang 1

References

1. Katoh, H.; Yamashita, K.; Enomoto, T.; Watanabe, M. Classification and general considerations of thyroid cancer. *Ann. Clin. Pathol.* **2015**, *3*, 1045.
2. James, B.C.; Mitchell, J.M.; Jeon, H.D.; Vasilottos, N.; Grogan, R.H.; Aschebrook-Kilfoy, B. An update in international trends in incidence rates of thyroid cancer, 1973–2007. *Cancer Causes Control* **2018**, *29*, 465–473. [CrossRef]

3.   Bray, F.; Ferlay, J.; Soerjomataram, I.; Siegel, R.L.; Torre, L.A.; Jemal, A. Global cancer statistics 2018: GLOBOCAN estimates of incidence and mortality worldwide for 36 cancers in 185 countries. *CA Cancer J. Clin.* **2018**, *68*, 394–424. [CrossRef]

4.   Davies, L.; Morris, L.; Hankey, B. Increases in thyroid cancer incidence and mortality. *JAMA* **2017**, *318*, 389–390. [CrossRef] [PubMed]

5.   National Cancer Institute. Surveillance, Epidemiology and End Results (SEER) Program. Available online: https://seer.cancer.gov/statfacts/html/thyro.htm (accessed on 4 October 2020).

6.   Fiore, M.; Oliveri Conti, G.; Caltabiano, R.; Buffone, A.; Zuccarello, P.; Cormaci, L.; Cannizzaro, M.A.; Ferrante, M. Role of emerging environmental risk factors in thyroid cancer: A brief review. *Int. J. Environ. Res. Public Health* **2019**, *16*, 1185. [CrossRef] [PubMed]

7.   Thun, M.; Linet, M.S.; Cerhan, J.R.; Haiman, C.A.; Schottenfeld, D. *Cancer Epidemiology and Prevention*; Oxford University Press: Oxford, UK, 2017.

8.   Pirahanchi, Y.; Jialal, I. Physiology, thyroid. In *StatPearls*; StatPearls Publishing: Treasure Island, FL, USA, 2018.

9.   Soundarrajan, M.; Kopp, P.A. Thyroid Hormone Biosynthesis and Physiology. In *Thyroid Disease and Reproduction*; Springer: Berlin/Heidelberg, Germany, 2019; pp. 1–17.

10. Romei, C.; Tacito, A.; Molinaro, E.; Piaggi, P.; Cappagli, V.; Pieruzzi, L.; Matrone, A.; Viola, D.; Agate, L.; Torregrossa, L. Clinical, pathological and genetic features of anaplastic and poorly differentiated thyroid cancer: A single institute experience. *Oncol. Lett.* **2018**, *15*, 9174–9182. [CrossRef] [PubMed]

11. Sipos, J.; Mazzaferri, E. Thyroid cancer epidemiology and prognostic variables. *Clin. Oncol.* **2010**, *22*, 395–404. [CrossRef] [PubMed]

12. Fugazzola, L.; Elisei, R.; Fuhrer, D.; Jarzab, B.; Leboulleux, S.; Newbold, K.; Smit, J. 2019 European Thyroid Association Guidelines for the Treatment and Follow-Up of Advanced Radioiodine-Refractory Thyroid Cancer. *Eur. Thyroid J.* **2019**, *8*, 227–245. [CrossRef]

13. Haugen, B.R.; Alexander, E.K.; Bible, K.C.; Doherty, G.M.; Mandel, S.J.; Nikiforov, Y.E.; Pacini, F.; Randolph, G.W.; Sawka, A.M.; Schlumberger, M. 2015 American Thyroid Association management guidelines for adult patients with thyroid nodules and differentiated thyroid cancer: The American Thyroid Association guidelines task force on thyroid nodules and differentiated thyroid cancer. *Thyroid* **2016**, *26*, 1–133. [CrossRef] [PubMed]

14. Grewal, R.K.; Ho, A.; Schöder, H. Novel approaches to thyroid cancer treatment and response assessment. *Semin. Nucl. Med.* **2016**, *46*, 109–118. [CrossRef]

15. Nikiforov, Y.E.; Ohori, N.P.; Hodak, S.P.; Carty, S.E.; LeBeau, S.O.; Ferris, R.L.; Yip, L.; Seethala, R.R.; Tublin, M.E.; Stang, M.T. Impact of mutational testing on the diagnosis and management of patients with cytologically indeterminate thyroid nodules: A prospective analysis of 1056 FNA samples. *J. Clin. Endocrinol. Metab.* **2011**, *96*, 3390–3397. [CrossRef]

16. Nikiforov, Y.E.; Carty, S.E.; Chiosea, S.I.; Coyne, C.; Duvvuri, U.; Ferris, R.L.; Gooding, W.E.; Hodak, S.P.; LeBeau, S.O.; Ohori, N.P. Highly accurate diagnosis of cancer in thyroid nodules with follicular neoplasm/suspicious for a follicular neoplasm cytology by ThyroSeq v2 next-generation sequencing assay. *Cancer* **2014**, *120*, 3627–3634. [CrossRef]

17. Santhanam, P.; Khthir, R.; Gress, T.; Elkadry, A.; Olajide, O.; Yaqub, A.; Driscoll, H. Gene expression classifier for the diagnosis of indeterminate thyroid nodules: A meta-analysis. *Med. Oncol.* **2016**, *33*, 14. [CrossRef]

18. Fagin, J.A.; Wells, S.A., Jr. Biologic and clinical perspectives on thyroid cancer. *New Engl. J. Med.* **2016**, *375*, 1054–1067. [CrossRef] [PubMed]

19. Liu, R.; Xing, M. TERT promoter mutations in thyroid cancer. *Endocr. Relat. Cancer* **2016**, *23*, R143–R155. [CrossRef]

20. Lathief, S.; Pothuloori, A.; Liu, X.; Chaidarun, S. Advances and practical use of the molecular markers for thyroid cancer. *Adv. Cell. Mol. Otolaryngol.* **2016**, *4*, 33948. [CrossRef]

21. Lin, C.; Yang, L. Long noncoding RNA in cancer: Wiring signaling circuitry. *Trends Cell Biol.* **2018**, *28*, 287–301. [CrossRef] [PubMed]

22. Agostini, M.; Ganini, C.; Candi, E.; Melino, G. The role of noncoding RNAs in epithelial cancer. *Cell Death Discov.* **2020**, *6*, 13. [CrossRef]

23. Choudhari, R.; Sedano, M.J.; Harrison, A.L.; Subramani, R.; Lin, K.Y.; Ramos, E.I.; Lakshmanaswamy, R.; Gadad, S.S. Long noncoding RNAs in cancer: From discovery to therapeutic targets. In *Advances in Clinical Chemistry*; Elsevier: Amsterdam, The Netherlands, 2020; Volume 95, pp. 105–147.

24. Sohel, M.M.H. Circulating microRNAs as biomarkers in cancer diagnosis. *Life Sci.* **2020**, 117473. [CrossRef]

25. Wright, M.W.; Bruford, E.A. Naming'junk: Human non-protein coding RNA (ncRNA) gene nomenclature. *Hum. Genom.* **2011**, *5*, 90. [CrossRef]

26. Esteller, M.; Pandolfi, P.P. The epitranscriptome of noncoding RNAs in cancer. *Cancer Discov.* **2017**, *7*, 359–368. [CrossRef]

27. Lü, L.; Sun, J.; Shi, P.; Kong, W.; Xu, K.; He, B.; Zhang, S.; Wang, J. Identification of circular RNAs as a promising new class of diagnostic biomarkers for human breast cancer. *Oncotarget* **2017**, *8*, 44096. [CrossRef]

28. Wang, W.-J.; Li, H.-T.; Yu, J.-P.; Han, X.-P.; Xu, Z.-P.; Li, Y.-M.; Jiao, Z.-Y.; Liu, H.-B. A competing endogenous RNA network reveals novel potential lncRNA, miRNA, and mRNA biomarkers in the prognosis of human colon adenocarcinoma. *J. Surg. Res.* **2019**, *235*, 22–33. [CrossRef] [PubMed]

29. Mai, D.; Ding, P.; Tan, L.; Zhang, J.; Pan, Z.; Bai, R.; Li, C.; Li, M.; Zhou, Y.; Tan, W. PIWI-interacting RNA-54265 is oncogenic and a potential therapeutic target in colorectal adenocarcinoma. *Theranostics* **2018**, *8*, 5213. [CrossRef]

30. Qin, X.-g.; Zeng, J.-H.; Lin, P.; Mo, W.-J.; Li, Q.; Feng, Z.-B.; Luo, D.-Z.; Yang, H.; Chen, G.; Zeng, J.-J. Prognostic value of small nuclear RNAs (snRNAs) for digestive tract pan-adenocarcinomas identified by RNA sequencing data. *Pathol. Res. Pract.* **2019**, *215*, 414–426. [CrossRef]

31. Zhang, P.; Wu, W.; Chen, Q.; Chen, M. Non-Coding RNAs and their Integrated Networks. *J. Integr. Bioinform.* **2019**, 16. [CrossRef]

32. Yang, L. Splicing noncoding RNAs from the inside out. *Wiley Interdiscip. Rev. RNA* **2015**, *6*, 651–660. [CrossRef]

33. Khorshidi, A.; Dhaliwal, P.; Yang, B.B. Noncoding RNAs in tumor angiogenesis. In *The Long and Short Non-Coding RNAs in Cancer Biology*; Springer: Berlin/Heidelberg, Germany, 2016; pp. 217–241.

34. Tang, X.J.; Wang, W.; Hann, S.S. Interactions among lncRNAs, miRNAs and mRNA in colorectal cancer. *Biochimie* **2019**, *163*, 58–72. [CrossRef] [PubMed]

35. Xu, S.; Gong, Y.; Yin, Y.; Xing, H.; Zhang, N. The multiple function of long noncoding RNAs in osteosarcoma progression, drug resistance and prognosis. *Biomed. Pharmacother.* **2020**, *127*, 110141. [CrossRef]

36. Rossi, M.; Gorospe, M. Noncoding RNAs Controlling Telomere Homeostasis in Senescence and Aging. *Trends Mol. Med.* **2020**, *26*, 422–433. [CrossRef]

37. Guzel, E.; Okyay, T.M.; Yalcinkaya, B.; Karacaoglu, S.; Gocmen, M.; Akcakuyu, M.H. Tumor suppressor and oncogenic role of long non-coding RNAs in cancer. *North. Clin. Istanb.* **2020**, *7*, 81. [CrossRef]

38. Dhamija, S.; Diederichs, S. From junk to master regulators of invasion: LncRNA functions in migration, EMT and metastasis. *Int. J. Cancer* **2016**, *139*, 269–280. [CrossRef]

39. Huang, Z.; Zhou, J.-K.; Peng, Y.; He, W.; Huang, C. The role of long noncoding RNAs in hepatocellular carcinoma. *Mol. Cancer* **2020**, *19*, 1–18. [CrossRef]

40. Liu, K.; Gao, L.; Ma, X.; Huang, J.-J.; Chen, J.; Zeng, L.; Ashby, C.R.; Zou, C.; Chen, Z.-S. Long non-coding RNAs regulate drug resistance in cancer. *Mol. Cancer* **2020**, *19*, 1–13. [CrossRef]

41. Li, Z.; Ruan, Y.; Zhang, H.; Shen, Y.; Li, T.; Xiao, B. Tumor-suppressive circular RNAs: Mechanisms underlying their suppression of tumor occurrence and use as therapeutic targets. *Cancer Sci.* **2019**, *110*, 3630. [CrossRef]

42. Shukla, G.C.; Singh, J.; Barik, S. MicroRNAs: Processing, maturation, target recognition and regulatory functions. *Mol. Cell. Pharmacol.* **2011**, *3*, 83.

43. Markopoulos, G.S.; Roupakia, E.; Tokamani, M.; Chavdoula, E.; Hatziapostolou, M.; Polytarchou, C.; Marcu, K.B.; Papavassiliou, A.G.; Sandaltzopoulos, R.; Kolettas, E. A step-by-step microRNA guide to cancer development and metastasis. *Cell. Oncol.* **2017**, *40*, 303–339. [CrossRef]

44. Dai, X.; Kaushik, A.C.; Zhang, J. The emerging role of major regulatory RNAs in cancer control. *Front. Oncol.* **2019**, *9*, 920. [CrossRef]

45. Zong, Y.; Zhang, Y.; Sun, X.; Xu, T.; Cheng, X.; Qin, Y. miR-221/222 promote tumor growth and suppress apoptosis by targeting lncRNA GAS5 in breast cancer. *Biosci. Rep.* **2019**, *39*. [CrossRef] [PubMed]

46. Ma, C.; Zhan, C.; Yuan, H.; Cui, Y.; Zhang, Z. MicroRNA-603 functions as an oncogene by suppressing BRCC2 protein translation in osteosarcoma. *Oncol. Rep.* **2016**, *35*, 3257–3264. [CrossRef]

47. Zeng, B.; Li, Y.; Jiang, F.; Wei, C.; Chen, G.; Zhang, W.; Zhao, W.; Yu, D. LncRNA GAS5 suppresses proliferation, migration, invasion, and epithelial-mesenchymal transition in oral squamous cell carcinoma by regulating the miR-21/PTEN axis. *Exp. Cell Res.* **2019**, *374*, 365–373. [CrossRef]

48. Calin, G.A.; Dumitru, C.D.; Shimizu, M.; Bichi, R.; Zupo, S.; Noch, E.; Aldler, H.; Rattan, S.; Keating, M.; Rai, K. Frequent deletions and down-regulation of micro-RNA genes miR15 and miR16 at 13q14 in chronic lymphocytic leukemia. *Proc. Natl. Acad. Sci. USA* **2002**, *99*, 15524–15529. [CrossRef]

49. He, H.; Jazdzewski, K.; Li, W.; Liyanarachchi, S.; Nagy, R.; Volinia, S.; Calin, G.A.; Liu, C.-g.; Franssila, K.; Suster, S. The role of microRNA genes in papillary thyroid carcinoma. *Proc. Natl. Acad. Sci. USA* **2005**, *102*, 19075–19080. [CrossRef]

50. Zembska, A.; Jawiarczyk-Przybyłowska, A.; Wojtczak, B.; Bolanowski, M. MicroRNA expression in the progression and aggressiveness of papillary thyroid carcinoma. *Anticancer Res.* **2019**, *39*, 33–40. [CrossRef]

51. Pishkari, S.; Paryan, M.; Hashemi, M.; Baldini, E.; Mohammadi-Yeganeh, S. The role of microRNAs in different types of thyroid carcinoma: A comprehensive analysis to find new miRNA supplementary therapies. *J. Endocrinol. Investig.* **2018**, *41*, 269–283. [CrossRef] [PubMed]

52. Czajka, A.A.; Wojcicka, A.; Kubiak, A.; Kotlarek, M.; Bakuła-Zalewska, E.; Koperski, Ł.; Wiechno, W.; Jażdżewski, K. Family of microRNA-146 regulates RARβ in papillary thyroid carcinoma. *PLoS ONE* **2016**, *11*, e0151968. [CrossRef]

53. Zhang, W.; Ji, W.; Zhao, X. MiR-155 promotes anaplastic thyroid cancer progression by directly targeting SOCS1. *BMC Cancer* **2019**, *19*, 1093. [CrossRef]

54. Zang, C.; Sun, J.; Liu, W.; Chu, C.; Jiang, L.; Ge, R. miRNA-21 promotes cell proliferation and invasion via VHL/PI3K/AKT in papillary thyroid carcinoma. *Hum. Cell* **2019**, *32*, 428–436. [CrossRef]

55. Wang, X.; Lu, X.; Geng, Z.; Yang, G.; Shi, Y. LncRNA PTCSC3/miR-574-5p governs cell proliferation and migration of papillary thyroid carcinoma via Wnt/β-catenin signaling. *J. Cell. Biochem.* **2017**, *118*, 4745–4752. [CrossRef] [PubMed]

56. Fang, L.; Xu, W.; Kong, D. Icariin inhibits cell proliferation, migration and invasion by down-regulation of microRNA-625-3p in thyroid cancer cells. *Biomed. Pharmacother.* **2019**, *109*, 2456–2463. [CrossRef]

57. Wang, X.-Z.; Hang, Y.-K.; Liu, J.-B.; Hou, Y.-Q.; Wang, N.; Wang, M.-J. Over-expression of microRNA-375 inhibits papillary thyroid carcinoma cell proliferation and induces cell apoptosis by targeting ERBB2. *J. Pharmacol. Sci.* **2016**, *130*, 78–84. [CrossRef]

58. Liu, N.; Zhou, Q.; Qi, Y.-H.; Wang, H.; Yang, L.; Fan, Q.-Y. Effects of long non-coding RNA H19 and microRNA let7a expression on thyroid cancer prognosis. *Exp. Mol. Pathol.* **2017**, *103*, 71–77. [CrossRef]

59. Huang, P.; Mao, L.-F.; Zhang, Z.-P.; Lv, W.-W.; Feng, X.-P.; Liao, H.-J.; Dong, C.; Kaluba, B.; Tang, X.-F.; Chang, S. Down-regulated miR-125a-5p promotes the reprogramming of glucose metabolism and cell malignancy by increasing levels of CD147 in thyroid cancer. *Thyroid* **2018**, *28*, 613–623. [CrossRef]

60. Fu, Y.; Zheng, H.; Zhang, D.; Zhou, L.; Sun, H. MicroRNA-1266 suppresses papillary thyroid carcinoma cell metastasis and growth via targeting FGFR2. *Eur. Rev. Med. Pharm. Sci.* **2018**, *22*, 3430–3438.

61. Fang, M.; Huang, W.; Wu, X.; Gao, Y.; Ou, J.; Zhang, X.; Li, Y. MiR-141-3p suppresses tumor growth and metastasis in papillary thyroid cancer via targeting Yin Yang 1. *Anat. Rec.* **2019**, *302*, 258–268. [CrossRef]

62. Liu, J.; Feng, L.; Zhang, H.; Zhang, J.; Zhang, Y.; Li, S.; Qin, L.; Yang, Z.; Xiong, J. Effects of miR-144 on the sensitivity of human anaplastic thyroid carcinoma cells to cisplatin by autophagy regulation. *Cancer Biol. Ther.* **2018**, *19*, 484–496. [CrossRef]

63. Sheng, W.; Chen, Y.; Gong, Y.; Dong, T.; Zhang, B.; Gao, W. miR-148a inhibits self-renewal of thyroid cancer stem cells via repressing INO80 expression. *Oncol. Rep.* **2016**, *36*, 3387–3396. [CrossRef]

64. Cheng, L.; Zhou, R.; Chen, M.; Feng, L.; Li, H. MicroRNA-150 targets Rho-associated protein kinase 1 to inhibit cell proliferation, migration and invasion in papillary thyroid carcinoma. *Mol. Med. Rep.* **2017**, *16*, 2217–2224. [CrossRef] [PubMed]

65. Yin, Y.; Hong, S.; Yu, S.; Huang, Y.; Chen, S.; Liu, Y.; Zhang, Q.; Li, Y.; Xiao, H. MiR-195 inhibits tumor growth and metastasis in papillary thyroid carcinoma cell lines by targeting CCND1 and FGF2. *Int. J. Endocrinol.* **2017**, *2017*, 6180425. [CrossRef] [PubMed]

66. Ren, L.; Xu, Y.; Qin, G.; Liu, C.; Yan, Y.; Zhang, H. miR-199b-5p-Stonin 2 axis regulates metastases and epithelial-to-mesenchymal transition of papillary thyroid carcinoma. *IUBMB Life* **2019**, *71*, 28–40. [CrossRef]

67. Liu, H.; Chen, X.; Lin, T.; Chen, X.; Yan, J.; Jiang, S. MicroRNA-524-5p suppresses the progression of papillary thyroid carcinoma cells via targeting on FOXE1 and ITGA3 in cell autophagy and cycling pathways. *J. Cell. Physiol.* **2019**, *234*, 18382–18391. [CrossRef]

68. Yue, K.; Wang, X.; Wu, Y.; Zhou, X.; He, Q.; Duan, Y. microRNA-7 regulates cell growth, migration and invasion via direct targeting of PAK1 in thyroid cancer. *Mol. Med. Rep.* **2016**, *14*, 2127–2134. [CrossRef]

69. Wang, Z.; Liu, W.; Wang, C.; Ai, Z. miR-873-5p Inhibits Cell Migration and Invasion of Papillary Thyroid Cancer via Regulation of CXCL16. *Oncotargets Ther.* **2020**, *13*, 1037. [CrossRef]

70. Guo, F.; Hou, X.; Sun, Q. MicroRNA-9-5p functions as a tumor suppressor in papillary thyroid cancer via targeting BRAF. *Oncol. Lett.* **2018**, *16*, 6815–6821. [CrossRef] [PubMed]

71. Sastre-Perona, A.; Santisteban, P. Role of the wnt pathway in thyroid cancer. *Front. Endocrinol.* **2012**, *3*, 31. [CrossRef]

72. Fuziwara, C.S.; Kimura, E.T. How does microRNA modulate Wnt/β-catenin signaling in thyroid oncogenesis? *Ann. Transl. Med.* **2020**, *8*, 266. [CrossRef]

73. Ely, K.A.; Bischoff, L.A.; Weiss, V.L. Wnt signaling in thyroid homeostasis and carcinogenesis. *Genes* **2018**, *9*, 204. [CrossRef]

74. Zhang, Z.-J.; Xiao, Q.; Li, X.-Y. MicroRNA-574-5p directly targets FOXN3 to mediate thyroid cancer progression via Wnt/β-catenin signaling pathway. *Pathol. Res. Pract.* **2020**, *216*, 152939. [CrossRef] [PubMed]

75. Zhang, Z.; Li, X.; Xiao, Q.; Wang, Z. MiR-574-5p mediates the cell cycle and apoptosis in thyroid cancer cells via Wnt/β-catenin signaling by repressing the expression of Quaking proteins. *Oncol. Lett.* **2018**, *15*, 5841–5848. [CrossRef]

76. Feige, J.; Cherradi, N. Serum miR-483-5p and miR-195 are predictive of recurrence risk in adrenocortical cancer patients. *Endocr. Relat. Cancer* **2013**, *20*, 579–594.

77. Liu, B.; Qu, J.; Xu, F.; Guo, Y.; Wang, Y.; Yu, H.; Qian, B. MiR-195 suppresses non-small cell lung cancer by targeting CHEK1. *Oncotarget* **2015**, *6*, 9445. [CrossRef]

78. Ujifuku, K.; Mitsutake, N.; Takakura, S.; Matsuse, M.; Saenko, V.; Suzuki, K.; Hayashi, K.; Matsuo, T.; Kamada, K.; Nagata, I. MiR-195, miR-455-3p and miR-10a* are implicated in acquired temozolomide resistance in glioblastoma multiforme cells. *Cancer Lett.* **2010**, *296*, 241–248. [CrossRef]

79. Liu, X.; Zhou, Y.; Ning, Y.-E.; Gu, H.; Tong, Y.; Wang, N. MiR-195-5p Inhibits Malignant Progression of Cervical Cancer by Targeting YAP1. *Oncotargets Ther.* **2020**, *13*, 931. [CrossRef]

80. Balmeh, N.; Tabatabaeian, H.; Asgari, M.; Mokhtarian, R.; Abharian, P.H.; Azadeh, M.; Ghaedi, K. miR-195 down-regulation is a distinctive biomarker of HER2 positive state in breast cancer. *Gene Rep.* **2020**, 100703. [CrossRef]

81. Jung, Y.-S.; Park, J.-I. Wnt signaling in cancer: Therapeutic targeting of Wnt signaling beyond β-catenin and the destruction complex. *Exp. Mol. Med.* **2020**, *52*, 183–191. [CrossRef]

82. Noorolyai, S.; Shajari, N.; Baghbani, E.; Sadreddini, S.; Baradaran, B. The relation between PI3K/AKT signalling pathway and cancer. *Gene* **2019**, *698*, 120–128. [CrossRef]

83. Nozhat, Z.; Hedayati, M. PI3K/AKT pathway and its mediators in thyroid carcinomas. *Mol. Diagn. Ther.* **2016**, *20*, 13–26. [CrossRef]

84. Zhang, J.-G.; Wang, J.-J.; Zhao, F.; Liu, Q.; Jiang, K.; Yang, G.-h. MicroRNA-21 (miR-21) represses tumor suppressor PTEN and promotes growth and invasion in non-small cell lung cancer (NSCLC). *Clin. Chim. Acta* **2010**, *411*, 846–852. [CrossRef]

85. Li, L.-Q.; Li, X.-L.; Wang, L.; Du, W.-J.; Guo, R.; Liang, H.-H.; Liu, X.; Liang, D.-S.; Lu, Y.-J.; Shan, H.-L. Matrine inhibits breast cancer growth via miR-21/PTEN/Akt pathway in MCF-7 cells. *Cell. Physiol. Biochem.* **2012**, *30*, 631–641. [CrossRef] [PubMed]

86. McClelland, A.D.; Herman-Edelstein, M.; Komers, R.; Jha, J.C.; Winbanks, C.E.; Hagiwara, S.; Gregorevic, P.; Kantharidis, P.; Cooper, M.E. miR-21 promotes renal fibrosis in diabetic nephropathy by targeting PTEN and SMAD7. *Clin. Sci.* **2015**, *129*, 1237–1249. [CrossRef]

87. Haugen, D.; Akslen, L.; Varhaug, J.A.; Lillehaug, J. Expression of c-erb B-2 protein in papillary thyroid carcinomas. *Br. J. Cancer* **1992**, *65*, 832–837. [CrossRef]

88. Montero-Conde, C.; Ruiz-Llorente, S.; Dominguez, J.M.; Knauf, J.A.; Viale, A.; Sherman, E.J.; Ryder, M.; Ghossein, R.A.; Rosen, N.; Fagin, J.A. Relief of feedback inhibition of HER3 transcription by RAF and MEK inhibitors attenuates their antitumor effects in BRAF-mutant thyroid carcinomas. *Cancer Discov.* **2013**, *3*, 520–533. [CrossRef]

89. Lv, Y.; Sui, F.; Ma, J.; Ren, X.; Yang, Q.; Zhang, Y.; Guan, H.; Shi, B.; Hou, P.; Ji, M. Increased expression of EHF contributes to thyroid tumorigenesis through transcriptionally regulating HER2 and HER3. *Oncotarget* **2016**, *7*, 57978. [CrossRef]

90. Zaballos, M.A.; Santisteban, P. FOXO1 controls thyroid cell proliferation in response to TSH and IGF-I and is involved in thyroid tumorigenesis. *Mol. Endocrinol.* **2013**, *27*, 50–62. [CrossRef]

91. Rena, G.; Prescott, A.R.; Guo, S.; Cohen, P.; Unterman, T.G. Roles of the forkhead in rhabdomyosarcoma (FKHR) phosphorylation sites in regulating 14-3-3 binding, transactivation and nuclear targetting. *Biochem. J.* **2001**, *354*, 605–612. [CrossRef]

92. Wong, K.; Di Cristofano, F.; Ranieri, M.; De Martino, D.; Di Cristofano, A. PI3K/mTOR inhibition potentiates and extends palbociclib activity in anaplastic thyroid cancer. *Endocr. Relat. Cancer* **2019**, *26*, 425–436. [CrossRef]

93. Coelho, R.G.; Fortunato, R.S.; Carvalho, D.P. Metabolic reprogramming in thyroid carcinoma. *Front. Oncol.* **2018**, *8*, 82. [CrossRef]

94. Suh, H.Y.; Choi, H.; Paeng, J.C.; Cheon, G.J.; Chung, J.-K.; Kang, K.W. Comprehensive gene expression analysis for exploring the association between glucose metabolism and differentiation of thyroid cancer. *BMC Cancer* **2019**, *19*, 1260. [CrossRef]

95. Miao, Y.; Zhang, L.-F.; Zhang, M.; Guo, R.; Liu, M.-F.; Li, B. Therapeutic Delivery of miR-143 Targeting Tumor Metabolism in Poorly Differentiated Thyroid Cancer Xenografts and Efficacy Evaluation Using 18F-FDG MicroPET-CT. *Hum. Gene Ther.* **2019**, *30*, 882–892. [CrossRef]

96. Hay, N. Reprogramming glucose metabolism in cancer: Can it be exploited for cancer therapy? *Nat. Rev. Cancer* **2016**, *16*, 635. [CrossRef]

97. Hamanaka, R.B.; Chandel, N.S. Targeting glucose metabolism for cancer therapy. *J. Exp. Med.* **2012**, *209*, 211–215. [CrossRef]

98. Yoshikawa, H.; Matsubara, K.; Qian, G.-S.; Jackson, P.; Groopman, J.D.; Manning, J.E.; Harris, C.C.; Herman, J.G. SOCS-1, a negative regulator of the JAK/STAT pathway, is silenced by methylation in human hepatocellular carcinoma and shows growth-suppression activity. *Nat. Genet.* **2001**, *28*, 29–35. [CrossRef]

99. Oshimo, Y.; Kuraoka, K.; Nakayama, H.; Kitadai, Y.; Yoshida, K.; Chayama, K.; Yasui, W. Epigenetic inactivation of SOCS-1 by CpG island hypermethylation in human gastric carcinoma. *Int. J. Cancer* **2004**, *112*, 1003–1009. [CrossRef]

100. Chen, C.Y.; Tsay, W.; Tang, J.L.; Shen, H.L.; Lin, S.W.; Huang, S.Y.; Yao, M.; Chen, Y.C.; Shen, M.C.; Wang, C.H. SOCS1 methylation in patients with newly diagnosed acute myeloid leukemia. *Genes Chromosomes Cancer* **2003**, *37*, 300–305. [CrossRef] [PubMed]

101. Liau, N.P.; Laktyushin, A.; Lucet, I.S.; Murphy, J.M.; Yao, S.; Whitlock, E.; Callaghan, K.; Nicola, N.A.; Kershaw, N.J.; Babon, J.J. The molecular basis of JAK/STAT inhibition by SOCS1. *Nat. Commun.* **2018**, *9*, 1–14. [CrossRef]

102. Danial, N.N.; Rothman, P. JAK-STAT signaling activated by Abl oncogenes. *Oncogene* **2000**, *19*, 2523–2531. [CrossRef]

103. Jin, S.; Borkhuu, O.; Bao, W.; Yang, Y.-T. Signaling pathways in thyroid cancer and their therapeutic implications. *J. Clin. Med. Res.* **2016**, *8*, 284. [CrossRef]

104. Braun, J.; Hoang-Vu, C.; Dralle, H.; Hüttelmaier, S. Downregulation of microRNAs directs the EMT and invasive potential of anaplastic thyroid carcinomas. *Oncogene* **2010**, *29*, 4237–4244. [CrossRef] [PubMed]

105. Cifuentes-Rojas, C.; Hernandez, A.J.; Sarma, K.; Lee, J.T. Regulatory interactions between RNA and polycomb repressive complex 2. *Mol. Cell* **2014**, *55*, 171–185. [CrossRef]

106. Guil, S.; Soler, M.; Portela, A.; Carrère, J.; Fonalleras, E.; Gómez, A.; Villanueva, A.; Esteller, M. Intronic RNAs mediate EZH2 regulation of epigenetic targets. *Nat. Struct. Mol. Biol.* **2012**, *19*, 664. [CrossRef]

107. Mohammad, F.; Mondal, T.; Guseva, N.; Pandey, G.K.; Kanduri, C. Kcnq1ot1 noncoding RNA mediates transcriptional gene silencing by interacting with Dnmt1. *Development* **2010**, *137*, 2493–2499. [CrossRef]

108. Merry, C.R.; Forrest, M.E.; Sabers, J.N.; Beard, L.; Gao, X.-H.; Hatzoglou, M.; Jackson, M.W.; Wang, Z.; Markowitz, S.D.; Khalil, A.M. DNMT1-associated long non-coding RNAs regulate global gene expression and DNA methylation in colon cancer. *Hum. Mol. Genet.* **2015**, *24*, 6240–6253. [CrossRef]

109. Arab, K.; Park, Y.J.; Lindroth, A.M.; Schäfer, A.; Oakes, C.; Weichenhan, D.; Lukanova, A.; Lundin, E.; Risch, A.; Meister, M. Long noncoding RNA TARID directs demethylation and activation of the tumor suppressor TCF21 via GADD45A. *Mol. Cell* **2014**, *55*, 604–614. [CrossRef]

110. Lister, N.; Shevchenko, G.; Walshe, J.L.; Groen, J.; Johnsson, P.; Vidarsdóttir, L.; Grander, D.; Ataide, S.F.; Morris, K.V. The molecular dynamics of long noncoding RNA control of transcription in PTEN and its pseudogene. *Proc. Natl. Acad. Sci. USA* **2017**, *114*, 9942–9947. [CrossRef] [PubMed]

111. Tripathi, V.; Ellis, J.D.; Shen, Z.; Song, D.Y.; Pan, Q.; Watt, A.T.; Freier, S.M.; Bennett, C.F.; Sharma, A.; Bubulya, P.A. The nuclear-retained noncoding RNA MALAT1 regulates alternative splicing by modulating SR splicing factor phosphorylation. *Mol. Cell* **2010**, *39*, 925–938. [CrossRef]

112. Zeng, T.; Ni, H.; Yu, Y.; Zhang, M.; Wu, M.; Wang, Q.; Wang, L.; Xu, S.; Xu, Z.; Xu, C. BACE1-AS prevents BACE1 mRNA degradation through the sequestration of BACE1-targeting miRNAs. *J. Chem. Neuroanat.* **2019**, *98*, 87–96. [CrossRef]

113. Yoon, J.-H.; Abdelmohsen, K.; Srikantan, S.; Yang, X.; Martindale, J.L.; De, S.; Huarte, M.; Zhan, M.; Becker, K.G.; Gorospe, M. LincRNA-p21 suppresses target mRNA translation. *Mol. Cell* **2012**, *47*, 648–655. [CrossRef]

114. Peng, W.-X.; Koirala, P.; Mo, Y.-Y. Y. LncRNA-mediated regulation of cell signaling in cancer. *Oncogene* **2017**, *36*, 5661–5667. [CrossRef]

115. Spizzo, R.; Almeida, M.I.; Colombatti, A.; Calin, G.A. Long non-coding RNAs and cancer: A new frontier of translational research? *Oncogene* **2012**, *31*, 4577–4587. [CrossRef]

116. Chen, Y.; Li, Y.; Gao, H. Long noncoding RNA CASC9 promotes the proliferation and metastasis of papillary thyroid cancer via sponging miR-488-3p. *Cancer Med.* **2020**, *9*, 1830–1841. [CrossRef]

117. Zhong, Z.; Wu, Y.; Luo, J.; Hu, X.; Yuan, Z.; Li, G.; Wang, Y.; Yao, G.; Ge, X. Knockdown of long noncoding RNA DLX6-AS1 inhibits migration and invasion of thyroid cancer cells by upregulating UPF1. *Eur. Rev. Med. Pharmacol. Sci.* **2019**, *23*, 10867–10873. [PubMed]

118. Song, B.; Li, R.; Zuo, Z.; Tan, J.; Liu, L.; Ding, D.; Lu, Y.; Hou, D. LncRNA ENST00000539653 acts as an oncogenic factor via MAPK signalling in papillary thyroid cancer. *BMC Cancer* **2019**, *19*, 297. [CrossRef]

119. Zhang, H.; Yu, Y.; Zhang, K.; Liu, X.; Dai, Y.; Jiao, X. Targeted inhibition of long non-coding RNA H19 blocks anaplastic thyroid carcinoma growth and metastasis. *Bioengineered* **2019**, *10*, 306–315. [CrossRef]

120. Li, M.; Chai, H.-F.; Peng, F.; Meng, Y.-T.; Zhang, L.-Z.; Zhang, L.; Zou, H.; Liang, Q.-L.; Li, M.-M.; Mao, K.-G. Estrogen receptor β upregulated by lncRNA-H19 to promote cancer stem-like properties in papillary thyroid carcinoma. *Cell Death Dis.* **2018**, *9*, 1–15. [CrossRef]

121. Liang, L.; Xu, J.; Wang, M.; Xu, G.; Zhang, N.; Wang, G.; Zhao, Y. LncRNA HCP5 promotes follicular thyroid carcinoma progression via miRNAs sponge. *Cell Death Dis.* **2018**, *9*, 1–13. [CrossRef]

122. Dai, Y.; Miao, Y.; Zhu, Q.; Gao, M.; Hao, F. Expression of long non-coding RNA H19 predicts distant metastasis in minimally invasive follicular thyroid carcinoma. *Bioengineered* **2019**, *10*, 383–389. [CrossRef] [PubMed]

123. Sun, Z.; Guo, X.; Zang, M.; Wang, P.; Xue, S.; Chen, G. Long non-coding RNA LINC00152 promotes cell growth and invasion of papillary thyroid carcinoma by regulating the miR-497/BDNF axis. *J. Cell. Physiol.* **2019**, *234*, 1336–1345. [CrossRef]

124. Li, X.; Zhong, W.; Xu, Y.; Yu, B.; Liu, H. Silencing of lncRNA LINC00514 inhibits the malignant behaviors of papillary thyroid cancer through miR-204–3p/CDC23 axis. *Biochem. Biophys. Res. Commun.* **2019**, *508*, 1145–1148. [CrossRef]

125. Gugnoni, M.; Manicardi, V.; Torricelli, F.; Sauta, E.; Bellazzi, R.; Manzotti, G.; Vitale, E.; de Biase, D.; Piana, S.; Ciarrocchi, A. Linc00941 is a novel TGFβ target that primes papillary thyroid cancer metastatic behavior by regulating the expression of Cadherin 6. *Thyroid* **2020**. [CrossRef]

126. Liu, J.; Dong, H.; Yang, Y.; Qian, Y.; Liu, J.; Li, Z.; Guan, H.; Chen, Z.; Li, C.; Zhang, K. Upregulation of long noncoding RNA MALAT1 in papillary thyroid cancer and its diagnostic value. *Future Oncol.* **2018**, *14*, 3015–3022. [CrossRef] [PubMed]

127. Wang, Y.; Bhandari, A.; Niu, J.; Yang, F.; Xia, E.; Yao, Z.; Jin, Y.; Zheng, Z.; Lv, S.; Wang, O. The lncRNA UNC5B-AS1 promotes proliferation, migration, and invasion in papillary thyroid cancer cell lines. *Hum. Cell* **2019**, *32*, 334–342. [CrossRef]

128. Liu, H.; Deng, H.; Zhao, Y.; Li, C.; Liang, Y. LncRNA XIST/miR-34a axis modulates the cell proliferation and tumor growth of thyroid cancer through MET-PI3K-AKT signaling. *J. Exp. Clin. Cancer Res.* **2018**, *37*, 1–12. [CrossRef]

129. Jiao, X.; Lu, J.; Huang, Y.; Zhang, J.; Zhang, H.; Zhang, K. Long non-coding RNA H19 may be a marker for prediction of prognosis in the follow-up of patients with papillary thyroid cancer. *Cancer Biomark.* **2019**, *26*, 203–207. [CrossRef] [PubMed]

130. Min, X.; Liu, K.; Zhu, H.; Zhang, J. Long noncoding RNA LINC003121 inhibits proliferation and invasion of thyroid cancer cells by suppression of the phosphatidylinositol-3-kinase (PI3K)/Akt signaling pathway. *Med. Sci. Monit. Int. Med. J. Exp. Clin. Res.* **2018**, *24*, 4592. [CrossRef]

131. Pellecchia, S.; Sepe, R.; Decaussin-Petrucci, M.; Ivan, C.; Shimizu, M.; Coppola, C.; Testa, D.; Calin, G.A.; Fusco, A.; Pallante, P. The long non-coding RNA prader willi/angelman region RNA5 (PAR5) is downregulated in anaplastic thyroid carcinomas where it acts as a tumor suppressor by reducing EZH2 activity. *Cancers* **2020**, *12*, 235. [CrossRef]

132. Wang, X.-M.; Liu, Y.; Fan, Y.-X.; Liu, Z.; Yuan, Q.-L.; Jia, M.; Geng, Z.-S.; Gu, L.; Lu, X.-B. LncRNA PTCSC3 affects drug resistance of anaplastic thyroid cancer through STAT3/INO80 pathway. *Cancer Biol. Ther.* **2018**, *19*, 590–597. [CrossRef]

133. Duan, Y.; Wang, Z.; Xu, L.; Sun, L.; Song, H.; Yin, H.; He, F. lncRNA SNHG3 acts as a novel Tumor Suppressor and regulates Tumor Proliferation and Metastasis via AKT/mTOR/ERK pathway in Papillary Thyroid Carcinoma. *J. Cancer* **2020**, *11*, 3492. [CrossRef] [PubMed]

134. Lv, M.; Zhong, Z.; Huang, M.; Tian, Q.; Jiang, R.; Chen, J. lncRNA H19 regulates epithelial–mesenchymal transition and metastasis of bladder cancer by miR-29b-3p as competing endogenous RNA. *Biochim. Biophys. Acta Mol. Cell Res.* **2017**, *1864*, 1887–1899. [CrossRef]

135. Zhou, W.; Ye, X.-L.; Xu, J.; Cao, M.-G.; Fang, Z.-Y.; Li, L.-Y.; Guan, G.-H.; Liu, Q.; Qian, Y.-H.; Xie, D. The lncRNA H19 mediates breast cancer cell plasticity during EMT and MET plasticity by differentially sponging miR-200b/c and let-7b. *Sci. Signal.* **2017**, *10*, eaak9557. [CrossRef]

136. Peng, F.; Li, T.-T.; Wang, K.-L.; Xiao, G.-Q.; Wang, J.-H.; Zhao, H.-D.; Kang, Z.-J.; Fan, W.-J.; Zhu, L.-L.; Li, M. H19/let-7/LIN28 reciprocal negative regulatory circuit promotes breast cancer stem cell maintenance. *Cell Death Dis.* **2018**, *8*, e2569. [CrossRef]

137. Vernucci, M.; Cerrato, F.; Besnard, N.; Casola, S.; Pedone, P.V.; Bruni, C.B.; Riccio, A. The H19 endodermal enhancer is required for Igf2 activation and tumor formation in experimental liver carcinogenesis. *Oncogene* **2000**, *19*, 6376–6385. [CrossRef]

138. Liu, W.; Xu, Q. Upregulation of circHIPK3 promotes the progression of gastric cancer via Wnt/beta-catenin pathway and indicates a poor prognosis. *Eur. Rev. Med. Pharm. Sci.* **2019**, *23*, 7905–7912.

139. Shen, Z.; Zhou, L.; Zhang, C.; Xu, J. Reduction of circular RNA Foxo3 promotes prostate cancer progression and chemoresistance to docetaxel. *Cancer Lett.* **2020**, *468*, 88–101. [CrossRef]

140. Sun, Y.; Li, X.; Chen, A.; Shi, W.; Wang, L.; Yi, R.; Qiu, J. circPIP5K1A serves as a competitive endogenous RNA contributing to ovarian cancer progression via regulation of miR-661/IGFBP5 signaling. *J. Cell. Biochem.* **2019**, *120*, 19406–19414. [CrossRef]

141. Zhou, G.; Zhang, G.; Yuan, Z.; Pei, R.; Liu, D. Has_circ_0008274 promotes cell proliferation and invasion involving AMPK/mTOR signaling pathway in papillary thyroid carcinoma. *Eur. Rev. Med. Pharm. Sci.* **2018**, *22*, 8772–8780.

142. Liu, F.; Zhang, J.; Qin, L.; Yang, Z.; Xiong, J.; Zhang, Y.; Li, R.; Li, S.; Wang, H.; Yu, B. Circular RNA EIF6 (Hsa_circ_0060060) sponges miR-144-3p to promote the cisplatin-resistance of human thyroid carcinoma cells by autophagy regulation. *Aging* **2018**, *10*, 3806. [CrossRef]

143. Ye, M.; Hou, H.; Shen, M.; Dong, S.; Zhang, T. Circular RNA circFOXM1 plays a role in papillary thyroid carcinoma by sponging miR-1179 and regulating HMGB1 expression. *Mol. Ther. Nucleic Acids* **2020**, *19*, 741–750. [CrossRef]

144. Cork, G.K.; Thompson, J.; Slawson, C. Real talk: The inter-play between the mTOR, AMPK, and hexosamine biosynthetic pathways in cell signaling. *Front. Endocrinol.* **2018**, *9*, 522. [CrossRef] [PubMed]

145. Allinen, M.; Beroukhim, R.; Cai, L.; Brennan, C.; Lahti-Domenici, J.; Huang, H.; Porter, D.; Hu, M.; Chin, L.; Richardson, A. Molecular characterization of the tumor microenvironment in breast cancer. *Cancer Cell* **2004**, *6*, 17–32. [CrossRef]

146. Moroishi, T.; Hayashi, T.; Pan, W.-W.; Fujita, Y.; Holt, M.V.; Qin, J.; Carson, D.A.; Guan, K.-L. The Hippo pathway kinases LATS1/2 suppress cancer immunity. *Cell* **2016**, *167*, 1525–1539. [CrossRef]

147. Locati, M.; Mantovani, A.; Sica, A. Macrophage activation and polarization as an adaptive component of innate immunity. In *Advances in Immunology*; Elsevier: Amsterdam, The Netherlands, 2013; Volume 120, pp. 163–184.

148. Baer, C.; Squadrito, M.L.; Iruela-Arispe, M.L.; De Palma, M. Reciprocal interactions between endothelial cells and macrophages in angiogenic vascular niches. *Exp. Cell Res.* **2013**, *319*, 1626–1634. [CrossRef]

149. Chen, Y.; Song, Y.; Du, W.; Gong, L.; Chang, H.; Zou, Z. Tumor-associated macrophages: An accomplice in solid tumor progression. *J. Biomed. Sci.* **2019**, *26*, 1–13. [CrossRef]

150. Adeegbe, D.O.; Nishikawa, H. Natural and induced T regulatory cells in cancer. *Front. Immunol.* **2013**, *4*, 190. [CrossRef]

151. Li, J.-H.; Zhang, S.-Q.; Qiu, X.-G.; Zhang, S.-J.; Zheng, S.-H.; Zhang, D.-H. Long non-coding RNA NEAT1 promotes malignant progression of thyroid carcinoma by regulating miRNA-214. *Int. J. Oncol.* **2017**, *50*, 708–716. [CrossRef]

152. Han, D.; Fang, Y.; Guo, Y.; Hong, W.; Tu, J.; Wei, W. The emerging role of long non-coding RNAs in tumor-associated macrophages. *J. Cancer* **2019**, *10*, 6738. [CrossRef]

153. Huang, J.K.; Ma, L.; Song, W.H.; Lu, B.Y.; Huang, Y.B.; Dong, H.M.; Ma, X.K.; Zhu, Z.Z.; Zhou, R. LncRNA-MALAT1 promotes angiogenesis of thyroid cancer by modulating tumor-associated macrophage FGF2 protein secretion. *J. Cell. Biochem.* **2017**, *118*, 4821–4830. [CrossRef]

154. Amit, M.; Rudnicki, Y.; Binenbaum, Y.; Trejo-Leider, L.; Cohen, J.T.; Gil, Z. Defining the outcome of patients with delayed diagnosis of differentiated thyroid cancer. *Laryngoscope* **2014**, *124*, 2837–2840. [CrossRef]

155. WHO. *Guide to Cancer Early Diagnosis*; World Health Organization: Geneva, Switzerland, 2017.

156. Cooper, D.S.; Doherty, G.M.; Haugen, B.R.; Kloos, R.T.; Lee, S.L.; Mandel, S.J.; Mazzaferri, E.L.; McIver, B.; Pacini, F.; Schlumberger, M. Revised American Thyroid Association management guidelines for patients with thyroid nodules and differentiated thyroid cancer: The American Thyroid Association (ATA) guidelines taskforce on thyroid nodules and differentiated thyroid cancer. *Thyroid* **2009**, *19*, 1167–1214. [CrossRef]

157. Sclabas, G.M.; Staerkel, G.A.; Shapiro, S.E.; Fornage, B.D.; Sherman, S.I.; Vassillopoulou-Sellin, R.; Lee, J.E.; Evans, D.B. Fine-needle aspiration of the thyroid and correlation with histopathology in a contemporary series of 240 patients. *Am. J. Surg.* **2003**, *186*, 702–710. [CrossRef]

158. Yang, J.; Schnadig, V.; Logrono, R.; Wasserman, P.G. Fine-needle aspiration of thyroid nodules: A study of 4703 patients with histologic and clinical correlations. *Cancer Cytopathol.* **2007**, *111*, 306–315. [CrossRef]

159. Cibas, E.S.; Ali, S.Z. The 2017 Bethesda system for reporting thyroid cytopathology. *Thyroid* **2017**, *27*, 1341–1346. [CrossRef]

160. Lum, J.N.M. Thyroid FNA: A retrospective audit of 1541 cases at NUH, Singapore. In *Proceedings of the NUH Cytopathology Workshop 2014, Singapore, 23–25 May 2014*; NUH: Singapore, 2014.

161. Sistrunk, J.; Alexander, S.; Marc, F.; Johnson, T.; Norman, F.; Philip, G.; Richard, G.; Edward, G. Clinical performance of multiplatform mutation panel microRNA risk classifier in indeterminate thyroid nodules. *J. Am. Soc. Cytopathol.* **2020**, *9*, 232–241. [CrossRef]

162. Albarel, F.; Conte-Devolx, B.; Oliver, C. From nodule to differentiated thyroid carcinoma: Contributions of molecular analysis in 2012. In *Annales D'endocrinologie*; Elsevier Masson: Paris, France; pp. 155–164.

163. Ferrari, S.M.; Fallahi, P.; Ruffilli, I.; Elia, G.; Ragusa, F.; Paparo, S.R.; Ulisse, S.; Baldini, E.; Giannini, R.; Miccoli, P. Molecular testing in the diagnosis of differentiated thyroid carcinomas. *Gland Surg.* **2018**, *7*, S19. [CrossRef] [PubMed]

164. Chakraborty, A.; Narkar, A.; Mukhopadhyaya, R.; Kane, S.; D'Cruz, A.; Rajan, M. BRAF V600E mutation in papillary thyroid carcinoma: Significant association with node metastases and extra thyroidal invasion. *Endocr. Pathol.* **2012**, *23*, 83–93. [CrossRef]

165. Guan, H.; Ji, M.; Bao, R.; Yu, H.; Wang, Y.; Hou, P.; Zhang, Y.; Shan, Z.; Teng, W.; Xing, M. Association of high iodine intake with the T1799A BRAF mutation in papillary thyroid cancer. *J. Clin. Endocrinol. Metab.* **2009**, *94*, 1612–1617. [CrossRef] [PubMed]

166. Jung, C.-K.; Im, S.-Y.; Kang, Y.-J.; Lee, H.; Jung, E.-S.; Kang, C.-S.; Bae, J.-S.; Choi, Y.-J. Mutational patterns and novel mutations of the BRAF gene in a large cohort of Korean patients with papillary thyroid carcinoma. *Thyroid* **2012**, *22*, 791–797. [CrossRef]

167. Network, C.G.A.R. Integrated genomic characterization of papillary thyroid carcinoma. *Cell* **2014**, *159*, 676–690.

168. Romei, C.; Fugazzola, L.; Puxeddu, E.; Frasca, F.; Viola, D.; Muzza, M.; Moretti, S.; Luisa Nicolosi, M.; Giani, C.; Cirello, V. Modifications in the papillary thyroid cancer gene profile over the last 15 years. *J. Clin. Endocrinol. Metab.* **2012**, *97*, E1758–E1765. [CrossRef]

169. Goh, X.; Lum, J.; Yang, S.P.; Chionh, S.B.; Koay, E.; Chiu, L.; Parameswaran, R.; Ngiam, K.Y.; Loh, T.K.S.; Nga, M.E. BRAF mutation in papillary thyroid cancer—Prevalence and clinical correlation in a South-East Asian cohort. *Clin. Otolaryngol.* **2019**, *44*, 114–123. [CrossRef]

170. Arita, T.; Ichikawa, D.; Konishi, H.; Komatsu, S.; Shiozaki, A.; Shoda, K.; Kawaguchi, T.; Hirajima, S.; Nagata, H.; Kubota, T. Circulating long non-coding RNAs in plasma of patients with gastric cancer. *Anticancer Res.* **2013**, *33*, 3185–3193.

171. Chen, X.; Ba, Y.; Ma, L.; Cai, X.; Yin, Y.; Wang, K.; Guo, J.; Zhang, Y.; Chen, J.; Guo, X. Characterization of microRNAs in serum: A novel class of biomarkers for diagnosis of cancer and other diseases. *Cell Res.* **2008**, *18*, 997–1006. [CrossRef] [PubMed]

172. Ferracin, M.; Veronese, A.; Negrini, M. Micromarkers: MiRNAs in cancer diagnosis and prognosis. *Expert Rev. Mol. Diagn.* **2010**, *10*, 297–308. [CrossRef] [PubMed]

173. Huang, Z.; Zhu, D.; Wu, L.; He, M.; Zhou, X.; Zhang, L.; Zhang, H.; Wang, W.; Zhu, J.; Cheng, W. Six serum-based miRNAs as potential diagnostic biomarkers for gastric cancer. *Cancer Epidemiol. Prev. Biomark.* **2017**, *26*, 188–196. [CrossRef]

174. Sartori, D.A.; Chan, D.W. Biomarkers in prostate cancer: What's new? *Curr. Opin. Oncol.* **2014**, *26*, 259. [CrossRef]

175. Labourier, E.; Shifrin, A.; Busseniers, A.E.; Lupo, M.A.; Manganelli, M.L.; Andruss, B.; Wylie, D.; Beaudenon-Huibregtse, S. Molecular testing for miRNA, mRNA, and DNA on fine-needle aspiration improves the preoperative diagnosis of thyroid nodules with indeterminate cytology. *J. Clin. Endocrinol. Metab.* **2015**, *100*, 2743–2750. [CrossRef]

176. Yang, F.; Zhang, H.; Leng, X.; Hao, F.; Wang, L. miR-146b measurement in FNA to distinguish papillary thyroid cancer from benign thyroid masses. *Br. J. Biomed. Sci.* **2018**, *75*, 43–45. [CrossRef]

177. Zhang, A.; Wang, C.; Lu, H.; Chen, X.; Ba, Y.; Zhang, C.; Zhang, C.-Y. Altered Serum MicroRNA Profile May Serve as an Auxiliary Tool for Discriminating Aggressive Thyroid Carcinoma from Nonaggressive Thyroid Cancer and Benign Thyroid Nodules. *Dis. Markers* **2019**, *2019*, 3717683. [CrossRef] [PubMed]

178. Rosignolo, F.; Sponziello, M.; Giacomelli, L.; Russo, D.; Pecce, V.; Biffoni, M.; Bellantone, R.; Lombardi, C.P.; Lamartina, L.; Grani, G. Identification of thyroid-associated serum microRNA profiles and their potential use in thyroid cancer follow-up. *J. Endocr. Soc.* **2017**, *1*, 3–13.

179. Zhang, Y.; Pan, J.; Xu, D.; Yang, Z.; Sun, J.; Sun, L.; Wu, Y.; Qiao, H. Combination of serum microRNAs and ultrasound profile as predictive biomarkers of diagnosis and prognosis for papillary thyroid microcarcinoma. *Oncol. Rep.* **2018**, *40*, 3611–3624. [CrossRef]

180. Graham, M.E.R.; Hart, R.D.; Douglas, S.; Makki, F.M.; Pinto, D.; Butler, A.L.; Bullock, M.; Rigby, M.H.; Trites, J.R.; Taylor, S.M. Serum microRNA profiling to distinguish papillary thyroid cancer from benign thyroid masses. *J. Otolaryngol. Head Neck Surg.* **2015**, *44*, 33. [CrossRef]

181. Ye, W.; Deng, X.; Fan, Y. Exosomal miRNA423-5p mediated oncogene activity in papillary thyroid carcinoma: A potential diagnostic and biological target for cancer therapy. *Neoplasma* **2019**, *66*, 516–523. [CrossRef]

182. Bonneau, E.; Neveu, B.; Kostantin, E.; Tsongalis, G.; De Guire, V. How close are miRNAs from clinical practice? A perspective on the diagnostic and therapeutic market. *Ejifcc* **2019**, *30*, 114.

183. Jackson, S.; Kumar, G.; Banizs, A.B.; Toney, N.; Silverman, J.F.; Narick, C.M.; Finkelstein, S.D. Incremental utility of expanded mutation panel when used in combination with microRNA classification in indeterminate thyroid nodules. *Diagn. Cytopathol.* **2020**, *48*, 43–52. [CrossRef] [PubMed]

184. Sistrunk, J.W.; Shifrin, A.; Frager, M.; Bardales, R.H.; Thomas, J.; Fishman, N.; Goldberg, P.; Guttler, R.; Grant, E. Clinical impact of testing for mutations and microRNAs in thyroid nodules. *Diagn. Cytopathol.* **2019**, *47*, 758–764. [CrossRef]

185. Schlosser, K.; Hanson, J.; Villeneuve, P.J.; Dimitroulakos, J.; McIntyre, L.; Pilote, L.; Stewart, D.J. Assessment of circulating LncRNAs under physiologic and pathologic conditions in humans reveals potential limitations as biomarkers. *Sci. Rep.* **2016**, *6*, 36596. [CrossRef]

186. Xu, W.; Zhou, G.; Wang, H.; Liu, Y.; Chen, B.; Chen, W.; Lin, C.; Wu, S.; Gong, A.; Xu, M. Circulating lncRNA SNHG11 as a novel biomarker for early diagnosis and prognosis of colorectal cancer. *Int. J. Cancer* **2020**, *146*, 2901–2912. [CrossRef] [PubMed]

187. Wang, L.; Duan, W.; Yan, S.; Xie, Y.; Wang, C. Circulating long non-coding RNA colon cancer-associated transcript 2 protected by exosome as a potential biomarker for colorectal cancer. *Biomed. Pharmacother.* **2019**, *113*, 108758. [CrossRef]

188. Abedini, P.; Fattahi, A.; Agah, S.; Talebi, A.; Beygi, A.H.; Amini, S.M.; Mirzaei, A.; Akbari, A. Expression analysis of circulating plasma long noncoding RNAs in colorectal cancer: The relevance of lncRNAs ATB and CCAT1 as potential clinical hallmarks. *J. Cell. Physiol.* **2019**, *234*, 22028–22033. [CrossRef]

189. Özgür, E.; Ferhatoğlu, F.; Şen, F.; Saip, P.; Gezer, U. Circulating lncRNA H19 may be a useful marker of response to neoadjuvant chemotherapy in breast cancer. *Cancer Biomark.* **2020**, *27*, 11–17. [CrossRef]

190. Liu, Y.; Feng, W.; Liu, W.; Kong, X.; Li, L.; He, J.; Wang, D.; Zhang, M.; Zhou, G.; Xu, W. Circulating lncRNA ABHD11-AS1 serves as a biomarker for early pancreatic cancer diagnosis. *J. Cancer* **2019**, *10*, 3746. [CrossRef]

191. Yu, J.; Ding, W.B.; Wang, M.C.; Guo, X.G.; Xu, J.; Xu, Q.G.; Yang, Y.; Sun, S.H.; Liu, J.F.; Qin, L.X. Plasma circular RNA panel to diagnose hepatitis B virus-related hepatocellular carcinoma: A large-scale, multicenter study. *Int. J. Cancer* **2020**, *146*, 1754–1763. [CrossRef] [PubMed]

192. Zhao, S.-Y.; Wang, J.; Ouyang, S.-B.; Huang, Z.-K.; Liao, L. Salivary circular RNAs Hsa_Circ_0001874 and Hsa_Circ_0001971 as novel biomarkers for the diagnosis of Oral squamous cell carcinoma. *Cell. Physiol. Biochem.* **2018**, *47*, 2511–2521. [CrossRef]

193. Liu, Y.-T.; Han, X.-H.; Xing, P.-Y.; Hu, X.-S.; Hao, X.-Z.; Wang, Y.; Li, J.-L.; Zhang, Z.-S.; Yang, Z.-H.; Shi, Y.-K. Circular RNA profiling identified as a biomarker for predicting the efficacy of gefitinib therapy for non-small cell lung cancer. *J. Thorac. Dis.* **2019**, *11*, 1779. [CrossRef]

194. Ortholan, C.; Puissegur, M.-P.; Ilie, M.; Barbry, P.; Mari, B.; Hofman, P. MicroRNAs and lung cancer: New oncogenes and tumor suppressors, new prognostic factors and potential therapeutic targets. *Curr. Med. Chem.* **2009**, *16*, 1047–1061. [CrossRef]

195. Huang, X.; Yuan, T.; Liang, M.; Du, M.; Xia, S.; Dittmar, R.; Wang, D.; See, W.; Costello, B.A.; Quevedo, F. Exosomal miR-1290 and miR-375 as prognostic markers in castration-resistant prostate cancer. *Eur. Urol.* **2015**, *67*, 33–41. [CrossRef]

196. Takamizawa, J.; Konishi, H.; Yanagisawa, K.; Tomida, S.; Osada, H.; Endoh, H.; Harano, T.; Yatabe, Y.; Nagino, M.; Nimura, Y. Reduced expression of the let-7 microRNAs in human lung cancers in association with shortened postoperative survival. *Cancer Res.* **2004**, *64*, 3753–3756. [CrossRef]

197. Hu, Y.; Tian, H.; Xu, J.; Fang, J.-Y. Roles of competing endogenous RNAs in gastric cancer. *Brief. Funct. Genom.* **2016**, *15*, 266–273. [CrossRef]

198. Donghi, R.; Longoni, A.; Pilotti, S.; Michieli, P.; Della Porta, G.; Pierotti, M.A. Gene p53 mutations are restricted to poorly differentiated and undifferentiated carcinomas of the thyroid gland. *J. Clin. Investig.* **1993**, *91*, 1753–1760. [CrossRef]

199. Handkiewicz-Junak, D.; Czarniecka, A.; Jarząb, B. Molecular prognostic markers in papillary and follicular thyroid cancer: Current status and future directions. *Mol. Cell. Endocrinol.* **2010**, *322*, 8–28. [CrossRef]

200. Passler, C.; Scheuba, C.; Prager, G.; Kaczirek, K.; Kaserer, K.; Zettinig, G.; Niederle, B. Prognostic factors of papillary and follicular thyroid cancer: Differences in an iodine-replete endemic goiter region. *Endocr. Relat. Cancer* **2004**, *11*, 131–139. [CrossRef] [PubMed]

201. Sondermann, A.; Andreghetto, F.M.; Moulatlet, A.C.B.; da Silva Victor, E.; de Castro, M.G.; Nunes, F.D.; Brandão, L.G.; Severino, P. MiR-9 and miR-21 as prognostic biomarkers for recurrence in papillary thyroid cancer. *Clin. Exp. Metastasis* **2015**, *32*, 521–530. [CrossRef]

202. Dai, L.; Wang, Y.; Chen, L.; Zheng, J.; Li, J.; Wu, X. MiR-221, a potential prognostic biomarker for recurrence in papillary thyroid cancer. *World J. Surg. Oncol.* **2017**, *15*, 11. [CrossRef]

203. Geraldo, M.V.; Nakaya, H.I.; Kimura, E.T. Down-regulation of 14q32-encoded miRNAs and tumor suppressor role for miR-654-3p in papillary thyroid cancer. *Oncotarget* **2017**, *8*, 9597. [CrossRef]

204. Haghpanah, V.; Fallah, P.; Tavakoli, R.; Naderi, M.; Samimi, H.; Soleimani, M.; Larijani, B. Antisense-miR-21 enhances differentiation/apoptosis and reduces cancer stemness state on anaplastic thyroid cancer. *Tumor Biol.* **2016**, *37*, 1299–1308. [CrossRef]

205. Frezzetti, D.; De Menna, M.; Zoppoli, P.; Guerra, C.; Ferraro, A.; Bello, A.; De Luca, P.; Calabrese, C.; Fusco, A.; Ceccarelli, M. Upregulation of miR-21 by Ras in vivo and its role in tumor growth. *Oncogene* **2011**, *30*, 275–286. [CrossRef]

206. Pennelli, G.; Galuppini, F.; Barollo, S.; Cavedon, E.; Bertazza, L.; Fassan, M.; Guzzardo, V.; Pelizzo, M.R.; Rugge, M.; Mian, C. The PDCD4/miR-21 pathway in medullary thyroid carcinoma. *Hum. Pathol.* **2015**, *46*, 50–57. [CrossRef]

207. Durante, C.; Haddy, N.; Baudin, E.; Leboulleux, S.; Hartl, D.; Travagli, J.; Caillou, B.; Ricard, M.; Lumbroso, J.; De Vathaire, F. Long-term outcome of 444 patients with distant metastases from papillary and follicular thyroid carcinoma: Benefits and limits of radioiodine therapy. *J. Clin. Endocrinol. Metab.* **2006**, *91*, 2892–2899. [CrossRef]

208. Schlumberger, M.; Tubiana, M.; Florent De, V.; Hill, C.; Paule, G.; Jean-Paul, T.; Philippe, F.; Jean, L.; Bernard, C.; Claude, P. Long-term results of treatment of 283 patients with lung and bone metastases from differentiated thyroid carcinoma. *J. Clin. Endocrinol. Metab.* **1986**, *63*, 960–967.

209. Buffet, C.; Wassermann, J.; Hecht, F.; Leenhardt, L.; Dupuy, C.; Groussin, L.; Lussey-Lepoutre, C. Redifferentiation of radioiodine-refractory thyroid cancers. *Endocr. Relat. Cancer* **2020**, *27*, R113–R132. [CrossRef]

210. Riesco-Eizaguirre, G.; Rodríguez, I.; De la Vieja, A.; Costamagna, E.; Carrasco, N.; Nistal, M.; Santisteban, P. The BRAFV600E oncogene induces transforming growth factor β secretion leading to sodium iodide symporter repression and increased malignancy in thyroid cancer. *Cancer Res.* **2009**, *69*, 8317–8325. [CrossRef] [PubMed]

211. Lakshmanan, A.; Scarberry, D.; Green, J.A.; Zhang, X.; Selmi-Ruby, S.; Jhiang, S.M. Modulation of thyroidal radioiodide uptake by oncological pipeline inhibitors and Apigenin. *Oncotarget* **2015**, *6*, 31792. [CrossRef]

212. Pak, K.; Shin, S.; Kim, S.-J.; Kim, I.-J.; Chang, S.; Koo, P.; Kwak, J.; Kim, J.-H. Response of retinoic acid in patients with radioactive iodine-refractory thyroid Cancer: A meta-analysis. *Oncol. Res. Treat.* **2018**, *41*, 100–104. [CrossRef]

213. Fu, H.; Cheng, L.; Jin, Y.; Cheng, L.; Liu, M.; Chen, L. MAPK Inhibitors Enhance HDAC Inhibitor-Induced Redifferentiation in Papillary Thyroid Cancer Cells Harboring BRAFV600E: An In Vitro Study. *Mol. Ther. Oncolytics* **2019**, *12*, 235–245. [CrossRef]

214. Rosenbaum-Krumme, S.J.; Freudenberg, L.S.; Jentzen, W.; Bockisch, A.; Nagarajah, J. Effects of rosiglitazone on radioiodine negative and progressive differentiated thyroid carcinoma as assessed by 124I PET/CT imaging. *Clin. Nucl. Med.* **2012**, *37*, e47–e52. [CrossRef]

215. Li, L.; Lv, B.; Chen, B.; Guan, M.; Sun, Y.; Li, H.; Zhang, B.; Ding, C.; He, S.; Zeng, Q. Inhibition of miR-146b expression increases radioiodine-sensitivity in poorly differential thyroid carcinoma via positively regulating NIS expression. *Biochem. Biophys. Res. Commun.* **2015**, *462*, 314–321. [CrossRef] [PubMed]

216. Riesco-Eizaguirre, G.; Wert-Lamas, L.; Perales-Patón, J.; Sastre-Perona, A.; Fernández, L.P.; Santisteban, P. The miR-146b-3p/PAX8/NIS regulatory circuit modulates the differentiation phenotype and function of thyroid cells during carcinogenesis. *Cancer Res.* **2015**, *75*, 4119–4130. [CrossRef] [PubMed]

217. Damanakis, A.I.; Eckhardt, S.; Wunderlich, A.; Roth, S.; Wissniowski, T.T.; Bartsch, D.K.; Di Fazio, P. MicroRNAs let7 expression in thyroid cancer: Correlation with their deputed targets HMGA2 and SLC5A5. *J. Cancer Res. Clin. Oncol.* **2016**, *142*, 1213–1220. [CrossRef]

218. Sahin Lacin, E.E.; Karakas, Y.; Yalcin, S. Metastatic medullary thyroid cancer: A dramatic response to a systemic chemotherapy (temozolomide and capecitabine) regimen. *Oncotargets Ther.* **2015**, *8*, 1039.

219. Chu, E.; Sartorelli, A. Cancer chemotherapy. *Basic Clin. Pharmacol.* **2004**, *10*, 878–907.

220. Stein, R.; Chen, S.; Reed, L.; Richel, H.; Goldenberg, D.M. Combining radioimmunotherapy and chemotherapy for treatment of medullary thyroid carcinoma: Effectiveness of dacarbazine. *Cancer* **2002**, *94*, 51–61. [CrossRef]

221. Wilson, P.; Millar, B.; Brierley, J. The management of advanced thyroid cancer. *Clin. Oncol.* **2004**, *16*, 561–568. [CrossRef] [PubMed]

222. Lessin, L.S.; Min, M. Chemotherapy for thyroid cancer. In *Thyroid Cancer*; Springer: Berlin/Heidelberg, Germany, 2000; pp. 179–182.

223. Gottlieb, J.A.; Hill, C.S., Jr. Chemotherapy of thyroid cancer with adriamycin: Experience with 30 patients. *New Engl. J. Med.* **1974**, *290*, 193–197. [CrossRef] [PubMed]

224. Brose, M.S.; Nutting, C.M.; Jarzab, B.; Elisei, R.; Siena, S.; Bastholt, L.; De La Fouchardiere, C.; Pacini, F.; Paschke, R.; Shong, Y.K. Sorafenib in radioactive iodine-refractory, locally advanced or metastatic differentiated thyroid cancer: A randomised, double-blind, phase 3 trial. *Lancet* **2014**, *384*, 319–328. [CrossRef]

225. Schlumberger, M.; Tahara, M.; Wirth, L.J.; Robinson, B.; Brose, M.S.; Elisei, R.; Habra, M.A.; Newbold, K.; Shah, M.H.; Hoff, A.O. Lenvatinib versus placebo in radioiodine-refractory thyroid cancer. *New Engl. J. Med.* **2015**, *372*, 621–630. [CrossRef]

226. Gianoukakis, A.G.; Dutcus, C.E.; Batty, N.; Guo, M.; Baig, M. Prolonged duration of response in lenvatinib responders with thyroid cancer. *Endocr. Relat. Cancer* **2018**, *25*, 699–704. [CrossRef]

227. Wells, S.A., Jr.; Robinson, B.G.; Gagel, R.F.; Dralle, H.; Fagin, J.A.; Santoro, M.; Baudin, E.; Elisei, R.; Jarzab, B.; Vasselli, J.R. Vandetanib in patients with locally advanced or metastatic medullary thyroid cancer: A randomized, double-blind phase III trial. *J. Clin. Oncol.* **2012**, *30*, 134. [CrossRef]

228. Schlumberger, M.; Elisei, R.; Müller, S.; Schöffski, P.; Brose, M.S.; Shah, M.H.; Licitra, L.F.; Jarzab, B.; Medvedev, V.; Kreissl, M. Final overall survival analysis of EXAM, an international, double-blind, randomized, placebo-controlled phase III trial of cabozantinib (Cabo) in medullary thyroid carcinoma (MTC) patients with documented RECIST progression at baseline. *Am. Soc. Clin. Oncol.* **2015**. [CrossRef]

229. Subbiah, V.; Kreitman, R.J.; Wainberg, Z.A.; Cho, J.Y.; Schellens, J.H.; Soria, J.C.; Wen, P.Y.; Zielinski, C.; Cabanillas, M.E.; Urbanowitz, G. Dabrafenib and trametinib treatment in patients with locally advanced or metastatic BRAF V600–mutant anaplastic thyroid cancer. *J. Clin. Oncol.* **2018**, *36*, 7. [CrossRef]

230. Hong, D.S.; DuBois, S.G.; Kummar, S.; Farago, A.F.; Albert, C.M.; Rohrberg, K.S.; van Tilburg, C.M.; Nagasubramanian, R.; Berlin, J.D.; Federman, N. Larotrectinib in patients with TRK fusion-positive solid tumours: A pooled analysis of three phase 1/2 clinical trials. *Lancet Oncol.* **2020**, *21*, 531–540. [CrossRef]

231. Doebele, R.C.; Drilon, A.; Paz-Ares, L.; Siena, S.; Shaw, A.T.; Farago, A.F.; Blakely, C.M.; Seto, T.; Cho, B.C.; Tosi, D. Entrectinib in patients with advanced or metastatic NTRK fusion-positive solid tumours: Integrated analysis of three phase 1–2 trials. *Lancet Oncol.* **2020**, *21*, 271–282. [CrossRef]

232. Robbins, R.J.; Srivastava, S.; Shaha, A.; Ghossein, R.; Larson, S.M.; Fleisher, M.; Tuttle, R.M. Factors influencing the basal and recombinant human thyrotropin-stimulated serum thyroglobulin in patients with metastatic thyroid carcinoma. *J. Clin. Endocrinol. Metab.* **2004**, *89*, 6010–6016. [CrossRef]

233. Schlumberger, M.; Hitzel, A.; Toubert, M.; Corone, C.; Troalen, F.; Schlageter, M.; Claustrat, F.; Koscielny, S.; Taieb, D.; Toubeau, M. Comparison of seven serum thyroglobulin assays in the follow-up of papillary and follicular thyroid cancer patients. *J. Clin. Endocrinol. Metab.* **2007**, *92*, 2487–2495. [CrossRef]

234. Spencer, C.A. Clinical utility of thyroglobulin antibody (TgAb) measurements for patients with differentiated thyroid cancers (DTC). *J. Clin. Endocrinol. Metab.* **2011**, *96*, 3615–3627. [CrossRef]

235. Giovanella, L.; Suriano, S.; Ceriani, L.; Verburg, F.A. Undetectable Thyroglobulin in Patients with Differentiated Thyroid Carcinoma and Residual Radioiodine Uptake on a Postablation Whole-Body Scan. *Clin. Nucl. Med.* **2011**, *36*, 109–112. [CrossRef]

236. Diesch, J.; Zwick, A.; Garz, A.-K.; Palau, A.; Buschbeck, M.; Götze, K.S. A clinical-molecular update on azanucleoside-based therapy for the treatment of hematologic cancers. *Clin. Epigenetics* **2016**, *8*, 71. [CrossRef]

237. Bainschab, A.; Quehenberger, F.; Greinix, H.T.; Krause, R.; Wölfler, A.; Sill, H.; Zebisch, A. Infections in patients with acute myeloid leukemia treated with low-intensity therapeutic regimens: Risk factors and efficacy of antibiotic prophylaxis. *Leuk. Res.* **2016**, *42*, 47–51. [CrossRef] [PubMed]

238. Slaby, O.; Laga, R.; Sedlacek, O. Therapeutic targeting of non-coding RNAs in cancer. *Biochem. J.* **2017**, *474*, 4219–4251. [CrossRef]

239. Wang, W.-T.; Han, C.; Sun, Y.-M.; Chen, T.-Q.; Chen, Y.-Q. Noncoding RNAs in cancer therapy resistance and targeted drug development. *J. Hematol. Oncol.* **2019**, *12*, 55. [CrossRef]

240. Xu, Y.; Han, Y.F.; Ye, B.; Zhang, Y.L.; Dong, J.D.; Zhu, S.J.; Chen, J. miR-27b-3p is Involved in Doxorubicin Resistance of Human Anaplastic Thyroid Cancer Cells via Targeting Peroxisome Proliferator-Activated Receptor Gamma. *Basic Clin. Pharmacol. Toxicol.* **2018**, *123*, 670–677. [CrossRef]

241. Liu, Y.; Yue, P.; Zhou, T.; Zhang, F.; Wang, H.; Chen, X. LncRNA MEG3 enhances 131I sensitivity in thyroid carcinoma via sponging miR-182. *Biomed. Pharmacother.* **2018**, *105*, 1232–1239. [CrossRef] [PubMed]

242. Liu, C.; Feng, Z.; Chen, T.; Lv, J.; Liu, P.; Jia, L.; Zhu, J.; Chen, F.; Yang, C.; Deng, Z. Downregulation of NEAT1 reverses the radioactive iodine resistance of papillary thyroid carcinoma cell via miR-101-3p/FN1/PI3K-AKT signaling pathway. *Cell Cycle* **2019**, *18*, 167–203. [CrossRef]

243. Xiang, C.; Zhang, M.-L.; Zhao, Q.-Z.; Xie, Q.-P.; Yan, H.-C.; Yu, X.; Wang, P.; Wang, Y. LncRNA-SLC6A9-5: 2: A potent sensitizer in 131I-resistant papillary thyroid carcinoma with PARP-1 induction. *Oncotarget* **2017**, *8*, 22954. [CrossRef]

244. Tan, H.-L.; Chan, K.-G.; Pusparajah, P.; Saokaew, S.; Duangjai, A.; Lee, L.-H.; Goh, B.-H. Anti-cancer properties of the naturally occurring aphrodisiacs: Icariin and its derivatives. *Front. Pharmacol.* **2016**, *7*, 191. [CrossRef] [PubMed]

245. Li, J.; Jiang, K.; Zhao, F. Icariin regulates the proliferation and apoptosis of human ovarian cancer cells through microRNA-21 by targeting PTEN, RECK and Bcl-2. *Oncol. Rep.* **2015**, *33*, 2829–2836. [CrossRef] [PubMed]

246. Zhang, Y.; Wei, Y.; Zhu, Z.; Gong, W.; Liu, X.; Hou, Q.; Sun, Y.; Chai, J.; Zou, L.; Zhou, T. Icariin enhances radiosensitivity of colorectal cancer cells by suppressing NF-κB activity. *Cell Biochem. Biophys.* **2014**, *69*, 303–310. [CrossRef] [PubMed]

247. Carrasco-Leon, A.; Ezponda, T.; Meydan, C.; Valcárcel, L.V.; Ordoñez, R.; Kulis, M.; Garate, L.; Miranda, E.; Segura, V.; Guruceaga, E. Role of lncRNAs as prognostic factor and potential therapeutic target in Multiple Myeloma. *Clin. Lymphoma Myeloma Leuk.* **2019**, *19*, e354–e355. [CrossRef]

248. Xiang, Z.; Song, S.; Zhu, Z.; Sun, W.; Sun, S.; Li, Q.S.; Yu, Y.; Li, K.K. LncRNAs GIHCG and SPINT1-AS1 are crucial factors for pan-cancer cells sensitivity to lapatinib. *Front. Genet.* **2019**, *10*, 25. [CrossRef]

249. Tsang, S.; Patel, T.; Yustein, J.T. Long non-coding RNAs regulation of therapeutic resistance. *Cancer Drug Resist.* **2019**, *2*, 550–567. [CrossRef]

250. Sheng, S.-R.; Wu, J.-S.; Tang, Y.-L.; Liang, X.-H. Long noncoding RNAs: Emerging regulators of tumor angiogenesis. *Future Oncol.* **2017**, *13*, 1551–1562. [CrossRef]

251. Zhou, Y.; Zhang, X.; Klibanski, A. MEG3 noncoding RNA: A tumor suppressor. *J. Mol. Endocrinol.* **2012**, *48*, R45–R53. [CrossRef]

252. Moradi, A.; Naiini, M.R.; Yazdanpanahi, N.; Tabatabaeian, H.; Nabatchian, F.; Baghi, M.; Azadeh, M.; Ghaedi, K. Evaluation of The Expression Levels of Three Long Non-Coding RNAs in Multiple Sclerosis. *Cell J.* **2020**, *22*, 165–170.

253. Shao, F.; Huang, M.; Meng, F.; Huang, Q. Circular RNA signature predicts gemcitabine resistance of pancreatic ductal adenocarcinoma. *Front. Pharmacol.* **2018**, *9*, 584. [CrossRef]

254. Kun-Peng, Z.; Xiao-Long, M.; Lei, Z.; Chun-Lin, Z.; Jian-Ping, H.; Tai-Cheng, Z. Screening circular RNA related to chemotherapeutic resistance in osteosarcoma by RNA sequencing. *Epigenomics* **2018**, *10*, 1327–1346. [CrossRef]

255. Yu, W.; Peng, W.; Sha, H.; Li, J. Hsa_circ_0003998 promotes chemoresistance via modulation of miR-326 in lung adenocarcinoma cells. *Oncol. Res. Featur. Preclin. Clin. Cancer Ther.* **2019**, *27*, 623–628. [CrossRef]

256. Shang, J.; Chen, W.-M.; Wang, Z.-H.; Wei, T.-N.; Chen, Z.-Z.; Wu, W.-B. CircPAN3 mediates drug resistance in acute myeloid leukemia through the miR-153-5p/miR-183-5p–XIAP axis. *Exp. Hematol.* **2019**, *70*, 42–54.e43. [CrossRef]

257. Huang, X.; Li, Z.; Zhang, Q.; Wang, W.; Li, B.; Wang, L.; Xu, Z.; Zeng, A.; Zhang, X.; Zhang, X. Circular RNA AKT3 upregulates PIK3R1 to enhance cisplatin resistance in gastric cancer via miR-198 suppression. *Mol. Cancer* **2019**, *18*, 71. [CrossRef]

258. Selvam, C.; Mutisya, D.; Prakash, S.; Ranganna, K.; Thilagavathi, R. Therapeutic potential of chemically modified si RNA: Recent trends. *Chem. Biol. Drug Des.* **2017**, *90*, 665–678. [CrossRef]

259. Steichen, S.D.; Caldorera-Moore, M.; Peppas, N.A. A review of current nanoparticle and targeting moieties for the delivery of cancer therapeutics. *Eur. J. Pharm. Sci.* **2013**, *48*, 416–427. [CrossRef] [PubMed]

260. Kramps, T.; Probst, J. Messenger RNA-based vaccines: Progress, challenges, applications. *Wiley Interdiscip. Rev. RNA* **2013**, *4*, 737–749. [CrossRef]

261. Tatiparti, K.; Sau, S.; Kashaw, S.K.; Iyer, A.K. siRNA delivery strategies: A comprehensive review of recent developments. *Nanomaterials* **2017**, *7*, 77. [CrossRef]

262. Dammes, N.; Peer, D. Paving the road for RNA therapeutics. *Trends Pharmacol. Sci.* **2020**, *41*, 755–775. [CrossRef]

263. Sugawa, H.; Smith, E.; Imura, H.; Mori, T. A thyroid cancer specific monoclonal antibody which recognizes cryptic epitope (s) of human thyroglobulin. *Mol. Cell. Endocrinol.* **1993**, *93*, 207–211. [CrossRef]

Publisher's Note: MDPI stays neutral with regard to jurisdictional claims in published maps and institutional affiliations.

Perspective

MicroRNAs and Their Targetomes in Tumor-Immune Communication

Sunglim Cho [1,†], Jesse W. Tai [1,†]  **and Li-Fan Lu [1,2,3,*]**

1   Division of Biological Sciences, University of California, San Diego, La Jolla, CA 92093, USA; limniim3324@gmail.com (S.C.); j2tai@ucsd.edu (J.W.T.)
2   Moores Cancer Center, University of California, San Diego, La Jolla, CA 92093, USA
3   Center for Microbiome Innovation, University of California, San Diego, La Jolla, CA 92093, USA
*   Correspondence: lifanlu@ucsd.edu
†   These authors contributed equally and are listed alphabetically.

Received: 24 June 2020; Accepted: 21 July 2020; Published: 24 July 2020

Abstract: The development of cancer is a complex and dynamically regulated multiple-step process that involves many changes in gene expression. Over the last decade, microRNAs (miRNAs), a class of short regulatory non-coding RNAs, have emerged as key molecular effectors and regulators of tumorigenesis. While aberrant expression of miRNAs or dysregulated miRNA-mediated gene regulation in tumor cells have been shown to be capable of directly promoting or inhibiting tumorigenesis, considering the well-reported role of the immune system in cancer, tumor-derived miRNAs could also impact tumor growth through regulating anti-tumor immune responses. Here, we discuss howmiRNAs can function as central mediators that influence the crosstalk between cancer and the immune system. Moreover, we also review the current progress in the development of novel experimental approaches for miRNA target identification that will facilitate our understanding of miRNA-mediated gene regulation in not only human malignancies, but also in other genetic disorders.

Keywords: miRNA; oncomiRNA; post-transcriptional regulation; immune regulation

1. Introduction

MicroRNAs (miRNA) are small, non-coding RNA molecules (~22 nucleotides) which play crucial roles in post-transcriptional regulation of gene expression. To date, more than 2600 mature human miRNAs have been registered at miRbase (Release 22.1: Oct. 2018) [1]. These mature miRNAs, incorporated together with Argonaute protein (Ago) to form the RISC (RNA-induced silencing complex), repress the expression of their targets by either inducing mRNA degradation or translational inhibition [2]. Since the first miRNA lin-4 was discovered in *Caenorhabditis elegans* in 1993 [3], miRNAs have been shown to control diverse biological pathways such as cell development, division, proliferation and differentiation, in both physiological and pathological conditions [4]. To date, many studies have reported that miRNA expression is dynamically regulated in different tumors. While dysregulation of miRNA biogenesis and function can directly contribute to tumorigenesis and malignant progression [5–7], considering the pivotal function of the host immune response in shaping the tumor microenvironment, the role of miRNA-mediated communication between tumors and the immune system involving exosomal miRNA, immunometabolites, and checkpoint regulators, has also begun to be appreciated. Previously, many research efforts in the identification of individual miRNA-mRNA pairs have helped shed some light on the importance of miRNAs in cancer [8]. However, as miRNAs generally repress a set of genes that are in a shared pathway or protein complex, to ensure their impact on gene regulation and the resultant biology, it is essential to obtain a comprehensive demonstration and validation of the targetome of the desired miRNAs.

To this end, moving from the complete reliance on computational prediction in the early days, in the past decade, many cutting-edge experimental techniques have been developed to afford unbiased examination of the interactions between miRNA and their target mRNAs in the selected cell types [9]. These technological advances have not only made the analysis of entire targetomes for specific miRNAs possible, but also allowed us to gain mechanistic insights into the biological impact of aberrantly expressed miRNAs on tumorigenesis.

2. MiRNAs as Mediators of Tumor–Immune Communication

Dysregulation of miRNAs can drive or promote carcinogenesis in a variety of fashions. In cancer, the expression of miRNAs may be dysregulated in various ways, including by mutations in the miRNA biogenesis machinery [10], changes in the epigenetic regulation of miRNA-transcribing genes [11], and altered expression of transcription factors involved in promoting or repressing miRNA expression [12,13]. Overexpression of miRNAs which target tumor-suppressive genes induce proliferative signaling, invasion and migration, resistance to apoptosis, etc. On the other hand, the loss of expression of miRNAs which target oncogenes also leads to such carcinogenic effects. It is also well known that cancer can co-opt the body's immune system to serve its needs, whether by generating inflammation and producing genotoxic damage, or by utilizing immunosuppressive regulatory cells and molecules to evade destruction. As such, aberrantly expressed miRNAs can also be employed by tumors to communicate with, and to deactivatethe body's defenses. Conversely, immune cells can also limit tumorigenesis through altering the expression of miRNAs in tumors (Figure 1).

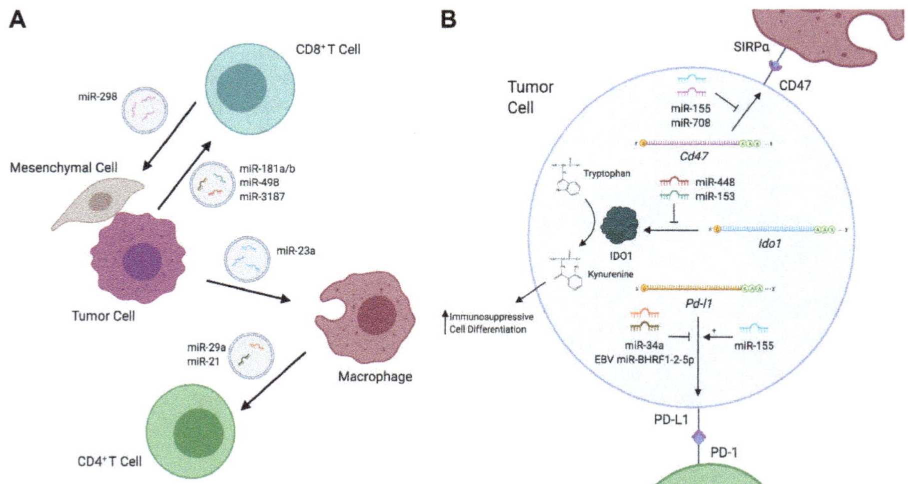

Figure 1. MicroRNAs (miRNAs) in tumor–immune communication. (**A**) Schematic of exosomal cross-talk between immune cells, stromal cells, and tumor cells in the tumor microenvironment (TME). (**B**) Representative diagram of miRNA-mediated control of immune-regulatory proteins in tumor cells. Created with BioRender.com.

2.1. Tumor-Derived Exosomal miRNAs

Exosomes are extracellular vesicles comprised of proteins, lipids, DNA, and RNA which are secreted by both healthy and cancerous cells. Crosstalk, mediated by exosomes between cancer cells and cells residing in the tumor microenvironment (TME), such as immune cells, fibroblasts, etc., has been implicated in driving tumorigenesis [14]. Thus, exosomes can serve as a mode of transportation by which tumor cells deliver immunosuppressive miRNAs directly to immune cell subsets (Figure 1A). To this end, a recent study found that exosomes secreted by melanoma cells altered cytokine secretion and T cell receptor signaling in CD8 T cells [15]. These melanoma-derived exosomes were enriched with miR-181a/b and miR-498, which directly bound to the 3'UTR of *TNF* and decreased TNFα secretion in CD8 T cells. Furthermore, another miRNA in melanoma-derived exosomes, miR-3187-3p, was also found to directly target *PTPRC*, a gene that encodes CD45, a key mediator of T cell receptor signaling. Such miRNAs likely serve to stymie the CD8 T cell response to melanoma and contribute to immune evasion. Notably, although the interaction was confirmed via the reporter assay, no consensus binding sequence on the *PTPRC* 3'UTR for miR-3187 was identified, highlighting the possible pitfalls of relying entirely on computational prediction.

In addition to their direct impact on T cells, tumor-derived exosomal miRNAs could also indirectly regulate T cell responses through targeting other immune cell subsets in the tumor microenvironment. For example, in liver cancer, miR-23a-3p released by endoplasmic reticulum-stressed hepatocellular carcinoma (HCC) cells has been reported to inhibit T cell function through targeting PTEN in macrophages from HCC tissues. Reduced PTEN expression led to elevated phosphorylated AKT, followed by increased expression of PD-L1, the ligand of PD-1 [16]. On the other hand, like tumor-derived exosomal miRNAs, miRNAs from tumor-associated macrophage (TAM)-secreted exosomes could also promote tumor growth and inhibit anti-tumor immunity. To this end, two miRNAs in TAM-derived exosomes, miR-29a-3p and miR-21-5p, have been shown to target the 3'UTR of *STAT3*, which plays a critical role in CD4 T cell differentiation into Th17 cells, thus inducing a higher regulatory T (Treg)/Th17 cell ratio [17]. It should be noted that while tumor- or TAM-derived exosomal miRNAs can suppress host anti-tumor responses, immune cells can secrete miRNA-containing exosomes to fight tumors as well. For example, Seo et al. have demonstrated that healthy, activated CD8 T cells can deplete mesenchymal stem cells (MSCs) and inhibit tumor invasion and metastasis in vivo through releasing cytotoxic extracellular vesicles [18]. The cytotoxic effect of the extracellular vesicles was attributed to miR-298-5p, a miRNA that was able to induce apoptotic depletion of MSCs, via the activation of caspase-3. This study not only highlights an anti-tumor role of immune cell derived exosomal miRNAs, but also demonstrates a novel effector mechanism by which CD8 T cells control tumor progression, independent from their canonical direct cytotoxicity against cancerous cells.

2.2. Immunometabolites Regulated by miRNAs

Tumors can also directly communicate with immune cells in the TME through the generation of immunomodulatory metabolites (Figure 1B). To this end, elevated expression of indoleamine 2,3-dioxygenase 1 (IDO1), an enzyme which metabolizes tryptophan into kynurenine, in the TME has been shown to drive the differentiation of several immunosuppressive cell types [19], including Treg cells [20–22], immunosuppressive dendritic cells [23,24] and macrophages [25,26], or to directly suppress anti-tumor immunity [27]. In colorectal cancer, overexpression of IDO1 suppressed the CD8 T cell responses, leading to enhanced tumor growth [27]. In this study, miR-448 was shown to be able to enhance the survival of CD8 T cells by directly attenuating the upregulation of IDO1 in colorectal tumor cells in response to IFNγ stimulation. Similarly, miR-153 has also been shown to target *IDO1* in colorectal cancer in response to IFNγ [27,28]. Moreover, when combining miR-153-mediated IDO1 inhibition and chimeric antigen receptor (CAR) T cell therapy, further enhanced in vitro T cell killing activities and reduced xenograft tumor growth in mice were reported [27,28].

2.3. Immune Checkpoint Regulators Targeted by miRNAs

Upon T cell activation, multiple co-inhibitory receptors such as CLTA4, PD-1 and LAG-3 are up-regulated. Through interacting with their corresponding ligands, these so-called immune checkpoint molecules are capable of modulating T cell responses to self proteins, as well as to tumor antigens [29]. In the past decade, great clinical success in cancer immunotherapy has been achieved by employing monoclonal antibodies targeting these immune checkpoints in patients with a variety of cancers [30,31]. In particular, therapeutic blockade of the PD-1 pathway has been considered as possibly one of the most important advances in the history of cancer treatment. Not only was PD-L1, the ligand of PD-1, found to be induced in TAM as discussed earlier, it is also highly expressed in multiple cancers [32]. PD-L1 expression can be regulated in a variety of ways, and recently the miRNA-mediated control of PD-L1 has become more apparent (Figure 1B) [33]. Indeed, disruption of the *Pdl1* 3′UTR led to elevated PD-L1 expression, implying that miRNAs play an active role in regulating PD-L1 expression [34]. To this end, recent investigation has uncovered multiple *Pdl1* mRNA-miRNA interactions in Epstein-Barr virus (EBV) associated B cell lymphomas, which are known to heavily rely on PD-L1 expression to evade immune defenses [35]. One report demonstrates that the co-localization of viral protein EBNA2, and B-cell-specific transcription factor EBF1 at the miR-34a promoter, leads to the repression of miR-34a, which targets the *PDl1* 3′UTR in Burkitt lymphoma (BL) and Diffuse Large B-Cell Lymphoma (DLBCL) cells [36]. Another group has found that a novel EBV-encoded miRNA, EBV miR-BHRF1-2-5p, could also target the 3′UTRs of *Pdl1* as well as another PD-1 ligand, *Pdl2*, in a model of EBV-positive DLBCL [37]. It should be noted that both miR-34a and EBV miR-BHRF1-2-5p coincided with LMP1 expression, which is known to drive PD-L1/2 expression. As such, the "fine-tuning" role for these counter-regulatory miRNAs may serve as attractive targets for therapeutics. On the other hand, while miR-155 expression was also higher in the serum of EBV-positive DLBCL patients [38], interestingly, miR-155-binding to the *Pdl1* 3′UTR actually served to upregulate PD-L1 expression, further demonstrating the complex nature of miRNA-mediated gene regulation [39].

In addition to the aforementioned immune checkpoint molecules that mainly act on controlling T cell responses, CD47, which is commonly expressed in blood cancers [40,41], sends a "don't eat me" signal to macrophages, preventing them from clearing cancerous cells via phagocytosis. Like PD-L1, CD47 was also recently shown to be targeted by miRNAs (Figure 1B). Specifically, binding of miR-708 to two sites in the *CD47* 3′UTR was capable of reducing CD47 levels in T cell acute lymphoblastic leukemia [42]. Functionally, enforced miR-708 in CCRF-CEM cell lines made the cells more vulnerable to phagocytosis, an effect that was synergistically strengthened with CD47 antibodies. Moreover, CD47 was also found to be targeted by miR-155 in multiple myeloma (MM). In advanced stages of the disease, miR-155 was downregulated [43]. When miR-155 was overexpressed in drug-resistant MM cell lines, reduced amounts of CD47 accompanied with an increase in phagocytosis by macrophages were observed.

3. Identification of miRNA Targetome

Identifying the targets of miRNAs is essential for understanding how aberrantly expressed miRNAs in tumors or tumor-associated immune cells could impact tumorigenesis. MiRNAs of vertebrates predominantly bind to partially complementary sequences in the 3′UTR of target mRNAs. Specificity of this binding is mostly determined by 7–8 nucleotides at the 5′ end of a miRNA molecule referred to as the seed sequence. As such, many computational miRNA target prediction algorithms such as TargetScan [44], as well as PicTar [45] and miRanda [46], were initially developed by relying on the seed rule that is dictated by Watson–Crick (WC) base pairing between the 5′ end of a miRNA molecule and the conserved complementary sequences in the 3′UTR of the target mRNA. Later, it became evident that there are exceptions to the seed rule [47,48]. For example, bulges, G:U wobbles, and seedless interaction can also affect miRNA–mRNA interaction. Furthermore, strong base pairing to the 3′ end of the miRNA canalso support seed pairing and structural accessibility into target sites [49,50]. Through further integrating the various aforementioned sequence or structure criteria, the performance and the accuracy of the target prediction have improved significantly over time. Nevertheless, while these computational prediction algorithms have been widely used in miRNA research and afforded the identification of many important miRNA targets, as discussed in the previous sections, the false positive rates remain very high [51]. As such, several experimental approaches have been developed to complement existing computational target prediction methods, allowinginvestigators to gain further insights into miRNA-mediated gene regulation in cancer and other biological processes.

3.1. CLIP-seq

CLIP-seq (cross-linking immunoprecipitation), also known as HITS-CLIP (high-throughput sequencing of RNAs isolated by UV crosslinking immunoprecipitation), is a method that was originally developed to identify functional protein–RNA interaction sites (Figure 2A) [52]. As only miRNA and mRNA incorporated into the RISC complex are pulled down for sequencing, not only does this method provide the opportunity to identify non-canonical miRNA–mRNA interactions, it has also helped exclude miRNA targets that are falsely predicted by computational means. Through taking this biochemical approach, unbiased analysis of specific miRNA–mRNA interactions has become possible [48,52]. For example, in hepatocellular carcinoma (HCC) patients, it has been previously reported that reduced expression of miR-122 in tumors correlates with metastasis and poor prognosis [53,54]. Through taking the CLIP-seq approach, many miR-122 targets, conserved in both humans and mice, and involved in thecell cycle, tight junction pathways, and cancer signaling pathways such as AMPK, PI3K/AKT, and WNT/β-catenin, were identified [53,54]. Among them, BCL9, a β-catenin co-factor which mediates transactivation of WNT target genes, was shown to be uniquely targeted by miR-122 at multiple sites. Further functional studies have established BCL9 as a conserved miR-122 target that impacts WNT-mediated progression of HCC, specifically through proliferation. Consistent with these findings, increased expression of BCL9 in tumors, along with other miR-122 targets such as STX6 and SLC52A2, are also significantly associated with poor patient survival. Similarly, like the aforementioned miR-122 in HCC, miR-203, the most highly expressed miRNA in the skin, was also found to be downregulated in the skin ofsquamous cell carcinoma (SCC) patients. Loss of miR-203 was shown to promote the initiation and development of SCC in both humans and mice [55]. Furthermore, CLIP-seq analysis revealed that miR-203 limits the proliferation of skin cells, particularly during the phase of tumor initiation, through targeting key components of the pro-proliferative Ras signaling pathway. Together, these studies demonstrate the power of the CLIP-seq approach in decoding disease-associated miRNA targeting networks and suggest that similar strategies could be applied to other tumor settings.

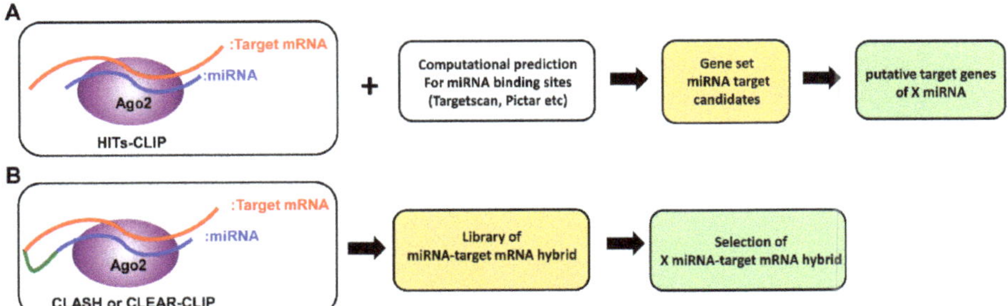

Figure 2. Comparison between HITS-CLIP (high-throughput sequencing of RNAs isolated by UV crosslinking immunoprecipitation) and CLASH (crosslinking, ligation and sequencing of hybrids). (**A**) In HITS-CLIP, RNA-Ago protein complexes are cross-linked with UV light and isolated by immunoprecipitation. MiRNAs and mRNAs purified from such complexes are then sequenced. MiRNA–mRNA interactions are matched computationally, and can be verified with additional assays. (**B**) In CLASH, following cross-linking, an additional ligation step generates miRNA-mRNA hybrids. Then, after isolation of RNA-Ago protein complexes by immunoprecipitation, miRNA-mRNA hybrids are sequenced to generate hybrid libraries.

Building upon the basis of HITS-CLIP, another CLIP-seq method, so-called PAR-CLIP (photoactivatable ribonucleoside-enhanced crosslinking and immunoprecipitation), was established [56]. Compared to HITS-CLIP, PAR-CLIP not only offers higher crosslinking efficiency, but also allows for a more precise localization of miRNA-containing Ago binding sites. To this end, by employing

PAR-CLIP analysis combined with RNA-seq in ovarian cancer cells with or without miR-450a expression, genes associated with the mitochondrial oxidative phosphorylation complex or glutamine metabolism, ACO2, TIMMDC1, ATP5B and MT-ND2, were identified as miR-450a targets [57]. Interestingly, while most of the Ago binding sites were located in the 3′UTR region, miR-450a seed complementary sequences were found exclusively in the coding sequences (CDS) of the miR-450a target genes. Nevertheless, compared to cells devoid of miR-450a, miR-450a expressing cells exhibited decreased mitochondrial membrane potential, as well as decreased glutamate production. Together, these data suggest that miR-450a regulates cellular energetic metabolism to limit tumor formation and progression.

3.2. CLASH (Crosslinking, Ligation and Sequencing of Hybrids)

Although CLIP-seq approaches have permitted the discoveries of many relevant miRNA targets in multiple studies, they have their own limitations [58]. Because biochemical identification of Ago binding sites came from CLIP-seq analysis of pooled mRNA after Ago immunoprecipitation, bioinformatics analysis is still required to identify the corresponding miRNAs responsible for Ago bindings. Moreover, when no obvious seed matches can be identified, there is no clear way to confirm whether the apparently seedless targeting was caused by non-canonical miRNA-target interactions or miRNA-independent mechanisms. To address this issue, building upon the original CLIP-seq approach, a new method, which is referred to as CLEAR (covalent ligation of endogenous Argonaute-bound RNAs)-CLIP [59] or CLASH (crosslinking, ligation and sequencing of hybrids) [60], was developed. In brief, miRNA–mRNA interactions are first ligated to generate miRNA–target mRNA chimeras while still bound to the Ago complex. Sequencing these miRNA–target mRNA chimeras allows for unambiguous mapping of canonical and non-canonical pairing (Figure 2B). Moreover, with recent advances in sequencing technology, this approach was further optimized to skip the clean-up steps that were previously required to remove unhybridized RNAs. As a result, this newly modified "quick" CLASH (qCLASH) not only is significantly faster, but can also be used to analyze patient biopsies which typically have much fewer cell numbers [61]. In this study, qCLASH was utilized to study the molecular mechanisms underlying Kaposi's sarcoma-associated herpesvirus (KSHV)-derived miRNAs that could drive the transformation of endothelial cells. KSHV, also known as human herpesvirus 8 (HHV8), is associated with malignant tumors such as Kaposi's sarcoma (KS), multicentric castleman's disease (MCD), and primary effusion lymphoma (PEL) [62–64]. Previously, KSHV-derived miRNAs have been shown to target tumor suppressor genes such as THBS1, TP53IPN1, and YWHAE [65,66]. By taking the qCLASH approach, 1433 gene targets were discovered and were involved in the cell cycle, glycolysis, and apoptosis pathways. Interestingly, 60% of the target sequences in KSHV-infected endothelial cells identified by the qCLASH hybrid were mapped to the CDS. In contrast, the majority of target sequences in KSHV-infected B cells are located in 3′UTR regions, indicating that miRNA–mRNA interactions can be cell type-specific. Moreover, 50% of the hybrids displayed non-seed pairing interaction, supporting the utility of generating miRNA–mRNA chimeras for the discovery of new canonical and non-canonical miRNA–mRNA interactions. Taken together, combining the qCLASH assay with pathway analysis provides an unbiased opportunity to understand the role of miRNAs in cancer biology.

4. Conclusions

Since the initial discovery of miRNAs back in the late 1990s, these small non-coding RNA species have been shown to be crucial for controlling almost every aspect of biological processes. In cancer, miRNAs can act both as "oncomirs" and tumor suppressors. Moreover, tumor-derived miRNAs can also impact cancer progression through directly, or indirectly, modulating immune responses, particularly in the tumor microenvironment. To this end, exosomal miRNAs released by tumor cells can directly nullify immune responses, while exosomal miRNAs released by immune cells may serve as a novel weapon in clearing cancer-associated cells, making this an exciting field for future research.

On the other hand, miRNAs can also indirectly control tumor–immune communication through regulating the expression of metabolites and immune checkpoint regulators, and so this field warrants further investigation given the growing importance of such molecules in cancer immunotherapy (Table 1). Moving forward, with continued technological advances made in miRNA target identification, functional miRNA–target mRNA interactions revealed by future studies will undoubtedly provide further mechanistic insights into the development of novel anti-cancer therapeutics.

Table 1. MiRNAs and their targets in tumor-immune communication.

miRNAs	Target Genes	Function	Cancer Type	Expression	Reference
miR-181a/b, miR-498	TNFα	Reduces the cytotoxicity of CD8 T cells	Melanoma		[15]
miR-3187-3p	PTPRC	Inhibition of TCR signaling			
miR-23a-3p	PTEN	Induces PD-L1 expression in macrophages	Hepatocellular carcinoma	Exosomes	[16]
miR-29a-3p, miR-21-5p	STAT3	Induces a higher Treg/Th17 cell ratio	TAM (Tumor-associated macrophages)		[17]
miR-298-5p	Unknown	Induces apoptosis of MSCs via Caspase-3	Fibrosarcoma		[18]
miR-448	IDO1	Enhance CD8 T cell survival	Colorectal cancer		[27]
miR-153					[28]
miR-34a	PDL1	Blocks the PD-1 pathway	Burkitt lymphoma, DLBCL	Tumor cells	[36]
miR-BHRF1-2-5p	PDL1, PDL2		EBV-positive DLBCL		[37]
miR-708	CD47	Promotes tumor cell elimination by phagocytosis	T-ALL		[42]
miR-155			MM		[43]
miR-122	BCL9, AMPK, PI3K/AKT, Wnt/β-catenin	Inhibits metastasis and proliferation	Hepatocellular carcinoma		[53,54]
miR-203	POLA1, HBEGF	Limits proliferation of skin cells	Squamous cell carcinoma		[55]
miR-450a	ACO2, TIMMDC1, ATP5B, MT-ND2	Limits tumor formation and progression	Ovarian cancer		[57]

Author Contributions: All authors contributed in writing, review and editing. All authors have read and agreed to the published version of the manuscript.

Funding: This work was supported by NIH grant AI108651 (L.-F.L.). L.-F.L. serves as the scientific consultant for Elixiron Immunotherapeutics.

Conflicts of Interest: The authors declare no conflict of interest.

References

1. Kozomara, A.; Birgaoanu, M.; Griffiths-Jones, S. MiRBase: From microRNA sequences to function. *Nucleic Acids Res.* **2019**, *47*, D155–D162. [CrossRef]
2. Ha, M.; Kim, V.N. Regulation of microRNA biogenesis. *Nat. Rev. Mol. Cell Biol.* **2014**, *15*, 509–524. [CrossRef] [PubMed]

3.   Lee, R.C.; Feinbaum, R.L.; Ambros, V. The C. elegans heterochronic gene lin-4 encodes small RNAs with antisense complementarity to lin-14. *Cell* **1993**, *75*, 843–854. [CrossRef]

4.   De Santa, F.; Iosue, I.; del Rio, A.; Fazi, F. MicroRNA biogenesis pathway as a therapeutic target for human disease and cancer. *Curr. Pharm. Des.* **2013**, *19*, 745–764. [CrossRef]

5.   Hamilton, M.P.; Rajapakshe, K.; Hartig, S.M.; Reva, B.; McLellan, M.D.; Kandoth, C.; Ding, L.; Zack, T.I.; Gunaratne, P.H.; Wheeler, D.A.; et al. Identification of a pan-cancer oncogenic microRNA superfamily anchored by a central core seed motif. *Nat. Commun.* **2013**, *4*, 2730. [CrossRef] [PubMed]

6.   Lu, J.; Getz, G.; Miska, E.A.; Alvarez-Saavedra, E.; Lamb, J.; Peck, D.; Sweet-Cordero, A.; Ebert, B.L.; Mak, R.H.; Ferrando, A.A.; et al. MicroRNA expression profiles classify human cancers. *Nature* **2005**, *435*, 834–838. [CrossRef] [PubMed]

7.   Iorio, M.V.; Ferracin, M.; Liu, C.G.; Veronese, A.; Spizzo, R.; Sabbioni, S.; Magri, E.; Pedriali, M.; Fabbri, M.; Campiglio, M.; et al. MicroRNA gene expression deregulation in human breast cancer. *Cancer Res.* **2005**, *65*, 7065–7070. [CrossRef]

8.   Hayes, J.; Peruzzi, P.P.; Lawler, S. MicroRNAs in cancer: Biomarkers, functions and therapy. *Trends Mol. Med.* **2014**, *20*, 460–469. [CrossRef]

9.   Lu, Y.; Leslie, C.S. Learning to predict miRNA-mRNA interactions from AGO CLIP sequencing and CLASH data. *PLoS Comput. Biol.* **2016**, *12*, e1005026. [CrossRef]

10.   Lin, S.; Gregory, R.I. MicroRNA biogenesis pathways in cancer. *Nat. Rev. Cancer* **2015**, *15*, 321–333. [CrossRef]

11.   Ramassone, A.; Pagotto, S.; Veronese, A.; Visone, R. Epigenetics and MicroRNAs in cancer. *Int. J. Mol. Sci.* **2018**, *19*, 459. [CrossRef] [PubMed]

12.   Chang, T.C.; Yu, D.; Lee, Y.S.; Wentzel, E.A.; Arking, D.E.; West, K.M.; Dang, C.V.; Thomas-Tikhonenko, A.; Mendell, J.T. Widespread microRNA repression by Myc contributes to tumorigenesis. *Nat. Genet.* **2008**, *40*, 43–50. [CrossRef] [PubMed]

13.   Li, X.L.; Jones, M.F.; Subramanian, M.; Lal, A. Mutant p53 exerts oncogenic effects through microRNAs and their target gene networks. *FEBS Lett.* **2014**, *588*, 2610–2615. [CrossRef] [PubMed]

14.   Othman, N.; Jamal, R.; Abu, N. Cancer-derived exosomes as effectors of key inflammation-related players. *Front. Immunol.* **2019**, *10*, 2103. [CrossRef] [PubMed]

15.   Vignard, V.; Labbe, M.; Marec, N.; Andre-Gregoire, G.; Jouand, N.; Fonteneau, J.F.; Labarriere, N.; Fradin, D. MicroRNAs in tumor exosomes drive immune escape in melanoma. *Cancer Immunol. Res.* **2020**, *8*, 255–267. [CrossRef]

16.   Liu, J.; Fan, L.; Yu, H.; Zhang, J.; He, Y.; Feng, D.; Wang, F.; Li, X.; Liu, Q.; Li, Y.; et al. Endoplasmic reticulum stress causes liver cancer cells to release exosomal miR-23a-3p and up-regulate programmed death ligand 1 expression in macrophages. *Hepatology* **2019**, *70*, 241–258. [CrossRef]

17.   Zhou, J.; Li, X.; Wu, X.; Zhang, T.; Zhu, Q.; Wang, X.; Wang, H.; Wang, K.; Lin, Y.; Wang, X. Exosomes released from tumor-associated macrophages transfer miRNAs that induce a Treg/Th17 cell imbalance in epithelial ovarian cancer. *Cancer Immunol. Res.* **2018**, *6*, 1578–1592. [CrossRef]

18.   Seo, N.; Shirakura, Y.; Tahara, Y.; Momose, F.; Harada, N.; Ikeda, H.; Akiyoshi, K.; Shiku, H. Activated CD8(+) T cell extracellular vesicles prevent tumour progression by targeting of lesional mesenchymal cells. *Nat. Commun.* **2018**, *9*, 435. [CrossRef]

19.   Munn, D.H.; Mellor, A.L. IDO in the tumor microenvironment: Inflammation, counter-regulation, and tolerance. *Trends Immunol.* **2016**, *37*, 193–207. [CrossRef]

20.   Munn, D.H.; Sharma, M.D.; Baban, B.; Harding, H.P.; Zhang, Y.; Ron, D.; Mellor, A.L. GCN2 kinase in T cells mediates proliferative arrest and anergy induction in response to indoleamine 2,3-dioxygenase. *Immunity* **2005**, *22*, 633–642. [CrossRef]

21.   Fallarino, F.; Grohmann, U.; You, S.; McGrath, B.C.; Cavener, D.R.; Vacca, C.; Orabona, C.; Bianchi, R.; Belladonna, M.L.; Volpi, C.; et al. The combined effects of tryptophan starvation and tryptophan catabolites down-regulate T cell receptor zeta-chain and induce a regulatory phenotype in naive T cells. *J. Immunol.* **2006**, *176*, 6752–6761. [CrossRef] [PubMed]

22.   Mezrich, J.D.; Fechner, J.H.; Zhang, X.; Johnson, B.P.; Burlingham, W.J.; Bradfield, C.A. An interaction between kynurenine and the aryl hydrocarbon receptor can generate regulatory T cells. *J. Immunol.* **2010**, *185*, 3190–3198. [CrossRef] [PubMed]

23. Quintana, F.J.; Murugaiyan, G.; Farez, M.F.; Mitsdoerffer, M.; Tukpah, A.M.; Burns, E.J.; Weiner, H.L. An endogenous aryl hydrocarbon receptor ligand acts on dendritic cells and T cells to suppress experimental autoimmune encephalomyelitis. *Proc. Natl. Acad. Sci. USA* **2010**, *107*, 20768–20773. [CrossRef] [PubMed]

24. Manlapat, A.K.; Kahler, D.J.; Chandler, P.R.; Munn, D.H.; Mellor, A.L. Cell-autonomous control of interferon type I expression by indoleamine 2,3-dioxygenase in regulatory CD19+ dendritic cells. *Eur. J. Immunol.* **2007**, *37*, 1064–1071. [CrossRef] [PubMed]

25. Liu, H.; Huang, L.; Bradley, J.; Liu, K.; Bardhan, K.; Ron, D.; Mellor, A.L.; Munn, D.H.; McGaha, T.L. GCN2-dependent metabolic stress is essential for endotoxemic cytokine induction and pathology. *Mol. Cell. Biol.* **2014**, *34*, 428–438. [CrossRef] [PubMed]

26. Ravishankar, B.; Liu, H.; Shinde, R.; Chaudhary, K.; Xiao, W.; Bradley, J.; Koritzinsky, M.; Madaio, M.P.; McGaha, T.L. The amino acid sensor GCN2 inhibits inflammatory responses to apoptotic cells promoting tolerance and suppressing systemic autoimmunity. *Proc. Natl. Acad. Sci. USA* **2015**, *112*, 10774–10779. [CrossRef] [PubMed]

27. Lou, Q.; Liu, R.; Yang, X.; Li, W.; Huang, L.; Wei, L.; Tan, H.; Xiang, N.; Chan, K.; Chen, J.; et al. MiR-448 targets IDO1 and regulates CD8(+) T cell response in human colon cancer. *J. Immunother. Cancer* **2019**, *7*, 210. [CrossRef]

28. Huang, Q.; Xia, J.; Wang, L.; Wang, X.; Ma, X.; Deng, Q.; Lu, Y.; Kumar, M.; Zhou, Z.; Li, L.; et al. MiR-153 suppresses IDO1 expression and enhances CAR T cell immunotherapy. *J. Hematol. Oncol.* **2018**, *11*, 58. [CrossRef]

29. Pardoll, D.M. The blockade of immune checkpoints in cancer immunotherapy. *Nat. Rev. Cancer* **2012**, *12*, 252–264. [CrossRef]

30. Topalian, S.L.; Drake, C.G.; Pardoll, D.M. Immune checkpoint blockade: A common denominator approach to cancer therapy. *Cancer Cell* **2015**, *27*, 450–461. [CrossRef]

31. Ribas, A.; Wolchok, J.D. Cancer immunotherapy using checkpoint blockade. *Science* **2018**, *359*, 1350–1355. [CrossRef]

32. Sun, C.; Mezzadra, R.; Schumacher, T.N. Regulation and function of the PD-L1 checkpoint. *Immunity* **2018**, *48*, 434–452. [CrossRef] [PubMed]

33. Wang, Q.; Lin, W.; Tang, X.; Li, S.; Guo, L.; Lin, Y.; Kwok, H.F. The roles of microRNAs in regulating the expression of PD-1/PD-L1 immune checkpoint. *Int. J. Mol. Sci.* **2017**, *18*, 2540. [CrossRef] [PubMed]

34. Kataoka, K.; Shiraishi, Y.; Takeda, Y.; Sakata, S.; Matsumoto, M.; Nagano, S.; Maeda, T.; Nagata, Y.; Kitanaka, A.; Mizuno, S.; et al. Aberrant PD-L1 expression through 3'-UTR disruption in multiple cancers. *Nature* **2016**, *534*, 402–406. [CrossRef] [PubMed]

35. Xu-Monette, Z.Y.; Zhou, J.; Young, K.H. PD-1 expression and clinical PD-1 blockade in B-cell lymphomas. *Blood* **2018**, *131*, 68–83. [CrossRef]

36. Anastasiadou, E.; Stroopinsky, D.; Alimperti, S.; Jiao, A.L.; Pyzer, A.R.; Cippitelli, C.; Pepe, G.; Severa, M.; Rosenblatt, J.; Etna, M.P.; et al. Epstein-Barr virus-encoded EBNA2 alters immune checkpoint PD-L1 expression by downregulating miR-34a in B-cell lymphomas. *Leukemia* **2019**, *33*, 132–147. [CrossRef]

37. Cristino, A.S.; Nourse, J.; West, R.A.; Sabdia, M.B.; Law, S.C.; Gunawardana, J.; Vari, F.; Mujaj, S.; Thillaiyampalam, G.; Snell, C.; et al. EBV microRNA-BHRF1-2-5p targets the 3'UTR of immune checkpoint ligands PD-L1 and PD-L2. *Blood* **2019**, *134*, 2261–2270. [CrossRef]

38. Zheng, Z.; Sun, R.; Zhao, H.J.; Fu, D.; Zhong, H.J.; Weng, X.Q.; Qu, B.; Zhao, Y.; Wang, L.; Zhao, W.L. MiR155 sensitized B-lymphoma cells to anti-PD-L1 antibody via PD-1/PD-L1-mediated lymphoma cell interaction with CD8+T cells. *Mol. Cancer* **2019**, *18*, 54. [CrossRef]

39. Vasudevan, S. Posttranscriptional upregulation by microRNAs. *Wiley Interdiscip. Rev. RNA* **2012**, *3*, 311–330. [CrossRef]

40. Chao, M.P.; Alizadeh, A.A.; Tang, C.; Jan, M.; Weissman-Tsukamoto, R.; Zhao, F.; Park, C.Y.; Weissman, I.L.; Majeti, R. Therapeutic antibody targeting of CD47 eliminates human acute lymphoblastic leukemia. *Cancer Res.* **2011**, *71*, 1374–1384. [CrossRef]

41. Majeti, R.; Chao, M.P.; Alizadeh, A.A.; Pang, W.W.; Jaiswal, S.; Gibbs, K.D., Jr.; Van Rooijen, N.; Weissman, I.L. CD47 is an adverse prognostic factor and therapeutic antibody target on human acute myeloid leukemia stem cells. *Cell* **2009**, *138*, 286–299. [CrossRef] [PubMed]

42. Huang, W.; Wang, W.T.; Fang, K.; Chen, Z.H.; Sun, Y.M.; Han, C.; Sun, L.Y.; Luo, X.Q.; Chen, Y.Q. MIR-708 promotes phagocytosis to eradicate T-ALL cells by targeting CD47. *Mol. Cancer* **2018**, *17*, 12. [CrossRef]

43. Rastgoo, N.; Wu, J.; Liu, M.; Pourabdollah, M.; Atenafu, E.G.; Reece, D.; Chen, W.; Chang, H. Targeting CD47/TNFAIP8 by miR-155 overcomes drug resistance and inhibits tumor growth through induction of phagocytosis and apoptosis in multiple myeloma. *Haematologica* **2019**. [CrossRef]

44. Lewis, B.P.; Shih, I.H.; Jones-Rhoades, M.W.; Bartel, D.P.; Burge, C.B. Prediction of mammalian microRNA targets. *Cell* **2003**, *115*, 787–798. [CrossRef]

45. Krek, A.; Grun, D.; Poy, M.N.; Wolf, R.; Rosenberg, L.; Epstein, E.J.; MacMenamin, P.; da Piedade, I.; Gunsalus, K.C.; Stoffel, M.; et al. Combinatorial microRNA target predictions. *Nat. Genet.* **2005**, *37*, 495–500. [CrossRef] [PubMed]

46. John, B.; Enright, A.J.; Aravin, A.; Tuschl, T.; Sander, C.; Marks, D.S. Human MicroRNA targets. *PLoS Biol.* **2004**, *2*, e363. [CrossRef]

47. Didiano, D.; Hobert, O. Perfect seed pairing is not a generally reliable predictor for miRNA-target interactions. *Nat. Struct. Mol. Biol.* **2006**, *13*, 849–851. [CrossRef]

48. Loeb, G.B.; Khan, A.A.; Canner, D.; Hiatt, J.B.; Shendure, J.; Darnell, R.B.; Leslie, C.S.; Rudensky, A.Y. Transcriptome-wide miR-155 binding map reveals widespread noncanonical microRNA targeting. *Mol. Cell* **2012**, *48*, 760–770. [CrossRef]

49. Chi, S.W.; Hannon, G.J.; Darnell, R.B. An alternative mode of microRNA target recognition. *Nat. Struct. Mol. Biol.* **2012**, *19*, 321–327. [CrossRef]

50. Bartel, D.P. MicroRNAs: Target recognition and regulatory functions. *Cell* **2009**, *136*, 215–233. [CrossRef]

51. Pinzon, N.; Li, B.; Martinez, L.; Sergeeva, A.; Presumey, J.; Apparailly, F.; Seitz, H. MicroRNA target prediction programs predict many false positives. *Genome Res.* **2017**, *27*, 234–245. [CrossRef] [PubMed]

52. Chi, S.W.; Zang, J.B.; Mele, A.; Darnell, R.B. Argonaute HITS-CLIP decodes microRNA-mRNA interaction maps. *Nature* **2009**, *460*, 479–486. [CrossRef] [PubMed]

53. Kutay, H.; Bai, S.; Datta, J.; Motiwala, T.; Pogribny, I.; Frankel, W.; Jacob, S.T.; Ghoshal, K. Downregulation of miR-122 in the rodent and human hepatocellular carcinomas. *J. Cell. Biochem.* **2006**, *99*, 671–678. [CrossRef]

54. Kojima, K.; Takata, A.; Vadnais, C.; Otsuka, M.; Yoshikawa, T.; Akanuma, M.; Kondo, Y.; Kang, Y.J.; Kishikawa, T.; Kato, N.; et al. MicroRNA122 is a key regulator of alpha-fetoprotein expression and influences the aggressiveness of hepatocellular carcinoma. *Nat. Commun.* **2011**, *2*, 338. [CrossRef]

55. Riemondy, K.; Wang, X.J.; Torchia, E.C.; Roop, D.R.; Yi, R. MicroRNA-203 represses selection and expansion of oncogenic Hras transformed tumor initiating cells. *Elife* **2015**, *4*, e07004. [CrossRef] [PubMed]

56. Hafner, M.; Landthaler, M.; Burger, L.; Khorshid, M.; Hausser, J.; Berninger, P.; Rothballer, A.; Ascano, M., Jr.; Jungkamp, A.C.; Munschauer, M.; et al. Transcriptome-wide identification of RNA-binding protein and microRNA target sites by PAR-CLIP. *Cell* **2010**, *141*, 129–141. [CrossRef]

57. Muys, B.R.; Sousa, J.F.; Placa, J.R.; de Araujo, L.F.; Sarshad, A.A.; Anastasakis, D.G.; Wang, X.; Li, X.L.; de Molfetta, G.A.; Ramao, A.; et al. MiR-450a Acts as a Tumor Suppressor in Ovarian Cancer by Regulating Energy Metabolism. *Cancer Res.* **2019**, *79*, 3294–3305. [CrossRef]

58. Lee, H.M.; Nguyen, D.T.; Lu, L.F. Progress and challenge of microRNA research in immunity. *Front. Genet.* **2014**, *5*, 178. [CrossRef]

59. Moore, M.J.; Scheel, T.K.; Luna, J.M.; Park, C.Y.; Fak, J.J.; Nishiuchi, E.; Rice, C.M.; Darnell, R.B. MiRNA-target chimeras reveal miRNA 3′-end pairing as a major determinant of Argonaute target specificity. *Nat. Commun.* **2015**, *6*, 8864. [CrossRef]

60. Helwak, A.; Kudla, G.; Dudnakova, T.; Tollervey, D. Mapping the human miRNA interactome by CLASH reveals frequent noncanonical binding. *Cell* **2013**, *153*, 654–665. [CrossRef]

61. Gay, L.A.; Sethuraman, S.; Thomas, M.; Turner, P.C.; Renne, R. Modified Cross-Linking, Ligation, and Sequencing of Hybrids (qCLASH) Identifies Kaposi's Sarcoma-Associated Herpesvirus MicroRNA Targets in Endothelial Cells. *J. Virol.* **2018**, *92*, e02138-17. [CrossRef] [PubMed]

62. Soulier, J.; Grollet, L.; Oksenhendler, E.; Cacoub, P.; Cazals-Hatem, D.; Babinet, P.; D'Agay, M.F.; Clauvel, J.P.; Raphael, M.; Degos, L.; et al. Kaposi's sarcoma-associated herpesvirus-like DNA sequences in multicentric Castleman's disease. *Blood* **1995**, *86*, 1276–1280. [CrossRef] [PubMed]

63. Cesarman, E.; Chang, Y.; Moore, P.S.; Said, J.W.; Knowles, D.M. Kaposi's sarcoma-associated herpesvirus-like DNA sequences in AIDS-related body-cavity-based lymphomas. *N. Engl. J. Med.* **1995**, *332*, 1186–1191. [CrossRef] [PubMed]

64. Chang, Y.; Cesarman, E.; Pessin, M.S.; Lee, F.; Culpepper, J.; Knowles, D.M.; Moore, P.S. Identification of herpesvirus-like DNA sequences in AIDS-associated Kaposi's sarcoma. *Science* **1994**, *266*, 1865–1869. [CrossRef] [PubMed]

65. Samols, M.A.; Skalsky, R.L.; Maldonado, A.M.; Riva, A.; Lopez, M.C.; Baker, H.V.; Renne, R. Identification of cellular genes targeted by KSHV-encoded microRNAs. *PLoS Pathog.* **2007**, *3*, e65. [CrossRef]

66. Gottwein, E.; Mukherjee, N.; Sachse, C.; Frenzel, C.; Majoros, W.H.; Chi, J.T.; Braich, R.; Manoharan, M.; Soutschek, J.; Ohler, U.; et al. A viral microRNA functions as an orthologue of cellular miR-155. *Nature* **2007**, *450*, 1096–1099. [CrossRef]

MDPI

St. Alban-Anlage 66

4052 Basel

Switzerland

Tel. +41 61 683 77 34

Fax +41 61 302 89 18

www.mdpi.com

Cancers Editorial Office

E-mail: cancers@mdpi.com

www.mdpi.com/journal/cancers

www.ingramcontent.com/pod-product-compliance
Lightning Source LLC
LaVergne TN
LVHW071524170726
843515LV00008B/2059